6th Vacuum and Surface Sciences Conference of Asia and Australia

(VASSCAA-6)

Journal of Physics: Conference Series Volume Volume 439

Islamabad, Pakistan
9-13 October 2012

ISBN: 978-1-62748-671-2
ISSN: 1742-6588

Printed from e-media with permission by:

Curran Associates, Inc.
57 Morehouse Lane
Red Hook, NY 12571

CURRAN ASSOCIATES INC. proceedings .com

Printed by Curran Associates, Inc. (2013)

Institute of Physics
Dirac House, Temple Back
Bristol BS1 6BE UK

Phone: 44 1 17 929 7481
Fax: 44 1 17 920 0979

techtracking@iop.org

Additional copies of this publication are available from:

Curran Associates, Inc.
57 Morehouse Lane
Red Hook, NY 12571 USA
Phone: 845-758-0400
Fax: 845-758-2634
Email: curran@proceedings.com
Web: www.proceedings.com

Papers

Surface Science and Engineering

Nanometer Structures

Thin Films

Electronic Material and Processing

Bio Interface

Plasma Science and Technique

Journal of Physics: Conference Series **439** (2013) 011001 doi:10.1088/1742-6596/439/1/011001

Preface

The Vacuum and Surface Sciences Conference of Asia and Australia (VASSCAA) conference series has been organized to create a new forum in Asia and Australia to discuss vacuum, surface and related sciences, techniques and applications. The conference series is officially endorsed by the International Union for Vacuum Science, Technique and Application (IUVSTA). The International Steering Committee of VASSCAA is comprised of Vacuum Societies in seven countries: Australia, China, India, Iran, Japan, South Korea and Pakistan. VASSCAA-1 was organized by the Vacuum Society of Japan in 1999 in Tokyo, Japan. VASSCAA-2 was held in 2002 in Hong Kong, VASSCAA-3 in Singapore in 2005. VASSCAA-4 was held in Matsue, Japan in 2008 and VASSCAA-5 in 2010 in Beijing, China.

The 6th Vacuum and Surface Sciences Conference of Asia and Australia (VASSCAA-6) was held from 9–13 October 2012 in the beautiful city of Islamabad, Pakistan. The venue of the conference was the Pak–China Friendship Centre, Islamabad. More than six hundred local delgates and around seventy delegates from different countries participated in this mega event. These delegates included scientists, researchers, engineers, professors, plant operators, designers, vendors, industrialists, businessmen and students from various research organizations, technical institutions, universities, industries and companies from Pakistan and abroad. The focal point of the event was to enhance cooperation between Pakistan and the international community in the fields of vacuum, surface science and other applied technologies.

At VASSCAA-6 85 oral presentations were delivered by local and foreign speakers. These were divided into different sessions according to their fields. A poster session was organized at which over 70 researchers and students displayed their posters. The best three posters won prizes. In parallel to the main conference sessions four technical short courses were held. The participants showed keen interest in all these courses. The most significant part of this event was an international exhibition of science, technology, energy and industry. In this international exhibition over 60 prominent international as well as local industrialists and vendors displayed their products. For the recreation of conference participants a cultural program and dinner was arranged. This entertaining program was fully enjoyed by all the participants especially the foreign guests. Recreational trips were also arranged for the foreign delegates. This mega event provided a unique opportunity to our scientific community to benefit from the rich international experience. The conference was a major forum for the exchange of knowledge and provided numerous scientific, technical and social opportunities for meeting leading experts.

Editors
Dr Javaid Ahsan Bhatti, Dr Talib Hussain, Dr Suleman Qaiser and Dr Wakil Khan
National Institute of Vacuum Science and Technology (NINVAST)
NCP Complex, Quaid-e-Azam University, Islamabad, Pakistan

Journal of Physics: Conference Series **439** (2013) 011001 doi:10.1088/1742-6596/439/1/011001

List of Delegates

Article No.	List of author names	Corresponding address
01-VASSCAA6	Vincenzo Ierardi, Francesca Ferrera, Enrico Millo, Gianluca Damonte, Gilberto Filaci and Ugo Valbusa	vincenzo.ierardi@unige.it
02-VASSCAA6	Salma Gul, Anwar-ul-Haq Ali Shah and Salma Bilal	salma.bilal@s2004.tu-chemnitz.de
03-VASSCAA6	Rahila Khalid, Francesc Salvat Pujol and Wolfgang S. M. Werner	khalid_rahila@yahoo.com
04-VASSCAA6	Asif Hussain Malik, Ajmal Shah and Shahab Khushnood	asifhussainmalik@yahoo.com
05-VASSCAA6	G Tasneem, C Tomastik, R Mroczyński and W S M Werner	ghazaltasneem@yahoo.com
06-VASSCAA6	A Hussain, A H Hamdani, R Akhter and M Aslam	pilo786@yahoo.com
07-VASSCAA6	Shoaib Ahmad	sahmad.ncp@gmail.com
08-VASSCAA6	M I Shahzad, M Giorcelli, D Perrone, A Virga, N Shahzad, P Jagdale, M Cocuzza and A Tagliaferro	imran.shahzad@polito.it
09-VASSCAA6	Shahid Nisar Ahmad, Saira Hakeem, Rashid Ahmed Alvi, Khawar Farooq, Naveed Farooq, Farida Yasmin, and Sadaf Saeed	s_n_ahmad@yahoo.com
10-VASSCAA6	M I Shahzad, M Giorcelli, N Shahzad, S Guastella, M Castellino, P Jagdale and A Tagliaferro	imran.shahzad@polito.it
11-VASSCAA6	A Quddos, N H Samtio and A M Syed	drabdulquddos@yahoo.com
12-VASSCAA6	N Shahzad, D Pugliese, A Lamberti, A Sacco, A Virga, R Gazia, S Bianco, M I Shahzad, E Tresso and C F Pirri	nadia.shahzad@polito.it
13-VASSCAA6	Syed Mansoor Ali, Syed Tajammul Hussain, Shahzad Abu Bakar, Jan Muhammad and Naeem ur Rehmand	mansoor_phys@yahoo.com
14-VASSCAA6	A Humaira and M Asghari	tpl.qau@usa.net
15-VASSCAA6	Erum Pervaiz and I H Gul	eramsyam@gmail.com
16-VASSCAA6	S Alam, M Irfan, B D Soomro, M Shahid and M Zeeshan	pitmaem.lhr@gmail.com
17-VASSCAA6	Sadaf Saeed, Saira Hakeem, Muhammad Faheem, Rashid Ahmed Alvi, Khawar Farooq, Syed Tajammul Hussain, and Shahid Nisar Ahmad	sadaf03_2000@yahoo.com
18-VASSCAA6	L F Oliveira, J Marques, P J G Coutinho, P Parpot and C J Tavares	ctavares@fisica.uminho.pt
19-VASSCAA6	M R Abdul Karim, M Pavese, EP Ambrosio, D Ugues, M Lombardi, S Biamino , C Badini and P Fino	muhammad.abdulkarim@polito.it
20-VASSCAA6	M Rashid, J A Bhatti, F Hussain, M Imran, I U Khawaja, K A Chaudhary and S A Ahmad	rapakistana@yahoo.com
21-VASSCAA6	M R Shahid and Musharaf Abbas	shahidmr786@hotmail.com
22-VASSCAA6	Nawazish A Khan and M Shamraiz	muhammadshamraiz6@gmail.com
23-VASSCAA6	K Ahmed, A N Khan, A Rauf, A Gul and M Aslam	drkamalahmed@yahoo.com
24-VASSCAA6	Sadia Sagar , Nadeem Iqbal and Asghari Maqsood	sadiasagar@scme.nust.edu.pk
25-VASSCAA6	M A Aleem, H Nawaz, M Shuaib, S Qaisar and M S Akbar	sam_qau@yahoo.com
26-VASSCAA6	Asif I Zia, A R Mohd Syaifudin, S C Mukhopadhyay, P L Yu, I H Al- Bahadly, Chinthaka P Gooneratne, Jűrgen Kose and Tai-Shan Liao	a.i.zia@massey.ac.nz
27-VASSCAA6	M Yousuf Zaman, S Ferrero, D Perrone, L Scaltrito, N Shahzad and D Pugliese	muhammad.zaman@polito.it

28-VASSCAA6	M Zafarullah Khan, M Mohsin Naveed and D M Akbar Hussain	mohsinnaveed30@gmail.com
29-VASSCAA6	M Aslam	m_aslammalik@yahoo.com
30-VASSCAA6	M Asghar, K Mahmood, M Faisal and M A Hasan	mhashmi@iub.edu.pk
31-VASSCAA6	M Asghar, K Mahmood, Faisal Malik and M A Hasan	mhashmi@iub.edu.pk
32-VASSCAA6	M Zafarullah Khan, M Mohsin Naveed and D M Akbar Hussain	mohsinnaveed30@gmail.com
33-VASSCAA6	V Ierardi, G Firpo and U Valbusa	vincenzo.ierardi@unige.it
34-VASSCAA6	G Y Hsiung, C C Chang, C L Chen, L H Wu, C M Cheng, C K Chan, Y C Yang, H P Hsueh, S N Hsu and J R Chen	hsiung@nsrrc.org.tw
35-VASSCAA6	M Maqsood, J Ali, A Usman and M Farooq	maqsood@ninvast.edu.pk
36-VASSCAA6	A Bahadur, M Zubair and M B Khan	principal.ces@nust.edu.pk
37-VASSCAA6	A Hai and Qurat-ul-Ain	haiabdul20@gmail.com
38-VASSCAA6	A Mazher	abeermazhar@yahoo.com
39-VASSCAA6	D Ahmed and A Ahmad	daniyal.ahmed@uettaxila.edu.pk
40-VASSCAA6	M Usman Saeed Khan, M Irfan Maqsood, Ehsan Ali, Shah Jamal and M Javed	usmansunshine@gmail.com
41-VASSCAA6	S Amna, N Samreen, B Khalid, A Shamim	shahzadi.bses169@iiu.edu.pk
42-VASSCAA6	T A Moughal	tauqir_mughal@yahoo.com
43-VASSCAA6	M Usman Saeed Khan, Ehsan Ali, M Irfan Maqsood and Haq Nawaz	usmansunshine@gmail.com
44-VASSCAA6	A Ahmed, A Akhtar, B Khalid and A Shamim	indus_10@hotmail.com
45-VASSCAA6	Akram Rashid and Ijaz Mansoor Qureshi	akram@mail.au.edu.pk
46-VASSCAA6	M Maqsood and M Nauman Nasir	maqsood@ninvast.edu.pk
47-VASSCAA6	Baoping Xin, Mengyue Shen, Hina Aslam and Feng Wu	xinbaoping@bit.edu.cn
48-VASSCAA6	S Iqbal and M M S Gualini	drsalmaniqbal@yahoo.com
49-VASSCAA6	F Amin and S Ahmed	faisal.msee226@iiu.edu.pk
50-VASSCAA6	T Hussain and M A Gondal	thminhas7@yahoo.com
51-VASSCAA6	A H Hamdani, A Nasir, S Sarwar, A Ansar, R Akhter and M Aslam	pilo786@yahoo.com
52-VASSCAA6	M Ayub Khan Yousuf Zai and Khusro Mian	ayubzai@yahoo.com

Journal of Physics: Conference Series **439** (2013) 012001 doi:10.1088/1742-6596/439/1/012001

Bioactive surfaces for antibody-antigen complex detection by Atomic Force Microscopy

Vincenzo Ierardi[1,2], Francesca Ferrera[3], Enrico Millo[3,4], Gianluca Damonte[3,4], Gilberto Filaci[3,5], Ugo Valbusa[1,2].

[1] Nanobiotechnologies, National Institute of Cancer Research (IST), Largo R. Benzi Genova, 16132, Italy.

[2] Nanomed lab , Physics Department, University of Genova, Via Dodecaneso,33, Genova, 16146, Italy.

[3] Centre of Excellence for Biomedical Research, University of Genova, Viale Benedetto XV, 7, Genova, 16146, Italy.

[4] Department of Experimental Medicine, University of Genova, Viale Benedetto XV, 1, Genova, 16132, Italy.

[5] Department of Internal Medicine, University of Genova, Viale Benedetto XV, 6, Genova, 16146, Italy

Corresponding author:Vincenzo Ierardi, vincenzo.ierardi@unige.it

Abstract. Recently there has been a great develop of new antibodies immobilization procedures, that keep antibodies to retain their orientation and functionality after the binding to a solid support. This allows the formation of immune-complexes useful for the detection of biomarkers from biological samples. We have developed a new method of functionalization for solid substrates that involves an initial surface activation, then a functionalization by means of 3-aminopropyltriethoxysilane-followed by another functionalization step with a layer of very small peptides, which have a high affinity to the Antibody Fc portion, acting as antibody linkers. These antibody binding peptides can immobilize the antibodies with a proper configuration that allows an unambiguous detection of antibody-antigen complexes by means of atomic force microscopy (AFM). The AFM can act as a powerful label free detection technique, which allows to detect, in principle, single molecule interactions, with the only limitation to use substrate with low-roughness surfaces; in this case, the roughness can be interpreted as background noise in the AFM analysis. Moreover, our functionalization method can be used to obtain bioactive surfaces on a wide range of solid supports, making them capable to suitably immobilize the antibodies for the antigenic binding.

1. Introduction

Recently, there has been a growing development of antibody immobilization procedures, which are able to stabilize the antibodies in order to retain an optimal orientation and functionality after the binding to a solid support, improving the antibody avidity for the specific antigen thus providing a new generation of analytical assays with increased sensitivity and specificity toward biomarkers present in biological samples. In fact in the last decade the Enzyme-Linked ImmunoSorbent Assay (ELISA), the conventional analytical assay largely used to measure protein biomarkers from biological fluids, has been partially replaced with the protein microarray technique, which can detect simultaneously several analytes [1]. In these assays each spot of the array contains antibodies for a single antigen, i.e. the antigens are captured specifically by antibodies and revealed usually by a fluorochrome-based detection system [2]. Although microarrays can be applied to the study of several aspects of cell activity, they still have a limited sensitivity since they cannot detect and measure any analyte below a specific cut-off [3]. Therefore, the only possibility for detecting biomarkers at very low concentrations and/or present in extremely small samples (such as could be the case for some biopsies, laser capture microdissected samples, forensics microspecimens, cell lysates or surnatants from few cell cultures), is to develop new high sensitive techniques able to detect the target at the single molecule scale. Certainly a further miniaturization of analytical assays could offer several advantages in terms of higher array density, smaller volumes, reduced amount of analyte [4-7]. However current ultramicroarrays and nanoarrays system are still proof of concept prototypes in which the spot size range from few nanometers to several microns [8].Moreover current conventional scanners used to detect fluorochrome-based recognition systems in microarrays are not capable to resolve nano-spots. All these considerations induce studies aimed at exploring the applicability of label-free techniques for single-molecule detection to analyze the cell activity in physiological and pathological conditions [9].

Ours idea is to use the Atomic Force Microscopy (AFM) as label-free revealing system [10]. This approach can allow us to detect the formation of antigen-antibody immune complexes without the need of fluorochrome labeled probe, since the AFM is able to visualize single molecules only if the substrate roughness does not interfere with the measure. Therefore the substrate functionalization method plays a central role in the process of detection of the antigen-antibody complexes by AFM imaging.

There are several functionalization methods that allow antibodies to retain their orientation and/or function [11]. The most common approach consists in the use of antibody binding proteins (protein A and protein G) that specifically bind the Fc region of antibodies. Although this method is simple, cheap and does not require antibody modification, it is not suitable for AFM imaging because it produces substrates presenting high superficial roughness (1-5 nm) [12]. Others approaches are time and cost consuming as they use tagged recombinant antibodies [13-15]. Indeed, the functionalization of a suitable substrate for AFM imaging should create a mono-layer of molecules that allows antibodies to retain their orientation and functionality without enhances the substrate roughness.

The antibody height detected by AFM varies from 4 to 6,5 nm [16, 17] [4]. These sizes are lower than that reported by crystallographic analyses: 14.5 x 8.5 x 4 nm [18]. This discrepancy is mainly due to the antibody high flexibility [19, 20]. Therefore, in order to permit unambiguous identification of the antibody by AFM imaging the substrate roughness must be very low. In this paper a new functionalization procedure that use a smooth solid substrate (e.g. mica, glass and silicon) and a short peptide as antibody linkers, which has a high affinity to the antibody Fc portion, is described. This procedure allows the production of a bioactive surface with a very low roughness (below 1 nm) able to bind properly the antibody, thus keeping its bioactivity. The bioactive surface preparation involves an initial surface activation, then a functionalization by means of 3-aminopropyltriethoxysilane

(APTES) followed by another functionalization step with a layer of the antibody binding peptide, which immobilize the antibodies with a proper configuration that allows an unambiguous detection of antibody-antigen complexes by means of atomic force microscopy (AFM).

2. Materials and Methods

2.1. Preparation of the antibody binding peptide monolayer

The antibody binding peptide could have 6-9 aminoacidic residues. We have manually synthesized three different peptides using the standard method of solid phase peptide synthesis which follows the 9-fluorenylmethoxycarbonyl (Fmoc) strategy with minor modifications.

The analysis of the crude powder was performed by liquid chromatography-electrospray mass-spectrometry (HPLC-ESI-MS) using an Agilent 1100 series LC/MSD ion trap instrument. The compound was purified by reverse phase high performance liquid chromatography (RP-HPLC) on a Shimadzu LC-9A preparative HPLC equipped with a Phenomenex C18 Luna column (21.20 x 250 mm). The molecular weight of the product was finally confirmed by electrospray ion trap mass spectrometry (ESI-TRAP-MS).

2.2. Preparation of the antibodies array for AFM and fluorescence investigation

The muscovite mica sheet was cut into (2 x 1) cm^2 pieces and cleaved in air. The cleaved mica was placed immediately into an oxygen plasma treatment chamber and then exposed for 2 min to oxygen RF plasma.

APTES-mica surface was obtained by placing the activated mica in a vacuum glass dessicator, which contained 50 μl of APTES 98% (Sigma-Aldrich) for 1h. The silanized substrate was then cured at 110°C for 1h.

To create a monolayer of antibody binding peptide on APTES-mica surface, 2 mg/ml peptide in DMF (Sigma-Aldrich) were treated with HATU and DIPEA (Sigma-Aldrich) to activate the carboxylic groups. The solution was then deposited on the APTES-mica substrate overnight, at room temperature, for the formation of the peptide bond between the amino groups of the APTES and the carboxylic groups of the peptide. The substrates were then washed in MilliQ water.

Five solutions with different concentrations (50 μg/ml, 1μg/ml, 10 ng/ml, 100 pg/ml, and 1 pg/ml) of anti-IL10 antibody (Biosource-Invitrogen, Carlsbad, CA) in PBS 30% Glicerol were spotted on the mica solid substrate by Piezoarray non-contact microarraying system (Perkin Elmer, Waltham, MA). The antibody array was incubated for 1 h at room temperature with a 1% mixture of aminoacids (Sigma-Aldrich) in PBS to block non-specific binding. The substrates were then washed in MilliQ water.

In order to test this procedure by means of fluorescence techniques, solutions of biotinylated recombinant human IL-10 (Peprotech, London, UK) in PBS, up to fifteen times more concentrated than the antibody solution, were deposited on the array and incubated for 1 h at room temperature, then washed with PBS. Streptavidin-Cy3 was then deposited on the array and incubated for 20 min in the dark and the fluorescence was detected by a standard microarray scanner system.

2.3. AFM imaging

The AFM analysis was performed using a Dimension 3100 Veeco AFM (Santa Barbara, CA) with a hybrid XYZ head, using Olympus OMCL-AC160 tips with nominal apical radius < 7 nm (Olympus corporate, Tokyo).

3. Results and Discussion

The idea to use AFM as detection technique induces some conditions on the characteristics of the substrate. In others words after each step of preparation the substrate must be smooth with very low roughness and must be able to bind antibodies with a proper orientation promoting the formation of the antibody-antigen complexes. Molecules such as protein A and G were not considered because of their elevated dimension. A review of the literature on this topic had shown that particular peptides could have antibody binding activity, so we focused our attention to some small peptides, which have the ability to recognize the Fc immunoglobulin portion [21]. Therefore, using peptides sequences with high affinity for the Fc region of anti-IL10 as model system, we have identified a class of peptides suitable for antibodies AFM investigation. Different peptides have been synthesized and tested. They are able to bind with high affinity the antibody Fc portion. In particular we have used the anti-IL10 monoclonal antibody as model. Indeed these peptides do not induce remarkable roughness on the substrate surface, reducing the AFM background signal, have high affinity for the antibody Fc portion and are able to link the antibody to the substrate with the proper orientation.

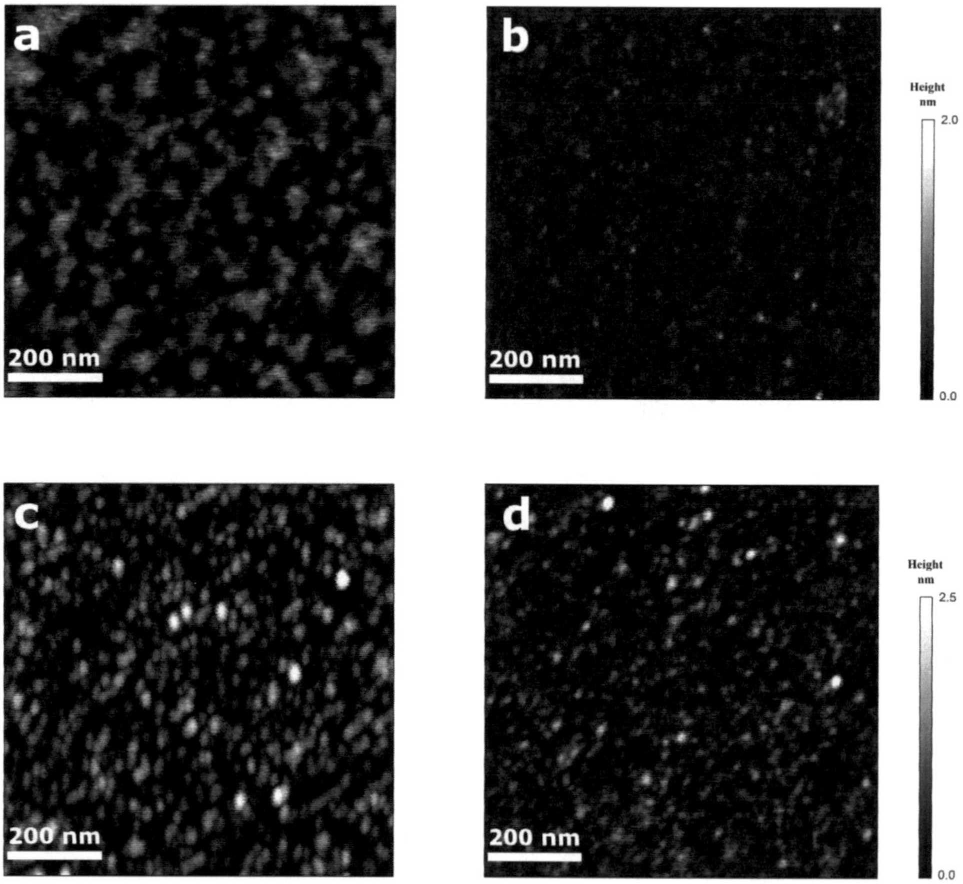

Figure 1: Tapping Mode AFM images of the four antibodies binding peptides investigated.

For this work we have analyzed four small peptides, which sequences are shown in table 1. They give us surfaces with very low roughness, even though with different morphology. In fact an AFM analysis of the surface obtained with these peptides shown a very different morphology, as it is possible to see in Figure 1. For instance, in figure 1c it is possible to identify globular structures, these structures disappear in the specimen a), but they are still present, although in less number, in the specimens b) and d). These structures very likely are due to the different value of the critical aggregation concentration (CAC) of these small peptides [22]. In spite of the morphological differences, the AFM-images show that all these peptides produced surfaces with very low roughness, as reported in table 1. In order to obtain the more suitable surface for our scope we have chosen the peptide that gives a surface with minor roughness. Therefore, five solutions with different concentrations of anti-IL10 antibody were deposited onto two solid supports of mica, previously coated with a layer of the peptide b). One support was used in order to verify the procedure by means of an immunofluorescence assay and the second one was used to perform AFM measurements.

Table 1. Peptides sequences and associated roughness. The roughness R_q^a was evaluated by Nanoscope 6.14 (Digital Instrument/Veeco) AFM software.

	Peptides Investigated	**Roughness R_q (nm)**
a)	$COOH\text{-}G\text{-}C_8\text{-}V\text{-}W\text{-}G\text{-}R\text{-}W\text{-}H\text{-}NH\text{-}COCH_3$	0.28
b)	$COOH\text{-}G\text{-}C_8\text{-}L\text{-}L\text{-}V\text{-}T\text{-}R\text{-}P\text{-}NH\text{-}COCH_3$	0.16
c)	$COOH\text{-}G\text{-}C_8\text{-}L\text{-}L\text{-}L\text{-}T\text{-}R\text{-}P\text{-}NH\text{-}COCH_3$	0.55
d)	$COOH\text{-}G\text{-}C_8\text{-}L\text{-}L\text{-}L\text{-}T\text{-}R\text{-}P\text{-}A\text{-}M\text{-}V\text{-}NH\text{-}COCH_3$	0.59

[a)] R_q is the root mean square (RMS) average of height deviations taken from the mean image data plane.

The specimen used in order to perform the immunofluorescence assay was treated with a 1% mixture of aminoacids in PBS in order to block non-specific binding. Afterward was incubated for 1 h at room temperature with a solution of biotinylated recombinant human IL-10 in PBS up to fifteen times more concentrated than the most concentrated antibody solution. After a gentle washing with PBS it was treated with a solution of Streptavidin-Cy3 and incubated for 20 min in the dark, next the fluorescence was immediately measured.

1 pg/ml →
100 pg/ml →
10 ng/ml →
1 μg/ml →
50 μg/ml →

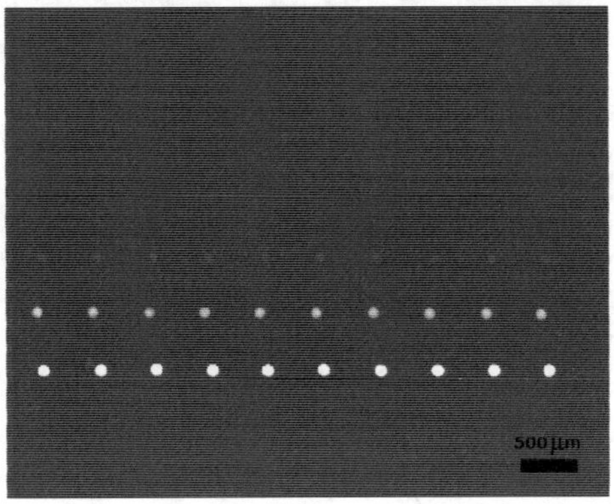

Figure 2: Fluorescence image of the immune assay experiment.

As it is possible to see in Figure 2, the signal is very clear for the spots with higher antibodies concentration while the signal disappears for the spots with a lower antibodies concentration. In spite of the scanner resolution limit, this experiment confirms that the antibodies maintain their bioactivity after the binding with the peptide.

The second specimen was analyzed by means of AFM in order to measure the height and the width of the antibodies. We have measured an antibody average height of 4.2 ± 0.3 nm and an antibody average width of 34.0 ± 7.0 nm. These values, within the tip convolution effects, are in accordance with literature data [20, 23]. Further details on the possible configurations that the antibodies can assume on the surface can be find in [20]. In addition, an important step in the immunoassay setting is the blocking of the binding sites of the antibodies in order to avoid unspecific interaction. In conventional immunoassay this is performed by various blocking agents with high molecular weight, such as bovine serum albumin. However, the large dimension of these agents makes them not suitable for an AFM based detection system because their presence would interfere with the antibody-antigen complex detection. We overcame this problem using a mixture of low molecular weight aminoacids, which did not produce relevant variation of both antibody height and surface roughness. It is also important to notice that the best AFM images of the antibodies were obtained analyzing the regions corresponding to the spots with the antibodies concentration of 100 pg/ml, whereas was impossible to identify unambiguously the antibodies in the regions corresponding to the spot with higher concentration because of the high surface density of antibodies. In order to measure the heights of these structures, several AFM images were collected. The height of every single antibody was measured taking the higher point of the antibody cross-section. In Figure 3 it possible to see an example of antibodies AFM image and the histogram of the antibodies height.

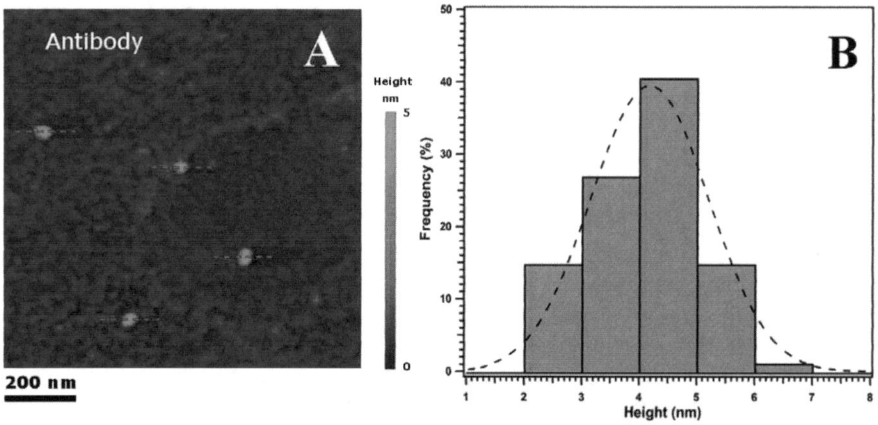

Figure 3: A) Tapping Mode AFM image of antibodies, the red lines above the antibodies are the tracks along which the antibodies heights were measured. B) Antibody height histogram. The height of antibodies is 4.2 ± 0.3 nm.

The same regions were analyzed by AFM after that the antibodies arrays were incubated with IL10. The AFM analysis shows an increase of the average height of the antibodies. On the average this increasing is around 2 nm, more precisely the average height of these structures is 6.6 ± 0.3 nm. This increase of the height is due to the formation of the antibody-antigen complexes, in fact this value of the height is in agreement with the expected one. The AFM analysis of the antigen IL-10, performed on a sample prepared by solution casting of the antigen solution on a surface similar to that one used for the preparation of the antibody array, indicates that the antigen has an average height of 2.3 ± 0.1 nm, as it possible to see in Figure 4 where it is shown an example of antigen AFM-image and the histogram of its height.

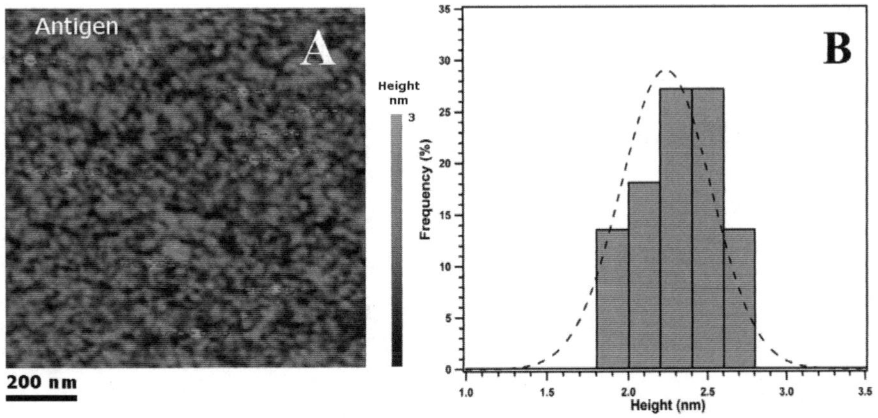

Figure 4: A) Tapping Mode AFM image of antigens, the red lines above the antigens are the tracks along which the antigens heights were measured. B) Antigen height histogram. The height of antigens is 2.3 ± 0.1 nm.

Accordingly the difference between the height of the antibody and the height of the antibody-antigen complex fits very well the height of the antigen. This is more evident in Figure 5, where is shown an example of the difference between the cross-section of an antibody and an antibody-antigen complex.

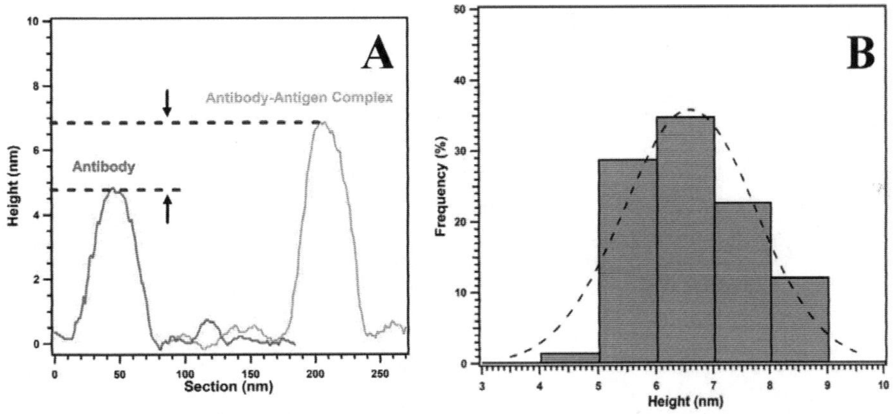

Figure 5: A) Cross-section of antibody and antibody-antigen complex. B) Antibody-antigen complexes height histogram. The average height of antibody-antigen complexes is 6.6 ± 0.3 nm.

4. Conclusions

In conclusion our procedure of surfaces functionalization is a clear advance on current methods. In fact we developed a bioactive surface that results to be efficient in binding properly the antibodies, keeping their bioactivity, as confirmed by the immunofluorescence test. Although further studies are needed in order to set up an analytical procedure to use with biological samples, we have the proof of principle that it is possible to detect the formation of antibody-antigen complexes using the AFM as label-free technique. This result was achieved taking advantage of small peptides with a peculiar and well defined aminoacidic sequence, which confer them high affinity for antibody Fc portion. In addition we have obtained an ultra-flat surface that allows us to use the AFM as label-free detection. Major advances can be attained by the described procedure in comparison with the current ones, if we consider that, thanks to the high resolution capability of the AFM can be possible to extend this analysis to the study of the interaction antigen-antibody in samples in which the analyte is present at extremely low concentration. Moreover, this procedure can be employed also to functionalize materials different of the mica, such as glass, silicon and some polymers as polydimethylsiloxane (PDMS).

Journal of Physics: Conference Series **439** (2013) 012001 doi:10.1088/1742-6596/439/1/012001

References

1. Dufva, M., Christensen C.B., *Diagnostic and analytical applications of protein microarrays.* Expert. Rev. Proteomics 2005. **2**(1): p. 41.
2. Pavlickova, P., Schneider E.M., Hug H. , *Advances in recombinant antibody microarrays.* Clin. Chim. Acta 2004. **343**(1-2): p. 17.
3. Borrebaeck, C.A., Wingren C., *Design of high-density antibody microarrays for disease proteomics: key technological issues.* J. Proteomics 2009. **72**(6): p. 928.
4. Lee, K.B., Park S.J., Mirkin C.A., Smith J.C., Mrksich M., *Protein nanoarrays generated by dip-pen nanolithography.* Science, 2002. **295**: p. 1702.
5. Nettikadan, S., Radke K., Johnson J., Xu J., Lynch M., Mosher C., Henderson E., *Detection and quantification of protein biomarkers from fewer than 10 cells.* Mol. Cell Proteomics, 2006. **5**: p. 895.
6. Sekula, S., Fuchs J., Weg-Remers S., Nagel P., Schuppler S., Fragala J., Theilacker N., Franzreb M., Wingren C., Ellmark P., *Multiplexed lipid dip-pen nanolithography on subcellular scales for the templating of functional proteins and cell culture.* Small 2008. **4**: p. 1785.
7. Zhang, G.J., Tanii T., Kanari Y., Ohdomari I., *Production of nanopatterns by a combination of electron beam lithography and a self-assembled monolayer for an antibody nanoarray.* J. Nanosci. Nanotechnol., 2007. **7**: p. 410.
8. Wingren, C., Borrebaeck C.A., *Progress in miniaturization of protein arrays--a step closer to high-density nanoarrays.* Drug Discov. Today, 2007. **12**: p. 813.
9. Ray, S., Mehta G., and Srivastava S., *Label-free deection techniques for protein microarrays: Prospect, merits and challenges.* Proteomics, 2010. **10**: p. 731.
10. Menotta, M., Crinelli R., Carloni E., Bianchi M., Giacomini E., Valbusa U., Magnani M. , *Label-free quantification of activated NF-kappaB in biological samples by atomic force microscopy.* Biosensors and Bioelectronics, 2010. **25**(11).
11. Nakanishi, K., Sakiyama T; Kumada Y., Imamura K., and Imanaka H., *Recent Advances in Controlled Immobilization of Proteins onto the Surface of the Solid Substrate and Its Possible Application to Proteomics.* Current Proteomics, 2008. **5**: p. 161.
12. Ohnishi, S., Murata M., Hato M., , *Correlation between surface morphology and surface forces of protein A adsorbed on mica.* . Biophys. J. , 1998. **74**(1): p. 455.
13. Steinhauer, C., Wingren C., Khan F., He M., Taussig M.J., Borrebaeck C.A., *Improved affinity coupling for antibody microarrays: engineering of double-(His)6-tagged single framework recombinant antibody fragments.* Proteomics, 2006. **6**: p. 4227.
14. Torrance, L., Ziegler A., Pittman H., Paterson M., Toth R., Eggleston I., *Oriented immobilisation of engineered single-chain antibodies to develop biosensors for virus detection.* J. Virol. Methods, 2006. **134**: p. 164.
15. Vallina-Garcia, R., del Mar Garcia-Suarez M., Fernandez-Abedul M.T., Mendez F.J., Costa-Garcia A., *Oriented immobilisation of anti-pneumolysin Fab through a histidine tag for electrochemical immunosensors.* Biosens Bioelectron, 2007. **23**: p. 210.
16. Lee, J.M., Park H.K., Jung Y., Kim J.K., Jung S.O., Chung B.H., *Direct immobilization of protein g variants with various numbers of cysteine residues on a gold surface.* Anal. Chem., 2007. **79**(7): p. 2680.
17. Lynch, M., Mosher C., Huff J., Nettikadan S., Johnson J., Henderson E.,, *Functional protein nanoarrays for biomarker profiling.* Proteomics, 2004. **4**(6): p. 1695.

18. Silverton, E.W., Navia M.A., Davies D.R., , *Three-dimensional structure of an intact human immunoglobulin.* Proc. Natl. Acad. Sci. U S A 1977. **74**(11): p. 5140.
19. Martinez, N.F., Lozano J.R., Herruzzo E. T., Garcia F., Richter C., Sulzbach T. and Garcia R., *Bimodal atomic force microscopy imaging of isolated antibodies in air and liquids.* Nanotechnology, 2008. **19**.
20. San Paulo, A., García R.,, *High-resolution Imaging of Antibodies by Tappin-Mode Atomic Force Microscopy: Attractive and Repulsive Tip-Sample Interaction Regimes.* Biophysical Journal, 2000. **78**: p. 1599.
21. Fassina, G., Verdoliva A., Odierna M. R., Ruvo M., and Cassini G., *Protein A Mimetic Peptide Ligand for Affinity Purificazione of Antibodies.* Journal of Molecular Recognition, 1996. **9**: p. 564.
22. Tiné, M.R., Alderighi M., Duce C., Ghezzi L., Solaro R., *Effect of temperature on self-assembly of an ionic tetrapeptide.* Journal of Thermal Analysis and Calorimetry, 2011. **103**(1): p. 75-80.
23. Browning-Kelley, M.E., Wadu-Mesthrige K., Hari V., and Liu G. Y., *Atomic Force Microscopic Study of Specific Antigen/Antibody Binding.* Langmuir, 1997. **13**: p. 343.

Synthesis and Characterization of Processable Polyaniline Salts

Salma Gul[a], Anwar-ul-Haq Ali Shah[b], Salma Bilal[a *]
[a]National Centre of Excellence in Physical Chemistry, University of Peshawar
[b]Institute of Chemical Sciences, University of Peshawar

[*]*Corresponding author: salma.bilal@s2004.tu-chemnitz.de*

Abstract
Polyaniline (PANI) is one of the most promising candidates for possible technological applications. PANI has potential applications in batteries, anion exchanger, tissue engineering, inhibition of steel corrosion, fuel cell, sensors and so on. However, its insolubility in common organic solvents limits its range of applications. In the present study an attempt has been made to synthesize soluble polyaniline salt via inverse polymerization pathway using benzoyl peroxide as oxidant and dodecylbenzenesulfonic acid (DBSA) as dopant as well as a surfactant. A mixture of chloroform and 2-butanol was used as dispersion medium for the first time. The influence of synthesis parameters such as concentration of aniline, benzoyl peroxide and DBSA on the yield and other properties of the resulting PANI salt was studied. The synthesized PANI salt was found to be completely soluble in DMSO, DMF, chloroform and in a mixture of toluene and 2-propanol. The synthesized polymer salt was also characterized with cyclic voltammetry, SEM, XRD, UV-Vis spectroscopy and viscosity measurements. TGA was used to analyze the thermal properties of synthesized polymer. The extent of doping of the PANI salt was determined from UV-Vis spectra and TGA analysis. The activation energy for the degradation of the polymer was calculated with the help of TGA.

Keywords: PANI-DBSA salt, common organic solvents, SEM, XRD, UV-Vis spectroscopy.

1. Introduction

In 1970s chemists began to prepare a new type of organic polymers which were good electronic conductors [1]. They were called intrinsically conducting polymers (ICPs). Since their discovery a tremendous research has been done in the field because of the interesting properties and numerous application possibilities of these ICPs. Among ICPs, polyaniline (PANI) is found to be the most promising because of its high stability toward air and moisture, high electrical conducting and unique redox properties [2]. It has been widely employed in various areas such as, fuel cells [3], batteries [4], anion exchanger [5], tissue engineering [6], inhibition of steel corrosion [7], sensors [8] and so on. It can be synthesized by oxidizing aniline either electrochemically or chemically. The latter one is of particular importance since this is the most feasible route for the production of PANI on large scale [9].

However, its poor process ability restricts not only its applications but also determination of molecular structure of this polymer. Therefore, several strategies were developed to overcome these shortcomings. These include copolymerization [10-12], use of substituted aniline as monomer [13] and synthesis by emulsion/inverse emulsion polymerization [14]. In first two cases processability increases but conductivity decreases [15]. So the only feasible route is emulsion/inverse emulsion polymerization. In this case aniline (monomer), a protonic acid and an oxidizing agent are combined with a mixture of water and non polar/ weakly polar solvent such as xylene, chloroform or toluene [16]. But it is seen that at the completion of the reaction the product remains entrain in the emulsion and it is very difficult to collect the PANI in pure form.

In the present work a new inverse emulsion polymerization pathway has been developed to resolve the current limitations of PANI. We introduced a mixture of chloroform and 2-butanol as a novel dispersing medium and polymerization was carried out in the presence of dodecyl benzenesulfonic acid (DBSA) surfactant. This system is unique and simple because after the completion of the reaction, two phases are separated from each other. The organic phase consists of the PANI-DBSA because of its high solubility in chloroform while the aqueous phase consists of byproducts of the reaction such as un-reacted DBSA, 2-butanol and benzoyl peroxide due to their solubility in this phase. The aqueous layer was readily removed from the highly concentrated organic phase. The latter phase was washed several times with acetone for further purification and then it was dried to get highly pure PANI-DBSA in powder form. The synthesized polymers were highly soluble various common organic solvents. The effects of reaction parameters on yield, intrinsic viscosity, morphology, structure, crystallinity of polymers were also investigated. Cyclic voltammetry confirmed the purity and good electroactivity of the polymer. TGA data showed the high thermal stability up to 500 ^{0}C.

1. Experimental

1.1. Materials
Aniline (Acros), Chloroform (Scharlau), Benzoyl Peroxide (Merck), 2-Butanol (Aldrich), DBSA (Acros), 2-Propanol (Merck), Toluene (Scharalau), Water (Milipore).

1.2. Methodology
1.2.1. Synthesis of Polyaniline Salt
In a typical experiment 0.29 mol of chloroform was taken in a 100 mL round bottom flask. Then 1.25 x 10^{-3} mol benzoyl peroxide was added to it under mechanical stirring. To the above solution 0.13 mol of 2-butanol, 3.73 x 10^{-3} mol DBSA, and 5.36 x 10^{-4} mol of aniline were added. To the resulting mixture 0.28 mol of water was added to form a milky white emulsion. The mixture turned green in 5 hours and polymerization reaction was allowed to proceed for 24 hours. In the end the organic phase containing Polyaniline salt was separated and washed four times with 50 mL of acetone. After thorough washing, a dark green highly concentrated polyaniline salt was obtained. Then it was dried at room temperature for 24 hours in a Petri dish. On addition small amount of acetone to the Petri dish the film broke into flakes. Then PANI-DBSA salt was separated by filtration and dried in a desiccator.

The polymers obtained with the varying amount of aniline were labeled as ANI 5, ANI 7, ANI 9, and ANI 11 where the aniline concentrations were 5.36 x 10^{-4} mol, 7.52 x 10^{-4} mol, 9.66 x 10^{-4} mol, 1.18 x 10^{-3} mol respectively. Similarly, the samples with different amounts of benzoyl peroxide were named as BP 101, BP 303, and BP 505 when the amounts of benzoyl peroxide were 4.17 x 10^{-4} , 1.25 x 10^{-3} and 2.08 x 10^{-3} mol, respectively. DB 12, DB 14 and DB16 represent polymers with 3.73 x 10^{-3} , 4.35 x 10^{-3} and 4.97 x 10^{-3} mol of DBSA respectively.

Percent yield of PANI-DBSA was calculated by using following formula [17].

$$\text{Percent yield} = \frac{\text{weight of PANI} - \text{DBSA}}{\text{weight of xM aniline} + \text{weight of xM DBSA}} \text{ X } 100 \ldots\ldots\ldots \text{(i)}$$

Viscosities of PANI-DBSA were determined using capillary viscometer. Intrinsic viscosity [η] was obtained by linearly extrapolating the line of specific viscosity divided by the c (η_{sp}/c) vs c and the line of the natural logarithm of relative viscosity divided by c (ln η_{rel}/ c) vs c to the same intercept at zero concentration [18].

UV-Vis spectra were obtained using Perkin Elmer 650 (UK) Spectrophotometer. A quartz cell of 1 cm path length was used for recording spectra of different samples of PANI-DBSA dissolved in chloroform. Cyclic voltammograms (CVs) were recorded with bipotentiostat 2323 (ALS) in 0.5 M H_2SO_4 at a scan rate of 50 mV/s in a three electrode cell at scan rate of 50 mV/s. PANI-DBSA was dissolved in chloroform and dip coated on gold disc which was used as a working electrode. The reference electrode was saturated calomel electrode and a gold wire was used as counter electrode. CVs were obtained by scanning the potential from E_{SCE} = -0.3 V to E_{SCE} = 1.0 V. The morphology of PANI-DBSA was investigated by using JSM5910 (JEOL Japan) scanning electron microscope. Powder material was deposited on adhesive tape fixed to stub and then coated with gold by sputter coater prior to SEM measurements. X-ray diffraction patterns were recorded by using Rigaku (Japan) X-ray diffractometer using Cu K radiations of wavelength 1.5405 A° with a continuous scan speed of 0.05°/s. Thermal analysis of the polymer was carried out by using Diamond Series TG/DTA Perkin Elmer (USA) at a heating rate 10°/min in N_2 atmosphere.

2. Results and Discussion

There have been several reports on the polymerization mechanism of polyaniline [19, 20]. In the first step oxidation of aniline into a radical cation place by an oxidant. The second step involves the formation of the intermediately oxidized nitrenium cation by losing one proton. Continuous combination of these intermediates leads to the formation of PANI, shown in scheme 1.

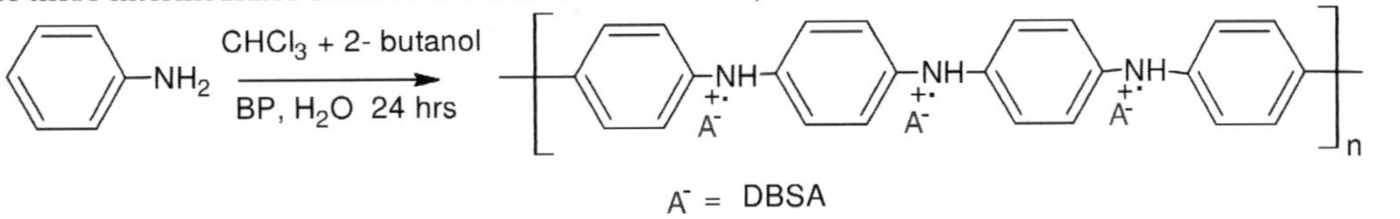

Scheme 1. Molecular structure of PANI-DBSA

3.1 . Yield of PANI-BDSA

The inverse emulsion polymerization reaction of aniline was carried out by varying the amount of benzoyl peroxide, aniline and surfactant. At first for low concentration of aniline no yield obtained, attained a maximum value at 5.36 x 10^{-4} mol and then decreased with further increasing the monomer concentration. Similar observations were obtained with benzoyl peroxide and surfactant concentration (Table 1). The observed change in the yield of the product with different parameters may be due to overoxidation process or may be that the kinetic effects were favored in the oxidation of aniline [21].

Table 1. % yield of PANI-DBSA

S.No.	Sample	% Yield
1	ANI 3	0
2	ANI 5	21
3	ANI 7	9.91
4	ANI 9	9.7
5	ANI 11	8.58
6	BP 101	0
7	BP 303	21
8	BP 505	0
9	DB 12	21
10	DB 14	15.3
11	DB 16	16.2

3.2. Solubility

It has been demonstrated that PANI doped by large sized organic sulfonic acids, such as DBSA have a good solubility and processibility in common organic solvents. It is due to the enlarge interchain distance, which means weaker interchain reaction, and strong interaction between dopant and solvent [22]. In the present work the synthesized PANI-DBSA is soluble in mixture of toluene and 2-propanol, chloroform, DMSO and DMF.

3.3. Intrinsic Viscosity

The intrinsic viscosities of PANI-DBSA prepared at different aniline concentrations are given in table 2. It was observed that intrinsic viscosity of PANI-DBSA decreased with increasing monomer concentration in the order of ANI 5 > ANI 7 > ANI 9 > ANI 11. The high intrinsic viscosity of PANI-DBSA was 32.09mL/g for ANI 5. The dropping trend in the intrinsic viscosity is viewed as a direct consequence of rapid lowering in initiator efficiency [23].

Table 2: Change of Intrinsic Viscosity with change in monomer concentration

S. No	Sample	Intrinsic viscosity [η] (mL/g)
1	ANI 5	33.43
2	ANI 7	24.96
3	ANI 9	21.77
4	ANI 11	19.31

3.4. UV-Vis Spectroscopy

Absorption spectroscopy is a valuable tool to detect the presence of polyaniline salt (conducting state) and its base (insulating state). Absorption spectra of PANI-DBSA dissolved in chloroform are shown in Fig. 1. All the spectra showed three peaks in the range of 349-360, 433-436, and 782-810 nm. The first absorption bands arises from π - π^* electron transition within benzoid (B) ring, the second and third absorption bands are attributed to the polaron transition (quinoid, Q) and bipolaron transition (exiton) respectively. The last two bands relate to the doping level and formation of polaron lattice, which represent the protonation stages of PANI chain. The last bands are usually ascribed to an exiton, located in the quinoid ring, ascribing from charge transfer from adjacent benzenoid rings with each side contributing half an electron on average [24, 25]. All these bands are characteristics of PANI salt.

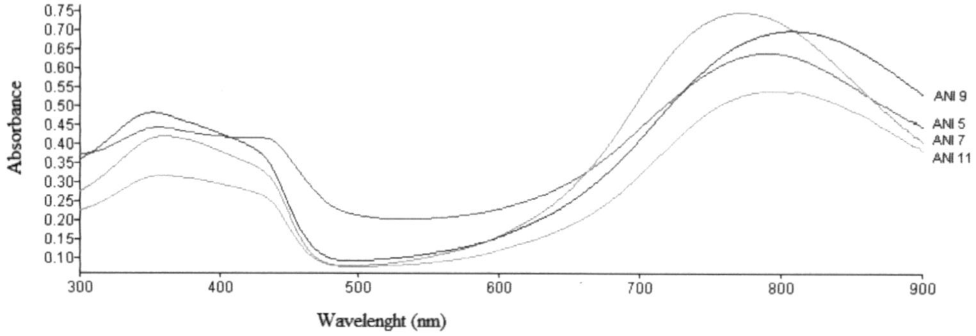

Fig. 1: UV-Vis spectra of PANI-DBSA salts recorded in chloroform solvent

3.4.1. Extent of doping

The extent of doping can be estimated from the absorption spectra of PANI in which the exciton (π-polaron transition)/ benzenoid (π-π^* transition) ratio indicates the doping level [26]. It is seen from table 3 that ANI 5 has the biggest ratio among all sample which means that doping level of ANI 5 is higher as compared to other samples of the synthesized PANI-DBSA. These results show that the concentration of aniline plays an important role in the synthesis of PANI-DBSA.

Table. 3: Absorption ratio exciton/benzenoid (E/B) for different samples of PANI-DBSA

1	ANI 5	2.27
2	ANI 7	2.24
3	ANI 9	2.24
4	ANI 11	2.25

3.5. Cyclic Voltammetry

The redox property of the synthesized PANI-DBSA was investigated by cyclic Voltammetry. Fig.2. shows the representative CV of ANI 5. It shows two redox peaks. The first redox peak around E_{SCE} = 0.2 V is assigned to the transition of leucoemeraldine to emeraldine state and the second peak around E_{SCE} 0.6V is due to the emeraldine to Pernigraniline transition. These peaks confirmed the electroactivity of the synthesized polymer [27].

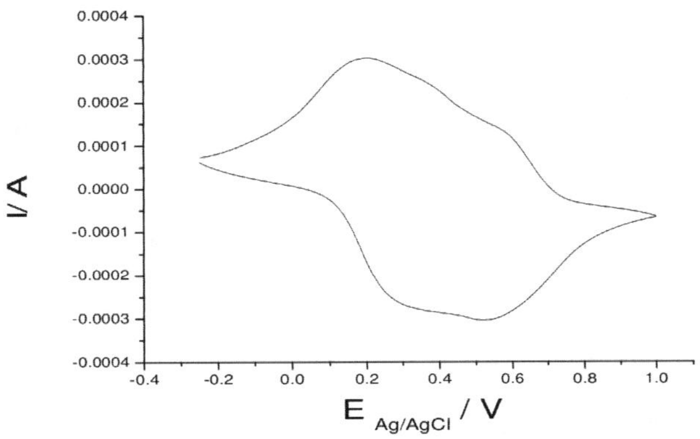

Fig.2. Cyclic voltammograme of ANI 5.

3.6. XRD of PANI-DBSA

XRD is extensively used in material identification. It is a non destructive, very simple and rapid technique for powder and other microcrystalline samples. The XRD patterns of ANI5 are shown in Fig. 3 and the value of 2Θ and d spaces are given in table 4. The peaks from 20° to 27° are ascribed to the momentum transfer and periodicity and perpendicular to the chain direction [28, 29]. The peaks at 2Θ ≈ 19° and another one at 2Θ ≈ 25° are characteristics of van der waals distances between aliphatic chain and between stakes of phenylene rings respectively [30, 31]. The peak at 2.5° showed the well organized layered structure in which stacks of charged aniline backbone are uniformly spaced by the alkyl tails of DBSA [32]. All peaks are observed at lower value of 2Θ indicating the better crystallinity [33].

Table 4. A comparison of the major peaks of XRD patterns for PANI-DBSA salts

Peak	ANI 5	
Number	2Θ (degree)	d (A°)
1	2.5	34.617
2	19	4.659
3	20	4.425
4	25.05	3.545
5	27	3.293

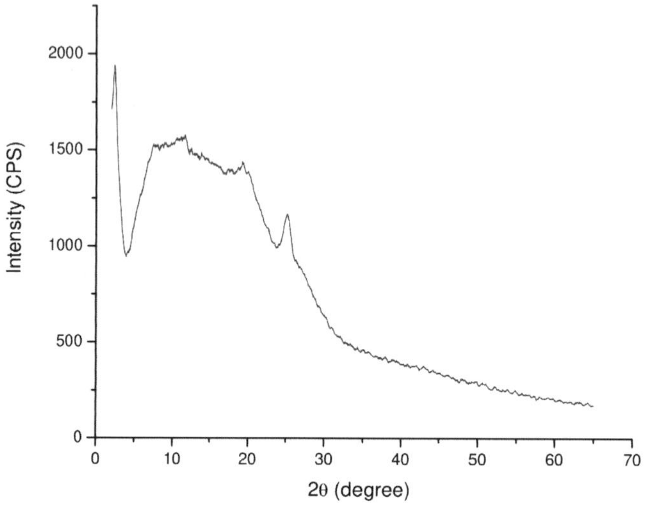

Fig. 3. XRD pattern of ANI 5

3.7. SEM of PANI-DBSA

Fig.4 demonstrates the scanning electron micrographs of PANI-DBSA prepared with different concentrations of aniline. These show granular morphology of the polymers, which is often been reported [34, 35]. It has been assumed that during polymerization phanazine nucleates associate randomly to produced granular structures of PANI. The reasons for such type of associations are high concentration and reduced solubility of these nucleates. Random associations also happen when the conversion rate of phanazine nucleate to initiation centers (to start PANI growth) is very fast and thus PANI chains grow before the organization of phanazine nucleates [36].

All these images show irregular shapes with no significant change in the morphology which indicate that change in concentration of monomer does not have an effect on the morphology, of resulting PANI. However, their porosity increased with increase in aniline concentration and can be used as biosensors [37].

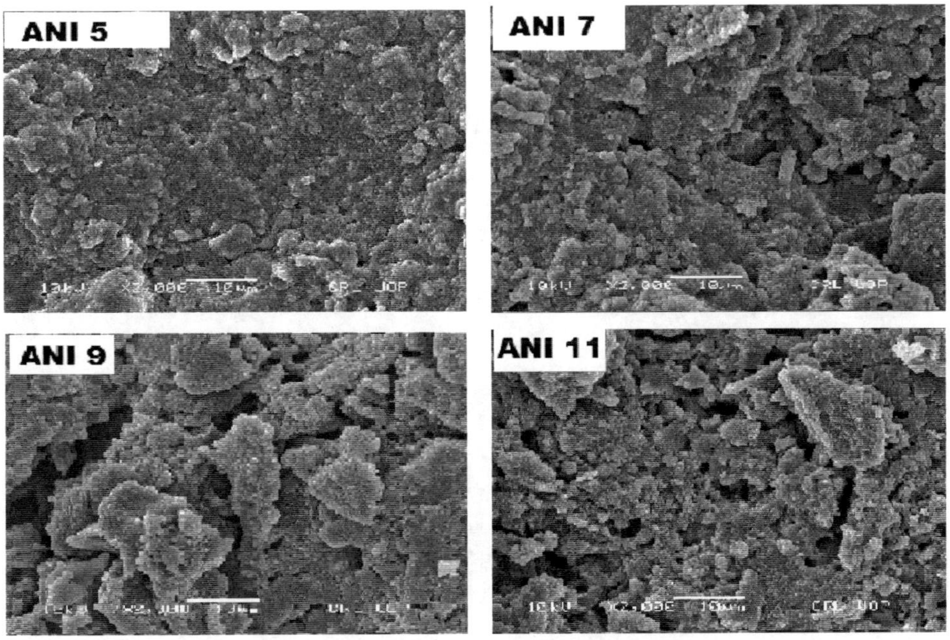

Fig. 4: SEM pictures of PANI-DBSA salts

3.8. Thermogravemetry of PANI-DBSA

The results of the TGA of PANI-DBSA are given in Fig. 5. The DBSA doped polyaniline salts synthesized at room temperature showed three main weight loss stages. The first step weight loss of about 5-12% is associated with the loss of moisture. PANI always shows high moisture loss because it is highly hygroscopic in nature and some moisture still remains even after vacuum drying [38]. The second step weight loss being in the range 48-66%, is attributed to destroy of the NH^+.....SO_3^- interaction between the PANI chain and the DBSA dopant with degradation of DBSA [39]. A slow and somewhat gradual weight loss was observed for these polymers at around 500 °C is due to the structure decomposition of PANI backbone. The weight loss of DBSA doped polymers are almost proportional to their percentage weight of dopant and attributed mainly to a dopant loss. The PANI-DBSA salts were not completely destroyed because in nitrogen atmosphere carbonization of polymer takes place leaving a marked residue [40]. These results show that DBSA doped PANI material is highly stable and can be used for various high temperature applications.

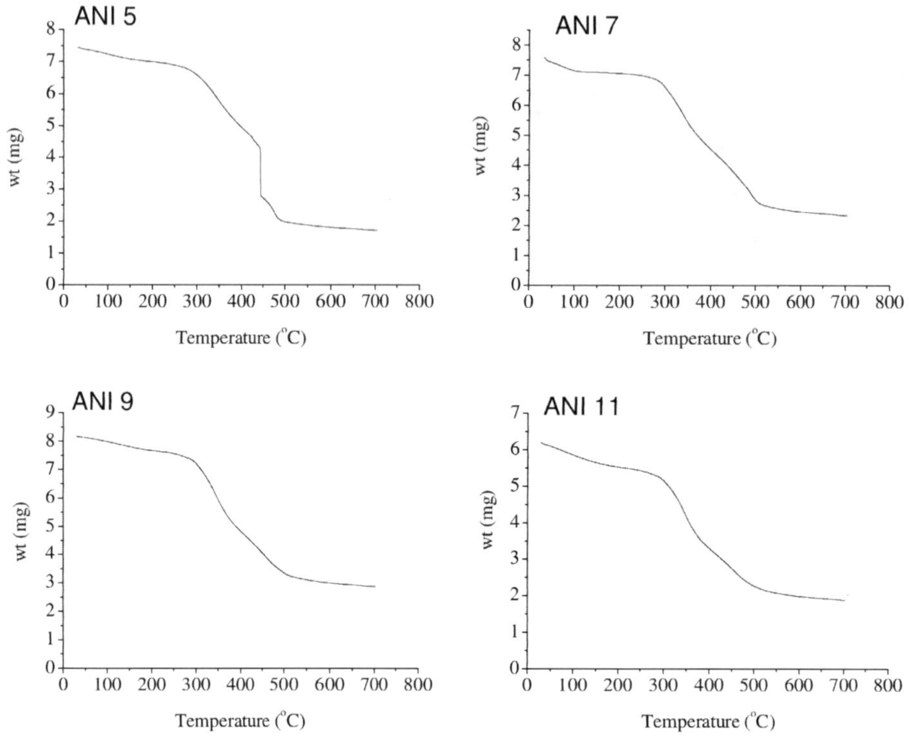

Fig. 5. TGA Thermograms of PANI-DBSA

3.8.1. Activation Energy for the Degradation of PANI-DBSA

The following method was used to evaluate kinetic parameters for degradation of PANI-DBSA by assuming first order kinetics [41].

3.8.1.1. Horowitz and Metzger method

Horowitz and Metzger [42] obtained following equation:

$$\ln[\ln\{(Wo-Wf)/(Wt-Wf)\}] = Ea\,\theta / RTs^2 \qquad (1)$$

Where
Wo = initial weight of the sample
Wf = final weight of the sample
Wt = the weight of the sample at temperature t
R = gas constant
θ = T - Ts
Ts is the reference temperature

From the slope (-Ea/R) of this plot activation energy can be calculated. Fig. 6 shows the plots obtained from kinetic analysis of TGA data of PANI doped with DBSA and activation energy calculated from these plots is tabulated (Table 5). The Ea for the degradation is the highest for ANI 9 and is thermally more stable.

Table 5. Activation energy (Ea in kJ/mol) for PANI-DBSA as a function of monomer (Aniline) concentration

S. No	Sample	Activation energy (kJ/mol)
1	ANI 5	46.16
2	ANI 7	47.77
3	ANI 9	47.94
4	ANI 11	38.91

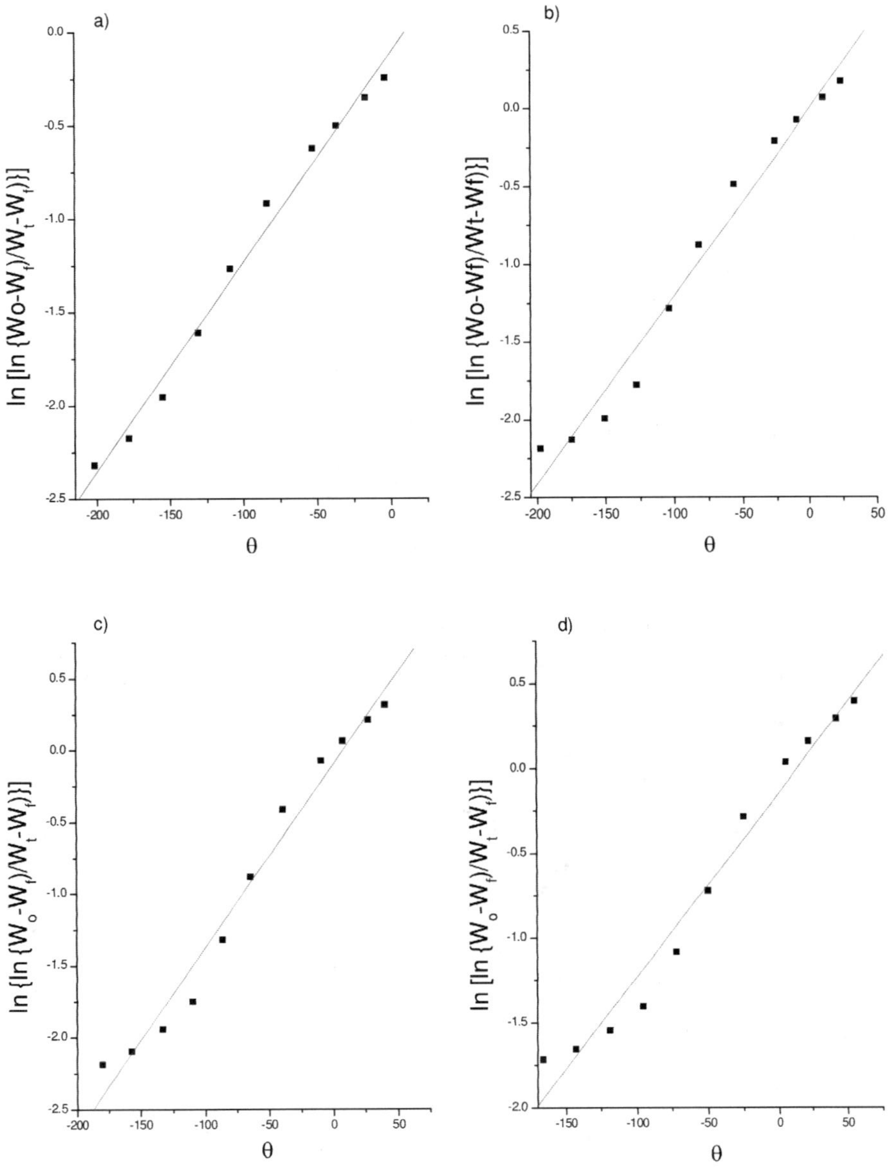

Fig. 6: Calculation of activation energy for a) ANI 5, b) ANI 7, c) ANI 9 d) ANI 11

Conclusions

PANI-DBSA salts were successfully prepared via inverse emulsion polymerization. These salts were soluble in so far highest number of common organic solvents. It was found that intrinsic viscosity decreases with increase in monomer concentration in the feed. The results of TGA reveal that the prepared polymers were thermally stable and can be used in various technologies.

Acknowledgement
National centre of Excellence in Physical Chemistry, University of Peshawar is gratefully acknowledged for research facilities and financial support.

References

[1] Palaniappan S 2004 *Polym. Adv. Technol.* **15** 111-117
[2] Roy B C, Gupta M D, Bhoumik L and Ray J K 2002 *Synth. Met.* **130** 27-33
[3] Huang Y F, Chang C S and Lin C W 2012 Int. J. Hydrogen Energy . **7** 11975-11983
[4] Lee M H, Luo Y C and Do J S 2005 *J. Power. Sourc.* **146** 340–344
[5] Ghorbani M, Esfandian H, Taghipour N and Katal R 2010 *Desalination* **263** 279–284
[6] Li M, Guo Y, Wei Y, MacDiarmid A G and Lelkes P I 2006 *Biomaterials* **27** 2705–2715
[7] Sakhri A, Perrin F X, Aragon E, Lamouric S and Benaboura A 2010 *Corr. Sci.* **52** 901-909
[8] Virji S, Huang J, Kaner R B and Weiller B H 2004 *Nano. Lett.* **4** 491-496
[9] Erdem E, Karakisla M and Sacak M 2004 *Eur. Polym. J.* **40** 785-791
[10] Shah A A and Holze R 2008 *Electrochim. Acta* **53** 4642-4653
[11] Bilal S and Holze R 2009 *Electrochim. Acta* **54** 4851-4856
[12] Shah A A, Bilal S and Holze R 2012 *Synth. Met.* **164** 356-363
[13] Bilal S, Shah A A and Holze R 2011 *Electrochim. Acta.* **56** 3353-3358
[14] Kim J, Kwon S and Ihm D 2007 *Curr. Appli. Physic.* **7** 205-210
[15] Palaniappan S and John A 2008 *Prog. Polym. Sci.* **33** 732-758
[16] Bhadra S, Khastgir D, Singha N K and Lee J H 2009 *Prog. Polym. Sci.* **34** 783-810
[17] Shreepathi S and Holze R 2006 *Langmuir* **22** 5196-5204
[18] Yang J and Weng B 2009 *Synth. Met.* **159** 2249-2252
[19] Sapurina I and Stejskal J 2008 *Polym. Int.* **57** 1295- 1325
[20] Ding Y, Padias A B, Hall H K and Jr 1999 *J. Polym. Sci. Part A: Polym. Chem.* **37** 2569-2579
[21] Palaniappan S and Amarnath C A 2006 *Rect. Func. Polym.* **66** 1741-1748
[22] Huang J and Wan M 1999 *J. Polym. Sci. Part A: Polym. Chem.* **37** 1277-1284
[23] Ghosh P, Siddhanta S K, Haque S R and Chakrabarti A 2001 *Synth. Met.* **123** 83-89
[24] Yang Y, Ding Y, Chen G and Li C 2007 *Eur. polym. J.* **43** 3337-3343
[25] Han Y G, Kusunose T and Sekino T 2009 *Synth. Met.* **159** 123-131
[26] Swaruparani H, Basavaraja S, Basavaraja C, Huh D S and Venkataraman A 2010 *J. Appl. Polym. Sci.* **117** 1350–1360
[27] Shreepathi S and Holze R 2005 *Chem. Mater.* **17** 4078-4085
[28] Chen J, Xu Y, Zheng Y, Dai L and Wu H 2008 *C. R. Chimie.* **11** 84-89
[29] Palaniappan S and Devi S L 2006 *Polym. Degrade. Stab.* **91** 2415-2422
[30] Sarvanan C, Palaniappan S and Chandezon F 2008 *Mat. Lett.* **62** 882-885
[31] Duffour B, Rannon P, Fedorko P, Djurado D and Travers J P 2001 *A. Pron. Chem. Mater.* **13** 4032-4040
[32] Han M G, Cho S K, Oh G S and Im S S 2002 *Synth. Met.* **126** 53-60
[33] Vijayan M and Trivedi D C 1999 *Synth. Met.* **107** 57-64
[34] Zhang X, Kolla H S, Wang X, Raja K and Monohar S K 2006 *Adv. Funt. Matter.* **16** 1145-1152
[35] Konyushenko E N, Stejskal J, Sedekova I, Trchova M, Sapurina I, Cieslar M and Prokes J 2006 *Polym. Int.* **55** 31-39
[36] Sapurina I and Stejskal 2008 *J, Polym. Int.* **57** 1295-1325
[37] Gaikwad P D, Shirale D J, Gade V K, Savale P A, Kharat H J, Kakde K P and Shirsat M D 2006 *Int. J. Electrochem. Sci.* **1** 425-434
[38] Bhadra S, Chattpadhyay S, Singa N K and Khastgir D 2008 *J. Appl. Polym. Sci.* **108** 57-64
[39] Babazadeh M 2009 *J. Appl. Polym. Sci.* **113** 3980-3984
[40] Brozova L, Holler P, Kovarova J, Stejskal J and Trchova M 2008 *Polym. Degrad. Stab.* **93** 592-600
[41] Rao P S, Subrahmanya S and Sathyanarayana D N 2002 *Synth. Met.* **128** 311-316
[42] Horowitz H H and Metzger G 1963 *Anal. Chem.* **35** 1464-1468

Secondary electron energy loss coincidence (e,2e) spectroscopy on Ag and Si surfaces

Rahila Khalid*, Francesc Salvat Pujol and Wolfgang S. M. Werner

Institute of Applied Physics, Vienna University of Technology, Wiedner Hauptstrasse

8-10/134, A-1040 Vienna, Austria

*E-mail: khalid_rahila@yahoo.com

Abstract. Qualitative information on the mechanism of secondary electron emission with respect to loss processes has been gathered by measuring the coincidences between secondary electrons and inelastically backscattered electrons on a polycrystalline silver (Ag) and an amorphised silicon (Si) sample using a specially designed (e,2e) coincidence spectrometer. Primary electrons impinge on the sample with an energy of 100 eV. Correlated emission of electron pairs was observed with a hemispherical mirror analyzer (backscattered electrons with losses up to 50 eV) and a time of flight analyzer (secondary electrons with energies up to 50eV). A comparison of these experimental data with a Monte Carlo (MC) simulation model is made. It is found that surface excitations play an important role in secondary electron emission mechanism. For Ag, low energy loss distributions were compared with calculations using linear response theory and agreed satisfactorily even though the expected limit of validity of this type of theories is for energies ≥100 eV.

1. Introduction

For electrons impinging on solid targets an important reaction channel involves a single collision event with a valence electron resulting in two electrons leaving the solid. Energy and angle resolved observation of these two electrons in coincidence is known as (e,2e) spectroscopy. The energy and momentum of the valence electron is determined from the known energies and momenta of the primary and the detected electrons using energy and momentum conservation. Early (e,2e) coincidence experiments were performed on gases [1]. The study of solids has always been more complicated. In a solid, before escaping from the surface, an unknown amount of energy can be lost in a "cascade" of subsequent scattering events. Besides this, the work function must be overcome. In case of transmission (e,2e) spectroscopy, high incident electron energy can help in reducing the additional scattering but it reduces the (e,2e) count-rate also and leads to poor statistics and long acquisition times. Also the transmission mode experiments are difficult to perform because of the need to prepare very thin samples. Experiments with thin foils and fast electrons in transmission geometry have the advantage that elastic processes are insignificant and the trajectory is more or less a straight line which can be interpreted entirely in terms of bulk scattering. An equally interesting application of (e,2e) spectroscopy is to use it in reflection geometry, first performed about 2 decades ago [2]. Use of a reflection mode involves electrons with low energies, where elastic scattering cannot be neglected. A mechanism of incoherent elastic reflection and inelastic collisions constitutes the framework for the interpretation of these experiments.

Secondary electron emission is a well-known phenomenon. This effect causes solid-state electrons to be emitted when a target is hit by an electron beam. This secondary emission effect was discovered a century ago [3] and has many important technological applications. In some areas (e.g. particle detectors) it is desirable to have large a secondary electron yield while in some cases (e.g. particle accelerators and storage rings [4]) a low secondary electron emission yield is desirable. A great effort has been made by Lin and Joy [5] in compiling measurements from over 80 years from more than hundred different groups.

Journal of Physics: Conference Series **439** (2013) 012003 doi:10.1088/1742-6596/439/1/012003

They developed a universal formula in which certain parameters are fitted for different materials with experimental data. But analysis of this huge dataset reveals significant discrepancies for the same dataset measured by different groups. Thus the phenomenon of secondary electron emission is still a subject of research.

Experimentally it is difficult to deal with slow secondary electrons. Reflection electron energy loss spectroscopy (REELS) is a well established tool to observe energy loss distribution of reflected electrons. But the interpretation of these spectra is more difficult at low energies since the secondary electron peak and loss features overlap. Furthermore the secondary electron spectrum is a superposition of many different excitation processes which are superpositioned in the emitted secondary electron energy distribution making it very difficult to study the emission mechanism of secondary electrons experimentally. Thus the information on the secondary electron emission mechanism one can obtain only by measuring the secondary electron spectrum is limited. Detailed information on the dynamics of low energy electrons in solids can be accessed by means of electron coincidence spectroscopy. To investigate the mechanism of secondary electron emission, (e,2e) spectroscopy has been utilized by measuring the secondary electrons in coincidence with the reflected/transmitted electrons with a specific energy loss[6, 7, 8, 9, 10]. In the present work double differential secondary electron spectra have been measured for a large energy loss range of backscattered electrons by using a specially designed (e,2e) coincidence spectrometer. The unique dataset, interpreted with the aid of state-of-the-art Monte Carlo (MC) simulations, gives detailed information on the emission mechanism of the secondary electrons from solids.

2. Experiment

The (e,2e) coincidence spectrometer consists of an electron gun and two electron analyzers flanged to a μ-metal shielded vacuum chamber with base pressure of ~ 10^{-10} mbar. The experimental setup is shown schematically in figure 1. A hemispherical mirror analyzer (HMA) with 5 detectors (channel electron multipliers CEMs) is used to record backscattered electrons with a polar opening angle of ±12° with an effective energy resolution of 1.25 eV for a pass energy of 50 eV. The secondary electrons are detected by a time of flight (TOF) analyzer with multichannel plate (MCP) detector. The acceptance angle of the TOF analyzer is 2π achieved by applying a positive bias to the outer two grids and the drift tube for fast acquisition (VG; VR; VD in figure 1). The plane of incidence describes an angle of 35°. The sample is mounted at the center of the vacuum chamber where it can be cleaned and in case of Si amorphised by sputtering with 4 keV Ar+ ions. The cleanliness of the sample is monitored by Auger electron spectroscopy before and after a coincidence measurement. This cleaning procedure was repeated every 24 h and subsequent a coincidence measurement was started.

The coincidences are detected by standard coincidence electronics. The signals from CEMs and MCP/DLA are processed by respective constant fraction discriminators (CFD). A time to digital convertor (TDC), started by a periodic signal from a PCI card, is used to store timing information of the events to computer hard disk. The evaluation software searches for coincidences by subtracting arrival times of electron events at both detectors. The flight times in the TOF analyzer are calibrated by a calibration measurement performed before each coincidence measurement. The time scale is converted to energy scale using the equations of motion of an electron inside the TOF analyzer. The outcome is a secondary electron coincidence spectrum for an energy loss of reflected electrons. These spectra have been measured for a range of energy losses between 0 and 50 eV in steps of 0.25 eV to obtain the double differential coincidence yield.

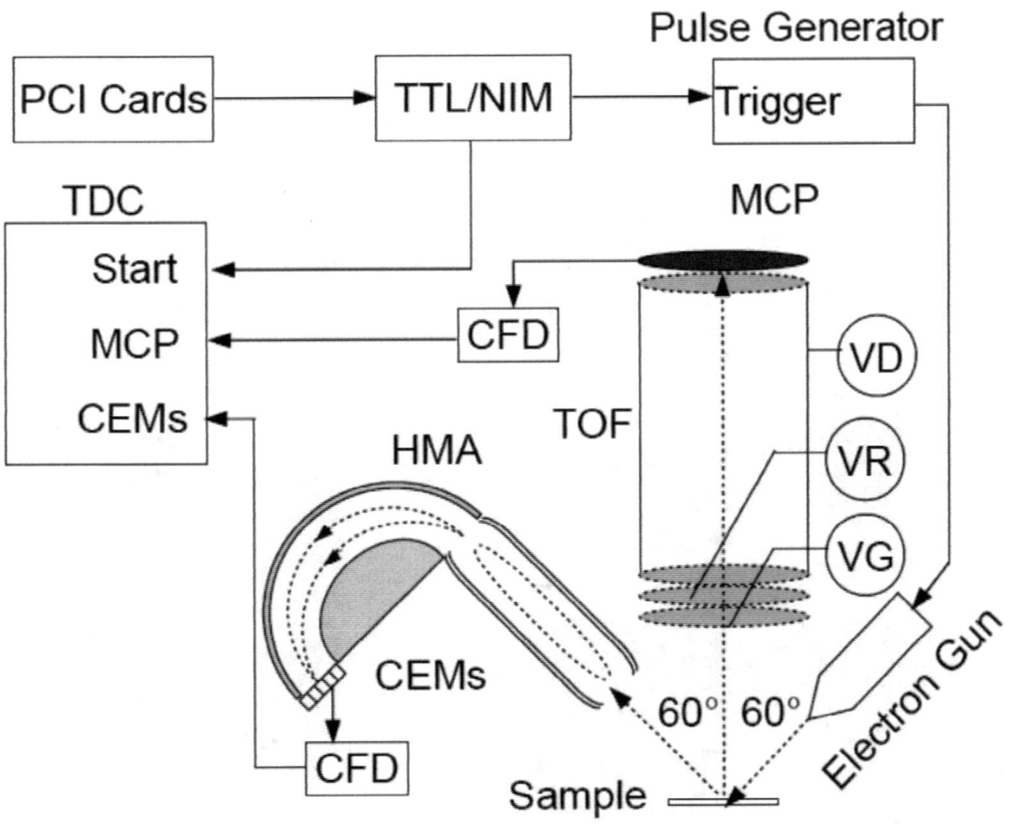

Figure 1. (e,2e) coincidence spectrometer

3. Results and Discussion

The measured double differential coincidence spectra for polycrystalline Si and Ag for primary energy of 100 eV are shown in figure 2. Panel (a) shows REELS for polycrystalline Si measured with 100 eV primary energy. The spectrum shows a peak at 12 eV corresponding to excitation of one surface plasmon (label "1s") and the peak seen at 17 eV corresponds to excitation of one bulk plasmon (label "1b"). The peak near 30 eV corresponds to reflected electrons that have excited a surface as well as a bulk plasmon excitation. Coincidence spectra for correlated electron pairs for Si is shown in panel (b) where the energy distribution of the secondary electrons is shown on a time of flight scale. The same dataset with the secondary electrons energy distribution on a kinetic energy scale is presented in panel (c). The red line in this curve represents the secondary electrons which are exactly at the Fermi edge originally and are excited by a collision with incoming primaries and escaping from the solid without any energy loss. The signal in these spectra represents the number of true coincidences obtained by subtracting at background of false coincidences. Panels (d)-(f) are the results of MC calculations based on the dielectric response theory of energy loss deep inside the solid and accounting for the influence of the vacuum-solid boundary (surface excitations)[11]. figure 2(d) shows MC results in which the surface excitations are neglected while panel (e) are MC simulations produced with surface excitations accounted only for the primary electrons . In the lower panel (f) surface excitations are accounted both for the primary and secondary electrons.

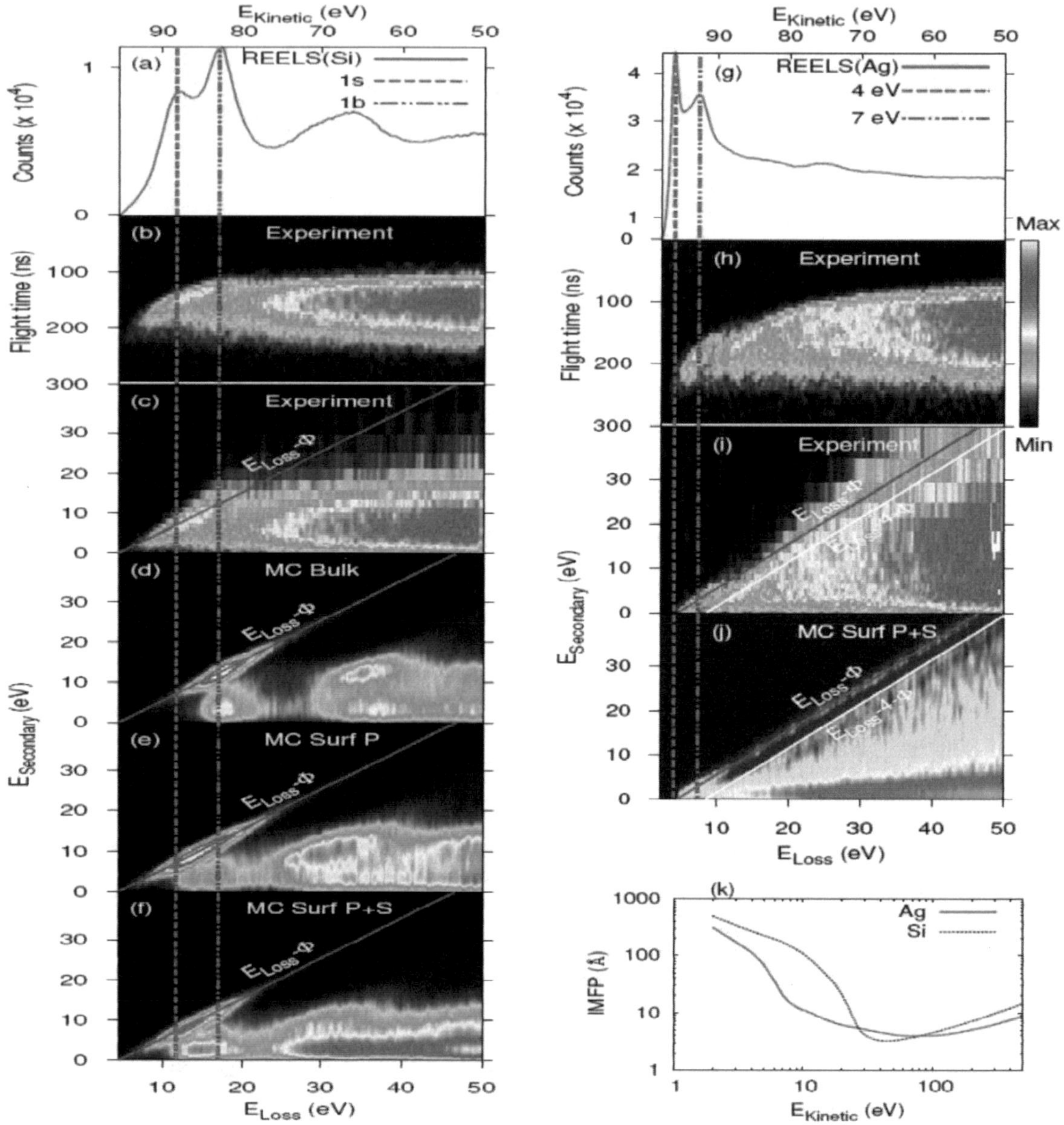

Figure 2. (a) Reflection Electron Energy Loss Spectrum (REELS) of Si; (b) Double differential coincidence yield for Si with the secondary electron energy distribution represented on a time of flight scale; (c) same as (b) using after conversion of the time of flight scale to an energy scale; (d) MC calculations in which surface excitations are neglected; (e) MC simulations in which surface excitations are taken into account for the primaries only; (f) MC results for which surface excitations of both primaries and secondaries are considered; (g) REELS of Ag; (h) same as (b) for Ag; (i) same as (c) for Ag; (j) MC results with surface excitations included both for the primary and secondary electrons; (k) Inelastic mean free path (IMFP) as a function of the energy for Ag and Si

Inspection of the measured spectra figure.2(b) and (c) reveals a region of high intensity between the red lines labelled "1s" and "1b". This region has its maximum exactly at the energy corresponding to excitation of a single surface plasmon ("1s"). For larger energy losses the intensity decreases rapidly reaching a minimum for $\Delta E \sim \hbar\omega_b$, as observed earlier in the case of Al [9, 10]. For the energy loss greater than twice the surface plasmon excitation energy, a broad region with gradually increasing intensity is seen. The narrow ridge between the lines labeled "1s" and "1b" can be understood by inspection of figure.2(k) which shows that for energies below the bulk plasmon excitation energy, the value of inelastic mean free path (IMFP) is very large. In consequence, when a primary liberates a solid-state electron by transferring an energy less than the bulk plasmon energy, the probability for the secondary to be scattered again is very small: in the majority of cases it will escape with the full energy transfer. This explains the ridge-like structure of this feature. The intensity of the secondary electrons for this region is due to the decay of the surface plasmons. For energies above the bulk plasmon loss the IMFP is small, therefore if a bulk plasmon decays, most probably the secondary electron can lose energy by multiple inelastic collisions in particular surface excitations. This also explains why the intensity of secondary electrons drops rapidly for secondary electron energies larger than the bulk plasmon energy (minus the work function).

Comparison of the simulated spectra with the experimental data clearly reveals the role of surface excitations played in secondary electron emission: when only volume scattering is accounted for in the simulations, the ridge at low energies appears at energy losses corresponding to the bulk plasmon energy instead of the surface plasmon energy (panel (d)). When surface excitations of the primaries are "switched on" the width of the ridge in the simulated data increases towards lower energies (panel (e)), but only when surface excitations of both the primaries and secondaries are accounted for (panel (f)), does the width of the ridge match the experimental data.

Essentially the same dataset is presented for polycrystalline Ag in figure.2 (g)-(j) (except panels (c) and (d)). The REELS of Ag presented in panel (g) has characteristic losses at ~4 and ~7 eV that correspond to surface plasmon excitations. Three distinct regions are observable in the double differential spectra presented in figure 2 (i) and (j): a narrow ridge-like region centered around the surface excitation energy (~7eV), a second narrow ridge extending from $\Delta E \sim 10$ - 20 eV and a broad region above $\Delta E \sim 20$ eV with extending up to the energy conservation line ($E_{sec} = E_{Loss} - \Phi$). There is no intensity around 4 eV in these spectra. The secondary electrons generated due to an energy loss of 4 eV cannot come out of the solid due to the work function barrier, which amounts to $\Phi = 4{:}5$ eV for Ag. The region around 7 eV corresponds to the secondary electrons generated due to a decay of surface plasmon without further inelastic collision on their way out from the solid as the IMFP at this kinetic energy is very large. The width of this region corresponds to the width of the surface excitations in the REELS of Ag. It is therefore concluded, just as in the case of Si (and Al, see Refs. [9, 10] that the sharp ridge-like structures in the double differential coincidence yield for energy losses just in excess of the work function are due to surface excitations. The intensity of the second region between 10 and 20 eV matches the white line $E_{Loss} - 4 - \Phi$ which corresponds to the secondary electrons originally created with an energy equal to the energy loss of the primary that are emitted after suffering exactly one inelastic collision before they come out of the solid. The IMFP for Ag is gradually decreasing till 50 eV kinetic energy which explains why the width of the coincidence spectra extends up to the line $E_{Loss} - \Phi$. The simulated spectra shown in panel (j) nicely reproduce the experimental results for Ag, at least qualitatively. Thus to correctly model the secondary electron coincidence spectra it is necessary to include the surface excitation loss channel both for the primary and the secondary electrons.

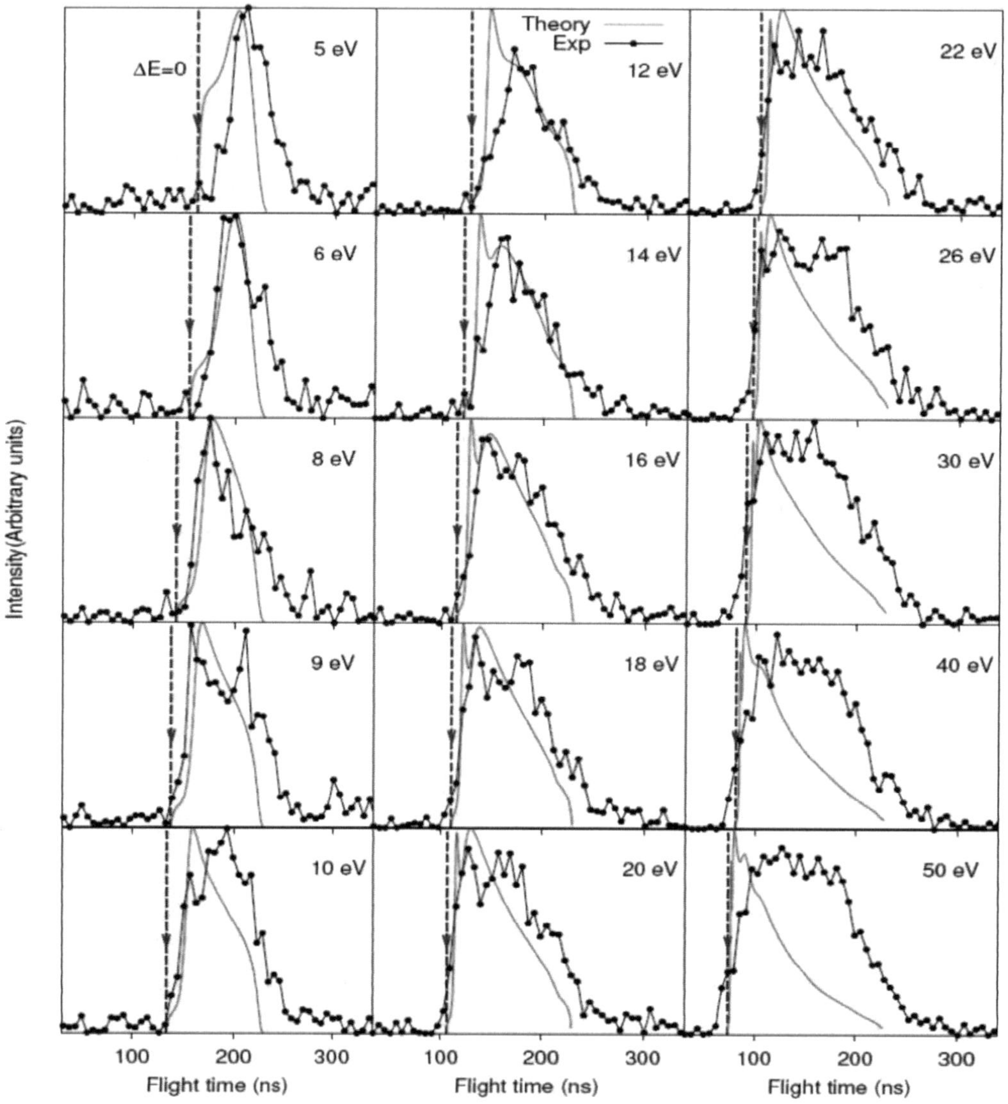

Figure 3. Qualitative comparison of experimental coincidence spectra for a specified secondary electron kinetic energy indicated in each panel with corresponding calculated differential inelastic inverse mean free paths (DIIMFP) for Ag. Line with points (black): Measured coincidence spectra, Line (red): Calculated DIIMFP, Dashed(blue): Zero energy loss without accounting for the work function

The inelastic cross section is an important parameter used for the quantitative understanding of the change in energy distribution of electrons as they move in solids. This quantity has been extracted experimentally [12, 13, 14] as well as theoretically [15, 16] for kinetic energies in the range 50 eV-10 keV. For kinetic energies below 50 eV, it is difficult to obtain experimental results for the differential inverse inelastic mean free path. Since the mean free path for secondary electrons in this energy range is very large, the coincidence data at any given energy loss should be very similar to the single scattering loss distributions for the secondary electrons traveling with this energy. This is confirmed by the coincidence spectra shown in figure 3 that are compared with the energy loss probability for single

Journal of Physics: Conference Series **439** (2013) 012003 doi:10.1088/1742-6596/439/1/012003

inelastic collision (differential inverse inelastic mean free paths, DIIMFP) calculated by the present model. This comparison indeed gives a good agreement with the experimental coincidence spectra for kinetic energies $\leq$ 20 eV. Discrepancies appear in the calculated DIIMFP and experimental distribution for energies $\geq$ 20 eV. The coincidence spectra at these energies become broader due to multiple scattering. The calculations based on dielectric theory correctly predict the single scattering loss distributions for very low energies at least qualitatively.

4. Conclusions

Double differential spectra of electron energy losses and secondary electrons emitted as a result have been measured and compared with MC calculations taking into account both volume and surface losses. Good qualitative agreement is found in general explaining the importance of surface excitation for secondary electron emission. Comparison of individual coincidence spectra for a given energy loss with calculations of the single scattering loss distributions gives reasonable agreement indicating that the employed theory is useful even for very low electron energies (~ 5 eV) being much lower than the assumed lower limit of validity of the theory (~100 eV).

Acknowledgments

The present work was supported by the Austrian Science Foundation FWF (Porject P20891- N20). A research scholarship of the Higher Education Commission of Pakistan via Österreichischer Austauschdienst in Vienna is gratefully acknowledged by the first author. F. Salvat-Pujol is grateful to Fundaci_on Caja madrid for PhD grant and a travel grant of the Fundacion Agusti Pedro i Pons.

References

[1] McCarthy I E and Weigold E 1988 *Rep. Prog. Phys.* **51** 299
[2] Kirschner O M A J and Terekhov A N 1992 *Phys. Rev. Lett.* **69** 1711
[3] Austin L and Starke H 1902 *Ann. Phys.* **4** 271
[4] Cimino R, Colins I R, Furman M A, Pivi M, Ruggiero F, Rumolo G and Zimmermann F 2004 *Phys. Rev. Lett.* 93 014801-1
[5] Lin Y and Joy D C 2005 Surf. Interface. Anal. **37** 895
[6] Pijper F J and Kruit P 1991 *Phys. Rev.* **B 44** 9192
[7] M ullejans H and Bleloch L 1992 *Phys. Rev.* **B 46** 8597
[8] Samarin S, Berakdar J, Suvorova A, Artamonov O M, Waterhouse D K, Kirschner J and Williams J F 2004 *Phys. Rev.* **B 70** 073403/1
[9] Werner W S M, Ruocco A, O_ F, Iacobucci S, Smekal W, Winter H and Stefani G 2008 *Phys. Rev.* **B 78** 233403
[10] Werner W S M, Salvat-Pujol F, Smekal W, Khalid R, Aumayer F, Stori H, Ruocco A and Stefani G 2011 *Appl. Phys. Lett.* **99** 184102
[11] Salvat-Pujol F 2012 Secondary electron emission from solids: coincidence experiments and dielectric formalism (PhD Thesis)
[12] Tougaard S and Chorkendorff I 1986 *Phys. Rev.* **B 35** 6570
[13] Werner W S M 2005 *J. Surf. Sci.* **588** 26
[14] Werner W S M 2006 *Phys. Rev.* **B 74** 075421
[15] Boutboul T, Akkerman A, Breskin A and Chechik R 1996 *J. Appl. Phys.* **79** 6714
[16] Salvat-Pujol F and Werner W S M *Surf. Intrf. Anal.* (Submitted for publication)

CFD analysis of heat transfer within a bottom heated vertical concentric cylindrical enclosure

Asif Hussain Malik[1]*, Ajmal Shah[2] and Shahab Khushnood[1]
[1] Department of Mechanical Engineering, University of Engineering and Technology, Taxila, Pakistan
[2] Department of Mechanical Engineering, Pakistan Institute of Engineering and Applied Sciences, Nilore, Islamabad, Pakistan

* Corresponding e-mail address: asifhussainmalik@yahoo.com

Abstract

The CFD analysis of heat transfer within a bottom heated vertical concentric cylindrical enclosure is important with respect to the process in the chemical and process industries. The research work focuses on the CFD analysis of the enclosure with respect to the machines used for the segregation of chemicals in the chemical industries. The CFD simulations are performed to study the effects of inner cylinder material and outer cylinder geometric configurations on the heat transfer mechanism in the enclosure. The CFD simulations are conducted at a bottom disc temperature of 393 K and compared with the published results at a temperature of 433 K. This research study depicts the behavior of bottom heated vertical concentric cylindrical enclosure at different bottom disc temperatures. This study also investigates the heat transfer mechanism of the enclosure using different inner cylinder materials and different configurations of the outer cylinder. In such enclosures a uniform temperature is required for the segregation of chemicals. A more uniform temperature is observed in the enclosure by using aluminum inner cylinder of the bottom disc and using two different diameter outer cylinders as compared to the mild steel and stainless steel.

Keywords: CFD simulations, heat transfer, enclosure, temperature,

1. Introduction

The past researchers studied heat transfer through cylindrical enclosures in the last couples of decades. It has many engineering applications in building machines, solar collectors, heat exchangers, materials processing, storage tanks, furnace designs and nuclear designs.

The past researchers conducted numerical study of heat transfer using different enclosures. Lemembre and Petit [1] numerically investigated two-dimensional laminar natural convection heat transfer in cylindrical enclosures of heated lateral walls at uniform heat flux and cooled at the same heat flux at the top surface by insulating the bottom surface. The convection heat transfer was enhanced with the increase of aspect ratio and Rayleigh number. Chen, Ho and Humphrey [2] numerically presented two-dimensional steady-state laminar natural convection in the rectangular enclosures. The flow and heat transfer in the enclosure were strongly affected by an asymmetric specification of side wall temperatures. The asymmetry induced a disproportionately rise in the Nusselt number of the hotter side wall relative to the decrease in Nusselt number of the colder wall. References [3-6] numerically investigated the convection heat transfer in the cylindrical enclosures with thermo-physical fluid properties assumed to be constant except density variation which varies according to Boussinesq approximation.

Kim and Viskanta [7] and Vargas et al. [8] presented the experimental and numerical study of steady natural convection in two-dimensional Newtonian and incompressible fluid-filled enclosures. They concluded that heat conduction reduced average temperature difference across the enclosure, partially stabilized the flow ultimately decreasing natural convection heat transfer. The heat transfer and fluid flow leading to thermal stratification inside the enclosure were also studied. Kee et al. [9] presented experimental and numerical study of steady natural convection of a heat generating tritium gas in a two-dimensional closed vertical cylinders and spheres with their bounding isothermal walls. They demonstrated experimental and analytical agreement for local temperatures over a range of Grashof numbers. Bohn and Anderson [10] experimentally investigated heat transfer between parallel and perpendicular vertical walls in a three-dimensional water-filled cubical enclosure with an adiabatic top and bottom and isothermal sides and discussed many aspects of natural convection flow.

Liaqat and Baytas [11] numerically analyzed laminar two-dimensional conjugate natural convection in an enclosure. Their research has shown a significant change in the buoyancy flow parameters as compared to convectional non-conventional investigations. Sharma et al. [3] numerically investigated the conjugate transient natural convection in a liquid sodium filled cylindrical enclosure with internal volumetric heat generating source. The lateral walls of the enclosure were isothermal, while the top and bottom walls were adiabatic. They presented the transient features of confined turbulent natural convection due to time varying generation of heat in the volumetric source and shown that the intensity of heat source exponentially decayed with time. Kuznetsov and Sheremet [12] numerically investigated the two-dimensional conjugate convective-conductive heat transfer in a rectangular enclosure due to the local heat and contaminant sources and convective radiative heat exchange with an environment. The effects of Grashof number, buoyancy ratio and transient factor on flow modes, heat and mass transfer are studied. Glakpe et al. [13] numerically investigated the laminar, three-dimensional natural convection in the air-filled vertical annular space formed by a square rod enclosed within a cylinder strongly affected by the Rayleigh number. They studied the effects of thermal radiation exchange between walls of an enclosure.
This research paper presents thermal behavior of the system, adjusts the cooling system as per requirements, gives insight of different configurations and promotes the understanding of natural convection heat transfer between two concentric cylinders and gives special attention to the material study not yet performed previously as per authors' knowledge.

2. Experimental set up

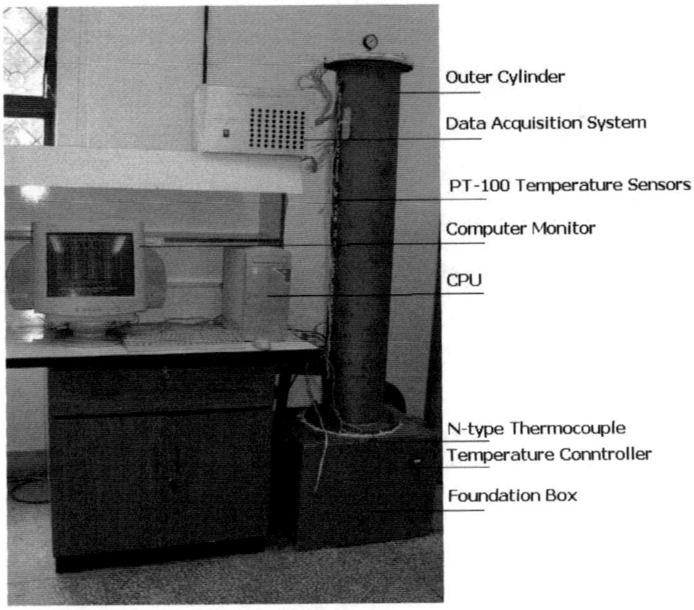

Figure 1: Experimental Apparatus

In this research paper, the CFD analysis of heat transfer within the bottom heated vertical concentric cylindrical enclosure (Figure 1) is numerically investigated at a specified bottom disc temperature of

393 K at the top centre of the bottom disc using two different configurations with three different materials of the inner cylinders and discs and compared with the published results at a temperature of 393 K [14]. Thermal analyses of the inner cylinders materials and two geometric configurations of enclosure (with aspect ratios A_1 = 6.016 and A_2 = 5.133 and Radius ratios RR_1 = 1.753 and RR_2 = 2.055 respectively) are performed.

3. Mathematical formulation

Mathematical formulation of the buoyancy flow effects in the fluid-filled enclosures includes the governing equations of conservation of mass, momentum and energy to solve the flow domain. To deal with the conduction within the solids and radiations from hot surfaces their mathematical formulation must be added. In the subsequent articles a brief introduction of the equations necessary for solving conjugate heat transfer processes is given.
The Navier-Stokes governing equations in their differential form in 2-D cylindrical coordinate system (r, z) can be written as under:

$$\frac{\partial\rho}{\partial t}+\frac{\partial}{\partial z}(\rho v_z)+\frac{\partial}{\partial r}(\rho v_r)+\frac{\rho v_r}{r}=S_m \qquad (1)$$

$$\frac{\partial}{\partial t}(\rho v_z)+\frac{1}{r}\frac{\partial}{\partial z}(r\rho v_z v_z)+\frac{1}{r}\frac{\partial}{\partial r}(r\rho v_r v_z)=-\frac{\partial p}{\partial z}+\frac{1}{r}\frac{\partial}{\partial z}\left(r\mu\left\{2\frac{\partial v_z}{\partial z}-\frac{2}{3}(\nabla.\boldsymbol{v})\right\}\right)+\frac{1}{r}\frac{\partial}{\partial r}\left(r\mu\left\{\frac{\partial v_z}{\partial r}+\frac{\partial v_r}{\partial z}\right\}\right)+F_z \qquad (2)$$

$$\frac{\partial}{\partial t}(\rho v_r)+\frac{1}{r}\frac{\partial}{\partial z}(r\rho v_z v_r)+\frac{1}{r}\frac{\partial}{\partial r}(r\rho v_r v_r)=-\frac{\partial p}{\partial r}+\frac{1}{r}\frac{\partial}{\partial z}\left(r\mu\left\{2\frac{\partial v_r}{\partial z}+\frac{\partial v_z}{\partial r}\right\}\right)+\frac{1}{r}\frac{\partial}{\partial r}\left(r\mu\left\{2\frac{\partial v_r}{\partial r}-\frac{2}{3}(\nabla.\boldsymbol{v})\right\}\right)-2\mu\frac{v_r}{r^2}+\frac{2}{3}\frac{\mu}{r}(\nabla.\boldsymbol{v})\rho\frac{v_z^2}{r}+F_r \qquad (3)$$

The stress tensor $\boldsymbol{\tau}$ is given by;

$$\boldsymbol{\tau}=\mu[(\nabla\boldsymbol{v}+\nabla\boldsymbol{v}^T)-\frac{2}{3}\nabla.\boldsymbol{v}\boldsymbol{I}] \qquad (4)$$

$$\frac{\partial}{\partial t}(\rho E)+\nabla.(\boldsymbol{v}(\rho E+p))=\nabla.(k_{eff}\nabla T-\textstyle\sum_j h_j\boldsymbol{J}_j+(\boldsymbol{\tau}_{eff}.\boldsymbol{v}))+S_h \qquad (5)$$

Where, r = radial coordinate, z = axial coordinate, v_r = radial velocity of the fluid, v_z = axial velocity of the fluid, ρ = density of fluid, S_m = mass source term within the differential volume, p = static pressure, $\boldsymbol{\tau}$ = stress tensor, $\boldsymbol{g}$ = acceleration due to gravity, $\boldsymbol{F}$ = external body forces including other model-dependent source terms such as porous medium and user-defined sources, μ = dynamic viscosity, $\boldsymbol{I}$ = unit tensor, k_{eff} = effective conductivity including the effect of turbulent thermal conductivity $\boldsymbol{J}_j$ = diffusion flux of species, and$j$. $\boldsymbol{\tau}_{eff}$ = effective stress tensor, S_h = energy source term including radiation source, chemical reaction source etc.

These equations (1-5) are given in detail in [15-18]. The first three terms on the right side of Eq. 5 give energy transfer due to conduction, species diffusion and viscous dissipation, respectively. The buoyancy forces can be designated by the ratio of Grashof and Reynolds numbers as shown below;

$$\frac{Gr}{Re^2}=\frac{g\beta\Delta TL}{v^2} \qquad (6)$$

There are strong buoyancy effects on the fluid flow when the above number approaches or exceeds unity. The strength of the buoyancy-induced flow in the pure natural convection is measured by the Rayleigh number;

$$Ra=\frac{g\rho\beta\Delta TL^3}{\mu\alpha} \qquad (7)$$

Where, β = coefficient of thermal expansion and α = thermal diffusivity.

$$\alpha=\frac{k}{\rho c_p} \qquad (8)$$

A buoyancy-induced laminar flow is assumed at $Ra<10^8$ and transition from laminar to turbulence is assumed in the range of $10^8<Ra<10^{10}$.

According to the Boussinesq approximation the density of the fluid is given by;

$$\rho = \rho_o[1 - \beta(T - T_o)] \quad (9)$$

Where, ρ_o = constant density of the fluid in the closed domain, T_o = operating temperature and β = coefficient of thermal expansion.

Fourth term of the energy equation (Eq. 5) ($\boldsymbol{S_h}$) also includes the heat source due to radiation. Rosseland radiation model is used which is faster and requires less memory as compared to other radiation models and deals with axisymmetric geometries and contains the effect of scattering. The radiative heat flux equation, the equation becomes;

$$q_r = -16\sigma\Gamma n^2 T^3 \nabla T \quad (10)$$

Both the specific radiative (q_r) and specific conduction heat fluxes (q_c) have the same form, therefore, these can be written as,

$$q = q_c + q_r = -(k + k_r)\nabla T \quad (11)$$

Where, k is the thermal conductivity, k_r is the radiative conductivity.

2.1 Boundary conditions

Appropriate boundary conditions are required to solve the Navier-Stokes equations. Boundary conditions are derived from the known parameters available at different physical boundaries of the domain as shown in Figure 2. The top and bottom walls of the enclosure are insulated (adiabatic), except the bottom disc, which act as a heat source for the enclosure. No slip boundary conditions are enforced at these walls of the enclosure. The outer side of the outer cylinder wall is exposed to natural convection at ambient conditions. The mathematical formulation of these boundary conditions is given below;

3.1.1. Bottom disc wall boundary

Heat source is applied at the bottom of the disc of radius, r_1. N-type thermocouple probe is mounted at the upper central location of the bottom disc and connected to the temperature controller. The temperatures are specified for different experiments at $z = 0$ and known. Its boundary conditions are given below;
At $z = 0;\ 0 < r < r_1;$
T is known; $v_r = v_z = 0$ (No slip condition);

3.1.2 Bottom insulated wall boundary

Bottom surface of enclosure is assumed to be insulated except the bottom disc heated by the heat source. Neither heat is entered into the bottom insulated wall nor does it come out of that wall. Its boundary conditions are given below;
At $z = 0;\ r_1 < r < r_2;$
$v_r = v_z = 0$ (No slip condition);
$q_{in} = q_{out} = 0$ for an adiabatic wall.

3.1.3 Side wall boundary

At the side wall of enclosure the natural convection heat transfer coefficient h_o is taken as a boundary of enclosure. Heat entering from the hot bottom disc after passing through the walls of inner cylinder comes out of the wall of outer cylinder of enclosure. Its boundary conditions are given below;
At $r = r_2;\ 0 < z < z_2;$

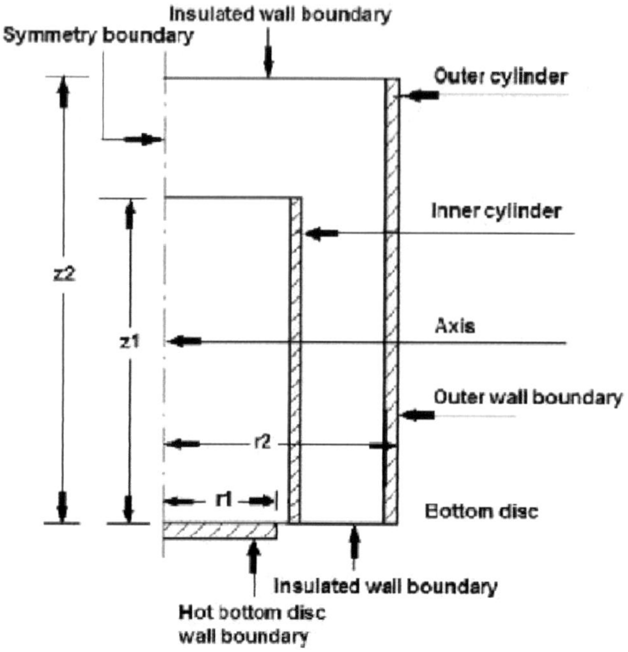

Figure 2: Boundary conditions of vertical concentric cylindrical enclosure

3.1.4 Symmetry boundary

In 2-D axisymmetric geometry the axis is used as a symmetry boundary and half of the geometry is taken for the analysis as shown in Figure 2 as used by the past researchers [1-5]. At the axis as a boundary conditions all the gradients of temperature and velocity are zero. Its boundary conditions are given below;

At $r = 0;\ 0 \leq z < z_2;$

$$\frac{\partial T}{\partial r} = 0\,;\ \frac{\partial v_z}{\partial r} = 0;\ \frac{\partial v_r}{\partial r} = 0;$$

3.1.5 Top insulated wall boundary

Top wall of enclosure is taken as insulated or adiabatic one. Neither heat is entered into the bottom insulated wall nor does it come out of that wall. Its boundary conditions are given below;

At $z = z_2,;\ 0 \leq r < r_2;$

$v_r = v_z = 0$ (No slip condition);

$q_{in} = q_{out} = 0$ for an adiabatic wall.

4. Numerical analysis

In this research work the CFD analysis of the experimental work [19] depicting heat transfer effects within the bottom heated vertical concentric cylindrical enclosure (Figures 1 and 3) in the bottom disc temperature range of 393 K at the upper centre of the bottom disc is performed. This analysis provided a thorough insight of the flow and buoyancy effects due to the change of inner cylinder wall material in the enclosure and outer cylinder diameter. FLUENT 6.3 software is used to study the heat, mass and momentum transfer mechanisms within the enclosure, compare and validate numerical results with the experimental ones and get insight of the flow and thermal energy enhancement mechanism within the enclosure as used by the past researchers [8, 19-20].

Two dimensional steady state analysis is carried out. The Rosseland radiation model is selected. Air is treated as a stationary, incompressible and laminar fluid.

The FLUENT have used the pressure-based solver, the second order scheme is used for pressure, while the second order upwind discretization schemes are used for momentum and energy [21]. The flow and energy equations are solved by the FLUENT [15] and the Green-Gauss cell-based gradient scheme is selected for this research work. Boussinesq approximation is used assuming the density as a constant, except in the buoyancy term of vertical momentum equation. In order to solve the pressure-velocity coupling SIMPLE algorithm is used as applied by the previous researchers [3, 6, 22-26]. The convergence criteria are different for different cases as shown in the previous studies by [27-29]. Iterations took place till the convergence is achieved to the required solution in the range of 765 ~ 4968 iterations. Three different materials of inner cylinders and discs in the enclosure are used in the analysis. Six experiments are performed at the specific temperatures of 393K on the upper central location of bottom discs. The mesh selected is square and total cells of both configurations (O_1 and O_2) are 213960 and 247400 respectively.

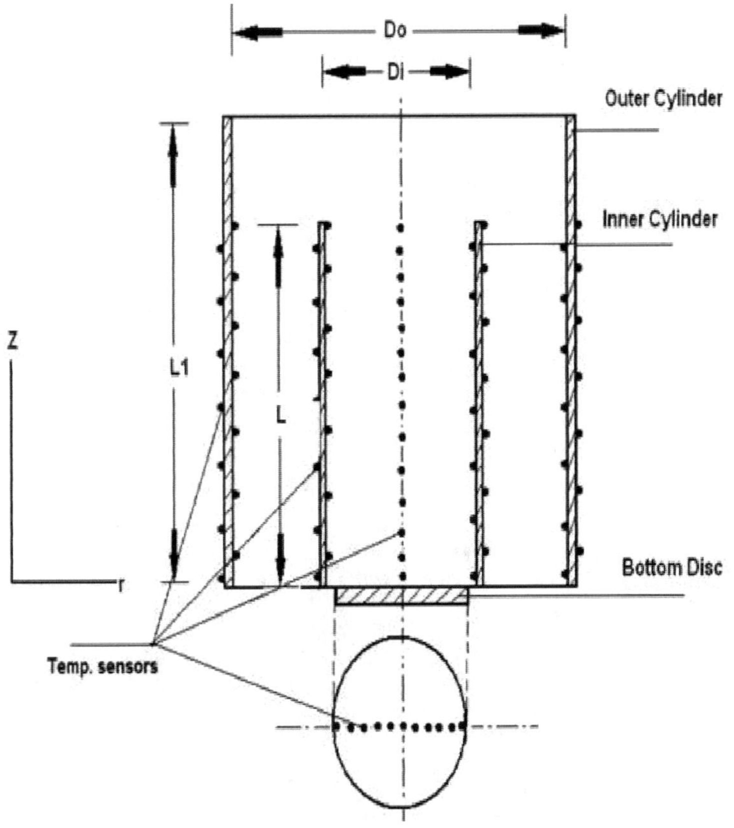

Figure 3: Cross-sectional view of concentric cylindrical enclosure

4.1 Geometry and meshing

In this research work the geometry considered for analysis is a 2-D axisymmetric one due to which the axis is used as a symmetry boundary. The half of the geometry is taken for the analysis due to its axisymmetric geometry as used by the past researchers [3, 19, 30-32]. Gambit 2.3.16 has been used as a preprocessor to construct the geometry and mesh.

The geometry is generated and meshed in the Gambit software as shown in the Figure 4. Square mesh is used to mesh the geometry. Along the axis of the geometry there are 1541 grid points, while at the top and bottom of the geometry there are 211 grid points. Due to change in diameters of the radial grid are 139 and 151 for outer cylinder, O_1 & O_2 respectively. The thermal energy enhancement about the axis of the geometry is assumed symmetric. Two-dimensional axisymmetric simulations are carried out. The bottom disc geometry of the enclosure being simulated is shown in the Figure 2. Top and bottom walls of the enclosure except the bottom disc are assumed to be adiabatic with no heat flux and no heat generation. No slip condition is considered for all walls of the enclosure. The walls within the enclosure are thermally coupled.

The convection heat transfer coefficient, h is selected as a boundary condition at the outer surface of the outer vertical cylinder by hit and trial method. The inner and outer cylinders walls are considered as conducting walls. The user defined function has incorporated the experimental temperature results of bottom disc and coupled with the FLUENT [15].The three inner cylinders of aluminum, mild-steel and stainless-steel having similar dimensions are used in different configurations.

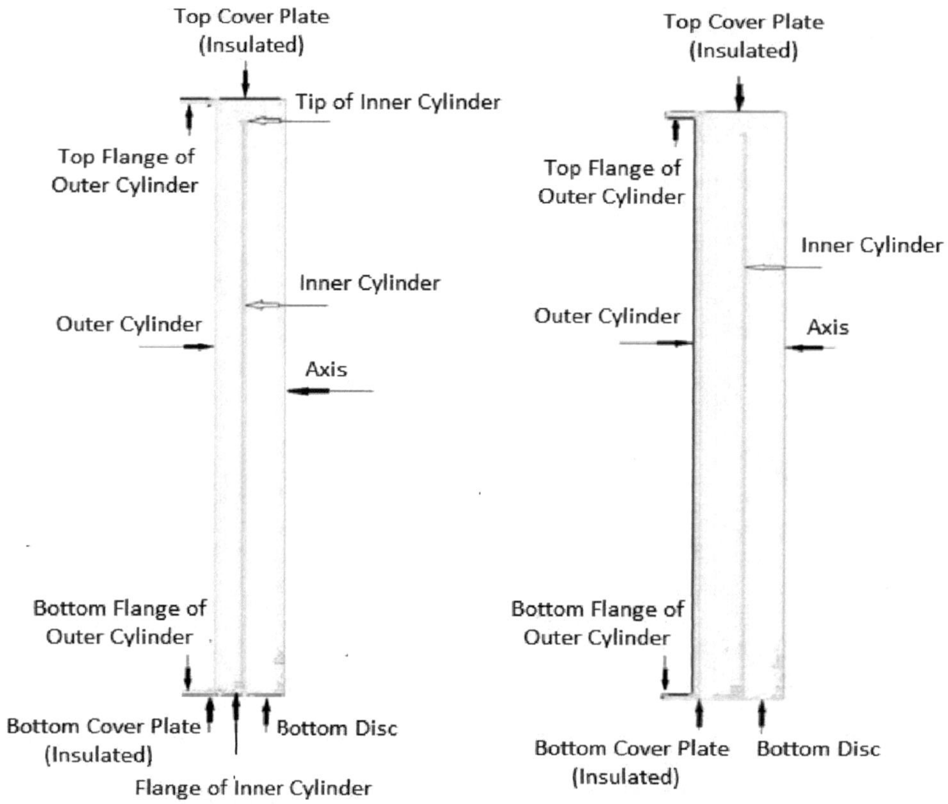

Figure 4: Meshing of enclosure geometries of outer cylinder, O_1 & O_2 outer cylinder

The outer surface of the simulated enclosure is far away from heat source, it was assumed that the pressure and temperature at the outer surface of the enclosure are equal to the ambient to allow free movement across this surface. There is an ambient temperature at the top of the vertical enclosure, so

Table 1: Properties of Fluid (Air)

Properties	Air
Density, kg/m^3	1.184
Specific heat capacity, J/kg.K	1007
Thermal conductivity, W/m.K	0. 02551
Dynamic viscosity, Pa.sec	1.8949E-5
Absorption coefficient, 1/m	0.49919, constant
Scattering phase function	Isotropic
Scattering coefficient, 1/m	0
Thermal Expansion Coefficient, 1/K	0.0033445
Refraction	1

Table 2: Properties of Materials

Properties	Aluminum	Mild Steel	Stainless Steel
Density, kg/m^3	2739	7833	8238
Specific heat capacity, J/kg.K	896	502	468
Thermal conductivity, W/m.K	222	45.3	13.4

Table 3: Geometric configurations of cylindrical enclosure

Parts of Geometry	Material	Diameter (m)	Thickness (m)	Height (m)
Bottom disc	Aluminum, Mild steel, Stainless steel	0.138	0.004	-
Inner cylinder	Aluminum, Mild steel, Stainless steel	0.146	0.005	1.48
Outer cylinder	Mild steel	O_1=0.256, O_2=0.300	0.01	1.54

the enclosure was assumed as an infinite enclosure. The properties of fluid (air) and materials [33] used in these simulations are given in the Tables 1 and 2 respectively.

4.2 Grid independence study

The grid independence study is carried out by taking the grids of cell sizes 2 mm, 1.5 mm, 1mm and 0.5 mm having 54260, 96135, 213960 and 855840 cells with percentage errors of 4.2, 2.9, 1.94 and 1.91 respectively. There is a prominent percentage error of cell sizes 2 mm, 1.5 mm and 1 mm, while cell sizes 1 mm and 0.5 mm have almost the same percentage error with the experimental results of [14]. Therefore, cell size of 1 mm is taken while meshing the enclosure under research.

4.3 Uncertainty analysis

Experimental measurements always contain some uncertainties in spite of using high precision measuring instruments. Such uncertainties come from the measuring instrument, the items being measured, the environment, the operator and many other sources. The researchers generally estimate such uncertainties by using statistical method presented by Moffat [34]. In this study the same method is used to estimate the uncertainties in experimental data.

The N-type thermocouple used for measuring the temperature at the center of bottom disc has an accuracy of ±0.75%, whereas, PT-100 temperature sensor has an accuracy of ±0.3%. The temperature controller 'CAL Series 9900' and contactor (solid state rely) have accuracies of ±0.25% and ±0.5%, respectively. The uncertainty in the reading of N-type thermocouple is estimated to be less than 1.5%.and that of PT-100 temperature sensor it is 1.1%, including the effect of the observed experimental data scatter.

5. Results and discussion

This research work also includes the study of effect of outer cylinder diameter on the heat transfer within the enclosure. The numerical analysis is carried out using Fluent code. Numerical simulation results of vertical concentric cylindrical enclosure are validated by comparing with the experimental data.

Thermal conductivity of inner cylinder material is thought to play a key role in affecting thermal behavior within the enclosure. For the material analysis three inner cylinders made of aluminum, mild steel and stainless steel are used. Thermal conductivity of aluminum is about 17 times greater than that of stainless steel and about 5 times greater than mild steel. To understand the heat transfer mechanism and buoyancy effects within the enclosure, the study of CFD results is required.

5.1 Axial thermal behavior

The temperature distribution along the axis and inner cylinder wall are very important to study the thermal behavior within the enclosure. The experimental and CFD results of axial temperature along the axis and inner cylinder of the enclosure are shown in Figures 6.1 and 6.2 respectively. The axial temperature distribution at axis of the geometry and inner cylinder are discussed in detail in the following articles.

5.1.1 Axis of the enclosure

Thermal behavior at the axis of enclosure gives true picture of heat transfer mechanism within the enclosure. The segregation of chemicals in the segregation machines takes place around the axis of the geometry and strongly depends on the thermal behavior along the axis.

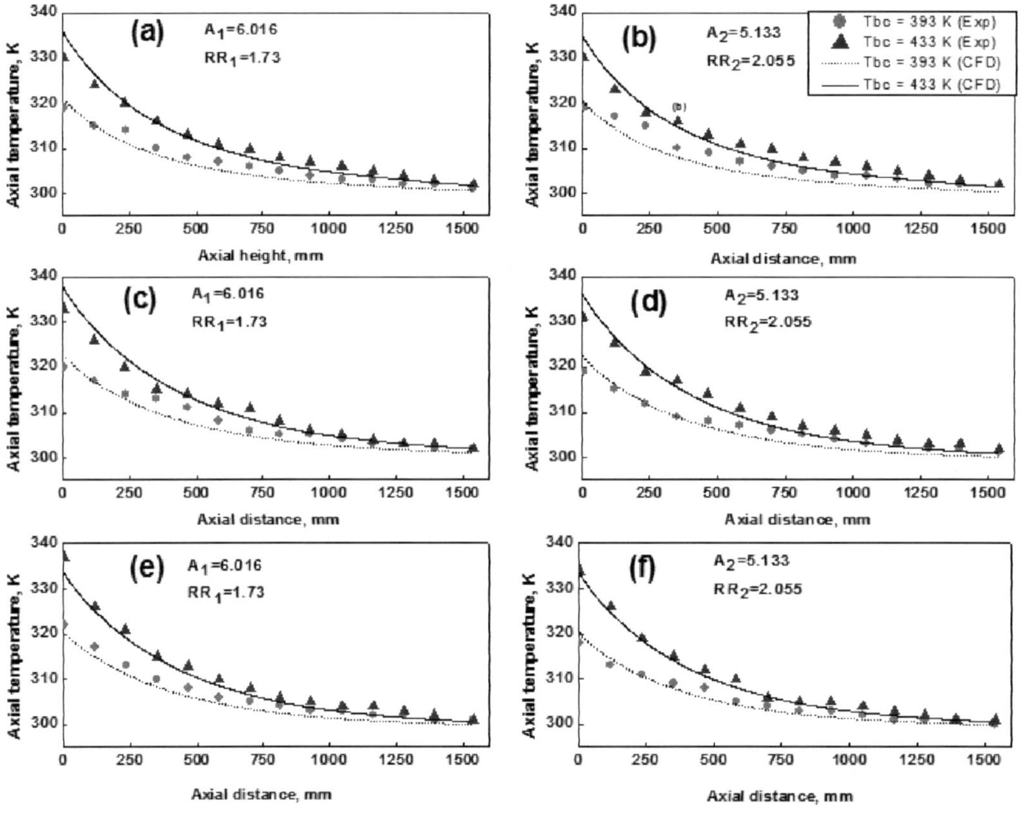

Figure 5: Axial temperature distribution along the axis of the enclosure with inner cylinder of aluminum (a, b), mild steel (c, d) and stainless steel (e, f).

Temperature at the axis is studied at bottom disc central temperature of 393 K, using inner cylinders of different materials (aluminum, mild steel and stainless steel) and outer cylinders of different diameter(O_1 and O_2) at bottom disc central temperature of 393K as shown in Figure 5. The experimental and numerical results of axial temperature match closely with each other for these six different cases and therefore validate the numerical simulations. The aspect ratio and radius ratio of the enclosure geometry with outer cylinder O_1 are 6.016 and 1.73, while with outer cylinder O_2 these values are 5.133 and 2.055 respectively.

The axial temperature decreases along the height of the enclosure for all the experiments performed due to the presence of heat source at the bottom of the enclosure. Figure 5 (a-b) shows that the axial temperature distribution is almost independent of the outer cylinder diameter. While, Figure 5 (c-f) shows that the axial temperature distribution is a function of the outer cylinder diameter especially at higher temperatures of the bottom disc. This response might be due to high thermal conductivity of aluminum as compared to mild steel and stainless steel.

5.1.2 Inner cylinder

The temperature distribution along the axis and inner cylinder tells the whole story of heat transfer in such enclosures. The experimental and the CFD results for axial temperature along the inner side of inner cylinder are shown in Figure 6. Figure 6 (a-b) shows the axial temperature distribution along the inner cylinder, using inner cylinder made of aluminum material, under the bottom disc central temperature of 393 K and using two different mild steel outer cylinders O_1 and O_2. It is observed that the axial temperature distribution along the inner cylinder is independent of the outer cylinder diameter.

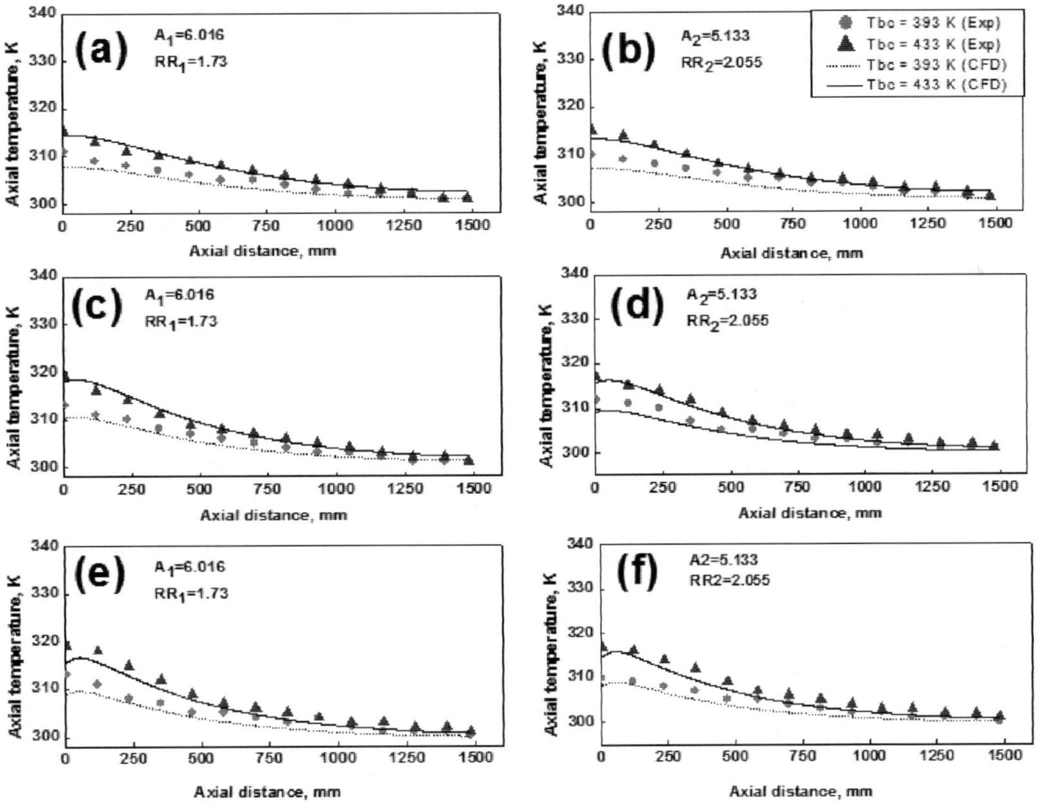

Figure 6: Axial temperature along the inner surface of inner cylinder of aluminum (a, b), mild steel (c, d) and stainless steel (e, f)

Figure 6 (c-d) and Figure 6 (e-f) show the same results using inner cylinders of mild steel and stainless steel respectively. It is observed that the axial temperature along the inner cylinder depends on the diameter of the outer cylinder.

Due to higher thermal conductivity of aluminum the heat is transferred easily to the outer enclosure irrespective of the diameter of the outer cylinder. While, in the case of mild steel and stainless steel inner cylinders the thermal conductivities of the inner cylinder are low and therefore the axial temperature along the inner cylinder is a function of the diameter of the outer cylinder as well. The spread in temperature distribution for various bottom disc temperatures is more in Figure 6 (c-f) as compared to Figure 6 (a-b). The reason might be due to higher thermal conductivity of aluminum as compared to mild steel and stainless steel. The close agreement between the experimental and the CFD results of axial temperature along inner cylinder support the numerical simulation of vertical concentric cylindrical enclosure.

5.2 Validation of the CFD simulation

Before discussing the CFD results an effort is made to validate the CFD simulations. The experimental data of six different experiments are used to validate the CFD results using aluminum, mild steel and stainless steel inner cylinders in the experiments. The CFD and experimental results have been compared in the following sections.

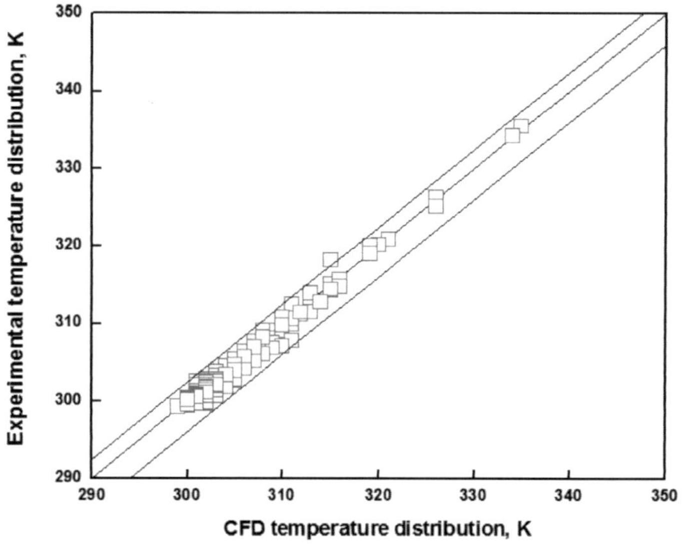

Figure 7: Comparison of experimental and the CFD results of enclosure with aluminum inner cylinder

The comparison of experimental results with the CFD simulation results of enclosure configurations (O_1, O_2) at different bottom disc central temperatures (353, 393 and 433 K) have been shown in Figure 7. The positive and negative errors of the CFD simulation results with the experimental ones are 0.8% and 1.3% respectively (Figure 7). This shows that the CFD simulation results are in best agreement with experimental ones. Hence, the CFD simulation results of enclosure configurations (O_1, O_2) with aluminum inner cylinder at different bottom disc central temperatures (353, 393 and 433 K) are validated with the experimental results.

5.3 The CFD simulation results

In order to study the heat transfer mechanism and buoyancy effects in vertical concentric cylindrical enclosure in more depth CFD results have been discussed in the following sections.

5.3.1 Contours of streamlines at 393 K

The contours of streamlines are obtained from CFD simulations for the enclosure configurations O_1 and O_2 at bottom disc central temperature of 393 K and are shown in Figure 8. Figure 8 (a-b) shows streamlines for two configurations (O_1, O_2) of aluminum inner cylinder. Similarly Figures 8 (c-d) and (e-f) show the corresponding results for mild steel and stainless steel inner cylinders. It is observed in Figure 8 (a-b) that the streamlines within the inner cylinder are same for both configurations (O_1, O_2) of aluminum inner cylinder. The same behavior is observed in case of mild steel (Figure 8 (c-d)) and

Journal of Physics: Conference Series **439** (2013) 012004 doi:10.1088/1742-6596/439/1/012004

stainless steel inner cylinders (Figure 8 (e-f)). The streamlines in the inner cylinder of Figure 8 (a-b) shows weak buoyancy effects as compared to Figure 8 (c-f). The buoyancy effects are stronger in the outer annulus with outer cylinder configuration O_2 as compared to configuration O_1 for all three inner cylinders used. The buoyancy effects in the outer annulus are stronger in case of aluminum inner cylinder as compared to mild steel and stainless steel for the same outer configuration. The same behavior is observed in case of mild steel inner cylinder as compared to stainless steel. At this temperature the heat added to the enclosure is small and buoyancy effects are relatively weak in Figure 8 (a-b) as compared to Figure 8 (c-f) due to difference of thermal conductivity.

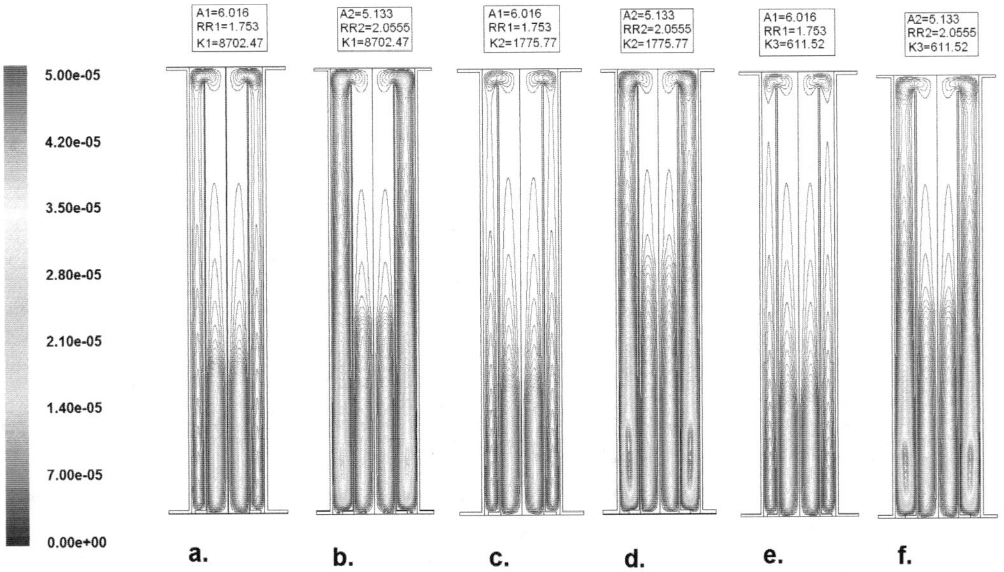

Figure 8: Streamlines of enclosure for configuration O_1 inner cylinders of a). aluminum, c). mild steel, e). stainless steel, for configuration O_2 inner cylinders of b). aluminum, mild steel, f). stainless steel at 393 K.

Due to increase in volume of outer cylinder configuration O_2 the buoyancy effects are stronger as compared to configuration O_1. In the outer annulus the buoyancy effects are stronger while using inner cylinder of aluminum as compared to other inner cylinders due to its high thermal conductivity.

5.3.2Comparison of contours of streamlines

At a bottom disc central temperature of 433 K the buoyancy effects are stronger [35] as compared to the buoyancy effects at 393 K for all three inner cylinders used. At 433 K the outer cylinder diameter also affects the heat transfer mechanism within the inner cylinder due to large quantity of heat transfer through the enclosure due to increase in volume as compared to configuration O_1 for three inner cylinders used while comparing the same streamlines at 393 K.

The stream lines at 393 K and 433 K suggest that due to high thermal conductivity of aluminum inner cylinder, the heat conduction in inner cylinder wall in the axial direction is high as compared to inner cylinders of mild steel and stainless steel. This behavior is more clearly visible at 433 K, where due to low thermal conductivity of mild steel and stainless steel double vortices are formed in the outer annulus region with outer cylinder configuration O_2.

6. Conclusion

In this study the CFD analysis of conjugate heat transfer within a bottom heated non-conventional cylindrical enclosure is performed. Thermal lines, streamlines and velocity vectors of the enclosure are studied at bottom disc temperature of 393K for the inner cylinder material (aluminum, mild steel, stainless steel) and outer cylinder diameter (O_1, O_2). In total six different simulations are performed. As a result of these experiments following conclusions are made;

1. A more uniform axial temperature is seen at the inner wall of aluminum cylinder at a bottom disc temperature of 393K and outer cylinder diameters (O_1, O_2) as compared to mild steel and stainless steel.
2. Stainless steel is more sensitive to outer cylinder diameter as compared to mild steel. Here, the outer cylinder diameter is the controlling parameter for the heat transfer mechanism rather along with the thermal conductivity. Aluminum is non-sensitive to the outer cylinder diameters (O_1, O_2) specifically for the inner cylinder thermal behavior.
3. In mild steel and stainless steel inner cylinders the temperature spread near the bottom disc at a temperature of the bottom disc 393K decreases with increasing outer cylinder diameters.
4. At the bottom disc temperature up to 393 K the streamlines within the inner cylinder are almost same for both configurations (O_1, O_2) of three inner cylinders of aluminum, mild steel and stainless steel being independent of outer diameter used.
5. Buoyancy effects are more prominent in the aluminum inner cylinder as compared to the mild steel and stainless steel inner cylinders showing its convection effects.
6. With outer cylinder configuration O_2 the buoyancy effects are stronger due to increase in volume as compared to configuration O_1 for three inner cylinders used.
7. In the outer annulus the buoyancy effects are stronger while using inner cylinder of aluminum as compared to other inner cylinders due to its high thermal conductivity.
8. With the increase in temperature from 393 K to 433 K the heat conduction in inner cylinder wall in the axial direction is high and more visible as compared to inner cylinders of mild steel and stainless steel due to high thermal conductivity of aluminum inner cylinder.

References

[1]. Lemembre, A.a.P., J. P., *Laminar Natural Convection in a Laterally Heated and Upper Cooled Vertical Cylindrical Enclosure.* International Journal of Heat and Mass Transfer, 1998. 41(16): p. 2437-2454.

[2]. Chen, S.A.H., J. R., Humphrey, J. A. C. , *Steady, Two-Dimensional, Natural Convection in Rectangular Enclosures With Differently Heated Walls* Journal of Heat Transfer, Transactions of the ASME, May 1987. 109: p. 400-406.

[3]. Sharma, A.K., Velusamy, K. and Balaji, C., *Conjugate Transient Natural Convection in a Cylindrical Enclosure with Internal Volumetric Heat Generation* Annals of Nuclear Energy, 2008. 35: p. 1502-1514.

[4]. Franke, M.E. and K.E. Hutson, *Effects of Corona Discharge on Free-Convection Heat Transfer Inside a Vertical Hollow Cylinder.* Journal of Heat Transfer, 1984. 106(2): p. 346-351.

[5] Roschina, N.A., Uvarov, A. V. and Osipov, A. I., *Natural Convection in an Annulus Between Coaxial Horizontal Cylinders with Internal Heat Generation.* International Journal of Heat and Mass Transfer October 2005. 48(21-22): p. 4518-4525.

[6]. Bairi, A., *Transient Natural 2D Convection in a Cylindrical Cavity with the Upper Face Cooled by Thermoelectric Peltier Effect Following an Exponential Law* Applied Thermal Engineering, March 2003. 23(4): p. 431-447.

[7]. Kim, D.M.a.V., R., *Effect of Wall Heat Conduction on Natural Convection Heat Transfer in a Square Enclosure.* Journal of Heat Transfer, Transactions of the ASME, February 1985. 107: p. 139-146.

[8]. Vargas, M., Sierra, F. Z., Ramos, E., and Avramenko, A. A., *Steady Natural Convection in a Cylindrical Cavity, .* International Communication of Heat and Mass Transfer, 2002. 29(2): p. 213-221.

[9]. Kee, R.J., Landram, C. S. and Miles, J. C., *Natural Convection of a Heat Generating Fluid within Closed Vertical Cylinders and Spheres* Journal of Heat Transfer, Transactions of the ASME, February 1976. 98(1): p. 55-61.

[10]. Bohn, M.S.a.A., R., *Temperature and Heat Flux Distribution in a Natural Convection Enclosure Flow.* Journal of Heat Transfer, Transactions of the ASME, May 1986. 108: p. 471-476.

[11]. Liaqat, A.a.B., A. C., *Conjugate Natural Convection in a Square Enclosure Containing Volumetric Sources* International Journal of Heat and Mass Transfer, September 2001. 44(17): p. 3273-3280.

[12]. Kuznetsov, C.V.a.S., M. A., *Conjugate Heat Transfer in an Enclosure under the Condition of Internal Mass Transfer and in the Presence of the Local Heat Source, ,* . International Journal of Heat and Mass Transfer, 2009. 52: p. 1-8.

[13]. Glakpe, E.K., Watkins, C.B. and Kurien, B.J., *Effect of Radiation and Specified Heat Flux on Natural Convection in a Vertical Region with a Rectangular Inner Boundary,* . AIAA and ASME, Joint Thermodynamics and Heat Transfer Conferences, 4th Boston, M.A. , June 2-4, 1986: p. 10 pages.

[14]. Malik, A.H., et al.,, *Experimental Study of Conjugate Heat Transfer within a Bottom Heated Vertical Concentric Cylindrical Enclosure.* International Journal of Heat and Mass Transfer, 2012. 55: p. 1154-1163.

[15]. Fluent, *Fluent 6.3 User's Guide.* Fluent Inc., 2006.

[16]. Rolf H. Sabersky, A.J.A., Edward G. Hauptmann, *Fluid Flow - A First Course in Fluid Mechanics.* Macmillan Publishing Co., Inc., New York, Collier Macmillan Publishers, London, 1971. Second Edition.

[17]. White, F.M., *Fluid Mechanics.* 1986. Second Edition.

[18]. Robert L. Daugherty, J.B.F.a.E.J.F., *Fluid Mechanics with Engineering Applications.* McGraw Hill Book Company, Singapore, 1996. S. I. Metric Edition.

[19]. Wrobel, W., Fornalik-Wajs, E., and Szmyd, J. S., *Experimental and Numerical Analysis of Thermo-magnetic Convection in a Vertical Annular Enclosure,* . International Journal of Heat and Fluid Flow, December 2010, 31(6): p. 1019-1031.

[20]. Covaro, F.a.P., Massimo *The Natural Convective Heat Transfer in a Partially Divided Enclosure: A study on the Influence of the Source Position.* Journal of Thermodynamics, 2009, 2009: p. 1-11.

[21]. Kuznetsov, G.V.a.S., Mikhail A., *Conjugate Natural Convection in an Enclosure with a Heat Source of Constant Heat Transfer Rate* International Journal of Heat and Mass Transfer, January 2011. 54(1-3): p. 260-268.

[22]. Aminossadati, S.M.a.G., B., *The Effects of Orientation of an Inclined Enclosure on Laminar Natural Convection* Heat and Technology, 2005. 23(2): p. 43-49.

[23]. Nazrul Islam, G., U.N., and Sharma, G.K., *Mixed Convection Heat Transfer in the Entrance Region of Horizontal Annuli* International Journal of Heat and Mass Transfer, June 2001. 44(11): p. 2107-2120.

[24]. Sezai, I.a.M.A.A., *Natural Convection Heat Transfer from a Discrete Heat Source on the Bottom of a Horizontal Enclosure* International Journal of Heat and Mass Transfer, 2000. 43: p. 2257-2266.

[25]. Mazumder, S., *On the Use of the Fully Compressible Navier Stokes Equations for the Steady-State Solution of Natural Convection Problems in Closed Cavities* Journal of Heat Transfer, Transactions of the ASME, March 2007. 129: p. 387-390.

[26]. Bouali, H., Mezrhab, A., Amouli, H. and Bouzidi, M., *Radiation – Natural Convection Heat Transfer in an Inclined Rectangular Enclosure* International Journal of Thermal Sciences, 2006. 45: p. 553-566.

[27]. Yu, E.a.J., Y.K., *Heat Transfer in Discretely Heated Side-Vented Compact Enclosures by Combined Conduction, Natural Convection and Radiation* Journal of Heat Transfer, Transactions of the ASME, November 1999. 121: p. 1002-1010.

[28]. Chang, T.S.a.T., Y.L., *Natural Convection Heat Transfer in an Enclosure with a Heated Background Step,* . International Journal of Heat and Mass Transfer, 2001. 44: p. 3963-3971.

[29]. Zhao, F.-Y., Liu, Di and Tang, Guang-Fa, *Natural Convection in an Enclosure with Localized Heating and Salting from Below* International Journal of Heat and Mass Transfer, 2008. 51: p. 2889-2904.

[30]. Lin, W.a.A., S. W., *Natural Convection Cooling of Rectangular and Cylindrical Containers, .* International Journal of Heat and Fluid Flow 2001. 22: p. 72-81.

[31]. Papanicolaou, E.a.B., V., *Transient Natural Convection in a Cylindrical Enclosure at High Rayleigh Numbers* International Journal of Heat and Mass Transfer, March 2002. 45(7): p. 1425-1444.

[32]. Khalilollahi, A.a.S., B., *Unsteady Natural Convection Generated by a Heated Surface Within an Enclosure* Numerical Heat Transfer, 1986. 9: p. 715-730.

[33]. Cengel, Y.A., *Heat Transfer a Practical Approach.* McGraw Hill Companies, Inc, 1221 Avenue of the Americas, New York, 2003. Second edition.

[34]. Moffat, R.J., *Contributions to the Theory of Single-Sample Uncertainty Analysis.* Journal of Fluids Engineering, 1982. 104(2): p. 250-258.

[35]. Malik A. H., e.a., *Numerical Study of Conjugate Heat Transfer within a Bottom Heated Cylindrical Enclosure.* Proceedings of 2012, 9th International Bhurban Conference on Applied Sciences & Technology (IBCAST), Islamabad, Pakistan, 9-12 January, 2012: p. 213-220.

SiON metrology using angular and energy distributions of photoelectrons

G Tasneem[1*], C Tomastik[2], R Mroczyński[3] and W S M Werner[4]

[1]Institut für Angewandte Physik, Vienna University of Technology, Wiedner Hauptstrasse 8-10, A-1040 Vienna, Austria
[2]Austrian Centre for Competence for Tribology, Viktor Kaplan-Straße 2, A-2700 Wiener Neustadt, Austria

[3]Institute of Microelectronics and Optoelectronics, Warsaw University of Technology, Koszykowa, Poland

[4]Surface and Microanalysis Science Division, National Institute of Standards and Technology, Gaithersburg, MD 20899-8370, USA

[*]*Email: ghazaltasneem@yahoo.com*

Abstract. Angle-resolved X-ray photoelectron spectroscopy (ARXPS) is a useful tool for non-destructive in-depth analysis of near surface regions. However, the reconstruction of depth profile from ARXPS data is an ill-posed mathematical problem. Thus, the main goal of this work was to develop a new, iterative algorithm based on the least square fitting which allows to solve this problem. The depth profiles were restored by dividing sample in thin virtual box shaped layers each with a different concentration. To extract information on the depth distribution, this algorithm is based on the analysis of the angular peak intensities along with the inelastic background. In addition, the physically trivial constraint of atomic fractions adding up to unity was imposed. The model takes into account the effect of elastic scattering and anisotropy of the photoelectric cross section. To test the algorithm, experimental spectrum for SiON samples on Si substrate were measured with a Thermo Theta Probe electron spectrometer for off-normal emission angles in the range between 25° and 75°. A very good agreement was found between the measured spectra and obtained spectra from the algorithm.

1. Introduction

Non-destructive methods for obtaining depth profile information from the topmost few nanometres of solids become increasingly important for many technological applications e.g. as the thickness of the nowadays gate dielectric layers used in ULSI CMOS devices, has reached the scale of a fraction of nanometres. Angle-resolved X-ray photoelectron spectroscopy (ARXPS) is widely used for such analysis in order to determine thickness of thin films and the variation of composition of different species with depth [1]. For ARXPS, transformation of the measured photoelectron intensities into a depth profile is not straightforward since inversion of ARXPS data is one of the notorious mathematically ill-posed problems. As much prior knowledge about the sample is utilized to constrain possible solutions and to obtain useful results [1].

In quantitative analysis by ARXPS, the area of the peak of signal electron that have not lost energy while escaping from the target is used for data interpretation after removing the broad background of inelastically scattered electrons. However, recently it was realized (e.g. [2-4]) that not only electrons that haven't lost their energy but the electrons having participated in a given number of inelastic collisions, also contain significant information about the depth distribution of the emitting species. Several methods have been proposed to extract the information about the depth profile from experimental spectra [2-4]. The numerical efforts involved in these methods are generally large and they are mostly based on some kind of least-squares fitting between the entire experimental and theoretical line shapes.

In present paper, a procedure to extract in-depth information from energy and angular distributions of photoelectrons, is introduced.

2. Theory

For a noncrystalline solid, the electron transport in solids may be formulated by a Boltzmann type quation [5-6]. Based on this approach, energy/angular spectrum of emitted electron is given by [5]

$$Y(\mu,E)=T(E)\otimes\sum_{j=0}^{n}A_j/\mu\sum_{n=0}^{\infty}\int_0^{\infty}dE'L_n(E-E')f(E')\times\int_0^{\infty}dzP_n(\mu,z)c(z) \quad (1)$$

In this expression, E is the electron energy, μ is the cosine of polar emission angle, θ (with respect to surface normal), f(E) is the true intrinsic spectrum, c(z) is the concentration of the specie at depth z and T(E) represents transmission function of the instrument. The symbol ⊗ stand for the convolution operation. Factor A describes the generation rate of signal electrons. The quantity L_n is the partial loss distributions

$$L_n(T)=\int L_{n-1}(T-T')w_b(T')dT' \quad (2)$$

Where $w(T)$ is the distribution of energy loss in a single scattering, E is the electron kinetic energy, and T is the energy loss. The second integral on the right hand side of equation (1) is partial intensity. Partial intensities are the number of electrons that have participated in n inelastic collisions and is denoted by the symbol Cn(μ) . Partial intensity can be defined as

$$C_{ni}(\mu)=\int_0^{\infty}dzP_n(\mu,z)c_i(z) \quad (3)$$

The quantity Pn (μ,z) represent the partial escape distribution (PED) which account for the inelastic background on the low kinetic energy side of the photo electron peak. PED is defined as [6- 7]: " The probability distribution for the process in which an electron generated in a certain depth interval will escape from the surface with a direction in a certain angular interval after experiencing a certain number of inelastic scattering processes in solids." It can be seen from equations 1-3 that partial intensities contain all the information regarding depth profile. Consequently, depth profile can be parameterized by variables X=(x1,x2,.....xi). The core of this analysis is determination of these quantities. It should be noted, that depth profile determination is usually greatly facilitated by measuring the yield at several emission angles. Taking above into account, following method acquired data at two different emission angles. Depth profile can be obtained by least square fitting of theoretical spectra to measured raw data by the following function

$$\sum_{i=1}^{n_i}A_i\left[\sum_{j=1}^{2}(Y_i^{Sim}(E,\theta_j)-Y_i^{Exp}(E,\theta_j))^2\right]=0 \quad (4)$$

Where A is the instrumental factor, i is the specie in the sample, j is angle, Y^{Exp} is the measured spectrum and Y^{Sim} is the simulated spectrum calculated by using equation (1) at two different angles θ_1

and θ_2 respectively. Equation (4) is the system of linear equations and the basis of proposed methods for retrieval of depth profile.

3. Experimental

Two-inch <100> *p* type silicon substrates of the resistivity 4-7 Ωcm were used in this study. The silicon oxynitride films were fabricated in an Oxford Plasma Technology PLASMALAB 80+ system. The silicon substrates before the experiments were cleaned using the standard RCA method. The parameters of plasma enhanced chemical vapor deposition (PECVD) process were optimized to allow repeatable formation of dielectric films. The PECVD process runs in parallel plate reactor in a RF-plasma (13.56 MHz). Process parameters necessary for ultra-thin oxynitride to be formed are presented in table 1. In this paper the following names are used to refer to the types of ultra-thin dielectric layers studied: 'SiOx' and 'SiOxNy' for sample with very low and significantly higher content of silicon nitride phase, respectively (see table 1).

Table 1. Process Parameters allowing formation of ultra-thin PECVD dielectric layers

	SiO_x	SiO_xN_y
Power [W]	10	10
Pressure [m Torr]	350	400
SiH_4 flow [sccm] (diluted 2% in N_2)	100	150
N_2O flow [sccm]	150	16
NH_3 flow [sccm]	-	32

The ARXPS measurements were collected using a Theta Probe instrument, a method of parallel ARXPS measurements. Monochromated Al Kα X-ray source (hν ≈ 1486.6 eV) was used throughout the work. The samples were introduced into the ultrahigh vacuum chamber that had a base pressure of ≈ 10^{-10} mbar and measurement pressure of ≈ 10^{-9} mbar (≈ 10^{-7} Pa) after cleaning by sputtering with 1 keV Ar^+ ions. The angular distributions were measured for emission angle (perpendicular to the surface normal) of 25° to 75° with a step size of 10°.

In the present study, surface losses were removed from each measured spectrum. The spectrum deconvolution procedure of Werner [7-9] was applied repeatedly to a spectrum until the background due to surface losses was eliminated in the considered energy range:

$$Y_{k+1} = Y_k(E) - q_{k+1}\int Y_k(E+T)L_{k+1}(T)dT \quad (5)$$

Where Y_k is the spectrum from which k-fold scattering has been eliminated. The coefficients q_k are functions of the reduced partial intensities, $c_n = C_n/C_0$, where the partial intensities C_n represent the number of electrons that reach the detector after n-fold inelastic scattering and $C0$ represents the number of electrons that reach the detector without inelastic scattering:

$$\begin{aligned} q_1 &= c_1 \\ q_2 &= c_2 - q_1^2 \\ q_3 &= c_3 - q_1 q_2 - q_1^3 \\ q_4 &= c_4 - q_1 q_3 - q_2^2 - q_1^2 q_2 - q_1^4 \end{aligned} \quad (6)$$

The kernels used in equation (5) for each stage of the deconvolution are shown in figure 1 and figure 2 for O1s and Si (2p, 2s) peaks. The differential surface excitation probability (DSEP) provide the function $ws(T)$ (for deconvolution of surface-inelastic scattering) and were taken from reference. [10].

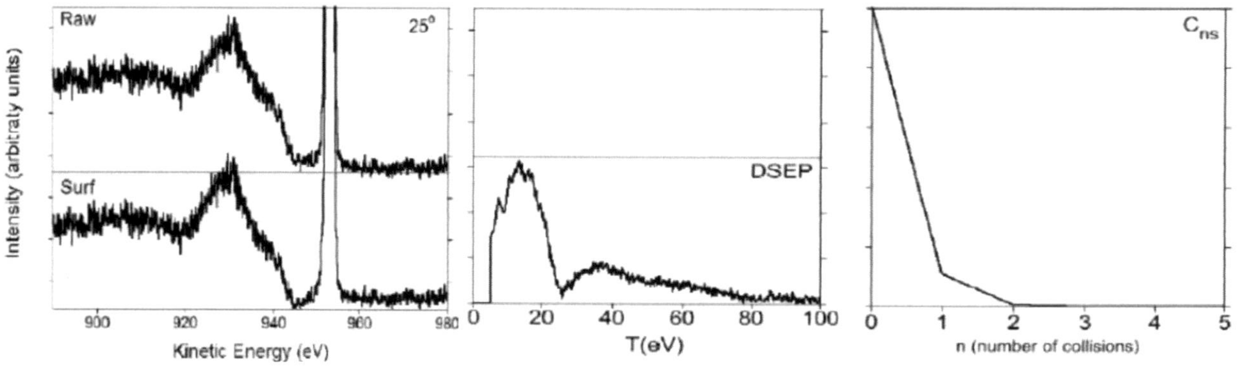

Figure 1. Kernel used for O1 s

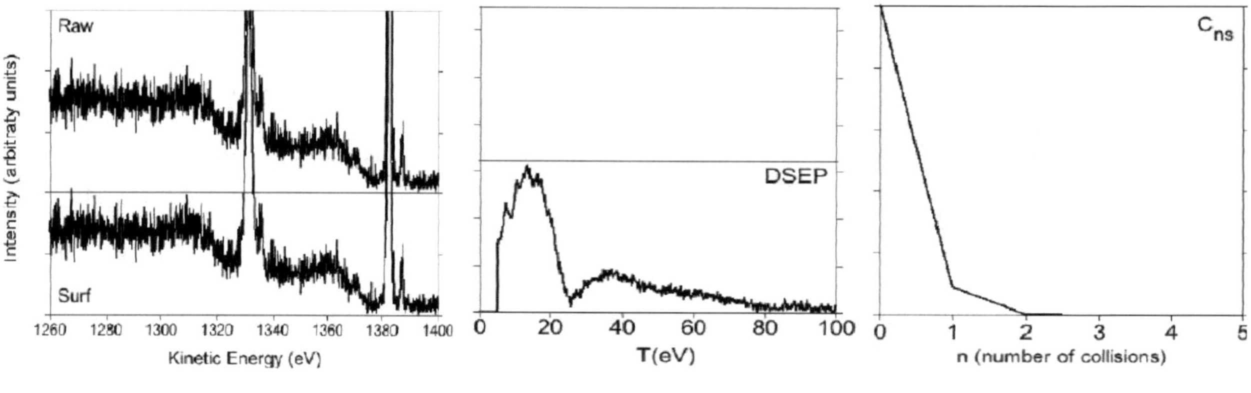

Figure 2. Kernel used for Si 2p

4. Results and Discussion

To analyze the performance of the proposed method described above, experimental data were used for SiOx and SiOxNy films on the Si substrate with C contamination at the top. Present model includes the effect of anisotropy in the photoionization cross-section and elastic scattering. The spectrums were calculated by using least square fitting at the emission angles of 25° and 55°. The partial intensities were calculated by using Simulation of Electron Spectra for Surface Analysis (SESSA) [11,12]. The obtained thickness of the films and concentrations of particular component in each layer is in table 2 and 3 for SiOx and SiOxNy respectively. Figure 3 and 5 shows the comparison of the calculated (dashed lines) and experimental spectrum (solid lines) after removal of surface excitations. There is a good agreement between experimental and calculated spectrum for all analyzed samples and the emission angles of 25° and 55°. Figure 4 and 6 shows the comparison of the experimental and calculated peak intensities after normalization. Experimental and calculated peak intensities generally agree well for all emission angles except the angles greater than 70°.

Table 2. Obtained thickness and concentration of particular component in each layer for SiO_x

Sample 1	**Thickness (nm)**	**Chemical Specie**
Layer 1	0.1	C
Layer 2	8.0	$Si_{0.4}O_{0.6}$
Layer 3	Substrate	Si

Table 3. Obtained thickness and concentration of particular component in each layer for SiO_xN_y

Sample 1	**Thickness (nm)**	**Chemical Specie**
Layer 1	0.15	C
Layer 2	1.0	$Si_{0.455}O_{0.545}$
Layer 2	3.5	$Si_{0.44}O_{0.34}N_{0.22}$
Layer 3	Substrate	Si

Journal of Physics: Conference Series **439** (2013) 012005 doi:10.1088/1742-6596/439/1/012005

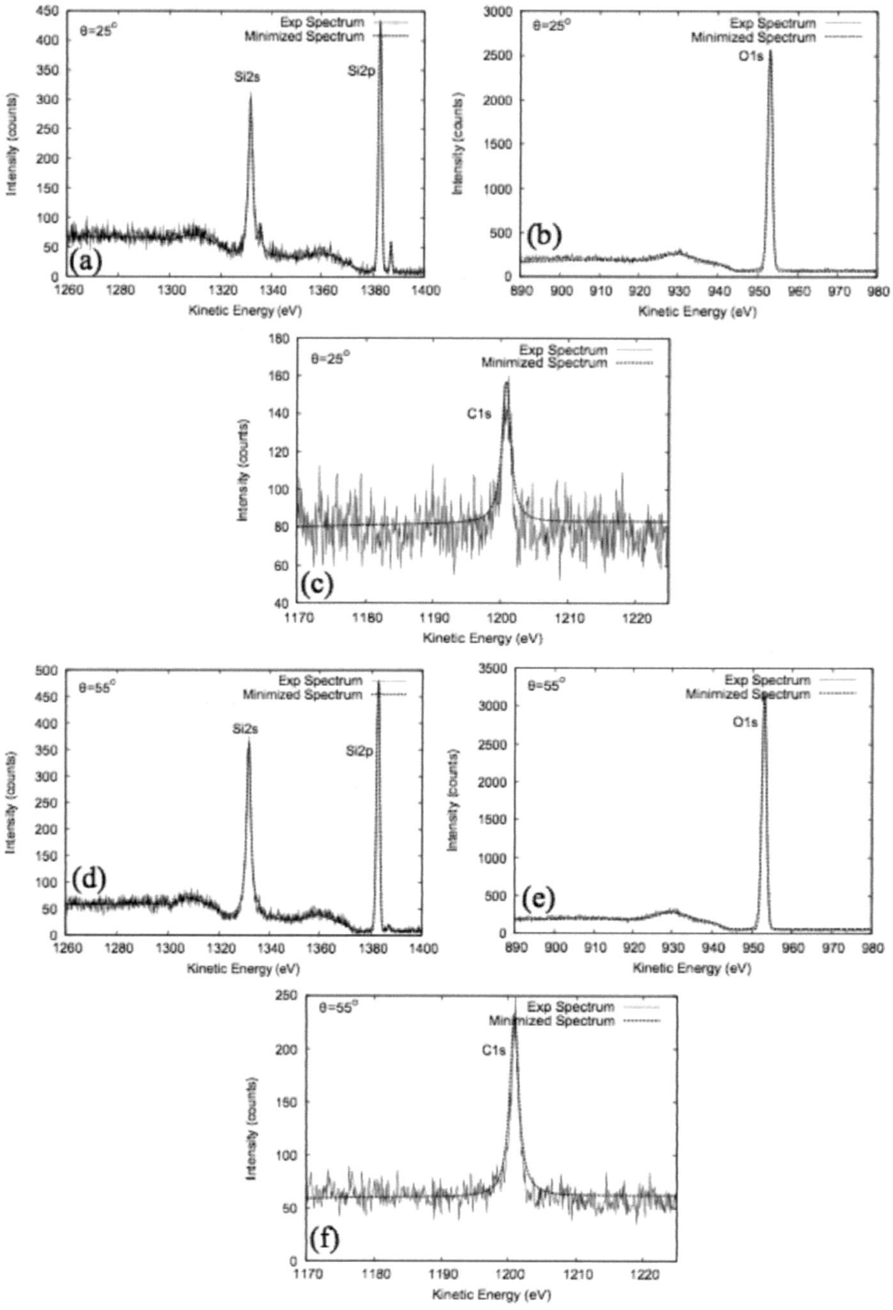

Figure 3. Fitting of Experimental spectrum and simulated spectrum at emission angle of 25° and 55° for Sample1. (a) Si 2p and Si 2s spectrum at 25°. (b) O1s spectrum at 25°. (c) C1s spectrum at 25°. (d) Si 2p and Si 2s spectrum at 55°. (e) O1s spectrum at 55°. (f) C1s spectrum at 55°

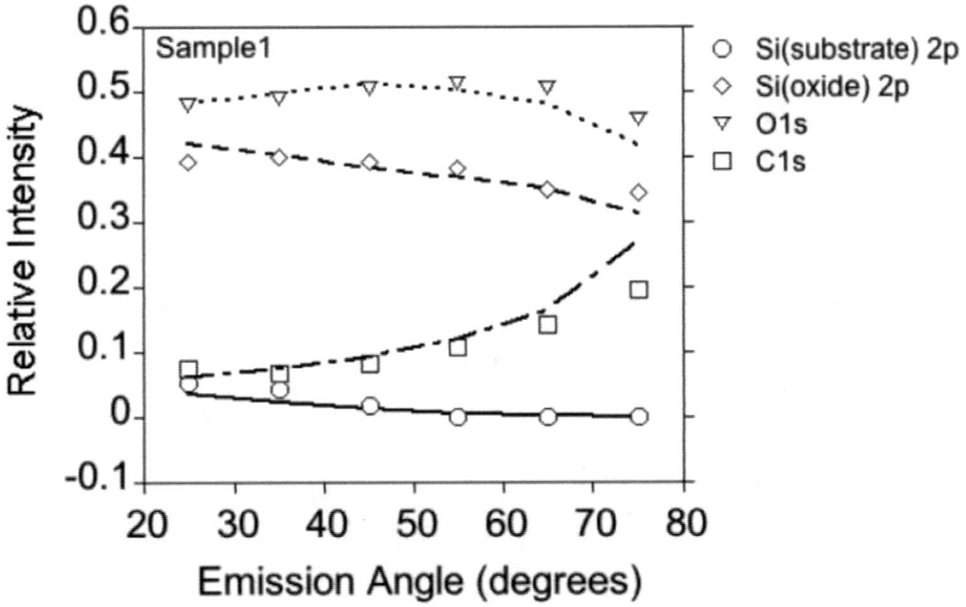

Figure 4. Comparison of experimental and calculated AR normalized peak intensities

Journal of Physics: Conference Series **439** (2013) 012005 doi:10.1088/1742-6596/439/1/012005

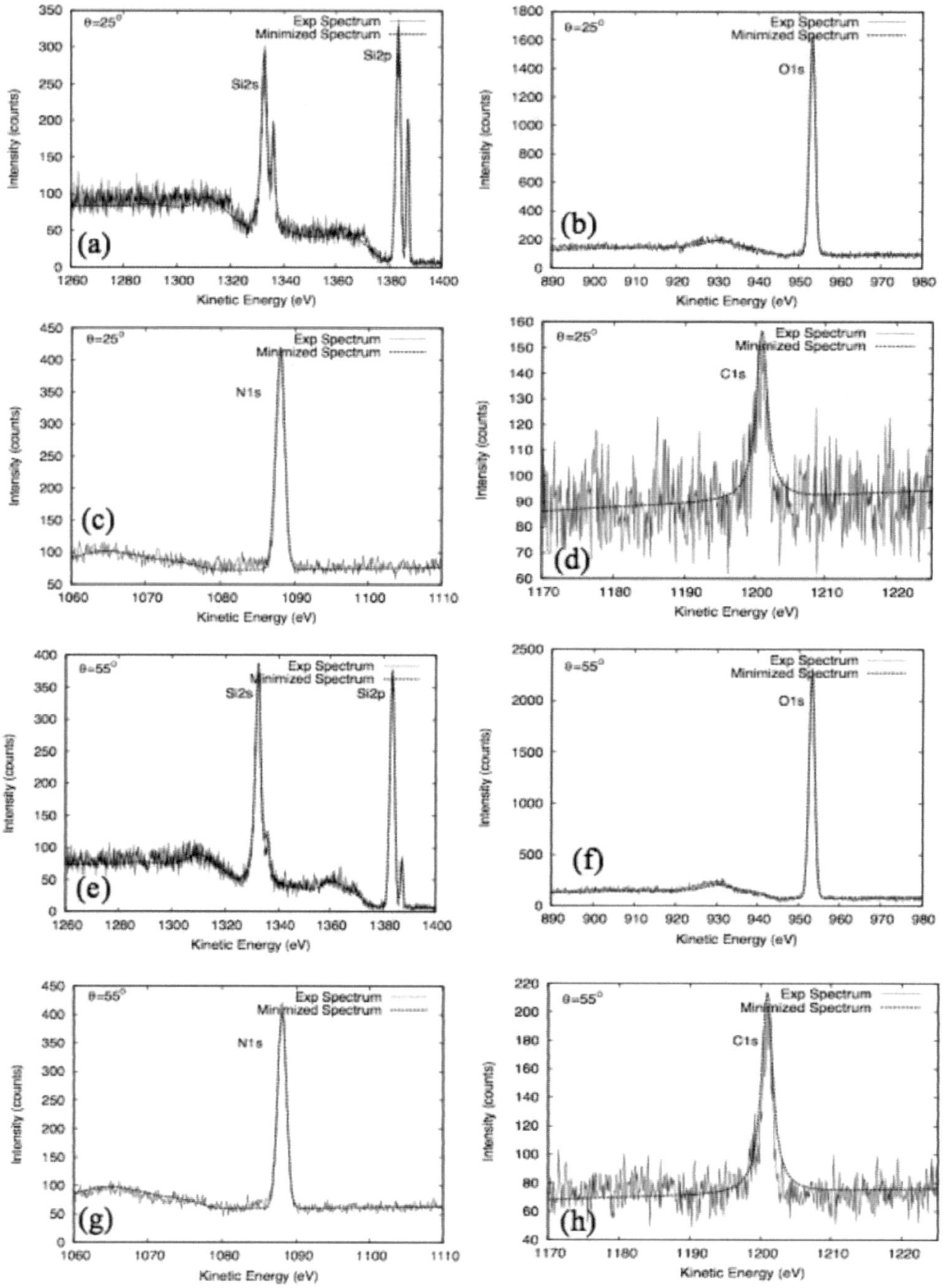

Figure 5. Fitting of Experimental spectrum and simulated spectrum at emission angle of 25° and 55° for Sample 2. (a) Si 2p and Si 2s spectrum at 25°. (b) O1s spectrum at 25°. (c) N1s spectrum at 25°. (d) C1s spectrum at 25°. (e) Si 2p and Si 2s spectrum at 55°. (f) O1s spectrum at 55°. (g) N1s spectrum at 55°. (h) C1s spectrum at 55°

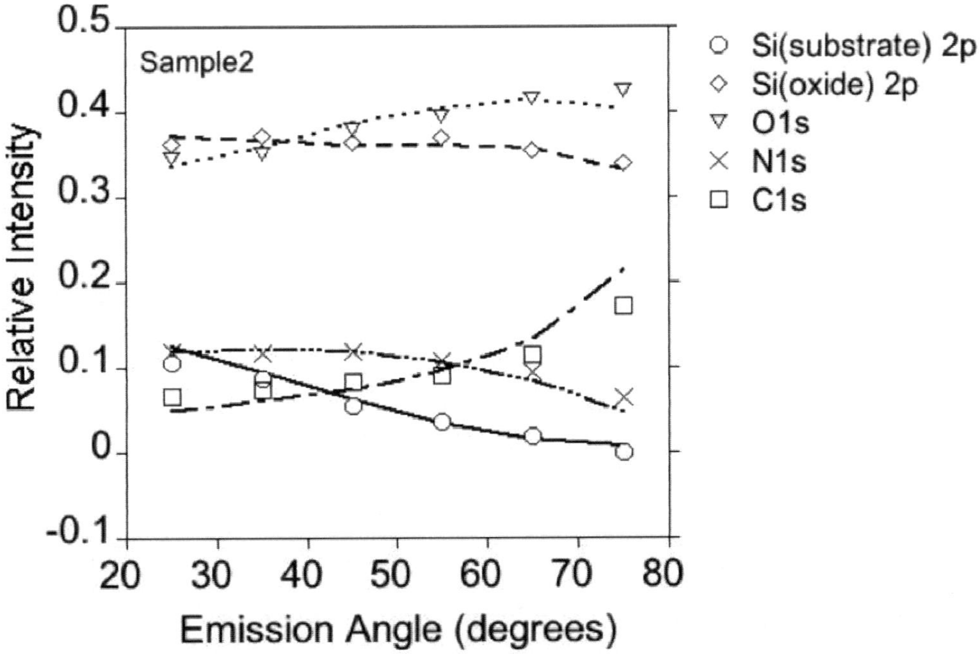

Figure 6. Comparison of experimental and calculated AR normalized peak intensities

5. Conclusions

Compositional depth profile can be retrieved accurately by using energy and angular distributions. Angular distribution in the Theta Probe configuration can only be quantitatively interpreted with a model that accounts for elastic scattering of the photoelectron as well as anisotropy of the photoionization cross section.

References

[1] Cumpson P J 2003 In Surface Analysis by Auger and X-ray Photoelectron Spectroscopy (Eds: Briggs D and Grant J T) *IM Publications, Chichester* 651
[2] Tougaard S 1986 *Surf. Interface Anal.* **8** 257
[3] Tougaard S and Hansen S H S 1989 *Surf. Interface Anal.* **14** 730
[4] Werner WSM 1995 *Surf. Interface Anal.* **B 23** 696
[5] Werner WSM, Tilinin I S, Beilschmid H and Hayek M 1994 *Surf. Interface Anal.* **B 21** 537
[6] Tilinin S and Werner WSM 1993 *Surf. Sci.* **290** 119
[7] Werner WSM, 1995 *Phys. Rev.* **B 52** 2964
[8] Werner WSM, 1995 *Surf. Interface Anal.* **23** 737
[9] Werner WSM 2001*Surf. Interface Anal.* **31** 141
[10] Werner WSM, Glantschnig K and Ambrosch-Draxl C 2009 *J. Phys. Chem. Ref. Data* **38** 1013
[11] http://www.nist.gov/srd/nist100.htm.
[12] Smekal W, Werner WSM and Powell CJ 2005 *Surf. Interface Anal.* **37** 1059

Laser Surface Hardening of Groove Edges

A Hussain[1*], A H Hamdani[1], R Akhter[1] and M Aslam[1]
Pakistan Institute of Lasers & Optics, P.O.Box 505, Rawalpindi, Pakistan

[*]E-mail: pilo786@yahoo.com

Abstract
Surface hardening of groove-edges made of 3Cr13 Stainless Steel has been carried out using 500 W CO_2 laser with a rectangular beam of 2.5x3 mm^2. The processing speed was varied from 150-500 mm/min. It was seen that the hardened depth increases with increase in laser interaction time. A maximum hardened depth of around 1mm was achieved. The microhardness of the transformed zone was 2.5 times the hardness of base metal. The XRD's and microstructural analysis were also reported.

Keywords: Surface hardening, CO_2 laser, Microhardness, Microstructure

1. Introduction

Laser surface hardening (LSH) is well known process in automotive industries. The various components require high hardness, low wear resistance, high elasticity and toughness [1, 2], since they are normally used in poor working environment. The high abrasion and wear enhance the friction which leads to premature failure. Thus it is important to improve the surface properties of the component. Extensive studies of LSH have been carried out for various components like the piston head in diesel engines [3], automotive shaft [2], bearing rings [4], different hollow parts [5], valves and valve seats of internal combustion engine [6] etc.

LSH process is applicable where conventional hardening process does not give the localized heating and low distortion associated with laser. A hardened layer of 0.5 to 1mm is sufficient to ensure the improvement in wear and fatigue resistance. Laser beam heats the surface to austenite temperature followed by self-quenching of the layer by the substrate. This allows the transformation of austenite to martensite. In practice, it is advisable to avoid surface melting which result in coarse structure and lower value of hardness on the surface. LSH process improves the fatigue life because cracks are usually initiated at the surface and propagated under continuous stresses. The developed compressive stresses in laser treated area must be over come before the crack can propagate in to the material.

The laser hardened layer contains high strength structure and residual compressive stresses which would slow down crack growth rate near early stage of fatigue crack propagation. This slowing down (retarding) effect is increased when hardened depth is increased [7].

A threefold increase in the surface hardness under optimum conditions has been reported [8]. Such a large increase is important for increasing the residual stresses, wear properties, corrosion resistance and fatigue life of the steel [9, 10].

In this work, LSH of groove-edges for V-type railing system of 3Cr13 SS has been carried out using 500 W CO_2 laser with a rectangular beam shape. The variation of case depth with speed, XRD's analysis and their microstructure were studied.

2. Material

The material was received in the form of component as shown in the Fig. 1. The microhardness hardness of the as received sample was 180-200Hv.

The composition of the base metal is given in the table 1.

Table-1. Composition of the material

Elements	C	S	Cr	Mn	Si	Fe
Composition (Wt.%)	0.308	0.0164	12.86	0.6	0.49	Balance

[*] Corresponding author

Fig. 1 shows (a) slide rail having V-shaped groove and (b) the surface area "A"to be hardened. The main objective was to harden the surface "A" as shown in the Fig. 1 up to a required hardness of 450-550Hv.

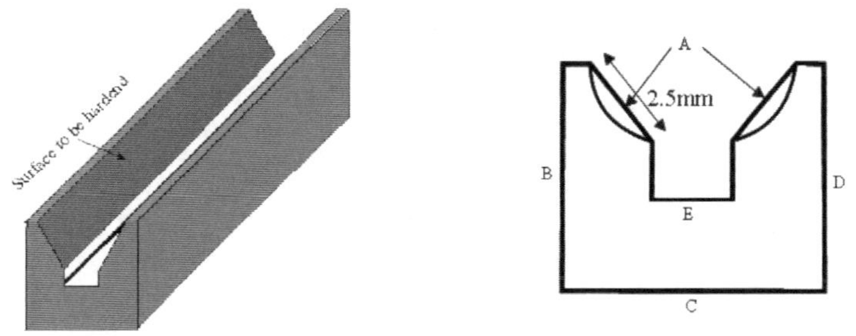

(a) (b)

Fig. 1 (a) slide rail and (b) the surface to be hardened (A)

3. Experimental setup

The experimental set up for CO_2 laser hardening process is shown in Fig. 2. A rectangular laser beam of size 3x 2.5 mm^2 was adjusted on the work piece with a focusing mirror of 280mm focal length, which was used to harden the required surface. The beam size was selected to cover the full width of the surface to be hardened, so that the required results were obtained in a single pass. The surface of the sample was black coated to increase the absorption of the laser beam. For laser treatment, the samples were fixed on a CNC table which moves under the laser beam along x-axis. The sample was positioned at an angle of 45° with the horizontal surface to access the area to be hardened. Shielding gas (nitrogen) was used to prevent the formation of oxide on the sample surface. The laser treated samples were cut and mounted along the cross-section. For metallography, the samples were ground and polished, and then etched in 2% Nital. The microstructure of the laser treated zone was examined under optical microscope and Vickers hardness was measured as a function of case depth. XRD's analysis was also carried out to compare the base metal and top surface of the laser treated specimens.

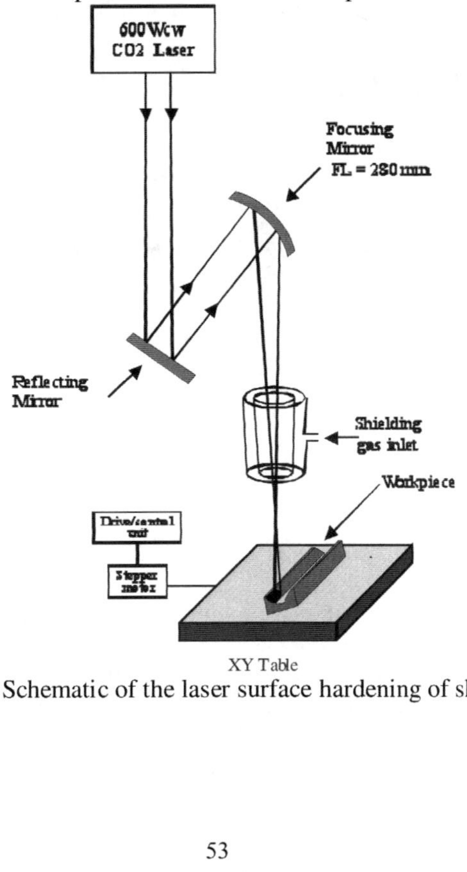

Fig. 2: Schematic of the laser surface hardening of slide rail

4. Results and Discussions

4.1 Effect of Laser Power and Traverse Speed on the Depth of Hardness: The important parameters in laser processing are laser power, working speed and spot size which determine the input power/energy to treat the materials. Fig. 3 shows a graph between the depth and width of the laser hardened zone versus the working speed. It depicts that the depth and width of the hardened zone decreases with increase of speed for fixed laser power because shorter laser interaction times lower the thermal input to the material. Fig. 4 shows that the depth and width of the hardened zone increases as the laser power is increased at constant speed due to higher heat input.

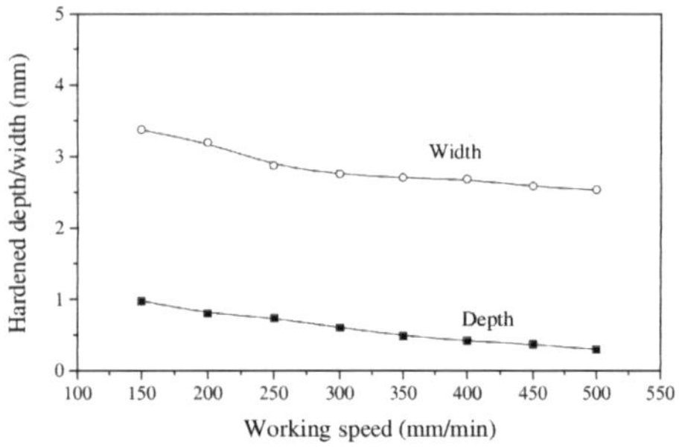

Fig. 3. Depth and width of hardened zone versus speed at laser power of 450W.

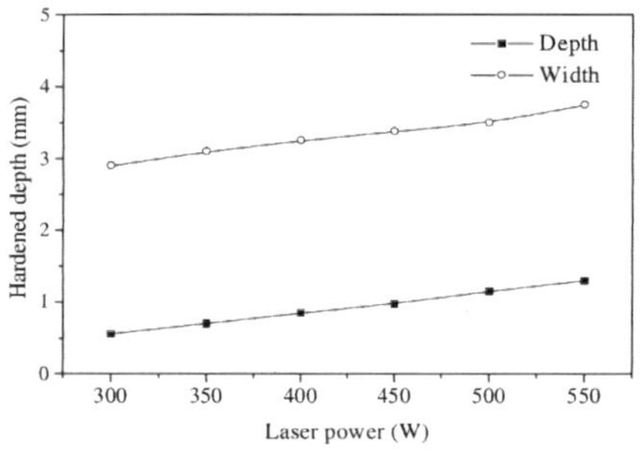

Fig. 4. Depth and width of hardened zone against laser power at constant working speed of 150mm/min

4.2 Microhardness and XRD's analysis: The microhardness as a function of case depth for various speeds at constant power level is presented in the Fig. 5. It shows that the case depth decreases with increasing the speed. When speeds are 150mm/min and 200mm/min, the surface melting occur which lowers the microhardness at the top surface of specimen. In order to attain such a large transformed depth i.e., about 1mm, the top surface melting of the order 0.1-0.2mm in depth could not be avoided. However, the surface grinding up to depth of 0.2mm can be used to completely remove the melted layer before final assembly.

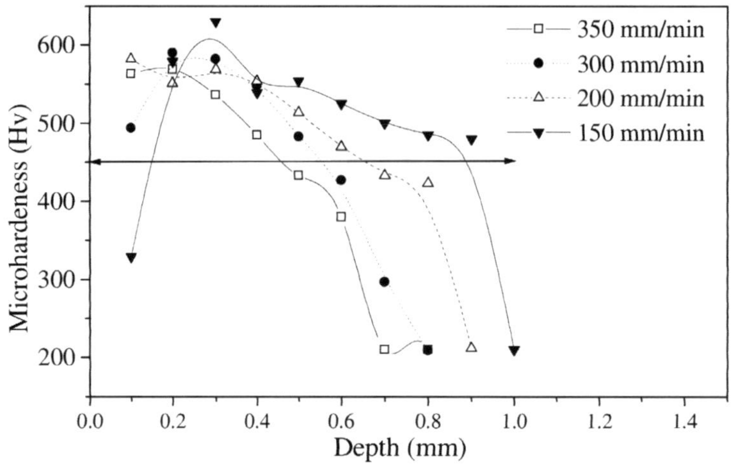

Fig. 5. Variation of microhardness with speed at constant laser power 450 W

Fig. 6(a & b) show the XRD's analysis of the base metal and laser treated samples, respectively. The samples were scanned with X-ray (CuKα) radiation source at the rate of 3°/min from 0 to 100°. In Fig. 6-a, the peaks of XRD's analysis correspond to the planes (110), (200) & (211) showing the BCC structure and represent the base metal is martensitic stainless steel. On the other hand the laser treated sample without any polishing/grinding was scanned. The XRD's analysis in Fig. 6-b shows that the iron carbides (Fe_3C) is the major phase present on the surface along with minor intensity peaks remained unidentified.

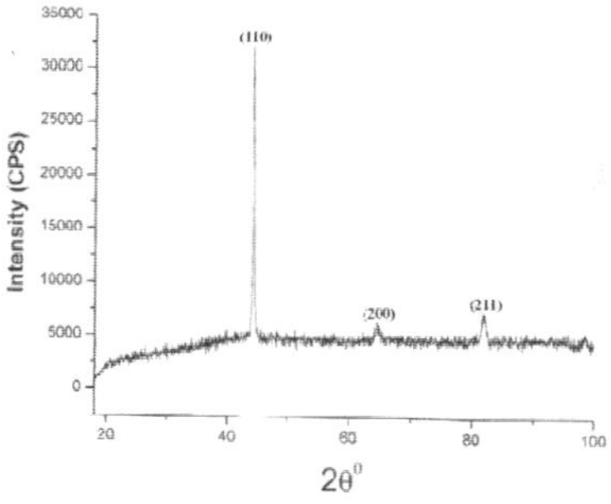

a)

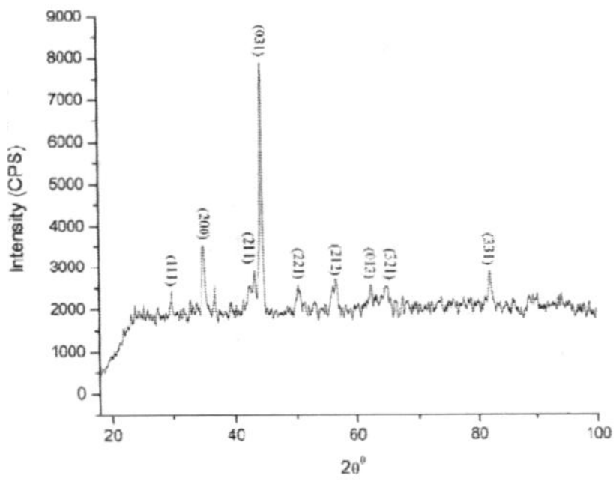

(b)

Fig. 6. XRD's analysis of a) as received sample and b) laser treated specimen at speed 150mm/min and power 450W.

4.3 Microstructure: Fig. 7(a-d) shows the microstructure of the laser hardened samples. In order to achieve hardened depth of 0.7mm to 1mm, the specimen was treated at low speed of 150mm/min and 450W laser power. The microstructure of the laser treated region is divided in to four zones. The regions consist of a) melted zone, b) hardened zone, c) partially transformed zone, and d) unaffected zone or base metal. In Figure 7-a, the top surface of the sample consists of a small melted layer/portion. This layer mostly consists of different types of carbides and retained austenite as confirmed by XRD's analysis. During melting, the material approaches to boiling point and carbon comes from lower to upper melted part and reacts with iron and makes iron carbides on the surface. The value of microhardness is lower in this zone due to coarse structure. In Fig.7-b, the microstructure is not resolved properly due to martensite formation and has the microhardness about 450 to 550Hv. In this case the different phases dissolved uniformly and formed fine structure due to self quenching process. This hardened layer is responsible for enhancing the wear resistance and lowers the crack propagation to the base metal and also increase the fatigue life of component. Fig.7-c shows the partially transformed zone which consists of martensite and base metal. Here the temperature was insufficient to dissolve all types of phases into austenite due to high cooling rate. Most of the heat is rapidly conducted to the adjacent bulk material. In this region the microhardness rapidly decreases as shown in the Fig. 5 .Fig. 7-d shows the microstructure of the base metal having hardness of about 200Hv.

The top surface melted layer has softening effect, since it contains different types of carbides and retained austenite which reduces the hardness, in addition, due to temperature distribution generated by laser beam, the grains gradually transform to fine from top surface to bottom. As subsurface has relatively lesser amount of retained austenite due to high cooling rate resulting in fine grain structure which gives high value of hardness. The top surface melted layer can be avoided, if the increased working speed is used at constant input laser power but at the cost of hardening depth. This melted layer provides an additional benefit by providing machining allowance for the final product.

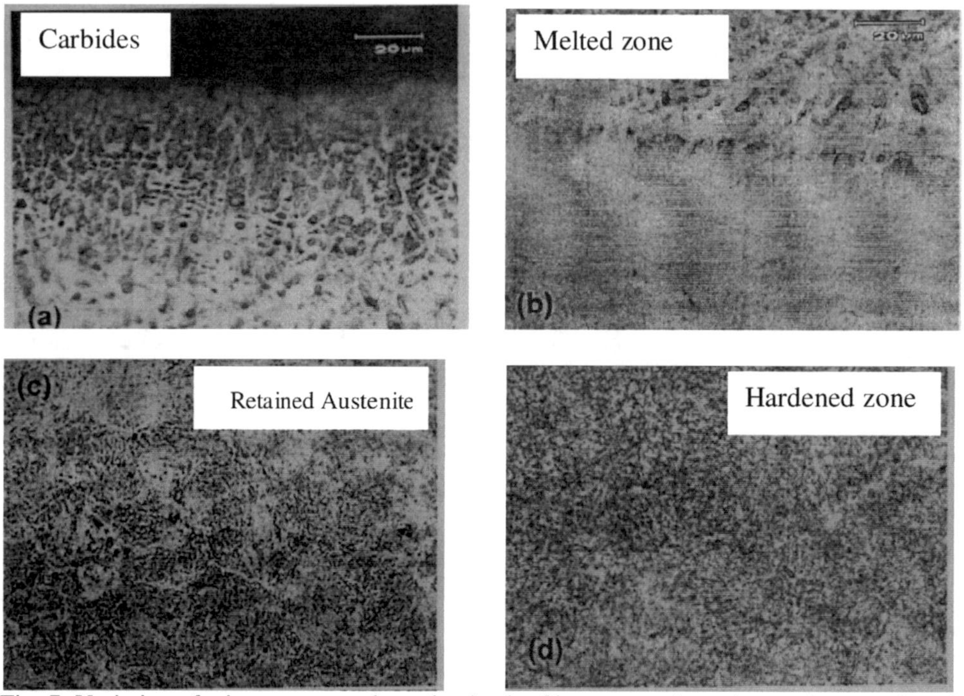

Fig. 7. Variation of microstructure along the depth of laser treated specimen: a) top surface (melted zone) ; b) Hardened zone; c) partially transformed zone; and d) base metal

5. Conclusions

- For a fixed laser power, the depth and width of the laser treated zone decreases as the working speed is increased. This is due to lower interaction time at higher working speed.
- For a fixed working speed, the depth and width of the laser treated zone increases with the increase of laser power. This is due to more heat input.
- 2.5 fold increase in the hardness was achieved by this process.
- The top surface melted layer can be avoided, if the working speed is increased for constant input laser power. However this melted layer provides an additional benefit by giving machining allowance for the final product.

References

[1] JM Beswick, Bearing Steel Technology-Advances and State of Art in Bearing Steel Quality and Assurance 2007 Vol. 1 No. 1465, ASTM International
[2] M S F de Lima, FA Goia, R Riva, A M E Santo, *Mater. Res.* 2007 Vol, 10, 4
[3] Q Liu, Y Song, Y Yang, G Xu and Z Zhao, *JMEPEG*, 1998 Vol. 7(3), 402
[4] R. Akhter, A.Hussain, W.A. Farooq and M. Aslam, Key Engin. Mater. 2010 Vol. 442, 130
[5] G Tani, A Fortunato, A Ascari, and G Campana, CIRP Annals – Manuf. Tech. 2010 Vol. 59(1) 207
[6] T Slatter, H Taylor, R Lewis and P King, Wear 2009 Vol. 267(5-8), 797
[7] D Ji-Liang, C Tsyr-Jang and T Yen-Hung, Engin. Fract. Mech., 1989 Vol 33(3), 483
[8] Hussain, R. Akhter, S. shahdin and M. Ashraf Atta, Proc. SPIE 2001 Vol. 4397 329
[9] Cerny, I Furbacher and V. Linhart, *JMEPEG*, 1998 Vol. 7, 361
[10] M L Escudero and J M Bello, *Mater. Sci and Eng.*, 1992 A158, 227

Journal of Physics: Conference Series **439** (2013) 012007 doi:10.1088/1742-6596/439/1/012007

Formation of carbon nano- and micro-structures on C_1^+ irradiated copper surfaces

Shoaib Ahmad

CASP, Government College University, Church Road, Lahore 54000, Pakistan

E- mail: sahmad.ncp@gmail.com

Abstract. A series of experiments has identified mechanisms of carbon nano- and micro-structure formation at room temperature, without catalyst and in the environment of immiscible metallic surroundings. The structures include threaded nano fibres, graphitic sheets and carbon onions. Copper as substrate was used due to its immiscibility with carbon. Energetic carbon ions (C_1^+) of 0.2-2.0 MeV irradiated Cu targets. Cu substrates, apertures and 3 mm dia TEM Cu grids were implanted with the carbon. We observed wide range of µm-size structures formed on Cu grids and along the edges of the irradiated apertures. These are shown to be threaded nano fibers (TNF) of few µm thicknesses with lengths varying from 10 to 3000 µm. Secondary electron microscopy (SEM) identifies the µm-size structures while Confocal microscopy was used to learn about the mechanisms by which C_1^+ irradiated Cu provides the growth environment. Huge carbon onions of diameters ranging from hundreds of nm to µm were observed in the as-grown and annealed samples. Transformations of the nanostructures were observed under prolonged electron irradiations of SEM and TEM. A mechanism for the formation of carbon nano- and micro-structures is proposed.

1. Introduction

A novel technique of producing carbon onions with unique characteristics is reported. Carbon atoms emerging out of the edges of C_1^+ irradiated Cu sheets are shown to form graphitic layered networks studded with carbon onions that are non-spherical, hollow and often contain smaller ones inside. Carbon onions, the multishelled fullerenes, were discovered by intense electron beam irradiation of soot[1] with their structural manipulation reported later[2], have also been produced by C_1^+ implantation[3-5] and other high temperature techniques[6-8]. Their remarkable spherical shelled nanostructures found few applications; the difficulty of their structural manipulation lies in the perfect shelled symmetry. In the experiments reported here we show that energetic C_1^+ irradiated Cu sheets with vertical holes allow C atoms at the end of their implantation range to emerge out and form nanostructures along the edges of the holes. This technique for the formation of graphitic nanostructures that include C onions[1] and networks of graphene[9,10]. It is different from similar C onion growing methods that utilize C_1^+ implantations[3-8] and the graphene formation techniques that include CVD[11,12] and ion implantation[13-16] where C addition is from outside onto a surface. We do not utilize the as-deposited C ions, but those C atoms which emerge out of the edges and sharp corners of the irradiated Cu sheets. These are shown to produce a variety of structures. This is the fundamental difference in our growth technique with that of the others. Carbon onions with various shapes and structures have been seen emerging with the under and over-lying graphene layers at all stages of C accretion and structure formation. The continuous C_1^+ irradiation at a steady rate is seen to produce stacks of graphene layers and nanotube-like structures. Discontinuous or variable C_1^+ irradiation rate may produce hollow onions that may be connected with each other or take the form that we identify as hollow onions. The hollow onions may contain smaller ones: the onion-inside-onions. Extremely low solubility of C in Cu at all temperatures and the immiscibility gap[17,18] is the key for the growth of C nanostructures on Cu surface or within. In almost all ion implantation experiments[3-5, 13-15] the C_1^+ dose at a given energy is the main parameter along with high substrate temperature ~700-1000 K. Their reported[3-5] critical C_1^+ dose for the formation of C onions ~10^{17} C cm^{-2}. On the other hand, we have implanted ~5×10^{14} C cm^{-2} at room temperature and found that the C_1^+ implantation rate rather than the dose is the most crucial parameter.

2. Experimental Method

C_1^+ is implanted into Cu grids for TEM (3 mm diameter) and 120 μm thick sheets. C_1^+ ions are delivered by GCU's 2 MV Pelletron. The energy of ions $E(C_1^+)$ is in the range 0.2 – 0.8 MeV. We have maintained the total dose at all irradiation energies ~ 5×10^{14} C_1^+ cm^{-2}. However, the rate of irradiation can be varied. The range of penetration of C_1^+ with these energies ~ 0.24-0.8 μm. The implanted C atoms at the end of the range, if allowed to emerge out from the sides of vertical holes in Cu sheets, are observed to form carbon nanostructures. We have exploited this peripheral aspect of C_1^+ irradiation whereby a fraction of C atoms escape the entrapment and emerges out of the Cu surface. These escaping C atoms are the feed of our nanostructures that are described in this letter. The carbon nanostructures so formed include multi-layered graphene stacks, nanotube-like and carbon onions. Initial investigations were done on 10 μm thick TEM Cu grids with square and circular holes. The later experiments were done on 3 mm diameter 120 μm thick Cu grids with ~ 1 mm inner holes for the nanostructure formation along the inner edges. The implantations were done at room temperature without any catalyst. Diagnostics were conducted with SEM (JEOL JSM-6480LV), Confocal microscope (Olympus FV1000) and TEM (JEOL JEM1230). One can generate a broad range of C nanostructures on 3 mm diameter, 10 μm thick Cu grids for TEM by irradiating with 0.2 MeV C_1^+ ions. In this paper, we present and discuss results obtained from 120 μm thick Cu rings with 3 mm OD and ~1 mm ID. Here, the initial batch of graphitic layers, sheets and shells are formed on the edges of the inner hole. These are lifted outwards by the next batch growing underneath by utilizing the C atoms emerging from within the Cu surface. The Confocal microscopic studies of the irradiated Cu grids and sheets reveal that all sharp surfaces of the holes, scratches and the edges of the grid bars are laced with C nano and microstructures. The emerging C atoms at the end of their implantation range with almost no kinetic energy are the basic ingredient and the requirement for C-C bonds to form. By exploiting this novel characteristic feature of the energetic C_1^+ irradiation of Cu our present technique was developed.

3. Results and discussion

Figure 1 shows TEM micrographs of graphitic layers with variable curvature and dimensions. Figure 1(a) has spherical layered structures that can be seen superimposed on a dense layered background. There is a continuity of the outer curved surfaces of ~25 nm. Figure 1(b) shows a half spherical, graphitic ball that grew earlier than the underlying graphene layers. Four peaks of a continuum of the curved graphene layers that seem to have adjusted to the variable rate of outward flow of C atoms from below can be seen in (c). Figure 1(a)-(c) show the Cu surface decorated with multiple graphene layers. The inner conical regions of variable densities are visible within all of the outward protruding structures. The inner regions point to the routes through which C accretion occurred. Much darker layers underneath may be due to the failure of the later C addition, most probably with different rates of C outflow, to follow the earlier pattern of growth.

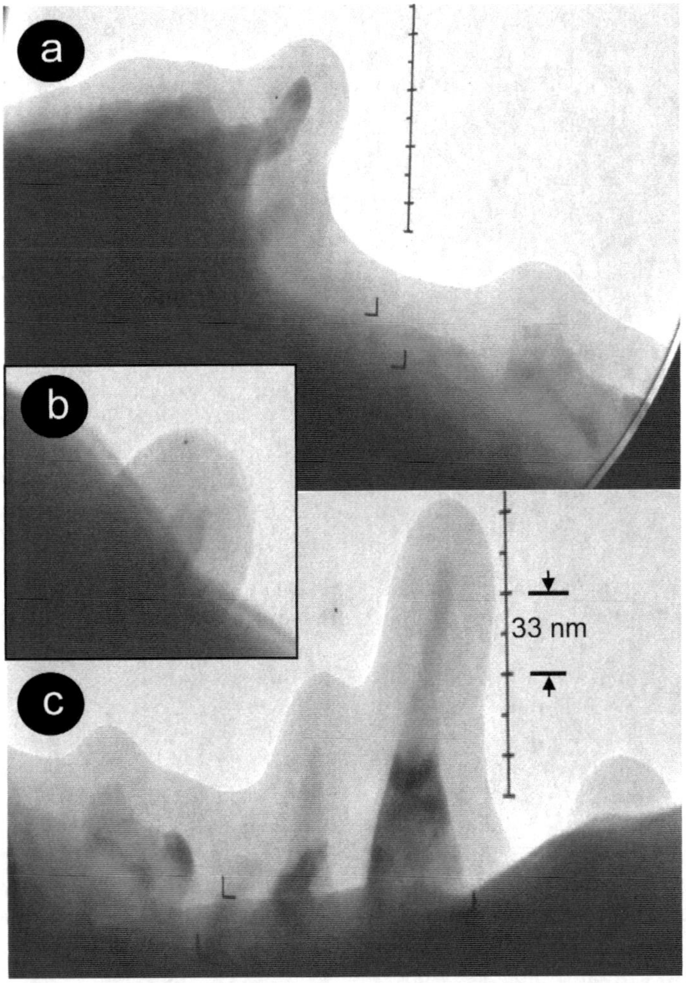

Figure 1. (a) Two partially spherical graphene layers emerging out of a dense background. There is a continuity of surfaces with varying heights and curvature. (b) A half spherical graphitic ball is shown that grew out of the underlying layers. (c) Four peaks of graphene layers; one on the extreme right is similar to (b).

A graphitic stack of irregular shape with spheroidal cages of dimensions between ~5 to 30 nm having different densities and curvature can be seen in Figure 2. Most of the cages have elliptic shapes. The cages are located on various layers of graphene with perforations around the cages especially around those that protrude out of the underlying graphitic network. The closed cages are identified as C onions with many having inner structures that have different curvatures and variable inner densities. Hollow C onions and networks of graphitic layers are shown with a heap of graphitic layers with different orientations, thicknesses and contours is shown studded with C onions of 5-30 nm diameters. Figure 2(a) elaborates the upper right hand portion of the main figure by a sample tilt to highlight the six larger (15-20 nm) and about a dozen smaller (≤ 5 nm) elliptic C onions. The perforated graphene layers between the outermost onions are also visible. Figure 2(b) shows overlapping and interconnected quasi-spherical mounds of graphene layers that are laced with onions

Journal of Physics: Conference Series **439** (2013) 012007 doi:10.1088/1742-6596/439/1/012007

grown inside and the one that is protruding outwards. The largest, outward protruding onion shows inner structure indicating variable shell density of the onion.

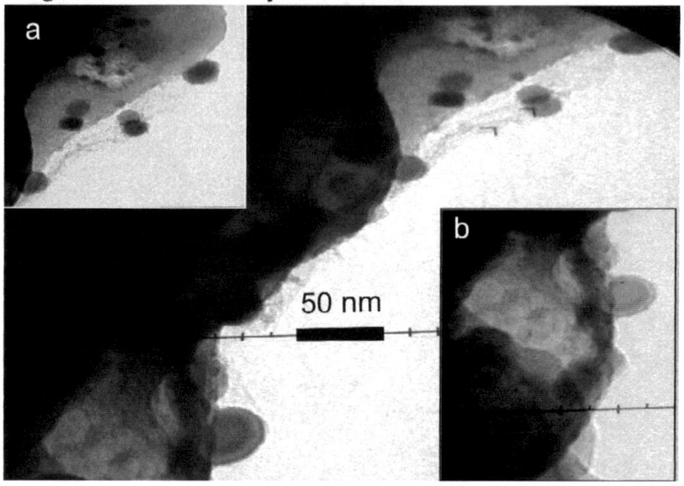

Figure 2. (a) The onions laced on various graphene layers in the upper right hand section are shown with sample tilt to highlight the six 10-15 nm elliptic onions and a number of smaller ones. (b) Overlapping and interconnecting mounds of graphene layers (~ 70 nm) with a protruding onion and some contained inside the mound are shown.

The diversity of the sizes and their respective positions on the graphene layers in Figure 2 can be contrasted with that shown in Figure 3(a) where a large area ~1 μm^2 is shown with various interconnected C structures including the onions, layers and curtain-like networks of graphene sheets. The darker regions in the upper and lower halves are the dense graphitic layers studded with few quasi-spherical objects (~100 nm) and lots of smaller onions spread all over. The two boxed areas are enlarged in (b) and (c), respectively. The range of the clearly visible onions is 20-40 nm in (b) and (c) where a much larger number of smaller ones are also present. The connectivity between the distinctly shaped and separately positioned onions is presumed to be through the broken and perforated graphene layers. The diffraction spots from the centre of (c) are shown in (d) that confirms the crystallinity of the graphitic shells, however, due to the inhomogeneous onion shapes and the associated variability of the curvature, the ordering of the spots is difficult.

We investigated with 120 keV TEM the possibility of transformations induced by intense energetic electron beam irradiations on graphitic structures with special emphasis on C onions. Intense electron beam induced changes on the arrays of onions shown in Figure 3(c). It is shown that the changes induced in onions that are spread on sheet-like graphitic networks occur via the reformation of the outer onions as well as those of the inner ones as a result of C removal. These changes were observed to occur continuously during the electron beam irradiations with many C onions retaining their identity as individual onions while a few did merge into each other.

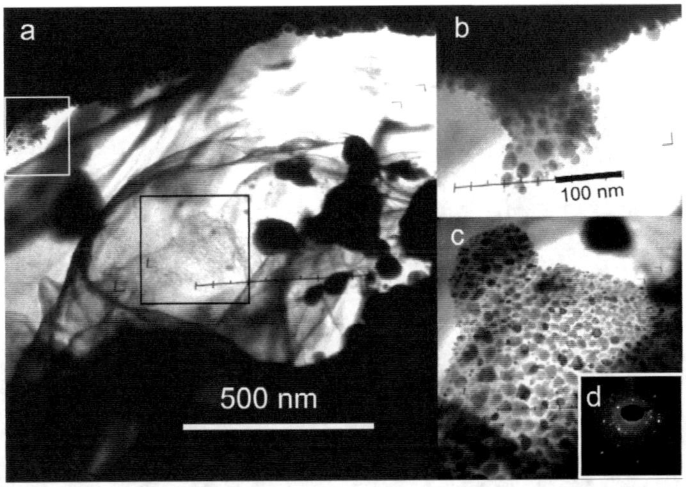

Figure 3. (a) An area of ~ 1μm^2 shows a rich variety of inter-connected graphitic layers, curtains and large and small C onions. Two boxed regions are further enlarged. Scale bar is 500 nm. (b) The boxed region shows a large number of C onions. (c) A large collection of ~100 C onions on relatively flatter layers with their shapes varying enormously. (d) Diffraction pattern from the centre of the onions sheet in (c).

On the other hand, the effect of C removal and the associated structural readjustments appeared significant in C onions that are connected with each other as depicted graphically in Figure 4 (a) and (b). The transformations induced in large ($\geq$100 nm) and small ($\leq$20-40 nm) cages is shown to occur via coalescence of the cages as is clear by comparing 4(a) with 4(b). In 4(a) two large cages and many smaller elliptic ones are connected with a much larger (~200 nm diameter x 500 nm length) graphitic caged structure. Under intense 120 keV electron beams the smaller onions on top of the structure in 4(a) coalesced into the larger ones shown in 4(b). Coalescence of smaller onions has produced comparatively larger spheroidal structures with increased densities. An observable feature of the reformed C onions is that these remain connected with each other. Major changes also occurred in a large kidney shaped cage that can be seen on the left in 4(a) prior to being subjected to intense 120 keV electron beam.

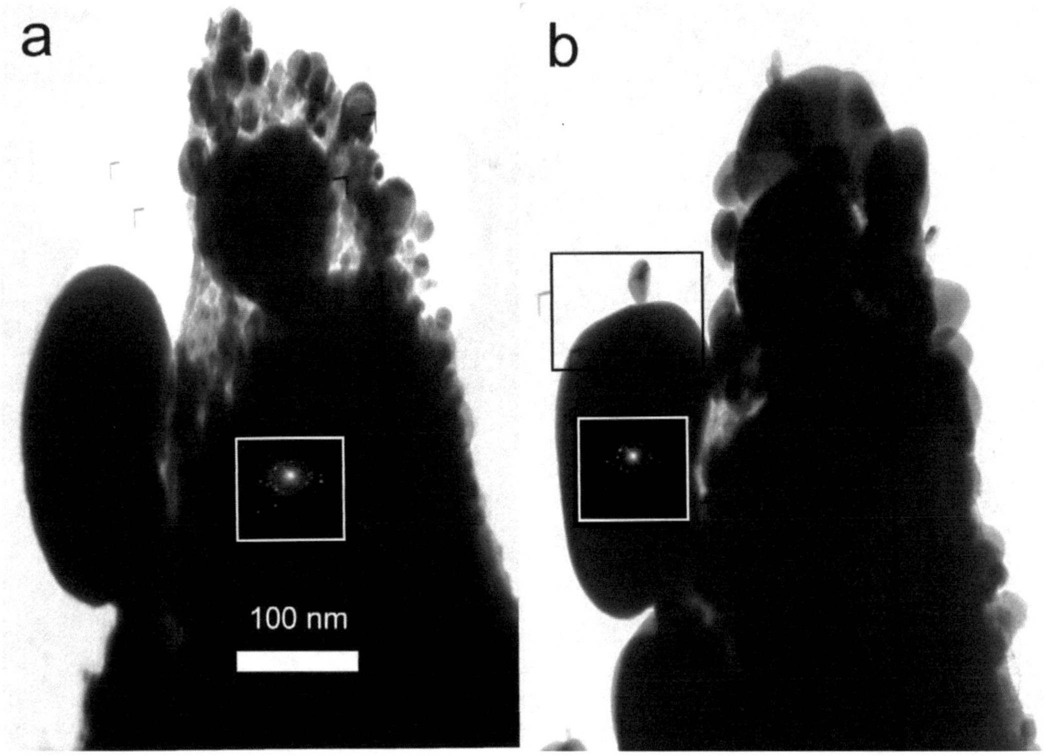

Figure 4. (a) A collection of three large (≥ 50 nm) and many smaller C onions, all sticking together in their pristine positions is shown in the first TEM micrograph. (b) After intense 120 keV electron beam irradiation in TEM for 1 minute the smaller ones coalesce into larger C onions that have relatively well defined surfaces. White boxes show the diffraction spots from the center of the graphitic caged regions. The scale bar is 100 nm.

The noticeable changes seen are in the outward convex curvature supplemented by the appearance of two types of nanostructures that are enlarged in Figure 5(a). The two largest structures in Figure 4 show diffraction spots from the beam that was focused at the point where the pattern is embossed. One can see that crystallinity of the objects is graphitic with wide range of overlapping planes in both the cages that seem to contain many smaller spheroidal, onion-like structures.

In Figure 5 the boxed portion of Figure 4(b) is enlarged where the central ellipsoidal that is a C onion-inside-onion in the making with ~30 nm x 45 nm dimensions. Six new hollow multiwalled nanotube-like objects with 5-7 shell thickness and different heights can be seen. This in situ observation of the different stages of C onion and nanotube growth out of the graphitic surface leads us to similar mechanisms of formation by the C atoms emerging from within the irradiated Cu surface. The nanotube-like features, onions and multiple layered stacks of graphene shown in Figures 1 and 2 can be understood by the comparing with the energetic electron induced transformations in Figure 5(a). In 5(a) two distinct but structurally related C nanostructures are formed by the C atoms that are released by 120 keV electrons from the underlying graphitic surface. The graphical representation of the growth mechanism is presented in four sequences 1-4 in Figure 5(b). The structure underneath the irradiated region and the rate of flow of the emerging C atoms determine the curvature and the lateral extent of the newly forming C shells (step 1). For a steady flow of C, shells-inside-shells grow without significant variations in diameter resulting in nanotube-like features (step 2). For increasing outward flow of C one is likely to obtain onion-like structure (step 3). Discontinuous or variable C flow rates are likely to yield onion-inside-onions (step 4). Majority of the C onions seen in Figure 3 belong to this category. Such an object can also be seen in the TEM micrograph 5(a). This in situ growth of an onion-inside-onion and the six hollow nanotubes is due to the C atoms released by the 120 keV electrons from the graphitic surface of the C cage. One finds similarities by comparing the C onions formed by C atoms that emerge from the irradiated Cu surfaces or the ones that are released by breaking the sp^2 bonds of the graphitic structures by 120 keV electrons. The emergence of C from under the Cu surface defines this mechanism of formation that is inward growth as opposed to the usual outward one.

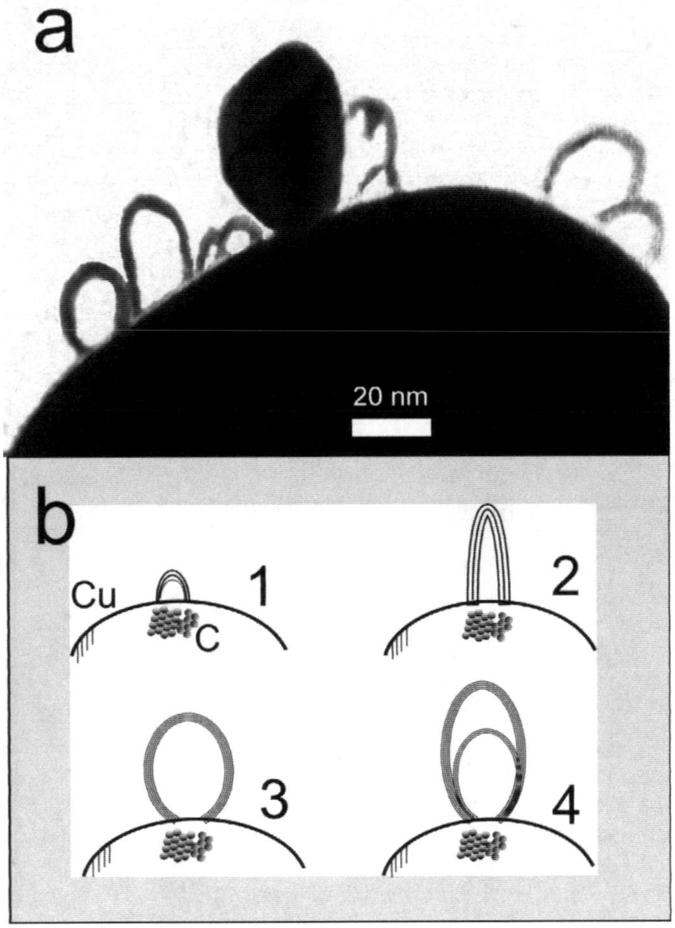

Figure 5. (a) The boxed region in Figure 4(b) is further enlarged to reveal six new, multi-walled nanotubes of ~2-3 nm thick shells. These emerged out of the reshaped, larger graphitic caged structure under energetic electron irradiations. (b) Four proposed stages (1-4) of the formation of multi-walled nanotubes or multi-shelled onion-inside-onions are graphically represented.

4. Conclusion

In conclusion I have shown that the sp^2-bonded nanostructures formed by C atoms emerging from within, onto an immiscible surface like Cu, will include all possible configurations including multiple layered graphene, nanotubes and onions. Their respective number densities and shapes, however, depend on the rate of emergence which is linked with the ion irradiation rate and of course, the local surface configurations at the emergence of C atoms. We have proposed a growth mechanism by demonstrating in situ growth of C cages. Carbon onions with well defined shell thickness and inner radii have been shown to be a possibility and the onion-inside-onion configuration can be exploited to yield desired physical properties. I show for the first time, the hollow, onion-inside-onion structures. These are formed by the C atoms emerging out of the C_1^+ irradiated Cu surfaces at room temperature. The rate of C_1^+ irradiation and the route that C atoms take to escape from within the Cu lattice determines the formation of C onions and stacks of graphene layers. Spherical and cylindrical curvature as discussed elsewhere[19] is seen as the essential feature by which the C onions and graphene layers accommodate variations in the C accretion rate and the Cu surface irregularities at the region of C emergence. Pentagons and heptagons may be forming along with the hexagonal networks to yield the diversity of curvature in the layers and shells of onions. Formation at room temperatures forbids the Stone Wales[20] transformation to smooth out the sharp corners by bond rearrangements.

References

[1] Ugarte D 1992 *Nature* **359** 707
[2] Banhart F and Ajayan, P. M. 1996 *Nature* **382** 433
[3] Cabioch T, Riviere J P and Delfond J 1995 *J. Mater. Sci.* **30** 4787
[4] Cabioch T *et al* 2000 *Coatings Technol.* **128-129** 43
[5] Abe H 2001 *Diamond and related Mater.* **10** 1201
[6] Sono N, Wang H, Chhowalla M, Alexandou I and Amaratunga G A 2001 *Nature* **414** 506
[7] Szerencsi M and Radnoczi G 2010 *Vacuum* **84** 197
[8] Blank V D, Kulnitskiy B A and Perezhogin I A 2009 *Scripta Materilia* **60** 407
[9] Meyer J C, Geim A K, Katsnelson M I, Novosolov K S, Booth T J and Roth S 2007 *Nature* **446** 60
[10] Geim A K 2009 *Science* **324** 1530
[11] Mathevi C, Kim H and Chhowalla M 2011 *J. Mater. Chem.* **21** 3324
[12] Reina A *et al* 2009 *Nano. Lett.* **9** 30
[13] Bangert U, Bleloch A, Gass M H, Seepujak A and van der Berg J 2010 *Phys. Rev. B* **81** 245423
[14] Garaj S, Hubbard W and Golovchenko J A 2010 *Appl. Phys. Lett.* **97** 183103
[15] Baraton L *et al.* 2011 *Nanotechnology* 22 085601
[16] Sun Z *et al* 2010 *Nature* **468** 549
[17] Fuks D *et al* 2001 *J. Mol. Structure (Theochem)* **539** 199
[18] Sun J and Zhang M D 1991 *J. Mater.Sci.* **26** 5762
[19] Ahmad S 2005 *Nanotechnology* **16** 1739
[20] Stone A J and Wales D J 1986 *Chem. Phys. Lett.* **128** 501

Growth of vertically aligned multiwall carbon nanotubes columns

M I Shahzad[1], M Giorcelli[1], D Perrone[2], A Virga[1], N Shahzad[1,2], P Jagdale[1], M Cocuzza[1] and A Tagliaferro[1]
[1]Department of Applied Science and Technology (DISAT), Polytechnic of Turin, Turin-10129, Italy.
[2]Center for Space Human Robotics, Italian Institute of Technology (IIT), Turin-10129, Italy.

E-mail: imran.shahzad@polito.it

Abstract. Capability of patterning carbon nanotubes (CNTs) growth is of tantamount importance for a number of applications ranging from thermal to electronic. This article reports on the columnar growth of vertically aligned multiwall carbon nanotubes (VA-MWCNTs) on patterned Silicon (Si) surface. We have developed procedures based on negative as well as positive masking approaches which allows the growth of predetermined MWCNTs patterns. We describe in detail the process steps leading to Si surface patterning. As quoted above, patterns are exploited to grow VA-MWCNTs. We have focused in particular on the growth of CNT pillars by chemical vapor despoition (CVD) technique at 850°C with camphor and ferrocene as carbon precursors and catalyst respectively. Field emission scanning electron microscopy (FESEM) is employed at low magnification to verify the correct patterning, and at high magnification to examine the surface morphology of CNTs pillars. The pillars are up to 2 mm high, their height being tailored through the deposition time. The diameter of each MWCNT is in the range 30-70 nm and the length is up to few hundred micrometers. The small CNT pillars produced, have several electrical and thermal applications. For instance they can be very useful for heat transfer systems as the lower thermal conductivity of fluids can be improved by the inclusion of nanotubes thanks to their peculiar 1-dimensional heat transfer characteristics.

1. Introduction

Nowadays Carbon based nanomaterials are among the most interesting materials. Following Ijima report [1], these materials have stimulated quite a lot of activity in different areas of science and engineering because of their extraordinary physical and chemical properties. Due to their high thermal and electrical conductivity, rigidity, strength, … [2], CNTs are ideal for a wide variety of applications. CNTs can be produced by various processes [3] including arc discharge, laser ablation and chemical vapor deposition (CVD). The most promising of these techniques for large scale production is CVD as it opens opportunities of controlled massive [4] as well as precise growth on patterned substrates [5].
The vertically aligned carbon nanotubes (VACNTs) uniformly grown by CVD on a patterned surface feature interesting unique mechanical, electrical and thermal properties which led to propose a variety

of potential application such as heat dissipation systems [5], biosensors [6], field emission sources [7], conductive electrodes and micro mechanical devices [8].

An efficient and economical method of patterning by soft lithography, which enables to produce highly dense VA-MWCNTs columns, is described in the present work. Exploiting such patterning we have grown a number of CNTs structures using precursors that allow fast growth and good CNT quality.

2. Surface patterning of silicon substrate

We have developed a procedure to pattern silicon wafer by soft photolithography technique in both positive and image reversal mode that ensure the growth of well defined structures of CNTs. The protocol we have used for the patterning of silicon surfaces involves the following steps;

2.1. Realization of the mask

The first important step for this patterning is to design the mask on a Mylar sheet. There are two types of photomasks: positive and negative photomasks (figure 1). Black colored parts represent-the areas which stop the UV radiation while these radiation can transmit through the white parts of the mask.

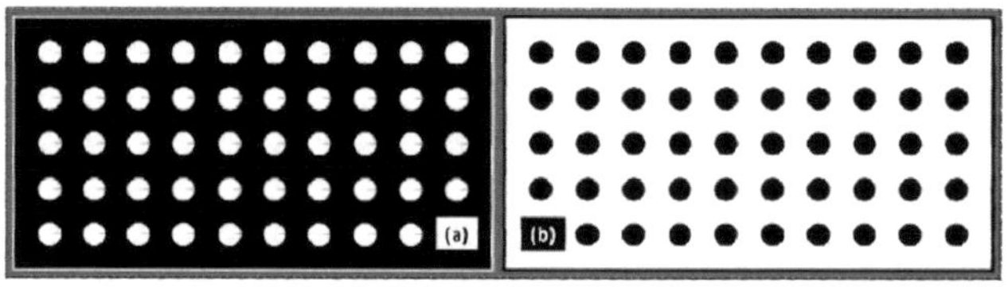

Figure 1. Soft Mylar photomasks a). Positive mask b). Negative mask

2.2. Mask transfer to the substrate

After the realization of photomasks, the substrates are prepared for the application of photoresist. The steps required to transfer the mask to the substrate are;

2.2.1. Wafer Cleaning. Silicon (100) substrates were cleaned with an acetone ultrasonic bath for 10 minutes followed by soaking in 2-Propanol for a few minutes to get rid of contaminations, debris and particulates. Furthermore, the wafers were washed in Piranha solution 3:1 (H_2SO_4:H_2O_2) for 5 minutes to enhance the hydrophilic behavior [9] as well as to remove organic residues from the surface. At the end of this procedure wafers were extensively rinsed in de-ionized water, dried under nitrogen flux and subsequently heated at 120°C for a few seconds to obtain moisture free cleaned Si substrates (figure 2-a).

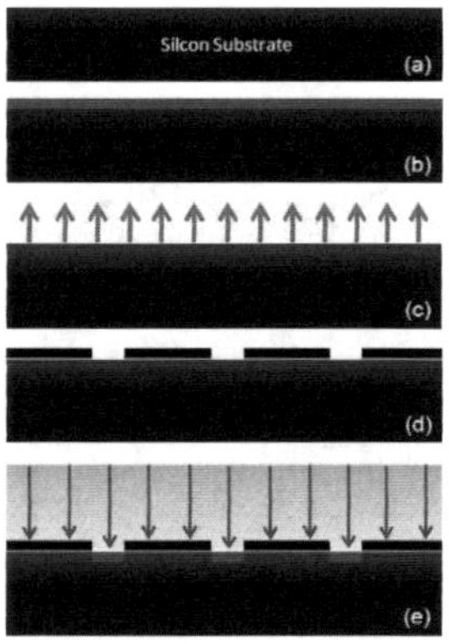

a). Cleaned Wafer

b).Uniform application of photoresist by spin coating

c). Soft baking at 120°C

d). In contact mask alignment on photoresist surface

e). Controlled UV exposure

Figure 2. Steps involved in the transferring of mask to the substrate

2.2.2. *Photoresist Application.* Commercially available photoresist AZ5214E from MicroChemicals has a capability to work for both positive and negative masks. The cleaned wafers were held on spinner's stage by creating a vacuum between substrate and sample holder. After dispensing enough quantity of photoresist on silicon wafer, a uniform layer of photoresist (figure 2-b) was obtained using a spinner programmed at 600 rpm for 5 seconds and at 4000 rpm for the subsequent 40 seconds.

2.2.3. *Soft Bake.* After the uniform application of the photoresist layer, the silicon substrate was soft baked at 120°C for 2 minutes (figure 2-c). The temperature and duration of soft baking are critical parameters in photo-imaging as it transfers the photoresist coatings to imageable layer. Soft baking for a too long time or at high temperatures can degrade the sensitivity of photoresist by destroying the sensitizer or reducing its solubility in the developer. It may also produce cracks in the photoresist film which ultimately leads to poor resolution. Shorter than needed baking may tend to affect the adhesion and exposure to the photoresist layer as both exposed and unexposed areas would be attacked by the developer. It could make the photoresist layer non uniform and patchy, resulting in the poor precision of photolithographic process [10].

2.2.4. *UV Exposure.* Before the UV exposure, the mask was aligned with the wafer, so that the pattern could be transferred onto the wafer surface precisely. It is important to remove any kind of particles, trapped between the resist and the mask, which may damage the mask and lead to defects in the pattern. The photoresist-coated silicon wafer was brought into physical contact with the photomask and held in vacuum to avoid the diffraction effect and to attain high resolution (figure 2-d). Once the mask and wafer were accurately aligned, the photoresist was exposed through the pattern with a high intensity ultraviolet light with UV light power density of $3 mW/cm^2$ (figure 2-e). The exposure to the UV light changes the chemical structure of the resist so that it becomes more soluble in the developer.

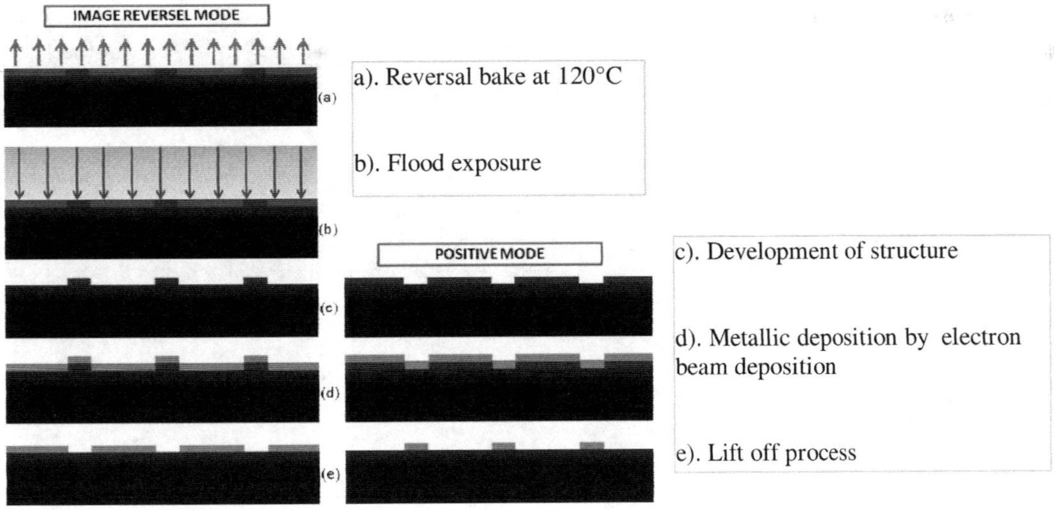

Figure 3. Fabrication of patterned structures in Positive and Image Reversal Mode

2.2.5. *Development of photomask.*

a) Development of positive mask. The next step was to dissolve soluble areas of photoresist by basic developer and to get visible patterns on wafer. The wafer with positive mask was directly developed in 1:1 solution of commercially available AZ 351 developer (from MicroChemicals) in de-ionized water.

The areas that were exposed to UV light were removed after developing, leaving windows exposing the underlying material. In other words, the mask, contains an exact copy of the pattern which will remain on the wafer (figure 3-c).
b) Development of negative mask. For the negative photomask (image reversal mode), few extra steps were involved after the UV exposure. After such step, the system is baked at 120°C for 2 minutes (figure 3-a) and a flood exposure in UV light without mask for 70 seconds (figure 3-b) is applied. These steps lead to an inversion in the chemistry of the photoresist that was not exposed to UV light at step 2.2.4. This resist become polymerized and more difficult to dissolve in 1:1 AZ Developer: de-ionized water solution. Hence it remains on the surface and the developer solution removed only the remaining resist portions (figure 3-c).

2.3. Fabrication of Patterned Structure:
After the successful imaging of the photomasks on the substrate the next step is to coat with a metal layer the areas where we don't want carbon nanotubes growth to occur. The following steps are required to such purpose.

2.3.1. *Metallic Deposition by Electron beam Evaporation.* CNTs can be grown on silicon substrate using an appropriate carbon source and a catalytically active metal. On the other end, non catalytic metals can be used to inhibit CNT growth. Hence, we used copper to cover the substrate where we don't wanted carbon nanotubes growth [11]. Metallic deposition was performed by electron beam evaporation technique in a vacuum chamber at 10^{-7} torr pressure. First a 20 nm thick titanium (Ti) layer was deposited and then a 200 nm thick copper (Cu) one added on top of it (figure 3-d). Ti is used as a buffer layer to overcome the poor adhesion of copper to silicon improving copper adhesion and stability.

2.3.2. *Lift off Process.* After deposition of metallic layers on the silicon substrate, a lift off process was performed in an ultrasonic bath at 60°C in acetone or normal methyl pirrolydinone (NMP) for 10-15 minutes (figure 3-e). During the lift off process the polymerized layer of photoresist under the metallic layer was detached from the silicon surface, leaving behind the required structures on the silicon substrate.

3. Growth of CNTs Columns by Thermal Chemical Vapor Deposition
The growth of vertical columns of multiwall carbon nanotubes on patterned silicon wafer was achieved by catalytic chemical vapor deposition of a reagent containing both the carbon and the catalyst sources. Ferrocene was used as catalytic source as it is a fine precursor to obtain iron nanoparticles, which leads to the formation of carbon nanotubes [12]. Commercially available camphor was used as carbon precursor as it is inexpensive and non-toxic. Furthermore, the arrangement of pentagonal and hexagonal carbon rings in camphor atoms makes it a good source to produce nanotubes [13].

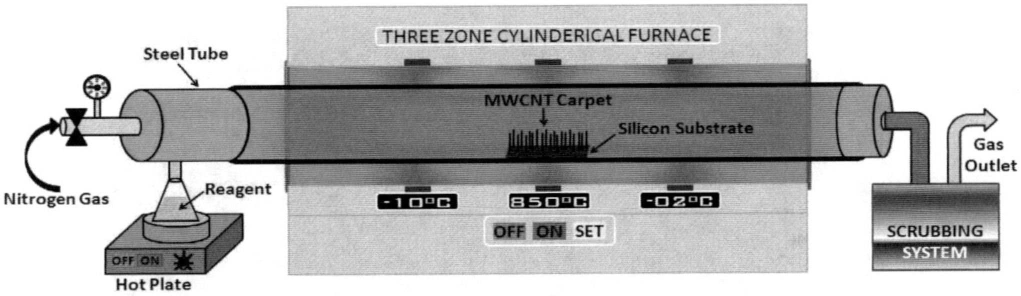

Figure 4. Thermal Chemical Vapor Deposition Growth System

The CVD unit used for CNTs growth is shown in Figure 4. A 100 cm long horizontal steel tube was placed in a 3-Zone furnace which has the ability to maintain a uniform temperature up to 1200°C throughout its length i.e. 60cm. The silicon substrates were placed in the central region of the furnace and inert atmosphere was maintained by nitrogen gas flux (420 ml/min). The pressure of the gas was kept slightly higher than the atmospheric pressure to avoid inward air leakages ensure. A pyrex conical flask containing the reagent mixture of camphor and ferrocene (20:1) was connected to the steel tube by a T joint between the furnace and nitrogen inlet. The pyrex flask assembly was rested on the hot plate.

The furnace temperature was maintained at 850°C and the pyrex flask was heated up to the vaporization of the reagent which starts above 200°C. The vaporized reagent was carried into the furnace by the nitrogen gas flux. The uniformity of the vapors flow was maintained by the temperature of the flask.

Because of the pyrolysis of the gases, the iron particles and carbon species were carried through the high temperature furnace. The iron particles works as catalyst and the deposition of carbon species on Si substrate resulted in the growth of carbon nanotubes. After a suitable time i.e.60 to 90 minutes, depending upon the required thickness of the CNTs carpet, the furnace was turned off. The similar method is described in further details by Musso et. al [4].

2. Surface Morphology

After the growth step is carried out, the produced structures were examined by FESEM. A ZEISS SUPRA-40 FESEM was used to study the structure, orientation, size and dimensions of nanotubes. In figure 5 (a,b), the circular patterning of different sizes on silicon substrate is shown. The vertical columns of carbon nanotubes grown by CVD shows that the length of circular columns is up to 1.5 mm. The diameters of the cylindrical columns are 250 μm (figure 5-c) and 500 μm (figure 5-d) and correspond to the mask features. The length of the columns, can be tailored by tuning the growth time in the range from a few micrometers to few millimetres.

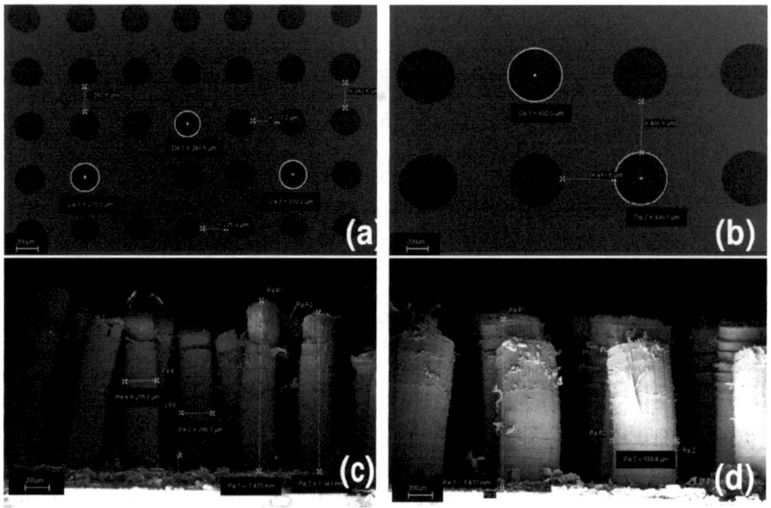

Figure 5. Patterned silicon substrates (a) & (b) and CNT columns (c) & (d)

Figure 6 shows some variety of structures based on MWCNTs varying in dimension and spacing as well as in structures type grown in our CVD system.

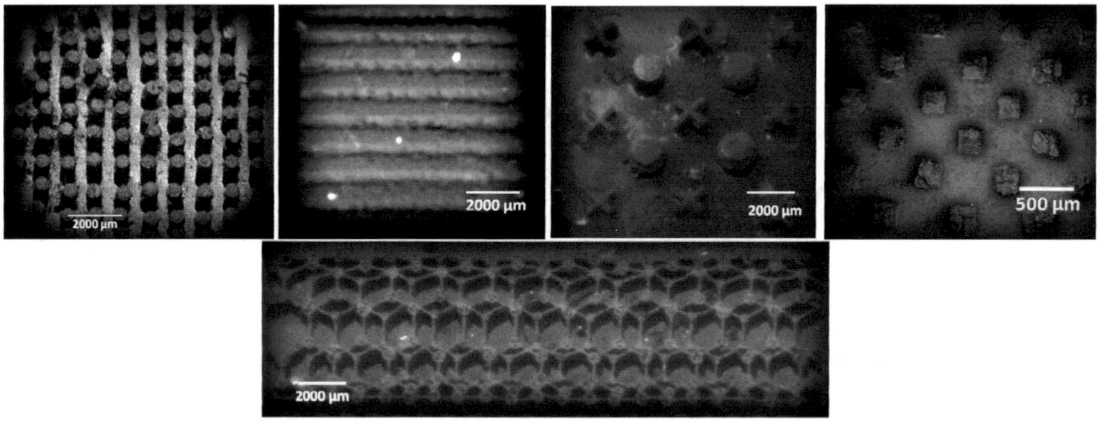

Figure 6. CNT based structures grown by CVD Technique

Figure 7 shows the structure, orientation, and dimension of carbon nanotubes inside the specific macroscopic structure. The diameter of a MWCNT is in the range of 30nm-70nm, only a few having diameter under 20 nm. The length of carbon nanotubes is up to few hundreds of micrometers. Most of the carbon nanotubes are well vertically aligned. Micrographs shown in figures 5 (a, b and c) are from the middle part of a column and micrographs in figure 5 (d, e and f) are from the top edge.

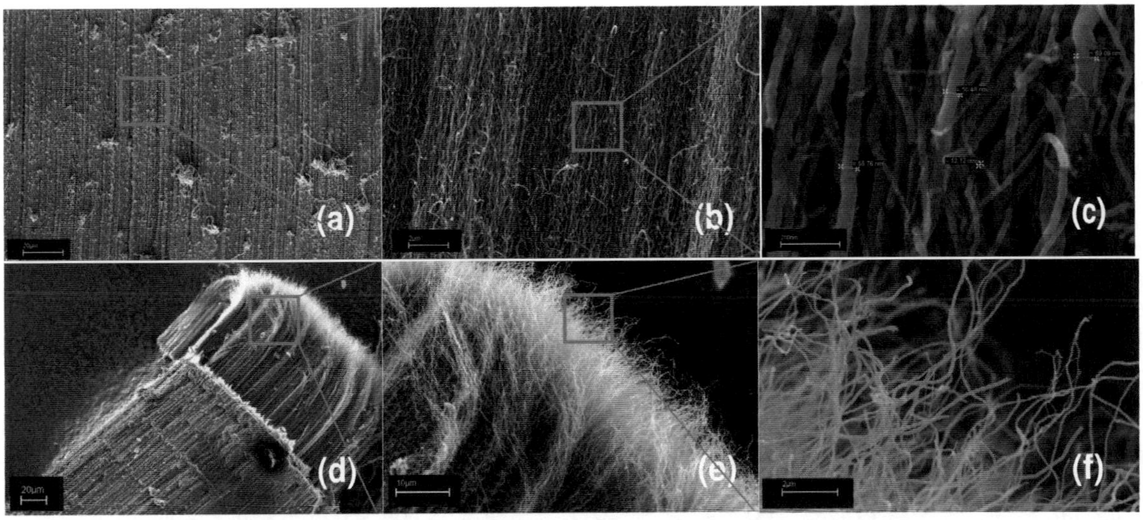

Figure 7. SEM micrographs of carbon nanotubes based structures

3. Structural Characterization

CNTs have been characterized by Raman spectroscopy to gain information about their structure. The Raman spectra of CNTs contains rich information about the electronic states and the phonon dispersion as it involves strong resonances of the incoming and outgoing photons and the vibrational states with the electronic energy levels of a tube [14].

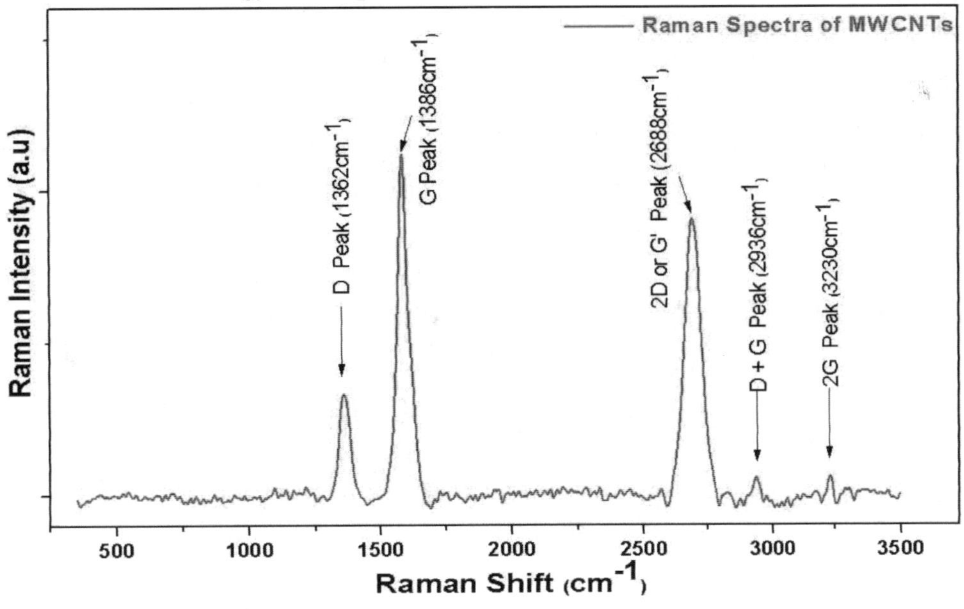

Figure 8. Raman Spectra of CNT based column

The D mode (breathing mode, A_{1g}band) is peaked in the 1300-1400 cm^{-1} range when excited with a visible laser. It is a typical disorder band observed in MWCNTs. The shorter intensity of D peak indicates the fineness of graphite layers. The G mode (Tangential Mode, E2g-band) is peaked in the 1550-1615 cm^{-1} range and corresponds to the stretching mode of sp_2 bonds in the graphite plane. A second-order-mode, often called G' mode is observed between 2620 and 2775 cm^{-1} [15, 16] which is based on double resonance mechanism. Its intensity increases with the decrease in the lattice disorder [17].

4. Conclusion and Future Viewpoints

An efficient protocol has been developed for the pattering of silicon substrate for different shapes and dimensions of carbon nanotubes columns. The patterning do not hinder the growth of CNTs as demonstrated by the results reported. The obtained vertically aligned multiwall carbon nanotubes based columns can be used for different applications including thermal dissipation system, microelectrodes etc.

Acknowledgment

The authors are thankful to Dr. Micaela Castellino, Dr. Salvatore Guastella, Diego Pugliese and Vijay Deep for useful discussion and assistance during experimentation and characterization. PA acknowledeges the support of the Italian Ministry of Research (FIRB grant RBFR10VZUG).

References

[1] Iijima S 1992 *Nature* **354** 56
[2] Saito R, Dresselhaus G and Dresselhaus M S 1999 *Physical Properties of Carbon Nanotubes* (London-Imperial College Press)
[3] Meyyappan M 2005 *Carbon Nanotubes: Science and Applications* (New York-CRC Press)
[4] Musso S, Fanchini G and Tagliaferro A 2005 *Diamond & Related Materials* **14** 784
[5] Rotkin S V, Perebeinos V, Petrov A G and Avouris P 2009 *Nano Letters* **9** 1850
[6] Balasubramanian K and Burghard M 2006 *Analytical and Bioanalytical Chemistry* **385** 452
[7] Bronikowski M J 2006 *Carbon* **44** 2822
[8] Yun Y H, Bange A, Heineman W R, Halsall H B, Shanov V N, Dong Z, Pixley S, Behbehani M, Jazieh A, Tue Y, Wong D K Y, Bhattacharya A and Schulz M J 2007 *Sensors and Actuators* **B 123** 177
[9] Seu K J, Pandey A P, Haque F, Proctor E A, Ribbe A E, and Hovis J S 2007 *Biophysical Journal* **92** 2445
[10] Anhoj T A, Jorgensen A M, Zauner D A and Hubner J 2006 *Journal of Micromechanics and Microengineering* **16** 1819
[11] Vander Wal R L, Ticich T M and Curtis V E 2001 *Carbon* **39** 2277
[12] Satishkumar B C, Govindaraj A and Rao C N R 1999 *Chemical Physics Letters* **307** 158
[13] Kumar M and Ando Y 2003 *Diamond and Related Materials* **12** 1845
[14] Dresselhaus M S, Dresselhaus G, Saito R and Jorio A 2005 *Physics Reports* **409** 47
[15] Zdrojek M, Gebicki W, Jastrzebski C, Melin T and Huczko A 2004 *Solid State Phenomena* **99**
[16] Delhaes P, Couzi M, Trinquecoste M, Dentzer J, Hamidou H and Vix-Guterl C 2006 *Carbon* **44** 3005
[17] Pelletier M J 1999 *Analytical Applications of Raman Spectroscopy* (London-Blackwell Science)

Journal of Physics: Conference Series **439** (2013) 012009 doi:10.1088/1742-6596/439/1/012009

Synthesis of multi-walled carbon nanotubes and their application in resin based nanocomposites

Shahid Nisar Ahmad[1,*], Saira Hakeem[2], Rashid Ahmed Alvi[2], Khawar Farooq[2], Naveed Farooq[2], Farida Yasmin[2], and Sadaf Saeed[2]
[1]National Center for Physics, QAU, Islamabad, Pakistan
[2] National Institute of Vacuum Science and technology, Islamabad, Pakistan

Corresponding author: s_n_ahmad@yahoo.com

Abstract. Multi-walled carbon nanotubes (MWCNTs) were synthesized by catalytic decomposition of hydrocarbon gas using chemical vapor deposition method. Synthesis was done at different growth temperatures and catalyst ratios. These MWCNTs were dispersed in epoxy resin (E-51) and their effect on mechanical strength of epoxy nanocomposites was studied. Increase in the mechanical strength of epoxy was observed with the addition of CNTs. The surface characterization was done by using optical microscope and scanning electron microscope (SEM). Mechanical properties were determined by the general tensile strength testing method.

1. Introduction

Carbon nanotubes (CNTs) are hollow nanometer size tubes of graphitic carbon. A carbon nanotube can be imagined as a graphene sheet rolled in the form of a seamless cylinder. The synthesis of CNTs can be back tracked to 1952 when Radushkevich et al. [1], a Russian group, published the first transmission electron microscopy (TEM) image supporting the tubular nature of the nanometric-sized carbon filaments. Iijima [2], from the NEC Laboratory in Tsukuba, Japan, was the first to identify the carbon nanotubes structure using High-Resolution Transmission Electron Microscopy (HRTEM). Afterwards, the field of carbon nanotubes was launched seriously. CNTs have exceptional mechanical [3-5], electrical [6-8] and thermal properties [9-11]. Their superior properties make them the most promising nanomaterials in the 21st century for a wide range of applications.

CNTs can mainly be divided into two types: single-walled carbon nanotubes (SWCNTs) and multi-walled carbon nanotubes (MWCNTs). A single-walled carbon nanotube can be considered as a single graphene sheet rolled into a tube, whereas, the multi-walled nanotubes can be described as a number of concentric SWCNTs having different diameters. The synthesis of CNTs can be done by various methods: arc-discharge [12], laser ablation [13], and chemical vapor deposition [14]. Chemical vapor deposition (CVD) is relatively a simple and economical technique for the synthesis of CNTs as compared to arc-discharge, and laser ablation methods. It can be done at low temperature and ambient pressure. This method offers higher purity, larger yield, and better control on the growth parameters and structure. To date, CVD is considered as a cost-effective method for the production of good quality CNTs. It has a potential to scale up the production of CNTs to the commercial level [15-18]

The most intriguing applications of CNTs are in CNTs based polymer nanocomposites. The effective utilization of CNTs in composites strongly depends upon their homogeneous dispersion in the host matrix, and an efficient load transfer to the CNTs [19-20]. Hence, good dispersion of CNTs is a fundamental feature to maximize their effect. In this paper, we report the synthesis of MWCNTs by catalytic decomposition of hydrocarbon gas (C_2H_2) using CVD method. The synthesis conditions were optimized in terms of growth temperature and catalyst ratio. These CNTs were heat treated for purification, and the purified CNTs were used for further studies. A low content of MWCNTs (~ 0.2 wt%) was used for these studies. The purified CNTs were dispersed in ethanol by ultrasonication for uniform dispersion in epoxy resin (E-51). A protocol was devised for homogeneous dispersion of MWCNTs in epoxy resin, and the CNTs based epoxy nanocomposites were prepared by following this protocol. The purpose of this study is to synthesize CNTs and to study their effect on mechanical strength of epoxy nanocomposites.

2. Experimental

2.1. Materials

The MWCNTs used in this study were synthesized at NCP, Islamabad. The synthesis was done by chemical vapor deposition method. Epoxy polymer matrix was prepared by mixing epoxy resin E-51, hardener (70-acid anhydride), and accelerator (*N*, *N*-dimethylbenzyl amine). Their chemical formulae are shown in figure 1. The precursors (Chromium acetate, Nickel Nitrate, Iron Oxide, conc. HCl, and Ethanol), carrier gas N_2, and source gas C_2H_2 used in this study were of high purity.

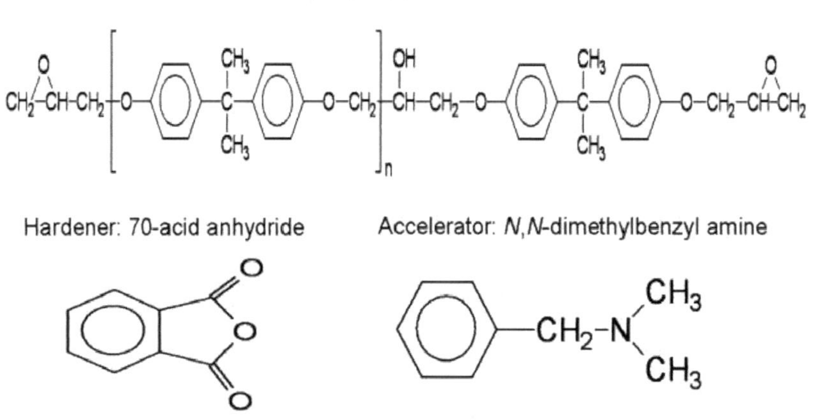

Figure 1. Epoxy resin, hardener, and accelerator used in this study.

2.2. CNTs synthesis

The synthesis of CNTs was done by catalytic decomposition of C_2H_2 using CVD method. A number of experiments were performed to optimize the experimental conditions like gas flow rate, reaction time, pre and post reaction temperature, reaction temperature, and mixing ratio of hydrocarbon gas and inert gas. A constant flow rate of nitrogen gas was maintained during heating. The schematic diagram of CVD setup is shown in the figure 2.

Journal of Physics: Conference Series **439** (2013) 012009 doi:10.1088/1742-6596/439/1/012009

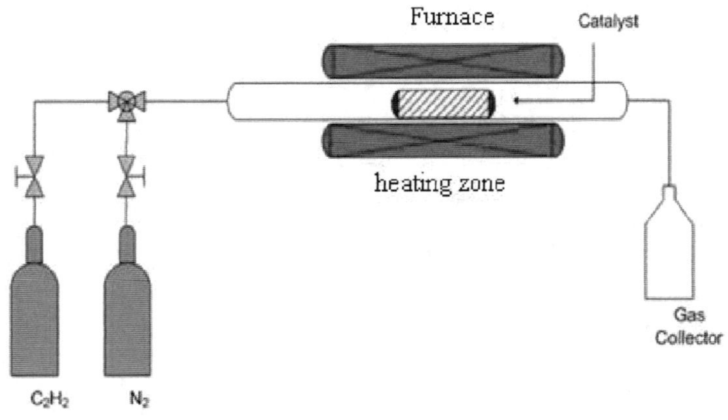

Figure 2. CVD set-up for the synthesis of CNTs.

A combination of Fe, Ni, and Cr was used as catalyst for the synthesis of CNTs. Liquid phase impregnation method was employed for catalyst preparation. Optimized amount of Chromium acetate, Nickel Nitrate, and Iron Oxide was dissolved in conc. HCl, and heating was done at 70°C for one hour. The catalyst substrate was placed at the center of horizontal quartz tube in the tube furnace. The furnace temperature was maintained at 850°C for half an hour in the N_2 flow to activate the catalyst. The synthesis was done at different growth temperatures (600-750°C), keeping the ratio of C_2H_2 : N_2 at 1:10 sccm. After the completion of reaction, the flow of C_2H_2 was stopped and the reaction temperature was maintained in the reactor for an hour in flowing nitrogen. Then the reactor was cooled down to the room temperature under the same flow rate of N_2.

2.3. Growth mechanism of carbon nanotubes

The growth mechanism of carbon nanotubes by catalytic decomposition of hydrocarbon gas can be described as follows. Hydrocarbon gas decomposes into carbon and hydrogen when it comes in contact with the high temperature catalyst metal particles. The decomposed carbon dissolves in the metal particles and reaches a supersaturated state. At this stage the dissolved carbon precipitates out and acquires an energetically stable state, by crystallizing in the form of seamless cylinders of carbon nanotubes. Depending upon the catalyst-substrate interactions, CNTs can grow by two different growth mechanisms in a CVD process. The first mechanism is called tip-growth mechanism and the second method is called base-growth mechanism. In the first method, the catalyst-substrate interactions are weak and the metal particle is pulled up at the top of grown CNTs. In the second method, the catalyst-substrate interactions are strong and the CNTs grow on the metal particle. The both growth mechanisms of CNTs are shown in figure 3.

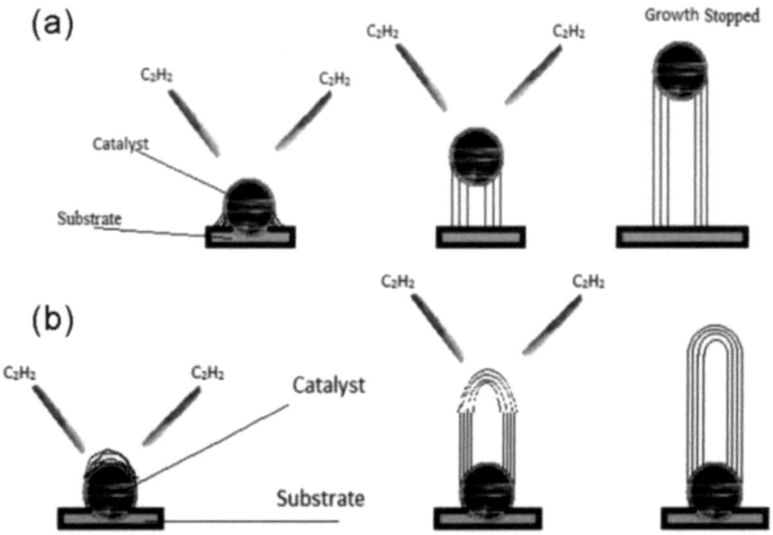

Figure 3. Growth mechanisms of CNTs: (a) Tip-growth mechanism, (b) Base-growth mechanism.

2.4. Preparation of nanocomposites

The protocol for preparation of CNTs dispersed epoxy nanocomposites is shown in the following flow chart figure 4.

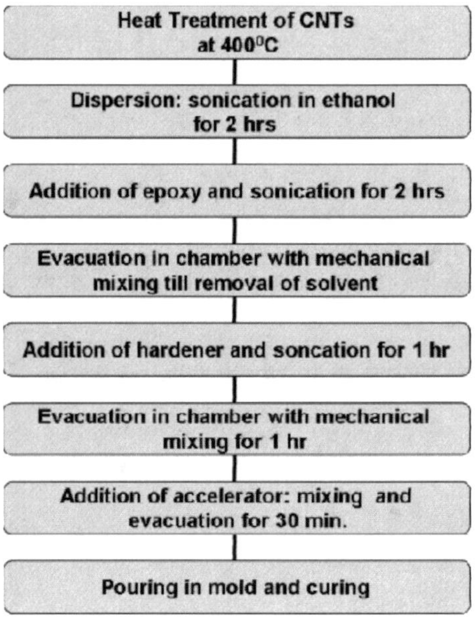

Figure 4. Protocol for the preparation of CNTs dispersed epoxy nanocomposites.

The synthesized MWCNTs were heat treated at 400°C for 1 hr in ambient atmosphere for purification. Ethanol was added to the purified CNTs and sonication was done for 2 hrs for de-bundling of agglomerated CNTs. The epoxy resin was added and the mixture was further sonicated for 2 hrs. After sonication, the solvent removal was done. This was done by mechanical mixing of the mixture under evacuation in a chamber. To ensure that the solvent is completely removed, the mixture was weighed before the addition of ethanol and after its removal. The hardener was added and sonication was done for 1 hr, followed by 1 hr mechanical mixing and evacuation. Finally accelerator was added and mechanical mixing and evacuation was done for 30 minutes. The uniform dispersion of CNTs in epoxy was checked by an optical microscope. The CNTs dispersed epoxy was poured in a mold for making tensile strength testing samples, and was cured in an oven. Neat epoxy resin samples were also prepared for the comparison of results. The cured dog bone shaped tensile strength testing samples of neat epoxy (orange) and CNTs doped epoxy (black) are shown in figure 5.

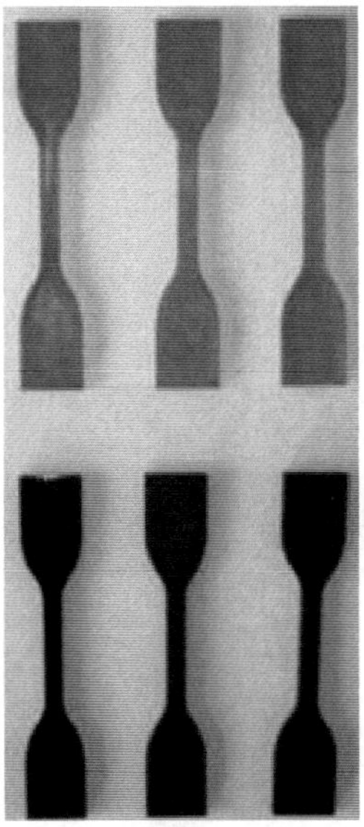

Figure 5. Tensile strength testing samples: Neat epoxy (orange) and CNTs doped epoxy (black) samples.

3. Result and Discussion

3.1. Synthesis

Parametric study was done for optimization of different parameters of the fixed-bed reactor. Catalyst composition, flow rates of nitrogen and acetylene gas, flow time of carbon source, growth temperature, and retention time were optimized. Synthesis temperature is the most important factor in our studies. C_2H_2 undergoes pyrolysis above 700°C [21]. The end products of C_2H_2 pyrolysis are polycyclic aromatic hydrocarbons (PAHs), which are thermodynamically stable [22]. These byproducts deplete the relative concentration of the CNTs carbon source. Also they can grow large enough to deactivate the catalyst by blocking active sites. These considerations restrict the large C_2H_2 concentration used for synthesis for the limited time period.

The SEM images of the systematic growth of CNTs at different reaction temperatures are shown in figure 6. At 600°C no growth of CNTs was observed (fig.6. a-1, a-2). As can be seen in the figure, uniform circular particles of carbon are grown instead of CNTs. These circular particles have diameter averaging around 100 nm. On increasing the growth temperature to 700°C, the growth of the CNTs is evidenced along with circular carbon particles (fig.6. b-1, b-2). These particles are larger than those observed at the previous growth temperature, but still they maintain the circular shape. More than half of the sample contains these particles. Further increase in growth temperature (750°C) results in better growth of CNTs (fig.6. c-1, c-2). Two type of CNTs are grown; one in the range of 150-100 nm and the other less than 100 nm. Still a significant part of the sample contains amorphous carbon. The Cr content was doubled keeping the amount of Ni and Fe same for the substrate preparation, and the experiment was repeated under the identical conditions. It can be evidenced in the figure 6 (d-1, d-2) that most of the CNTs grown in this case are less than 100 nm, only a few CNTs lie in the range of 150-100 nm. Also these CNTs are clean as compared to those grown in the previous cases and almost no amorphous carbon is observed in this case.

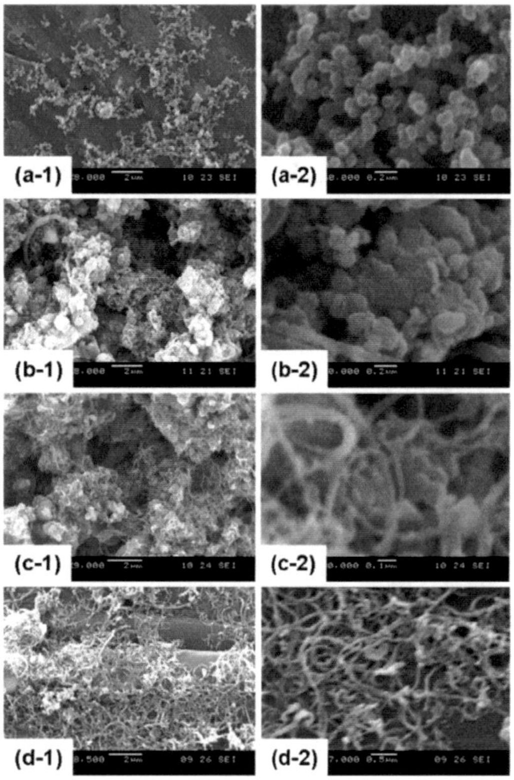

Figure 6. Comparison of SEM images of CNTs growth at different reaction temperatures (T_R). (a-1, a-2) T_R = 600°C, (b-1, b-2) T_R = 700°C, (c-1, c-2) T_R = 750°C, (d-1, d-2), T_R = 750°C and Cr content is doubled.

3.2. CNTs dispersed epoxy nanocomposites

Dispersion of synthesized CNTs in epoxy resin was done following the flow chart discussed in section 2.4 (figure 3). The dispersion of CNTs in epoxy was checked visually and by an optical microscope. A comparison of dispersion is shown in figure 7.

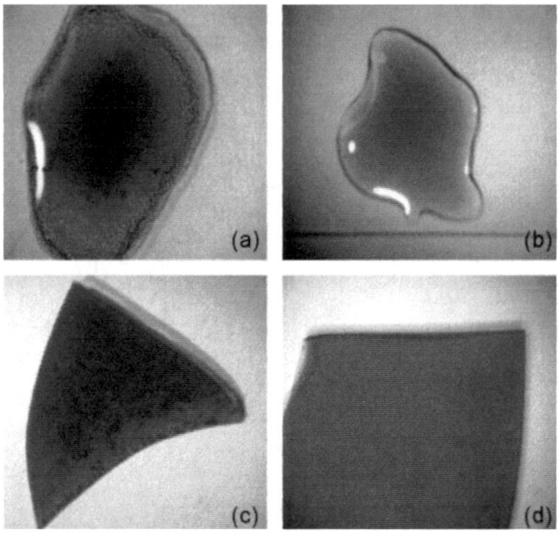

Figure 7. Comparison of optical micrographs of homogeneous and poor dispersion of CNTs in epoxy resin. (a) Direct mechanical mixing of CNTs in epoxy resin, (b) Homogeneous dispersion of CNTs in epoxy resin following the protocol developed in section 2.4, (c) Cured films of mechanically mixed CNTs in epoxy, (d) Cured films of CNTs homogeneously dispersed in epoxy following the protocol developed in section 2.4.

The optical micrographs of homogeneous and non-homogeneous dispersion of CNTs are shown in the figure 7 (a, b). Agglomerates of CNTs can be seen in figure 7 (a). In this case CNTs were directly dispersed in the epoxy resin by mechanical mixing. No solvent was used for dispersion. After 4 hrs of mechanical mixing the mixture was poured in the mold and curing was done. After mixing, no agglomerates were observed on visual inspection. A drop of this mixture was placed on a glass slide and after a few hours it was showing two shades. When the slide was checked under a microscope the clustering of CNTs was apparent. On curing CNTs agglomerates were observed at the bottom of the cured film, as shown in the figure 7 (c). This shows that by mechanical mixing CNTs were not completely dispersed in the epoxy. Only partial dispersion was achieved and the van der Waal forces were strong enough to make the agglomerates of CNTs after mechanical mixing. On the other hand, when the dispersion was done using a solvent, following the protocol discussed in section 2.4, the dispersion was homogeneous and no agglomeration was observed in cured and uncured samples. This is shown in figure 7 (b, d). This effect was further confirmed when the tensile strength samples were prepared and tested.

To measure the effect of CNTs on the strength of epoxy samples with specific size were prepared, as shown in figure 5. Tensile testing (ASTM D 638) was done by Instron tensile tester at a cross speed of 2mm/min at 25°C. A batch of five samples was tested and the average value was reported.

Three types of samples were prepared;

1. Direct mixing of CNTs in epoxy without solvent
2. Mixing of CNTs following the protocol discussed in section 2.4
3. Neat samples of epoxy without CNTs (for comparison)

The CNTs doped epoxy samples (~ 0.2 wt%) show greater strength (84 MPa) as compared to the pure epoxy samples (72 MPa). On the other hand, the samples prepared by direct mixing of CNTs showed a decrease in the strength and their value (45 MPa) was even less than the neat epoxy samples. This improvement in UTS can be attributed to the high specific mechanical properties and high aspect ratio of the CNTs, when it is homogeneously dispersed in the epoxy matrix. This shows that addition of carbon nanotubes enhances the stress transfer effect of epoxy on application of a load, resulting in overall enhancement in the strength and plasticity of the resin matrix. The same, even better, effect has been observed in the strength of carbon fiber based epoxy composites on addition of CNTs (results not shown here). On the contrary, agglomerates of CNTs act as impurity sites in the epoxy which result in reduction in the strength.

This confirms that the effective utilization of CNTs in composites strongly depends upon their homogeneous dispersion in the host matrix. If a homogeneous dispersion may lead to enhancement in the properties, then a poor dispersion of CNTs may destroy the existing properties. Hence purification of CNTs and their good dispersion in epoxy resin play a key role to maximize the effect of CNTs on the overall properties of epoxy based composite system.

4. Conclusions

Carbon nanotubes having diameter in the range of 50 - 150 nm and less were successfully synthesized by the catalytic chemical vapor deposition of acetylene gas over Ni–Cr–Fe. It has been observed that certain parameters like growth temperature, catalyst composition, and flow time of acetylene have strong influence on the synthesis of carbon nanotubes. The affect of addition of these CNTs on mechanical properties of epoxy nanocomposites was also studied. When the CNTs are homogeneously dispersed, a small addition of CNTs (~ 0.2 wt%) may increase the strength of epoxy resin. This On the other hand, a poor dispersion of the same amount of CNTs may decrase the existing properties. The comparison of these result shows that improvement in the strength of epoxy resin with doping of CNTs strongly depends upon better dispersion CNTs in epoxy resin.

5. References

[1] Radushkevich L V and Lukyanovich V M 1952 *Zurn. Fisic. Chim.* **26** 88
[2] Iijima S 1991 *Nature.* **354** 56
[3] Poncharal P, Wang Z L, Ugarte D, and Heer W A 1999 *Science* **283** 1513
[4] Tracy M M J, Ebbesen T W and Gibson J M 1996 *Nature.***381** 678
[5] Wong E W, Sheehan P E and Lieber C N 1997 *Science.* **277** 1971
[6] Zhu H W, Xu L, Wu D, Wei B Q, R Vajtai and Ajayan P M 2002 *Science.* **296** 884
[7] Li Y, Kinloch I A and Windle A H 2004 *Science.* **304** 276

[8] Zhou W, Vavro J, Guthy C, Winey K I, Fisher J E and Smalley R E 2004 *J. Appl. Phys.* **95** 649
[9] Yosida Y 2000 *J. Appl. Phys.* **8** 3338
[10] Berber S, Kwon Y K and Toma`nek D 2000 *Phys. Rev. Lett.* **84** 4613
[11] Foygel M, Morris RD, Anez D, French S and Sobolev V L 2005 *Phys. Rev.* B **71** 104201
[12] Journet C, Maser W, Bernier P, Loiseau A, Lamy D, Chapelle M, Lefrant S, Deniard P, Lee R and Fischer J 1997 *Nature.***388** 756
[13] A Thess A, Lee R, Nikolaev P, Dai H, Petit P, Robert J, Xu C, Lee Y H, Kim S G, Rinzler A G, Colbert D T, Scuseria G E, Tomek D, Fischer J E and Smalley R E 1996 *Science.* **273** 483
[14] S Bethune D S, Klang C H, de Vries M S, G Gorman G, Savoy R, Vazquez J and Beyers R 1993 *Nature.* **363** 605
[15] Govindaraj A and Rao C N R 2002 Pure. Appl. Chem. 74 1557
[16] Rao C N R and Govindaraj A 2001 Proc. Indian Acad. Sci. (Chem. Sci.) 113 375
[17] Fonseca A, Hernadi K, Piedigrosso P, Colomer J F, Mukhopadhyay K, Doome R, Lazarescu S, Biro L P, Lambin Ph, Thiry P A, Bernaerts D and Nagy J B 1998 *Appl. Phys.* A **67** 11
[18] Lee C J, Lyu S C, Kim H W, Park C Y and Yang C W 2002 *Chem. Phys. Lett.* **359** 109
[19] Gojny F H, Nastalczyk J, Roslaniec Z and Schulte K 2003 *Chem. Phys. Lett.* **370** 820
[20] Villoria R G, Miravete A, Guartero J, Chiminelli A and Tolosana N 2006 *Composites Part B: Eng.* **37** 273
[21] Cui H, Zhou O and Stoner B R *J. Appl. Phys.* **88** 6072
[22] Okai M, Muneyoshi T, Yaguchi T and Sasaki S 2000 *Appl. Phys. Lett.* **77** 3468

Study of carbon nanotubes based Polydimethylsiloxane composite films

M I Shahzad[1], M Giorcelli[1], N Shahzad[1,2], S Guastella[1], M Castellino[1], P Jagdale[1] and A Tagliaferro[1]

[1]Department of Applied Science and Technology (DISAT), Polytechnic of Turin, Turin-10129, Italy

[2]Center for Space Human Robotics, Italian Institute of Technology (IIT), Turin-10129, Italy

E-mail: imran.shahzad@polito.it

Abstract. Thanks to their remarkable characteristics, carbon nanotubes (CNTs) have fields of applications which are growing every day. Among them, the use of CNTs as filler for polymers is one of the most promising. In this work we report on Polydimethylsiloxane (PDMS) composites with different weight percentages (0.0% to 3.0%) of multiwall carbon nanotubes (MWCNTs) having diameter 10-30 nm and length 20-30 µm. To achieve optimum dispersion of CNTs in PDMS matrix, high speed mechanical stirring and ultrasonication were performed. By using the doctor blade technique, 70 µm thick uniform films were produced on glass. They were subsequently thermally cured and detached from the glass to get flexible and self standing films. The surface morphological study done by FESEM, shows that CNTs are well dispersed in the PDMS. Raman spectroscopy and FTIR were used to investigate the possible structural changes in the polymer composite. To examine the optical behavior UV-VIS spectroscopy was employed in both specular and diffused modes. A linear increase in absorption coefficient is found with the increasing percentage of CNTs while the transmittance decreases exponentially. The results confirm the dependence of optical limiting effect on the quantity of MWCNTs. Based on optical study, MWCNTs/PDMS composite films can be a promising material to extend performances of optical limiters against laser pulses, which is often required in lasing systems.

1. Introduction

The breakthrough of carbon nanotubes (CNTs) has got remarkable progress in various fields of research and applications due to their peculiar structural, electrical, mechanical, optical and magnetic properties [1]. CNTs diameters are in the 1-100 nm range and they can be up to millimeters long [2]. These dimensions lead to very high aspect ratios as compared to that of carbon fibers. This combination of properties allows better interaction in composite matrices, resulting in improved properties and performance [3].

The incorporation of CNTs into the polymer can enhance the properties of material by increasing mechanical strength and electrical conductivity. However the formation of aggregates and low dispersions of CNTs in the polymer matrix are the major cause of poor and inhomogeneous composites [4]. The interface between the CNTs and the polymer matrix plays as well a vital role in achieving good dispersion. Successful incorporation of CNTs in polymer matrices could result in different types of lightweight and strong materials for flexible electronic devices and sensors [5].

So, for the exploitation of the potential of CNTs, the issues regarding economical and controlled fabrication of well dispersed reinforced composite materials has to be tackled and solved. We report here an efficient and inexpensive process of incorporating carbon MWCNTs into a PDMS (polydimethylsiloxane) matrix. Furthermore, we focused on structural and optical properties of CNTs and summarize the current issue and future viewpoint on this subject.

2. Fabrication of MWCNT/PDMS composite films

The commercial MWCNTs used in this work were purchased from CHEAPTUBES®. These nanotubes are produced by catalyzed chemical vapor deposition and have a 99.9% purity. Their diameter is in the 10-30 nm range and their length is in the 20-30 μm range (figure 1). The FESEM micrographs show that MWCNTs are highly entangled due to Van der Waals attractive forces.

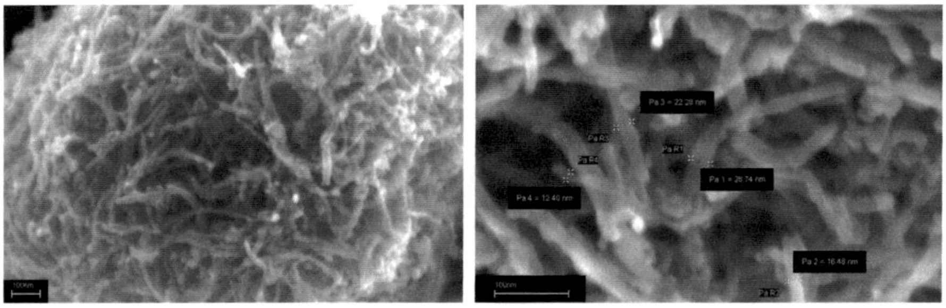

Figure 1. Field emission scanning electron microscopy of MWCNTs used in our work

In order to prepare MWCNTs-PDMS composites with different percentages of carbon nanotubes, the desired quantity of fillers has been added to the PDMS monomer and dispersed by means of high speed mechanical stirring (1000 rpm for 10 min). After the addition of hardener, the matrix is mixed for 10 more minutes. The final step to achieve a uniform dispersion of MWCNTs in the composite solution is sonication (ultrasonic frequency 37 KHz for 15 min) at room temperature. The composite films with an average thickness of 70 micrometers and different weight percentages of MWCNTs (from 0.0 % to 3.0%) were prepared on glass by Doctor Blade Technique as described in the figure 2. The films were then degassed in vacuum to get rid of entrapped bubbles. Subsequently, films were thermally cured at 70°C for 4 hours and finally detached from the glass surface to get flexible and self standing films.

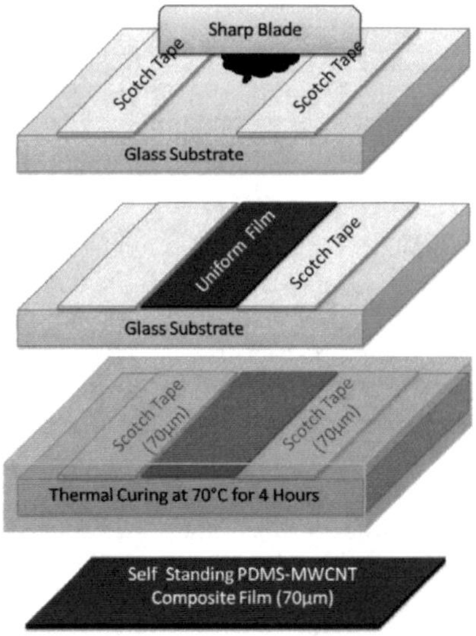

Figure 2. Schematic of steps involved in the fabrication of MWCNT/PDMS film

3. Surface morphology of films

After the preparation, these films were cryofractured and characterized by field emission scanning electron microscopy (ZEISS SUPRA 40, FESEM). This technique allows to investigate the dispersion of MWCNTs inside the polymer composite. The samples were coated with a thin metal layer before measurement to avoid charge accumulation and obtain a better insight on the CNTs distribution in the matrix. Figure 3 shows the cross sectional view of PDMS with different percentages (1wt%, 2wt% and 3wt %) of MWCNTs at 100kx resolution.

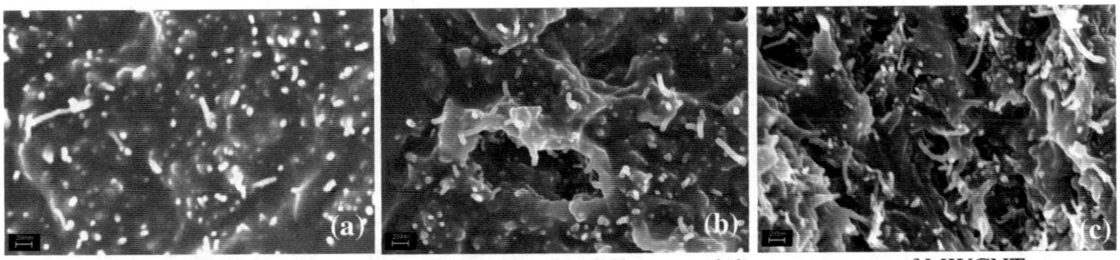

Figure 3. FESEM micrographs of PDMS with different weight percentages of MWCNTs a). 1.0 wt% b). 2.0 wt% c). 3.0 wt% at 100kx magnification

In figure 4, micrographs of the PDMS composite with 3.0 wt% of MWCNTS are shown at different magnifications. These micrographs show that most of the CNTs are well dispersed and randomly oriented into the polymer. The high magnification micrograph (figure 4-c) shows that the diameter of a single nanotubes-like structure is larger than that of CNTs (figure 1), even duly taking into account the increase due to metallization. This is attributed to the strong linkage between nanotubes and polymer matrix that leads to a decoration of CNTs by polymer molecules.

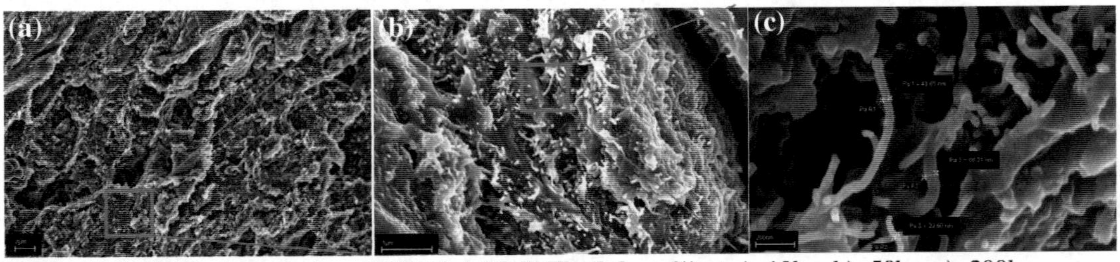

Figure 4. FESEM micrographs of PDMS+MWCNTs (3.0 wt %) at a). 10kx b). 50kx c). 200kx

4. Raman Spectroscopy

Raman spectroscopy is a non destructive technique which gives relevant structural information on carbon based materials.

Raman spectroscopy of PDMS-MWCNTs composite films (figure 5) was performed using a green laser source (wavelength 532 nm). The D-peak (located between 1300-1400 cm^{-1}) and the G-peak (located between1550-1615cm^{-1}) are the characteristic peaks of MWCNTs [6, 7]. The intensity of D and G peaks of carbon structures are increasing proportionally with the increase in CNTs inside the polymer, proving the fact that they are well dispersed and the films are enough uniform.

The other peaks in Raman spectra are attributed to the PDMS structure, in agreement with the characteristic peaks of well polymerized polydimethylsiloxane PDMS [8]. There are significant and gradual changes in the intensity and broadening of PDMS characteristic peaks which appears after the inclusion of CNTs shown in the figure above. Their investigation is however outside the purpose of the present work and will be discussed elsewhere.

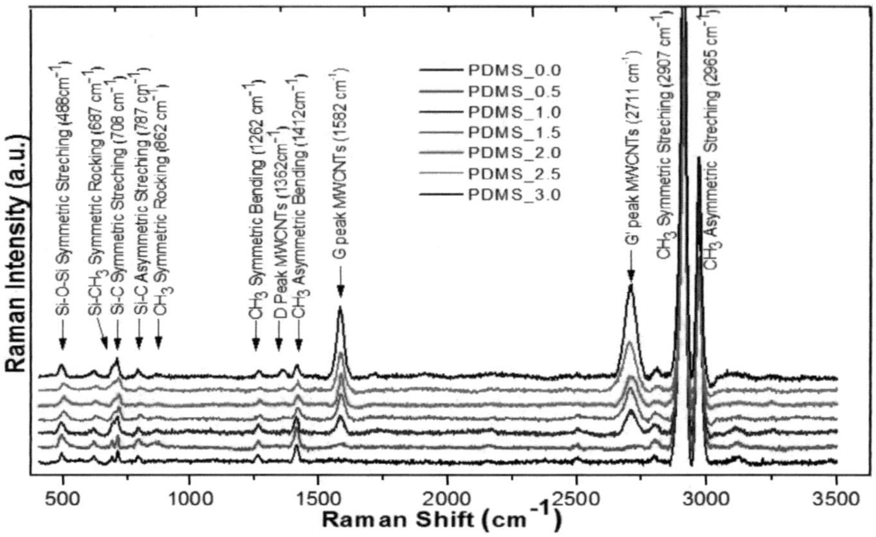

Figure 5. Raman spectra for PDMS composite films for different wt % of MWCNTs

5. FTIR Spectroscopy

Figure 6 summarizes the positions of IR absorption bands of the chemical groups found in the PDMS-CNT composites. -CH_3 wagging related peaks are located at 1410 cm^{-1} and at 1258 cm^{-1}. A wide multi-component peak ranging from 1000 cm^{-1} to 1100 cm^{-1} corresponding to symmetrical Si–O–Si stretching is also present. There is a minor decrease in the peak at 930 cm^{-1} which shifts to lower wave numbers as the concentration of MWCNTs increase in PDMS. In addition the ratio between the two transmissions values at 900 cm^{-1} and 930 cm^{-1} decrease with increasing CNT content as highlighted in the inset of figure 6. This is in good agreement with the literature [9] as this effect is also observed for other carbon materials based composites. Si–C bands and rocking peaks for $Si(CH_3)_2$ are observed in 835–855 cm^{-1} and 785–815 cm^{-1} regions, respectively [10,11].

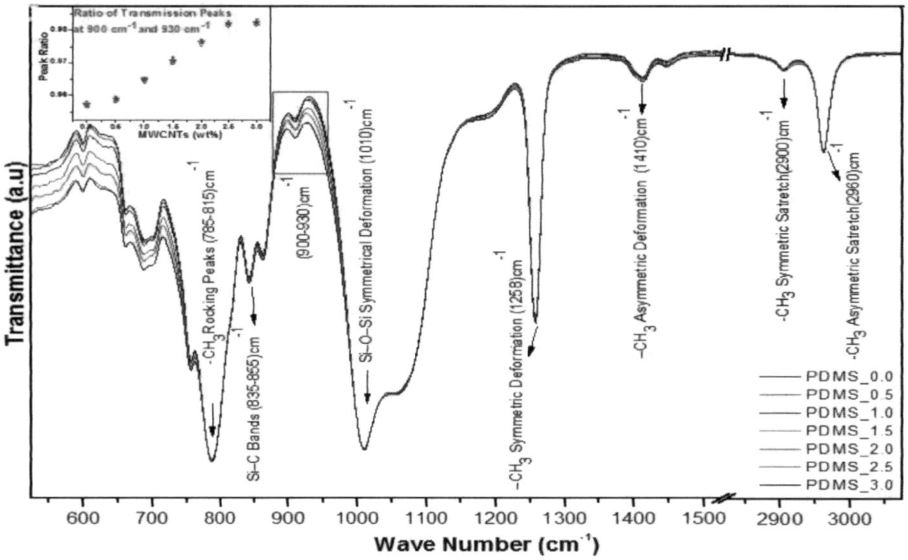

Figure 6. FTIR Spectra for Pure PDMS composite with different weight percentages of MWCNTS

6. UV-VIS Spectroscopy

For the comprehensive study of optical behavior of these polymeric films, optical characteristics were measured both for direct as well as diffused light. In figure 7, total reflectance (a) and transmittance (b) spectra are shown for the various composites. As expected, the addition of nanotubes to PDMS has reduced the transmittance as well as reflectance of the films. For the inclusion of 3wt% carbon nanotubes, the films have an almost negligible transmission.

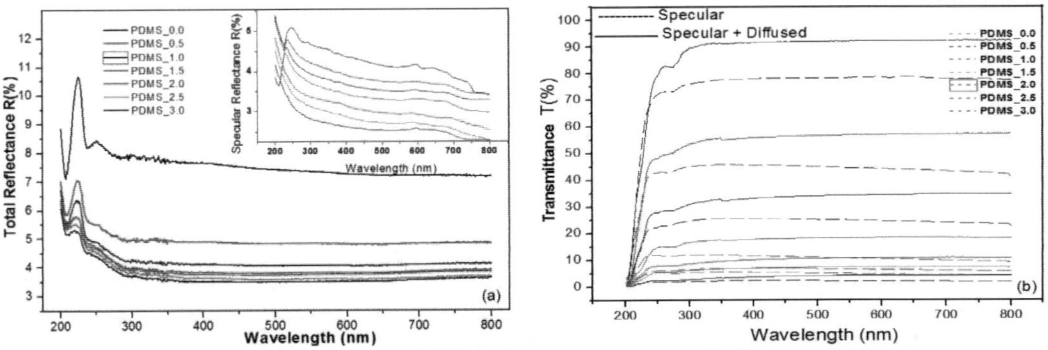

Figure 7. Comparison of specular and diffused measurements: a) Reflectance b) Transmittance

The overall absorptance (absorption) of each film is evaluated as

$$A=100-(R_d+R_s)-(T_d+T_s) \qquad (1)$$

where R and T denote reflectance and transmittance, while $_d$ and $_s$ stand for diffused and specular respectively [12]. The results concerning A values are shown in the figure 8-a. The absorption coefficient α has been calculated as

$$\alpha=1/t * \log(1-R)/T \qquad (2)$$

where t is the thickness of the film and R and T are the total reflectance and transmittance respectively [12]. The absorption coefficient increases with increasing content of CNTs as shown in figure 8-b.

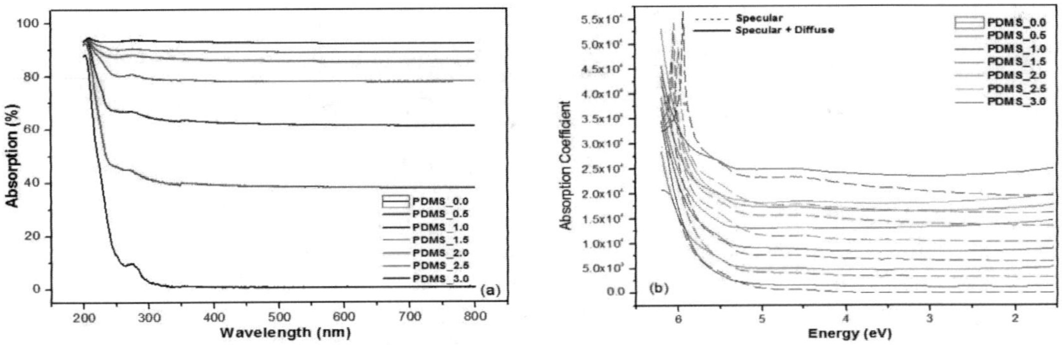

Figure 8. The Absorption (a) and the Absorption Coefficient (b) of the PDMS-MWCNTs Composites

7. MWCNTS/PDMS composite film for optical limitation

The absorption coefficient increases linearly with the increasing content of carbon nanotubes (figure 9-a) while the transmittance decreases exponentially (figure 9-b).

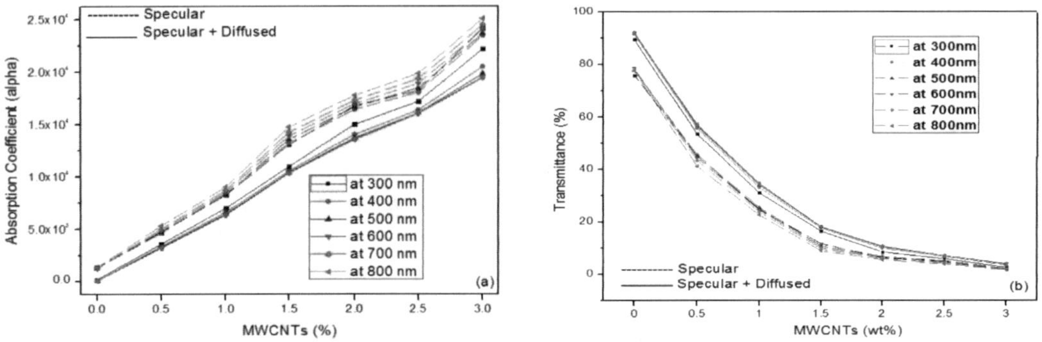

Figure 9. Dependence of absorption coefficient (a) and transmittance (b) on MWCNTs (wt%)

Optical density spectra, calculated [12] as

$$OD = log\ (1/T) \quad (3)$$

for the different percentages of MWCNTs are compared in figure 10-a. The optical density is increased by increasing the weight percentage of MWCNTs. However, it remains almost constant throughout the visible range for a given CNTs content.

Diffusivity is defined as

$$D = (T_d+R_d)/(1-A) \quad (4)$$

and describes the rate at which light diffuses in the substance. The diffusivities of the PDMS-MWCNTs films are shown in figure 10-b. While in pure PDMS diffusivity is almost wavelength independent, the presence of CNTs lead to diffusivity increasing with increasing wavelength in composites.

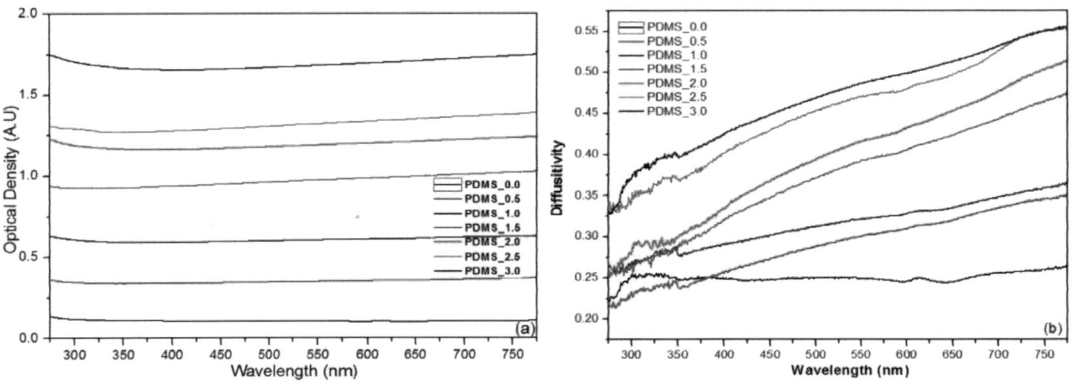

Figure 10. The spectra of (a) Optical Density and (b) Diffusivity

The above results suggest that, through a tuning of the specular transmittance, these films might be of interest in laser technology as optical absorbers to tune the laser beam power density.

8. Conclusion

Embedding CNTs in the PDMS matrix can open new fields of application for this well established polymer. The morphological study proved reasonably well dispersion and random orientation of CNTs into the PDMS composite films. The Raman Spectroscopy as well as FTIR analysis demonstrated the bonding between CNTs and polymer. The optical characteristics established that PDMS-MWCNTs composite film are promising materials to extend performances of optical limited devices. CNTs proved excellent candidate material for optical limiting and the optical limiting effect depends on the quantity of CNTs.

Acknowledgement

I highly acknowledge Dr. A. Virga for his valuable support in morphological characterizations.

Journal of Physics: Conference Series **439** (2013) 012010 doi:10.1088/1742-6596/439/1/012010

References

[1] Lau K T, Lu M and Hui D 2006 *Composites: Part B* **37** 437
[2] Deck C P, Vecchio K 2006 *Carbon* **44** 267
[3] Coleman J N, Khan U, Blau W J and Gun'ko Y K 2006 *Carbon* **44** 1624
[4] Harris P J F 2004 *International Materials Reviews* **49** 31
[5] Bakshi S R, Lahiri D and Agarwal 2010 *International Materials Reviews* **55** 51
[6] Dresselhaus M S, Dresselhaus G, Saito R and Jorio A 2005 *Physics Reports* **409** 47
[7] Zdrojek M, Gebicki W, Jastrzebski C, Melin T and Huczko A 2004 *Solid State Phenomena* **99**
[8] Bae S C, Lee H, Lin Z and Granick S 2005 *Langmuir* **21** 5685
[9] Nour M, Berean K, Griffin M J, Matthews G I, Bhaskaran M, Sriram S, Zadeh K K 2012 *Sensors and Actuators B: Chemical* **161** 982
[10] Maji D, Lahiri S K and Dasa S 2011 *Surface and Interface Analysis* **44** 62
[11] Agarwal R, Tandon P and Gupta V D 2006 *Journal of Organometallic Chemistry* **691** 2902
[12] Dahm D J and Dahm K D 2007 *Interpreting Diffuse Reflectance and Transmittance* (Chichester, NIR Publications)

Journal of Physics: Conference Series **439** (2013) 012011 doi:10.1088/1742-6596/439/1/012011

Characterization and the Pattern of Surfaces of Sealant with nano size Composite Materials

A Quddos[1], N H Samtio[2] and A M Syed[3]
[1]Institute Of Industrial Control System (IICS) P. O. Box 1398 Rawalpindi, Pakistan

Corresponding author: [1]drabdulquddos@yahoo.com

Abstract
Nano composite sealant is low viscosity, room temperature cured, opaque and flowable nature. They have variety of uses such as potting, pressure sealant and shock resistant. Most important factor influencing use of fillers in polymer composites is their ability to effectively transfer the applied load in the matrix. The effective utilization of fillers in composites for structural applications depends strongly on the ability to disperse the nano fillers homogeneously in the matrix without damaging them. R-Belite supper epoxy adhesive (RBSEA) were formulated with different nano fillers (KCl, Al_2O_3, ZrO_2, SiO_2, ZrO_2) at room temperature. The composite were prepared with the 0.02 to 0.10 weight ratios to promote the nucleation of the nanoparticles in the applied sealant. Two main problems which arise in improving the properties are poor dispersion of the fillers in the composite and weak bonding between nano fillers and the matrix. These problems are solved by mechanical and chemical means. It was observed that mechanical properties like tensile strength, elongation hardness etc and thermal properties were also improved with incorporation of nanofillers in the working applied polymer matrix. The dispersion of nano fillers in polymer matrix is studied by Scanning electron microscopy (SEM). The results confirm the presence of nanomaterial in RBSEA/fillers nanocomposites. SEM is also used to characterize the pattern of surfaces with nano size composite materials.

1. Introduction
Composite resins, first introduced in the 1970s, were typically quartz-filled and did not meet the compressive and tensile force requirements for composite applications. They had relatively high surface roughness and poor bonding at the interface which, in combination with polymerization shrinkage resulted in margin degradation [1].
Early resin composites were difficult to handle due to their manual mixing of the resin paste (A) and the catalyst (B). They were available only in tubs or syringes of pastes (A and B separately), and were always self-cured with a limited working time. Over the time, the strength characteristics of composite resins were improved and the resins were introduced for use in different applications [2-6].
The focus was still on etching and bonding, and early attempts at strong bonding resulted in weak bonding to the layer and a rapidly inside bonding. It was also recommended that, after curing of the resin, the interface over this surface to seal it and help to seal and wear [7-9]. These enabled a more conservative approach during preparation and they were used with improved bonding systems for improved marginal integrity at the interface.
Current composite materials (and bonding systems) are vastly superior to earlier versions due to several types of fillers, advanced enamel and uniform adhesive systems. They offer strength, lower polymerization shrinkage and stress, wear resistance, and excellent bonding [10-12].
Sealant nano composite (SNC) contain crushed particles ranging in size from 0.04 to 1 micron. Filler load is lower in micro filled composite resins, and the internal bonding between the matrix and the

filler is strong, resulting in high strength. This is an important consideration in stress-bearing areas [13, 14].
Nanofilled composite resins have high filler load in order to obtain strength and wear resistance similar to that of micro hybrid composite resins. Nanofilled composite resins contain smaller particles of filler in the range of 0.02 to 0.1 microns. One nanofilled composite contains nanofiller particles that are approximately 0.02 microns in diameter, sintered into nanoclusters of 0.6 to 1.4 microns that contain zirconium particles, in order to improve physical characteristics.
The recent paper in this research work has included reductions in polymerization shrinkage and in polymerization stress. We also report the effect of nano fillers on different properties of SNC. These have been achieved through increased filler loads and novel chemical technology. Polymerization shrinkage, and therefore stress, has been reduced by increasing conversion rate of the monomer [15]. Direct reduction of polymerization stress has now also been achieved through the use of a polymerization process.

2. Experiment

2.1 Materials
R-Belite super (RBS) 2-part epoxy resins were supplied by USA. The chemicals used in this study were Ethyl alcohol C_2H_5OH 99% Aldrich, Zirconium oxide ZrO_2 99.99 % Merck, Aluminum oxide Al_2O_3 99.99 % Merck, Potassium chloride KCl 99.999% Merck and Silicon dioxide SiO_2 99.99 % Merck added to RBS.

2.2 Preparation of Material
All-solid ingredients were dried in an oven at 100 ± 2 °C for minimum 4 h (moisture content < 0.1% by Karl Fischer). R-Belite super 2-part epoxy resins were added in a vertical mixer for uniform distribution in 30 min. Fillers were added in two equal installments and mixed for 10 min. The mixing was continued for 60 min without vacuum and then for 60 min in vacuum. All these operations were carried out under controlled humidity 55 ± 5 per cent.

2.3 Characterization of Material
Mechanical properties of the composite formulations were tested according to ASTM- D-638 using tensile testing machine (Instron 1185). Shore hardness was measured by Shore hardness tester (SHR-4 111). SEM (Jeol-JSM-5910LV) and Thermo gravimetric analyses (TGA) were also used for characterizations.

2.4 Processing of composites
The first step of processing nanocomposites is to determine the right proportion of resin curing agent and nanoparticles, and weighed them carefully as shown in Table-1. A correct ratio of resin and curing agent, when mixed, will make sure that a complete reaction would occur between them. In most of the cases, it would not be possible for a complete reaction to occur between the resin and curing agent where monomers are not mixed in right proportions. Because of the incomplete reaction any unreacted resin or curing agent will remain in the matrix and this will affect the final properties of the composite sealant.

2.5 Curing of Nanocomposites
The mold is prepared by applying the releasing agent. The dispersed mixture of resin, nanoparticles and curing agent should be kept in vacuum before pouring into the mold to make sure that there are no air bubbles in it. The curing of the composite is carried out in two stages. They are cured at room and high temperature.
It was also observed during many experiments that in spite of pouring the mixture into the mold very slowly there were still some air trapped in it. The mold was again kept in vacuum to remove the few air bubbles which might have trapped during the time of pouring of the mixture into the mold. Curing was carried out at room temperature for 30 minutes. The mold and all the parts were tightly fixed with all the screws in position to avoid leakage of the mixture through any of the loosely connected parts.

The properties of the composite to an extent depend on the curing temperature. In all the experiments curing of the nanocomposite samples was carried out at room temperature for 4 hours.

3. Results and discussion

3.1 Formulation designation

Polymer system suitable for sealing purpose must have a low viscosity, a long pot life time, and a cured polymer resin possessing good mechanical properties [6]. R-Belite super (RBS) 2-part epoxy resins have been widely used in many industrial fields because of its good mechanical and relatively low cost properties. Thus, in this study, we selected the RBS as the matrix for the composite sealant. At room temperature, this polymer system has excellent properties because of their low viscosity. The curing agent in this study had to meet all the following requirements: (1) a liquid state at room temperature and low viscosity, (2) low cost and low toxicity, and (3) a long pot life (at least 1 h under 25 °C). Among the many liquid amine curing agents, it was difficult to select one for the epoxy system suitable for this study because many curing agents meet only some of the requirements mentioned previously [4].

Nano composite sealant formed the inner-lining-layer and the outer-wound-layer in the polymerization process. Sealant composites often must endure various inner pressures ranging from 2.5 to 40 MPa according to the different types encountered during long-term use. Therefore, to improve the strength of the products, a toughened polymer sealant nano matrix was required in this study. The toughening agent nano size materials over comes the brittleness of the RBS as the strength property of the layer would depend on the polymer net work. In this study, we used RBS resin with nano materials as the toughening agent. Nano size particles could react with the polymer matrix and become a part of the network. Thus, the cured nano sealant networks obtained better molecule movement capability than the polymer system without nano particles.

Generally, viscosity diluter is often added to the epoxy system. However, the strength is reduced on the addition of diluter.

The formulations of the polymer system for the sealant nano composite is listed in Table-I.

Table-1. Compositions of The Nanoparticles Based Composite Materials.

Ingredient (%)	sealant nano composite (SNC-1)	sealant nano composite (SNC-1I)	sealant nano composite (SNC-1II)	sealant nano composite (SNC-1V)	sealant nano composite (SNC- V)	sealant nano composite (SNC-VI)	sealant nano composite (SNC-VII)
RBS resin	49	49	49	49	49	47	45
RBS hardener	49	49	49	49	49	47	45
KCl	2	-	-	-	-	-	-
Al_2O_3	-	2	-	-	-	-	-
ZrO_2	-	-	2	-	2	6	10
SiO_2	-	-	-	2	-	-	-

Nano particles of ZrO_2, Al_2O_3, KCl and SiO_2 were added in the polymer matrix sealant formulations (Table-1). The shear strength (MPa) and nano particles distribution were improved. A possible reason could be attributed to the fact that the incorporation of nano particles into the polymer system introduced a few flexible chains containing chemical bonds inside the network structure of the sealant. During the formulation of RBS, nano particles reacted with the matrix nearly at the same time. Thus, many flexible chain segments were expected cross linked to the nano composite sealant structure and finally embed into the structure of the polymer system. The improvement in the shear strength could be due to fact that few flexible chains containing chemical bonds were introduced into the nano sealant composite.

3.2 Mechanical properties

The mechanical properties of the sealant nano composites (SNC I-VII) were evaluated through shear strength, hardness, density and thermal properties (Table-2). It has been reported [6] that low-interfacial interaction between the embedding nano particles and the matrix results in good mechanical properties. This observation can be seen in the sample SNC-I to IV in the range of 1.04 to 1.05 MPa. The maximum property of the shear strength was obtained in the sample no SNC-VI.

Table-2. Mechanical Properties of Nanoparticles base composite materials.

Compositions	Thickness (mm)	Particles size (Micron)	Hardness (Shore A)	Density (gcm^{-3})	Tensile shear strength (MPa)	Thermal Temperature range (°C)
SNC-1	4	500 pass	92	1.04	1.0	280-570
SNC-1I	4	500 pass	90	1.04	1.0	280-570
SNC-1II	4	500 pass	91	1.04	1.2	280-570
SNC-1V	4	500 pass	89	1.05	0.9	280-570
SNC -V	4	500 pass	90	1.06	1.4	280-570
SNC -VI	4	500 pass	90	1.10	1.6	280-570
SNC -VII	4	500 pass	90	1.09	1.4	280-570

This was an indication of bonding with polymeric matrix and the nano materials were improved. On the basis of these results, this formulation SNC-VI was chosen as the candidate because of its good overall mechanical properties. Viscosity and pot life of this formulation is practically feasible during the application of the sealant. With respect to the sealing of the component in the applied process, the viscosity and pot life of the polymer system are crucial because a low viscosity and a long pot life make the nano sealant process easier.

It is due to the formulation that the viscosity of the polymer system has to be kept as low as possible to seal the assembly completely. If the pot life is too short, it is difficult to finish the sealing operation because the parts cannot be wetted fully by the sealant; the short pot life causes many air bubbles to be trapped inside the sealing during the wet process if the viscosity increases quickly.

3.3 Morphological studies

Figure 1(a, b, c) presents the SEM images of the SNC V to VII system respectively. The images show irregular boundaries in the range of 10 to 100 nm. In the case of the polymer Fig. 1(a) the pattern was found to be non uniform type, whereas in the case of the Fig. 1(b) nano composite sealant, the micrographs show regular patterns. Dispersed particles and some large agglomerates in the matrix were also observed. The spherical particles can clearly be seen in the polymer matrix. From these various observations, it can be deduced that the formation of the regular is probably the result of uniform distribution of the nano particles.

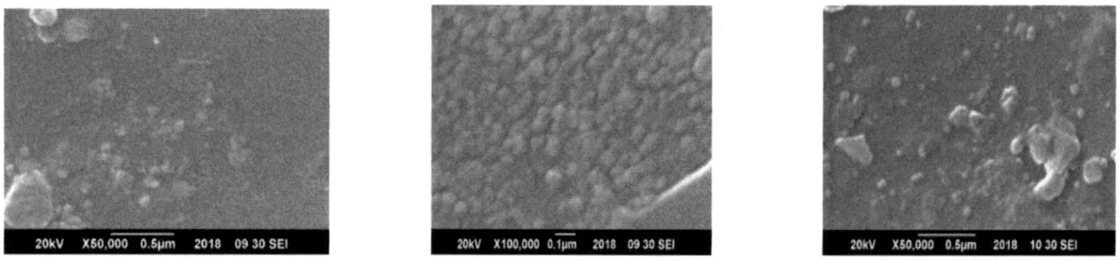

Figure 1. Scanning electron micrographs **(a)** Sealant nano composite (SNC- V) **(b)** Sealant nano composite (SNC- VI) and **(c)** Sealant nano composite (SNC- VII).

The interface plays a vital role in stress transfer from nano particles to the polymer nano composite matrix. Optimal interfacial adhesion is required for taking advantage of the excellent mechanical properties of nano composite sealant.

The addition of nano particles to the polymer matrix has an effect on the mechanical properties of nano composites through the interface because the toughness, shear strength, and viscosity of the resin matrix are affected by the bonding strength and interfacial behavior between the matrix and nano particles.

4. Conclusion

Sealant nano composites were successfully prepared by one-step blending. The results show that polymeric materials effectively improves the compatibility between filler and matrix, forming chemical bonding between the active end of monomer and the surfaces of nano particles, which make them well distributed in matrix. The results of mechanical properties show that nanocomposite improves shear strength. When the content of nano-ZrO_2 is 2%, the shear strength is the highest, improving 1 to 1.6 MPa, compared with traditional composite systems, nano-composite systems.

Scanning electron micrographs (SEM) have demonstrated the smoother surface resulting in a more uniform pattern, predicting the uniform distribution of nano particles in the host matrix.

5. References

[1] Small B W 2002 Direct resin composites for 2002 and beyond. Gen Dent. **50** (1):30-3
[2] Saito R and Tobe T 2005 Polymers for advanced technologies **16** 232
[3] Ca Y M, Sun J and Yu D H 2002 Journal of Applied Polymer Science **83** 70
[4] Chiang P C, Whang W T, Wu S C and Chuang K R 2004 Polymer **45** 465
[5] Sanchez C and mez-Romero P Go 2004 Functional Hybrid Materials, Wiley VCH, Weinheim
[6] Rivaa A, Zanetti M, Braglia M, Camino G and Falqui L 2002 Polymer Degradation and Stability **77** 299
[7] Gao P X and Wang Z L 2005 Journal of Applied Physics **97** 44304
[8] Kong Y and Hay J N 2004 J Polym Sci Part B: Polym. Physics **42** 2129
[9] Wang Q, Xia H S and Zhang C H 2001 J. Appl. Poly. Sci. **80** 478
[10] Quddos A, Khan M B, Wadood A, Malik M A and O'Brien P 2010 "Deposition and Characterization of CdSe Nanoparticles in Polymeric Materials" Key Engineering Materials, Trans Tech Publications, Switzerland Vol 442 pp 364-371
[11] Yang Y H and Dan Y 2003 Colloid Polym Sci 281 94
[12] Marchese P, Celli A and Fiorini M 2002 Macromol Chem. Phys **203**, 695
[13] Zhou W Y, Qi S H and Tu Chin C C 2005 J Northwest Polytech. Univ. **23** 365
[14] Zhou W Y, Qi S H and Zhao H Z 2006 China Plast **20** 34
[15] Peng L, Luo Y, Dan Y, Zhang L, Xia S and Zhang X 2006 Colloid Polym. Sci.285, 153

Journal of Physics: Conference Series **439** (2013) 012012 doi:10.1088/1742-6596/439/1/012012

Monitoring the dye impregnation time of nanostructured photoanodes for dye sensitized solar cells

N. Shahzad[a,b] , D. Pugliese[a,b], A. Lamberti[a,b], A. Sacco[a,b], A. Virga[b], R. Gazia[a], S. Bianco[a,b], M. I. Shahzad[b], E. Tresso[a,b] and C. F. Pirri[a,b]

[a] Center for Space Human Robotics @PoliTo, Istituto Italiano di Tecnologia, Corso Trento 21, Torino, IT-10129, Italy
[b] Applied Science and Technology Department (DISAT), Politecnico di Torino, Corso Duca degli Abruzzi 24, Torino, IT-10129, Italy

Corresponding Author: **Nadia Shahzad**
DISAT, Politecnico di Torino,
Corso Duca Degli Abruzzi-24, Torino-10128, Italy
Tel: +39 011 090 7380-48, Fax: +39 011 090 7399
Email: nadia.shahzad@polito.it

Abstract. Dye-sensitized solar cells (DSSCs) are getting increasing attention as low-cost, easy-to-prepare and colored photovoltaic devices. In the current work, in view of optimizing the fabrication procedures and understanding the mechanisms of dye attachment to the semiconductor photoanode, absorbance measurements have been performed at different dye impregnation times ranging from few minutes to 24 hours using UV-Vis spectroscopy. In addition to the traditional absorbance experiments, based on diffuse and specular reflectance on dye impregnated thin films and on the desorption of dye molecules from the photoanodes by means of a basic solution, an alternative in-situ solution depletion measurement, which enables fast and continuous evaluation of dye uptake, is presented. Photoanodes have been prepared with two different nanostructured semiconducting films: mesoporous TiO_2, using a commercially available paste from *Solaronix*, and sponge-like ZnO obtained in our laboratory from sputtering and thermal annealing. Two different dyes have been analyzed: Ruthenizer 535-bisTBA (N719), which is widely used because it gives optimal photovoltaic performances, and a new metal-free organic dye based on a hemisquaraine molecule (CT1). Dye sensitized cells were fabricated using a customized microfluidic architecture. The results of absorbance measurements are presented and discussed in relation to the obtained solar energy conversion efficiencies and the incident photon-to-electron conversion efficiencies (IPCE).

1. Introduction

Dye-Sensitized Solar Cells (DSSCs) have attracted widespread interest since their first description by Grätzel and O'Regan [1] as an influential and low cost solar energy harvester. In a conventional DSSC, the photoanode consists of a TiO_2 layer (8-15 μm) deposited on a Transparent Conducting Oxide (TCO) covered glass substrate and sensitized with dye molecules (usually Ru complexes). An electrolyte containing I^-/I_3^- redox couple acts as hole conductor and electrically regenerate the dye molecules. Another TCO covered substrate with a thin Pt layer (few nm) serves as counter electrode, to promote the reduction of the triiodide [2, 3]. In these cells, the sensitizer plays an important role: it is responsible for the electric current generation, by harvesting the solar radiation and injecting the excited electrons into the semiconductor conduction band. Consequently, the dying process has an elemental impact on both cell performance and scale-up viewpoint. Existing dye uptake measurements are based on dye desorption from photoanode after specified time intervals using a basic (*e.g.* NaOH or KOH) solution, and the subsequent UV-Vis spectroscopy. Although this procedure is quantitative, it is destructive of the devices and most of all *in-situ* measurements are impracticable. In alternative, the so-called solution depletion method is a powerful way to monitor dying processes [4, 5].

Over the last 20 years, ruthenium complexes gifted with thiocyanate ligands have achieved power conversion efficiencies beyond 11% and showed good stability [6-8]. Nevertheless, metal-free organic dyes have attracted considerable attention for practical applications because of the high cost and rarity of Ru metal. In comparison to the ruthenium sensitizers, the obtained lower efficiencies of organic dyes are mainly due to the formation of dye aggregates on the semiconductor surface and to their narrow dye absorption bands in the visible region. Therefore, lots of efforts have been made to study organic dye molecules with broader spectral responses and higher molar extinction coefficients. [9-11].

Another key component, which is responsible of the dye absorption in the DSSCs is the nanostructured semiconductor oxide. The highest photovoltaic conversion efficiencies have been obtained for TiO_2-photoanodes impregnated not only with Ru-based dyes but also organic sensitizers [6-11]. ZnO photoanodes, having the interesting property of higher specific surface area available for dye loading, have also been suggested. Many research groups have tried to investigate different ZnO nanostructures, such as nanowires, nanorods, platelets and branched structures, which exhibit great surface-to-volume ratio [12]. Moreover ZnO has attractive electronic properties [13] and nowadays the conversion efficiencies of ZnO based DSSCs, employing the Ru-based N719 dye, have been improved exceeding 6% [14, 15].

Aim of this work is to monitor the dye absorption in view of understanding the mechanisms of dye attachment to the semiconductor photoanode and of optimizing the assembly procedures in DSSCs. We have chosen to investigate two different semiconductor oxides: the traditionally employed nanostructured TiO_2 and a sponge-like ZnO film fabricated by sputtering in our laboratory. We have decided to compare the dye absorption properties of two different dyes: the largely used Ru-based N719 and a recently proposed organic dye [16] based on the hemisquaraine molecule (CT1).

2. Experimental details

2.1. Preparation of Electrodes

Fluorine-doped tin oxide (FTO) covered glasses (7 Ω/sq, Solaronix) were firstly cleaned by immersing in acetone in an ultrasonic bath for 10 minutes and rinsing with isopropanol. Afterwards the glasses were dipped for other 10 minutes in a 1:3 (Hydrogen Peroxide: Sulphuric Acid) piranha solution, so as to remove organic residues, rinsed with deionised water and finally dried using Nitrogen-shower. Then, a TiO_2 layer (Ti-Nanoxide D37, Solaronix) with a circular shape (nearly 1cm in diameter) was deposited onto FTO covered glasses by doctor blading technique described elsewhere [2].The deposited films were then dried at 50 °C for 10 minutes on a hot plate and left for half an hour at ambient temperature. A subsequent sintering process at 450 °C for 30 minutes enabled the formation of nanoporous TiO_2 film with a mean thickness of 8.5μm, as measured by a profilometer (P.10 KLA-Tencor Profiler). Post-treatment with $TiCl_4$ was performed according to the method described in literature [17, 18]. Briefly, an aqueous stock solution of 2 M $TiCl_4$ was diluted to 0.05 M. Freshly sintered TiO_2 films were immersed into this solution and placed on a hot plate at 70°C for 30 min in a closed hood. After $TiCl_4$ treatment, the films were again sintered at 450°C for 30 minutes.

Zn nanostructured layers (5 μm) were deposited on glasses coated with FTO by RF magnetron sputtering technique. The geometry of the sputtered layer was defined by covering the substrate with a hard circular mask. In order to obtain sponge-like ZnO layers, the deposited films were placed on a hot plate at 380 °C for 60 minutes in ambient air [19].

Photoanodes were immersed into a dye solution for different time periods at room temperature and then rinsed in ethanol to remove the unadsorbed dye molecules. For each impregnation time, eight anodes were prepared: five for cell fabrication and the remaining three for dye desorption procedure. Table 1 gives parameters for four different combinations of semiconducting oxide and dye.

Table 1: Photoanode parameters.

	Semiconducting Oxide	Dye	Solvent	Concentration of dye solution
Set 1	TiO_2	N719 (Ruthenizer 535bis-TBA)	Ethanol	0.33mM
Set 2	TiO_2	CT1 (Hemisquaraine)	Acetonitrile	0.17mM
Set 3	ZnO	N719 (Ruthenizer 535bis-TBA)	Ethanol	0.25mM
Set 4	ZnO	CT1 (Hemisquaraine)	Acetonitrile	0.17mM

For the fabrication of counter electrodes, two small pin-holes as inlet/outlet ports for electrolyte filling were drilled in the FTO glass substrate using powder blasting technology. Substrates were then cleaned with the same cleaning method described above and thermal evaporation was employed to deposit a 5 nm Pt thin film.

2.2. Cell Fabrication

Dye sensitized cells were fabricated employing a customized microfluidic architecture [20] consisting of a Poly-DiMethylSiloxane (PDMS) thin membrane reversibly sealed between the two electrodes with an external PolyMethylMethAcrylate (PMMA) clamping system. The structure of the cell is presented in figure 1. The electrolyte (Iodolyte AN 50, Solaronix) filling was performed with a syringe connected to the O-Ring PDMS interconnections. Copper foils (50 µm thick, 1.5-cm^2 area) were used as electric contacts with FTO, dielectrically insulated by the PDMS membrane. Active area of the cells was 0.86 cm^2 and measurements were performed with a 0.22-cm^2 black rigid mask.

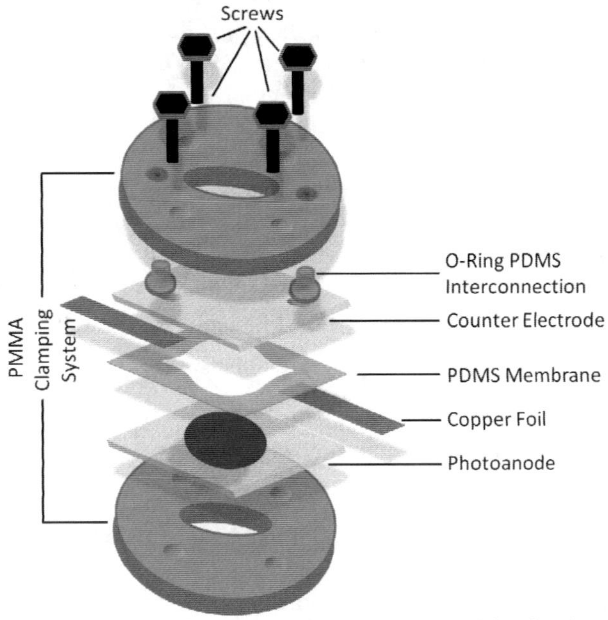

Figure 1: Scheme of the microfluidic dye-sensitized solar cell.

Journal of Physics: Conference Series **439** (2013) 012012 doi:10.1088/1742-6596/439/1/012012

3. Characterizations

3.1. Optical Measurements

3.1.1. *Photoanode Absorbance.* Photoanodes were optically characterized by UV-Vis spectrophotometer (Agilent, Varian Cary 5000) measuring Kubelka-Munk function, F(R), utilizing integrating sphere as shown in figure 2a. The Kubelka-Munk theory is normally used for the analysis of diffuse reflectance spectra obtained from diffusing samples. It provides a correlation between reflectance and absorbance: $F(R) = (1-R)^2/2R = k/s = Ac/s$ where R is the reflectance, k is the absorption coefficient, s is the scattering coefficient, A is the absorbance and c is the concentration of absorbing species. As c and s remain same, spectrum obtained for Kubelka-Munk function can be considered as absorbance.

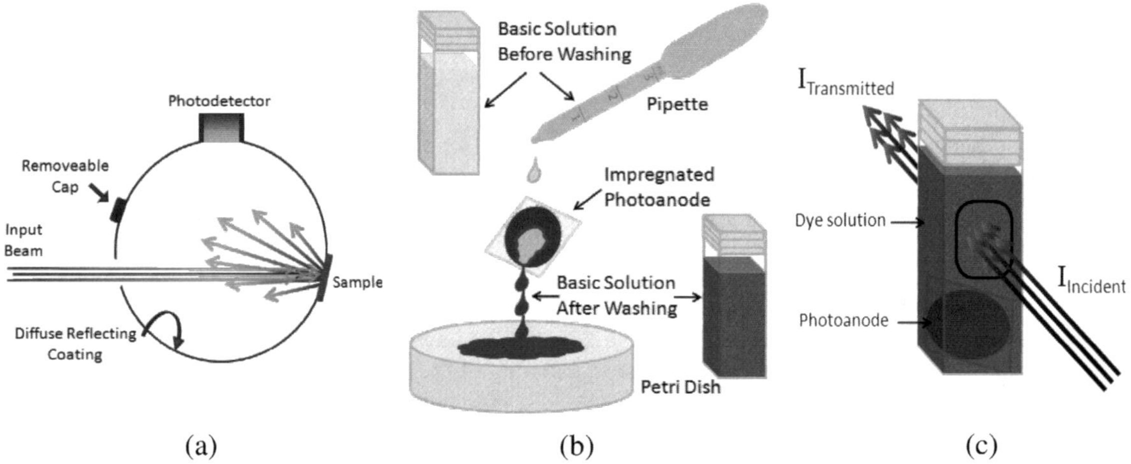

Figure 2: Schematic diagram of: (a) The integrating sphere for reflectance measurements, (b) washing procedure for dye loading measurements and (c) the solution depletion method to study the in situ uptake of dye molecules on the photoanode.

3.1.2. Washing Procedure. Investigation of dye loading amount was determined by desorbing the dye molecules from photoanode using a basic solution followed by measurement of the optical absorbance in optical density units with a UV-Vis spectrophotometer in the wavelength region from 300 to 800 nm. A simple but depictive scheme of the procedure is shown in figure 2b.

3.1.3. Solution Depletion Method. The sintered photoanode was immersed in a spectrophotometer quartz cuvette containing the dye solution consisting of 0.27mM N719 (Ruthenizer535bis-TBA, Solaronix) dye in ethanol (0.17mM hemisquaraine dye in acetonitrile). The cuvette with the immersed photoanode was placed in the UV-Vis spectrophotometer at room temperature for more than 24 hours in order to record the absorbance spectra. The photoanode was kept at the cuvette bottom as shown in the schematic diagram of the setup (figure 2c). This allowed monitoring of the depletion of dye solution with time, due to continuous adsorption of dye molecules by the photoanode. During the whole experiment cuvette was kept tightly covered with a cap to avoid the evaporation of solvent to make sure that the volume of solution remained constant. Experiment was repeated for three times to enhance the certainty of results.

3.2. Electrical Measurements

I-V electrical characterizations were performed under AM1.5G illumination (1000 W/m^2) using a class A solar simulator (91195A, Newport) and a Keithley 2440 source measure unit. Incident photon-to-electron conversion efficiency (IPCE) measurements were carried out employing a 100-W QTH lamp (Newport) as light source and a 150-mm Czerny Turner monochromator (Omni-l 150, Lot-Oriel).

4. Results and Discussions

4.1. Nanostructured TiO_2: absorption measurements

Figure 3 shows the absorption spectra of the dye-impregnated photoanodes of Set 1 (8.5um TiO_2 films sensitized with N719) for different impregnation times, ranging from 3 to 24 hours. The spectra exhibit a typical trend, with the N719 absorption peaks located at 420 and 540 nm and superimposed to the TiO_2 absorption (essentially located in the UV region). The dye-related peaks increase with the impregnation time as it was expected (except 24 hours). In order to have better evaluation of the dye absorption, the areas enclosed by the F(R) curves before and after dye incubation (which correspond respectively to the absorbance of the TiO_2 and of the TiO_2/dye films) have been computed: these areas are a sort of "integrated" dye absorbance in the visible region. The results are shown in figure 3b, and they evidence that the increased light absorption of the photoanodes is due to an increased number of adsorbed dye molecules on the TiO_2 film.

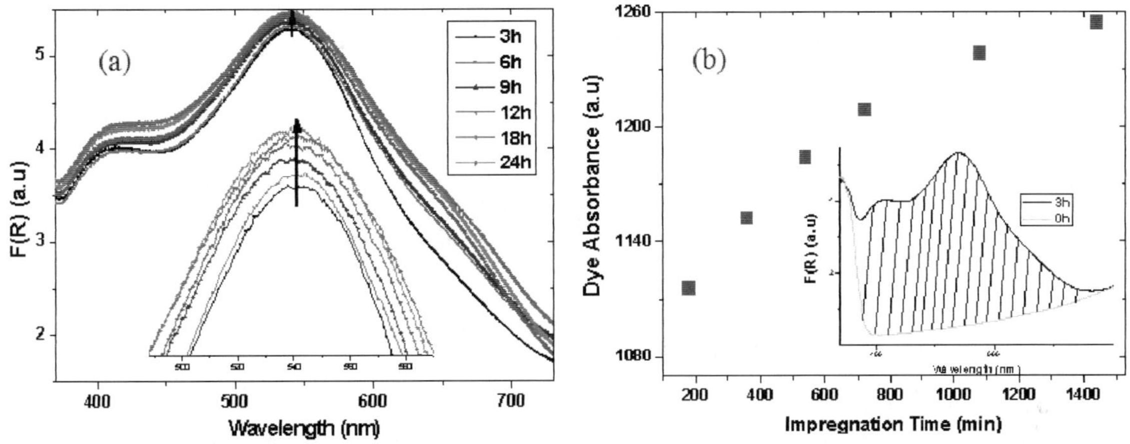

Figure 3: (a) Absorption spectra of photoanodes after dye impregnation for different time periods, (b) Area between F(R) curves after and before dye incubation (shown in inset only for 3h to have clear view).

For a quantitative analysis of dye loading, the washing procedure for desorption of dye from the anodes has been performed using a known volume of 0.1M NaOH aqueous solution, as implemented for different TiO_2 nanostructures in literature [21-24]. The absorbance spectra of the resulting solutions are shown in the figure 4a. In addition, the depletion solution experiment has also been performed and the spectra are shown in figure 4b. When N719 is dissolved in a solvent it presents two narrow and well defined peaks, which are located at 386 and 530 nm when the solvent is ethanol, and at 372 and at 500 nm in aqueous solution of NaOH. By comparison with figure 3a it can be observed that the peaks are broadened and red-shifted when the dye is attached to the TiO_2 film, this is due to the anchoring and to the chemical interaction of the dye molecules to the semiconductor surface.

From the spectra of figures 4a and 4b the concentration and hence the total number of adsorbed dye moles were calculated employing the Lambert-Beer law and the pre-determined extinction coefficient values: 12717 M^{-1} cm^{-1} at 500 nm for the basic solution and 14614 $M^{-1}cm^{-1}$ at 529 nm for the ethanolic solution. The two dye incubation measures are compared in figure 4c and a strong correlation is observed in the obtained values of dye loading. The number of moles obtained from desorption is little lower than the one obtained from depletion. This dissimilarity can be due to an incomplete removal of dye molecules by the basic solution, or also to the fact that, after impregnation, the photoanodes are rinsed with ethanol. Such rinsing procedure removes also the physisorbed dye molecules which are not anchored to the TiO_2 surface but are computed with the depletion method. In any case, both the methods clearly show that there is an upper limit in the number of absorbed dye molecules and indicate that the saturation is reached at about 16 hours of impregnation time.

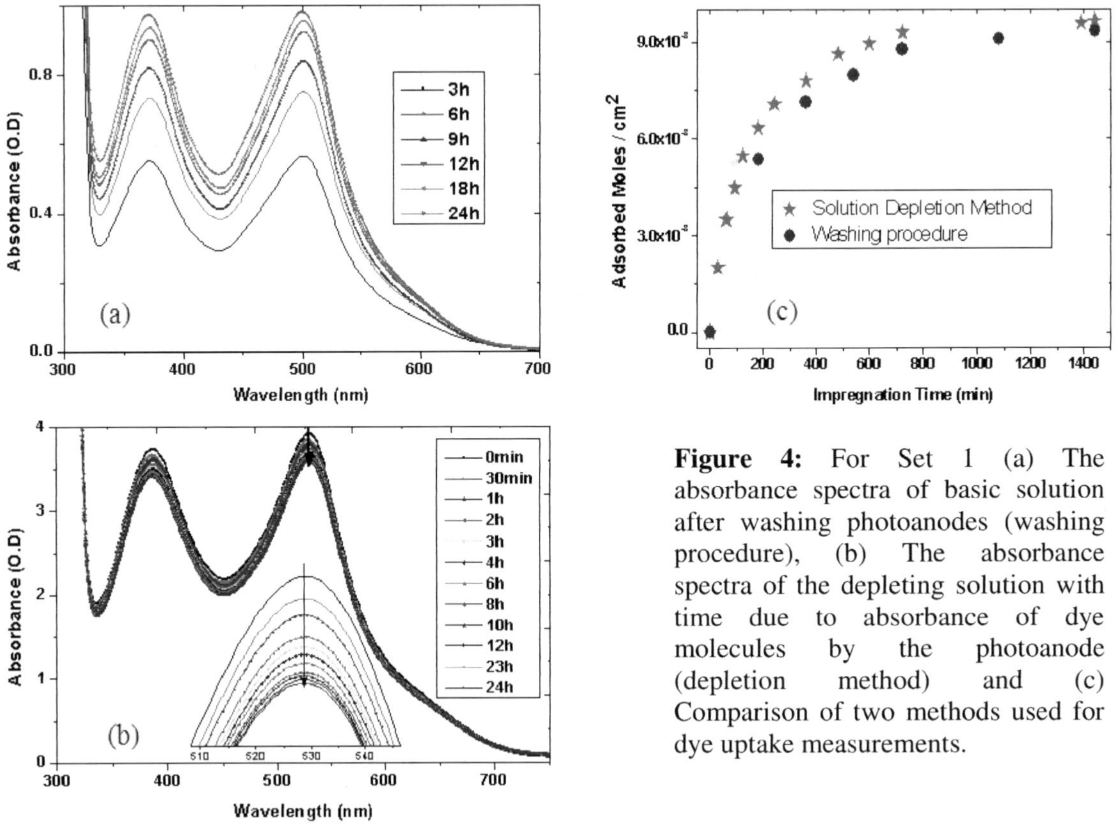

Figure 4: For Set 1 (a) The absorbance spectra of basic solution after washing photoanodes (washing procedure), (b) The absorbance spectra of the depleting solution with time due to absorbance of dye molecules by the photoanode (depletion method) and (c) Comparison of two methods used for dye uptake measurements.

The obtained results evidenced a well known behaviour of the N719 dye when attached to the TiO_2 surface (in fact the typical impregnation time suggested for N719-based DSSCs is overnight [3, 21]), and they confirm that our experimental procedures for absorption evaluation are correct. We decided to apply these procedures to study the recently introduced hemisquaraine CT1 dye. This dye is prepared following a simple, easy and inexpensive three steps synthesis: quaternarization, coupling and hydrolysis, according to a classical procedure for hemisquaraine dyes [25]. The CT1 absorbance spectrum in acetonitrile solution is reported in figure 5a (Set 2). It exhibits a wide band composed by three neighboring peaks centered at 388nm, 419nm and 434nm with an average molar extinction coefficient of 15220 $M^{-1}cm^{-1}$. CT1 shows promising property of good light absorption, even if – compared to N719- the absorption extends only till 500 nm and does not cover the red region of the visible spectrum. The absorption spectra obtained on the CT1/TiO_2 photoanodes are compared in the inset of figure 5b for impregnation times ranging from 2 min to 24 hours, where again a broadening and a red- shift of the peaks due to dye anchoring to the TiO_2 surface is observed. Such a red-shift extends till 600 nm the CT1 region of light absorption. Moreover it can be observed that the peak at 480nm shows very small variation in intensity with increasing incubation period, while the 414 nm peak intensity strongly increases till 3 hours of impregnation, and remains nearly constant for longer times. Another shoulder-like peak appears only for impregnation times higher than 2-3 hours and it can be seen increasing up to 15 hours. The integrated dye absorbance of Set 2 (CT1 hemisquaraine dye/TiO_2 photoanodes) shown in figure 5b has been obtained, as previously described, by calculating the area between the F(R) curves before and after dye impregnation.

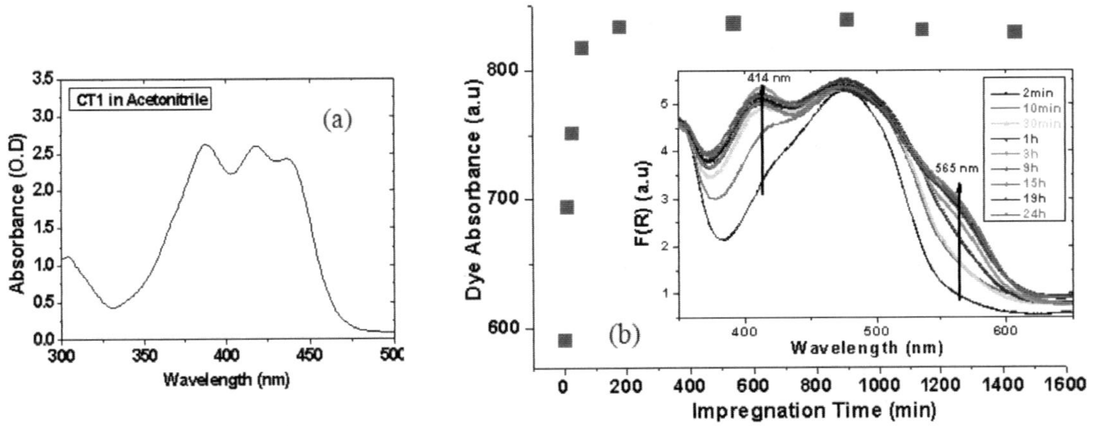

Figure 5: (a) Optical absorbance of hemi-squaraine dye in acetonitrile (b) Dye absorbance obtained by calculating area between the F(R) curves before and after dye impregnation. Inset compares the absorption spectrum of Set 2 (CT1 absorbed on TiO_2 film).

For obtaining a quantitative analysis different tests have been performed for washing procedures of hemisquaraine dye from the TiO_2 photoanodes: the best performing solution was found to be a 0.5M aqueous solution of KOH. Absorbance spectra of such basic solution after rinsing the photoanodes are shown in the figure 6a. The desorbed dye moles numbers were calculated using the extinction coefficient value (25715 M^{-1} cm^{-1} at 447 nm) of dye in the corresponding basic aqueous solution.

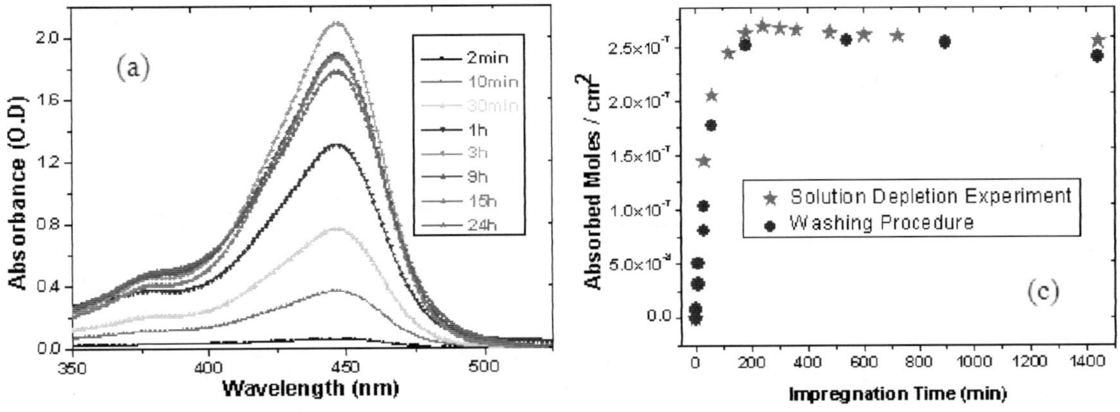

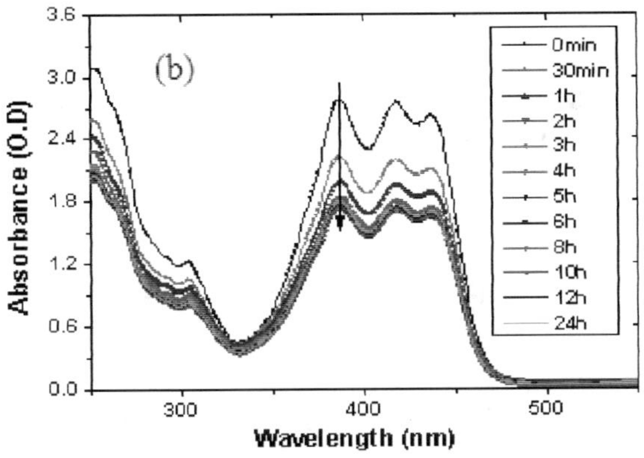

Figure 6: For Set 2 (a) The absorbance spectra of basic solution after washing CT1 impregnated photoanodes (washing procedure), (b) the absorbance spectra of the depleting dye solution with time (depletion method) and (c) comparison of adsorbed number of moles calculated from two dye uptake measures.

Journal of Physics: Conference Series **439** (2013) 012012 doi:10.1088/1742-6596/439/1/012012

The absorbance spectra obtained from solution depletion experiment (shown in figure 6b) has been used to calculate the number of moles using molar extinction coefficient value of CT1 solution in acetonitrile which is 15930 $M^{-1}cm^{-1}$ at 388 nm. Comparing the two measurements (figure 6c), it is found that the dye loading results are in agreement, with a little dissimilarity as it was also observed for the traditional dye N719. Both the methods reveal saturation at around 3 hours.

4.2 - Nanostructured TiO_2: Comparison with electrical measurements

The dye impregnated photoanodes of Set1 (N719/TiO_2) and Set 2 (CT1/TiO_2) have all been employed for fabrication of DSSCs. The dependence of the cell efficiency on dye incubation time was studied by I-V electrical characterization and incident photon-to-electron conversion efficiency (IPCE) measurements. The current density-voltage curves of Set 1 and Set 2 are reported in (a) and (b) parts of figure 7, respectively, as a function of incubation time. A monotonic behavior, with a saturating incubation time around 16 hours clearly appears for N719, while a non-monotonic trend, with favorable incubation time around 1 hour, is observed with CT1. Sections (c) and (d) of the same figure compare the IPCE curves of the N719 and hemisquaraine based DSSCs, respectively. For Set 2 (CT1 impregnated), the peak at about 460nm is blue shifted, lowered and broadened for impregnation times longer than 1h. This fact indicates degradation in the photon-to-current conversion, it can be related to the observed trend in the F(R) peak at 414 nm (figure 5a) and hence it can lead to a lower efficiency of the solar cell [26].

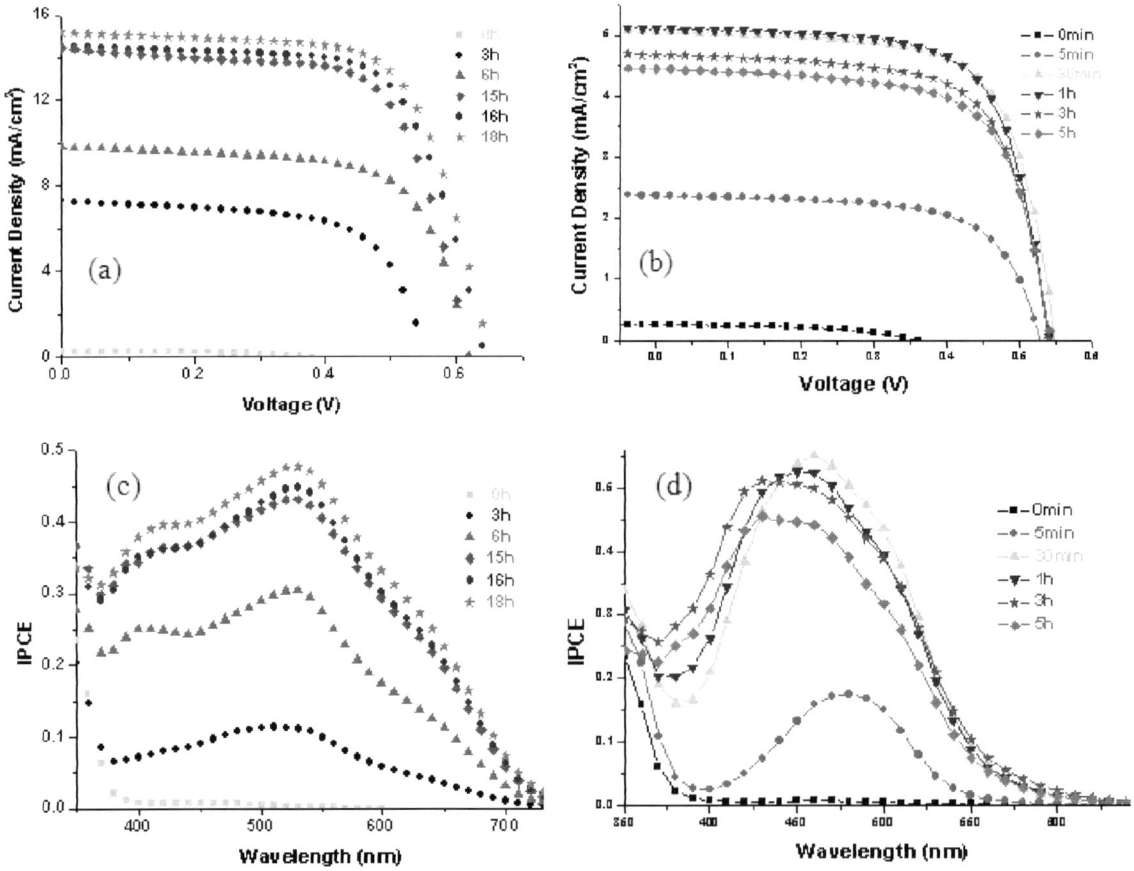

Figure 7: Current density-voltage curves for cells based on (a) Set 1(N719 adsorbed on TiO_2) and (b) Set 2 (CT1 adsorbed on TiO_2). IPCE curves of DSSCs based on (c) Set 1 and (d) Set 2, for different impregnation times.

In figure 8 the absorbance and the electrical results are compared. Panels (a) and (b) compare the cell efficiency and the absorbed number of moles as a function of impregnation time for Set 1(N719/TiO_2 photoanodes) and Set 2 (CT1/TiO_2 photoanodes). In case of N719, the efficiency of the cell is increasing with increasing number of absorbed moles till the saturation point is reached. On the other hand efficiency of the cell based on hemisquaraine has been observed increasing up to 30min and then started decreasing regardless of increasing number of absorbed moles. This indicates that, for impregnation times higher than 1h, only a part of the hemisquaraine molecules present in the photoanode contribute to electric current generation. A consistent fraction of dye molecules is not able to transfer the photogenerated electron to TiO_2, because they tend to aggregate each other or because their anchorage to TiO_2 surface is not effective. Moreover the non monotonic behavior of absorption spectra of the CT1/TiO_2 photoanodes (inset of figure 5b) evidence changes in the CT1attachment to TiO_2, in particular the peak at 565nm, appearing only after 1h, can be related to dye molecules aggregation.

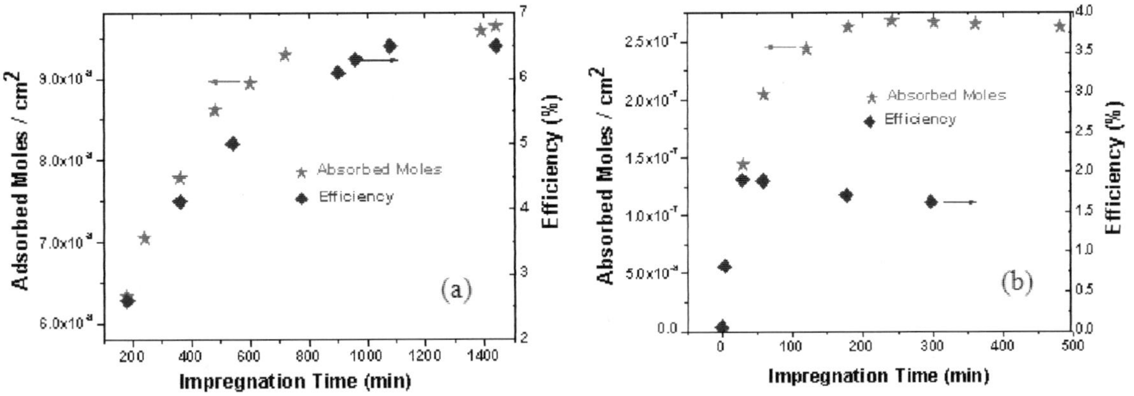

Figure 8: Number of absorbed moles has been compared with the efficiency of DSSCs based on: (a) Set 1, (b) Set 2.

4.2. Sponge-like ZnO: absorption and electrical measurements

We performed impregnation time experiments also on ZnO-based photoanodes. A simple and low cost two-step method, based on RF magnetron sputtering deposition and thermal oxidation, for obtaining sponge-like ZnO nanostructured films has been recently proposed. These films exhibited a high density of branches, and a specific surface area of 14.1 m^2/g. A photoconversion efficiency value of 6.67% was obtained with an 18 μm thick film incubated with N719 dye molecules [15]. In this work 5 μm thick sponge-like ZnO films were immersed in 0.25mM ethanolic solution of N719 dye (Set 3) and characterized for optical absorption (figure 9a). The results depict the strong absorption properties of ZnO films, up to 24 hours of incubation. The typical N719 peak (located at 530 nm) monotonically amplifies with increasing immersion time from 10 min to 5 hours. Only one very long (24h) impregnation time was measured and in the resultant absorption spectrum a radical change is observed: two new shoulder-like peaks appear at about 405nm and 480nm, and hence the peak at 530 nm is strongly broadened. This suggests an alteration in the N719 anchorage to ZnO surface for longer impregnation times. This hypothesis is also confirmed by the electrical characterizations (current density-voltage and IPCE curves) of DSSCs fabricated with these N719/ZnO photoanodes, given in figure 9 (b and c respectively). The best result is obtained for 2 hours of impregnation: it exhibits a short circuit current density (J_{SC}) of 9.6 mA/cm^2 and an open circuit voltage (V_{OC}) of 592 mV, leading to a maximum Photo Conversion Efficiency (PCE) value of 3.21%.

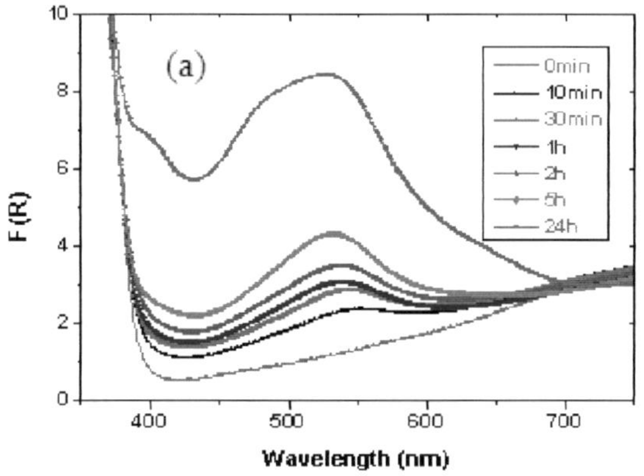

Figure 9: (a) Absorption spectrum of Set 3 (N719 dye absorbed on ZnO films), (b) Current density-voltage curve and (c) IPCE curves of Set 3 (solar cells based on sponge-like ZnO anode immersed in N719 solution) for different impregnation times.

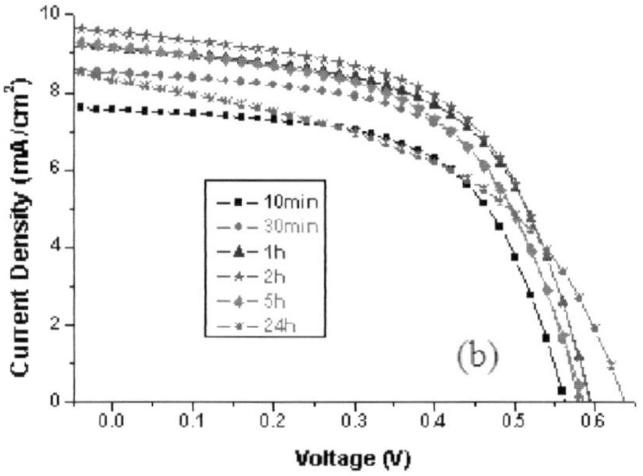

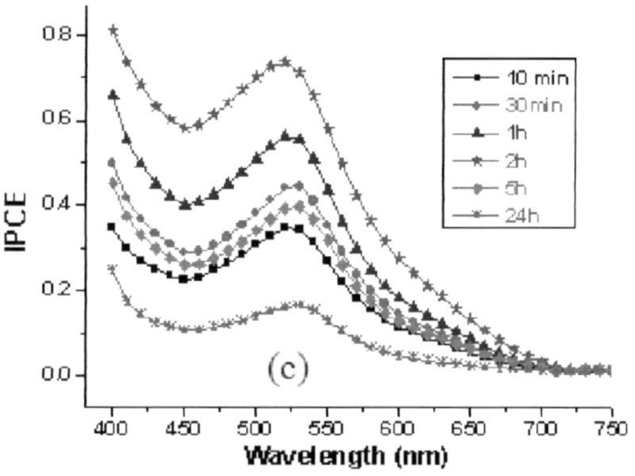

The quantitative analysis results, obtained from depletion solution experiment, are reported in the inset of figure 10a. The number of absorbed dye molecules reaches a maximum at 1h and 45min, then it decreases and finally starts again to increase, with a different slope, at 2h and 20min. This indicates that nearly at 2h of impregnation the ZnO film stopped absorbing, and it released a few molecules to the solution, then it started absorbing molecules anew. The integrated dye absorbance reported in figure 10a, evaluated from F(R) results, confirms such increase in dye absorption up to 24h. In figure 10a the DSSC efficiency values are also reported: the most favorable incubation time results to be 2h: exceeding this time a decrease in efficiency is observed despite the increasing amount of dye loading, which is evident from both integrated dye absorbance and absorbed dye molecules. This can be related to the formation of molecular aggregates between dye molecules and dissolved Zn^{2+} ions instigating from the ZnO surface, as reported by several researchers [27, 28]. This dissolution of Zn surface atoms is caused by the acidic carboxylic groups of the N719 dye [13]. Drastic change in absorbance behavior after 24 hours of incubation also suggests the possible formation of new aggregates. Such molecular complexes show strong adsorption properties but are clearly unable to inject electrons in the conduction band of the semiconductor [19, 29].

Finally, some sponge-like ZnO substrates were also immersed in CT1 hemisquaraine dye solution (Set 4) and dye sensitized solar cells were fabricated using microfluidic approach to investigate their photon to electron conversion efficiency. As shown in figure 10b, this set remained unable to achieve efficiency higher than 0.65%. Such a low efficiency range was not enough encouraging for further investigations. The results obtained for all the different combinations of semiconducting oxide and dye are summarized in table 2.

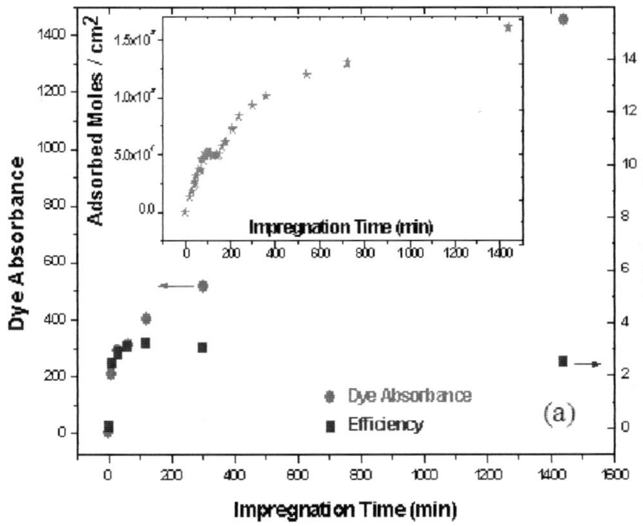

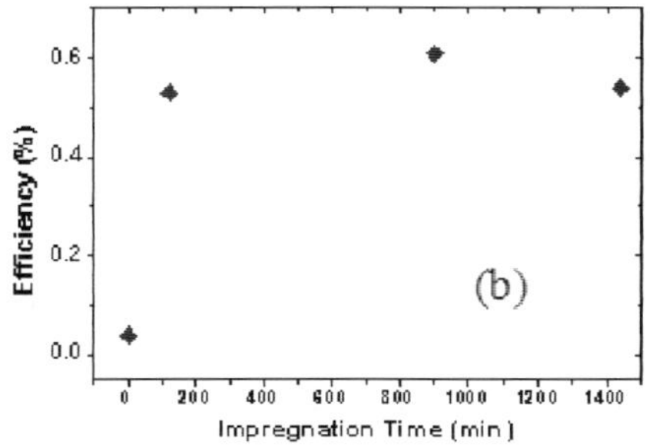

Figure 10: (a) Dye absorbance has been compared with the efficiency of DSSCs based on ZnO anodes (Set 3). Inset shows adsorbed moles calculated from solution depletion experiment, (b) Photon conversion efficiency of cells based on ZnO films immersed in hemisquaraine dye solution (Set 4).

Table 2: Summarized results.

	Favorable Impregnation time (hours)	Maximum Efficiency (%)	Absorption Behavior	Adsorbed Moles at Favorable time	Max Adsorbed Moles
Set 1	16	6.5	monotonic	9.51×10^{-8}	9.63×10^{-8}
Set 2	1	1.9	Non-monotonic	14.5×10^{-8}	25.5×10^{-8}
Set 3	2	3.2	Non-monotonic	5.15×10^{-8}	16.0×10^{-8}
Set 4	---	0.6	---	---	---

5. Conclusions

- We have employed three different UV-Vis absorbance measurements for monitoring the dye impregnation time of nanostructured photoanodes for dye sensitized solar cells.
- Two different dyes, the largely used Ru-based N719 and a recently proposed organic dye based on the hemisquaraine molecule, CT1, have been tested on two different semiconductor oxides: the traditionally employed nanostructured TiO_2 and a sponge-like ZnO film fabricated in our laboratory.
- Four different sets of dye impregnated semiconducting films have been used for dye sensitized solar cells fabrication, employing a microfluidic architecture; the cells have been electrically characterized by I-V and IPCE measurements.
- It has been shown that the analysis of the dye loading amount in photoanodes is helpful for interpretation of the conversion efficiency trend versus impregnation time, in view of optimization of the dying procedure.

References:

[1] O'Regan B and Grätzel M 1991*Nature* **353** 737-740
[2] Nazeeruddin M K, Kay A, Rodicio I, Humphry-Baker R, Mueller E, Liska P, Vlachopoulos N and Graetzel M 1993 *J. Am. Chem. Soc.* **115** 6382-6390
[3] Nazeeruddin M K, Klein C, Liska P and Gratzel M 2005 *Coord. Chem. Rev.* **249** 1460-1467

[4] Waston T, Holliman P and Worsley D 2011 *J. Mater. Chem.* **21** 4321-4325
[5] Concina I, Frison E, Braga A, Silvestrini S, Maggini M, Sberveglieri G, Vomiero A and Carofiglio T 2011 *Chem. Commun.* **47** 11656-11658
[6] Gratzel M 2009 *Accounts of Chem. Res.* **42** 1788
[7] Jang S R , Yum J H, Klein C, Kim K J, Wagner P, Officer D, Gratzel M and Nazeeruddin M K 2009 *J. Phys. Chem. C* **113** 1998-2003
[8] Gao F, Wang Y, Shi D, Zhang J, Wang M, Jing X, Humphry-Baker R, Wang P, Zakeeruddin S M, and Gratzel M 2008 *J. Am. Chem. Soc.* **130** 10720-10728
[9] Choi H, Raabe I, Kim D, Teocoli F, Kim C, Song K, Yum J H, Ko J, Nazeeruddin M K and Gratzel 2010 *M Chem. Eur. J.* **16** 1193-1201
[10] Cai N, Moon S J, Cevey-Ha L, Moehl T, Humphry-Baker R, Wang P, Zakeeruddin S M and Gratzel M 2011 *Nano Lett.* **11** 1452-1456
[11] Im H, Kim S, Park C, Jang S H, Kim C J, Kim K, Park N G and Kim C 2010 *Chem. Commun.* **46** 1335-1337
[12] Zhang Q, Dandeneau C S, Zhou X, and Cao G 2009 *Adv. Mater.* **21** 4087-4108
[13] Ambade S B, Mane R S, Ghule A V, Takwale M G, Abhyankar A, Chod B W and Han S H 2009 *Scripta Mater.* **61** 12-15
[14] Saito M and Fujihara S 2008 *Energy Environ. Sci.* **1** 280-283
[15] Sacco A, Lamberti A, Gazia R, Bianco S, Manfredi D, Shahzad N, Cappelluti F, Ma S and Tresso E 2012 High efficiency Dye-sensitized Solar Cell exploiting sponge-like ZnO nanostructures. *Phys. Chem. Chem. Phys.* In Press, DOI: 10.1039/C2CP42705B
[16] Cicero G *et. al.* 2012 Combined experimental and theoretical investigation of the hemi-squaraine/TiO_2 interface in a dye sensitized solar cell. Energy & Environmental Science (submitted)
[17] Demopoulos G P, Charbonneau C, Lee K E, Shan G B, Gomez M A and Gauvin R 2009 *ECS Trans.* **21** 23-34
[18] Manmeeta, Saxena D, Sharma G D and Roy M S 2012 *Research Journal of Chemical Sciences* **2** 61-71
[19] Lamberti A, Gazia R, Sacco A, Bianco S, Quaglio M, Chiodoni A, Tresso E and Pirri C F 2012 Coral-shaped ZnO nanostructures for dye-sensitized solar cell photoanodes *Prog. Photovolt: Res. Appl.* In Press, DOI: 10.1002/pip.2251
[20] Lamberti A, Sacco A, Bianco S, Giuri E, Quaglio M, Chiodoni A and Tresso E 2011 *Microelectron. Eng.* **88** 2308-2310
[21] Katoh R, Yaguchi K and Furube A 2011 *Chem. Phys. Lett.* **511** 336-339
[22] Hwang K J, Yoo S J, Jung S H, Park D W, Kim S I, and Lee J W 2009 *Bull. Korean Chem. Soc.* **30** 172-176
[23] Stergiopoulos T, Ghicov A, Likodimos V, Tsoukleris D S, Kunze J, Schmuki P and Falaras P 2008 *Nanotechnology* **19** 235602-235609
[24] Koo H J, Kim Y J, Lee Y H, Lee W I, Kim K, and Park N G 2008 *Adv. Mater.* **20** 195-199
[25] Yum J H, Walter P, Huber S, Rentsch D, Geiger T, Nüesch F, Angelis F D, Grätzel M, and Nazeeruddin M K 2007 *J. Am. Chem. Soc.* **129** 10320-10321
[26] Andersen A R, Halme J, Lund T, Asghar M I, Nguyen P T, Miettunen K, Kemppainen E and Albrektsen O 2011 *J. Phys. Chem. C* **115** 15598-15606
[27] Keis K, Lindgren J, Lindquist S E and Hagfeldt A 2000 *Langmuir* **16** 4688-4694
[28] Memarian N, Concina I, Braga A, Rozati S M, Vomiero A and Sberveglieri G 2011 *Angew. Chem. Int. Ed.* **50** 12111-12366
[29] Zhang Q, Chou T P, Russo B, Jenekhe S A and Cao G 2008 *Angew. Chem. Int. Ed.* **47** 2402-2406

Effect of doping on the Structural and Optical Properties of SnO_2 Thin Films fabricated by Aerosol Assisted Chemical Vapor Deposition

Syed Mansoor Ali[1], Syed Tajammul Hussain[2], Shahzad Abu Bakar[2,3], Jan Muhammad[1], ,Naeem ur Rehman[4]

[1] Department of Physics, The Islamia University of Bahawalpur, Pakistan
[2] NS& CD, National Center for Physics Islamabad, Pakistan
[3] Department of Chemistry, Quaid-i-Azam University Islamabad, Pakistan
[3] National Center for Nanotechnology, Department of Chemical and Material Engineering, Pakistan Institute of Engineering and Applied Sciences (PIEAS), Islamabad, Pakistan

E.mail: mansoor_phys@yahoo.com

Abstract. In order to achieve high conductivity and transmittance of transparent conducting oxide (TCO), we attempted to fabricate Mg doped SnO_2 ($Mg_xSn_{1-x}O_2$) thin films and characterized them for their structural and optical properties. The $Mg_xSn_{1-x}O_2$ thin films have been deposited on glass substrate by using aero-sole assisted chemical vapor deposition (AACVD). The molar concentration of Mg concentration was changed from 0 to 8%. The confirmation of tetragonal structure and particle size (32 to 87nm) has been calculated of thin films by XRD. The surface roughness is decreased with the increase of the dopant concentration, which has been investigated by atomic force microscopy (AFM). The optical transmission has increased from 54 to 78% and the band gape of pure SnO_2 has been found to be in the range of 3.76eV and it is shifted to 3.69eV for 6Wt % Mg doping and then increase on further increasing the Mg doping.

Keywords: Mg Doped SnO_2; Thin Films; AACVD; RBS; XRD

1. Introduction

Nanometre size materials are great insert in the recent year due to the promising technological application in various fields. Among these materials, tin oxide (SnO_2) is a very important material, due to its attractive properties like optical transmittance, uniformity, low resistivity, mechanical hardness, stability to heat treatment and piezoelectric behaviour [1-2]. These properties make SnO_2 a suitable candidate for device such as for gas sensor application, Li- ion batteries, photovoltaic, super capacitor, Light emitting diode, display devices and solar cell [3-4].

SnO_2 thin film has been prepared by various techniques such as RF magnetron sputtering [5-6], electron beam evaporation [7], sol gel coating [8], chemical vapour deposition [9], etc. Among them AACVD is very simple technique for the growth of SnO_2 due to its simplicity and low cost. In comparison of other deposition techniques which required to maintains high vacuum during deposition, in AACVD the deposition is performed under atmospheric pressure.

To improve the quality of film as well as the physical and chemical properties of thin films, the addition of some metal ions as impurities is expected to play an important role in changing the charge carriers concentration of the metal oxide matrix, catalytic activity, the surface potential, the phase composition, the size of crystallites, and so on [10-11]. In literature we have found that Cu [12],

Fe [13], Mn [14], Sb [15], Cs [16] and F [17] were doped to improve the properties of SnO_2. In the literature, no one has reported the deposition of $Mg_xSn_{1-x}O_2$ thin films by AACVD. The Mg^{+2} has ionic radius of 0.67Å, which is close to the ionic radius of Sn^{+4} that is 0.71Å. Therefore replacement of Sn by Mg does not change in lattice constants and crystal structure. It is possible to change the band gape by doping SnO_2 with suitable ratio of Mg.

In this study, we have grown SnO_2 thin film with different concentration of Mg (2-8 wt%) and study the effect of Mg doping on the structure and optical properties of the growth film using X-ray diffraction (XRD), Fourier transform infra-red spectroscopy (FTIR), Rutherford back scattering (RBS), UV-Vis transmission spectroscopy and atomic force microscopy (AFM).

2. Materials and Methods

The schematic diagram of AACVD apparatus is shown in figure 1. The 0.2 M solution of $Mg_xSn_{1-x}O_2$ (x=0-0.08in) ethanol was prepared and stirred with magnetic stirrer on hotplate for 30 minutes. The glass slides (1 x 2 cm) were washed ultrasonically prior to deposition with ethanol, propanol and mixture of acetone-propanol before deposition. The slides were placed in distilled water. Before the deposition, the glass slides were taken from the distilled water and placed inside the quartz tube and heated to the required temperature (450 °C). The temperature of the glass slides were continuously monitored by the K-type thermocouple. The 0.2 M solution of $Mg_xSn_{1-x}O_2$ (x = 0-0.08) in a two neck flask was placed on the humidifier. The aero-sol was generated at room temperature were transported to the heated substrate by an inert gas (Argon gas). The deposition was performed for 30 mints and then deposited thin films were allowed to cool down to the room temperature under the inert gas passage.

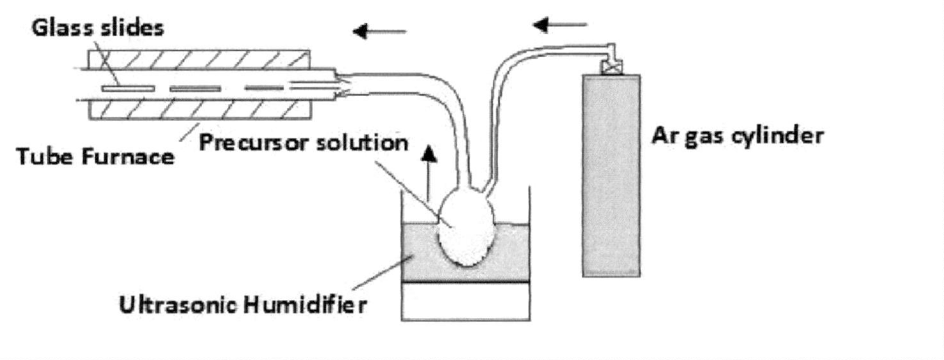

Figure 1. Schematic diagram of AACVD facilities.

The phase crystallinity and structural analysis of the deposited $Mg_xSn_{1-x}O_2$ was analyzed by X-ray diffraction patterns using X' pert PRO (PAN analytical) with Cu-K_α source radiation. The optical transmission of the thin films were recorded at room temperature by a Perkin Elmer UV/VIS/NIR Lambda19 spectrometer in the wave length 300-850nm. FTIR transmission spectra were recorded at room temperature on a NICOLET 6700, Thermo electron Co. USA in the range of 500-1800 cm^{-1}. Surface morphology of the thin films were investigated by atomic force microscopy (Quesant Universal SPM, Ambios Technology) in non connected mode. An AFM tip of silicon nitride was used having an approximate radius of curvature of 10nm. Rutherford backscattering (RBS) was performed to confirm the layer structure across the thickness of the thin films. Incident particles were the He^{+4} ions accelerated by 5UDH Tandem Accelerator to energy of 2.84MeV. The scattering angle was 170°, which is most common for conventional RBS. However, required depth resolution was achieved by maintaining the surface normal of the sample to 0° with respect to incident beam and 10° with respect to the scattered beam.

3. Result and Discussion

Non-destructive X ray diffraction technique was used to determine the phase crystallinity and structural analysis of the $Mg_xSn_{1-x}O_2$ deposited thin films as shown in figure 2(a) and 2(b). As shown in figure 2(a), the d_{hkl} (inter planer distance) values of their corresponding <hkl> have been identified as (110), (101), (200), (211), (220), (310), and (321) of the pure SnO_2 thin film which indicate the tetragonal structure of SnO_2. The lattice parameter are a = b = 4.67Å, c = 3.19 Å respectively with unit cell volume 69.57$Å^3$. The lattice parameters were not change by Mg doping because Sn^{+4} have the close radii to the Mg^{+2}, might be due to the substitution of metal ions in the crystallite structure and therefore no structure strain is generated. These results comply with the standard SnO_2 XRD pattern of the JCPDS line [18]. The $Mg_xSn_{1-x}O_2$ XRD spectrum shows in figure 2(b) that the thin film retain the tetragonal structure but peaks slightly shifted towards the lower angle, this is due to the substitution of Sn^{4+} ions at the lattice sites with the Mg^{2+} ion.

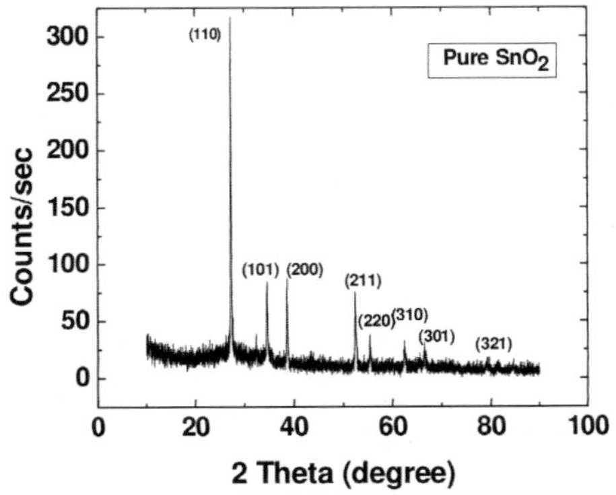

Figure 2 (a). XRD spectra for pure SnO_2 thin film.

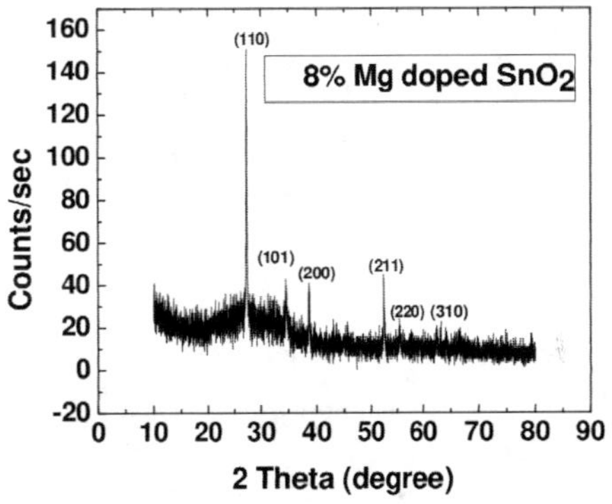

Figure 2(b). XRD spectrum of $Mg_xSn_{1-x}O_2$ thin film.

For calculate the crystallite size of the deposited thin films, the Debye Scherer formula was employed on the XRD spectra which is given by

$$D = \frac{0.9\lambda}{\beta \cos\theta}$$

Where λ is the wave length of the X-rays employed which in this case is 0.1548 nm for Cu-K_{α}, 0.9 is the correction factor in the Debye Scherer equation, β is the FWHM (full width and half maxima) in radian and θ is the Bragg's angle in degree. The calculated crystalline size from the Scherer formula is in the range of 32 nm for the deposited thin films.

The depth profile study and the elemental concentration of the deposited $Mg_xSn_{1-x}O_2$ thin films by AACVD have been examined by Rutherford Back scattering analysis. Figures 3(a), (b), (c) and (d) show the simulation and experimental data of the thin films with the energy of the incident He^{4+} beam was 2.084MeV. The counts come from the Mg in the thin films overlap with the substrate Si because of minor difference in their atomic weight. The O, Mg and Sn are simulated at the channel energies corresponding to 0.82, 1.05 and 1.69MeV respectively. The elemental composition of the thin films and their thickness examined by RBS simulated spectra are presented in Table 1.

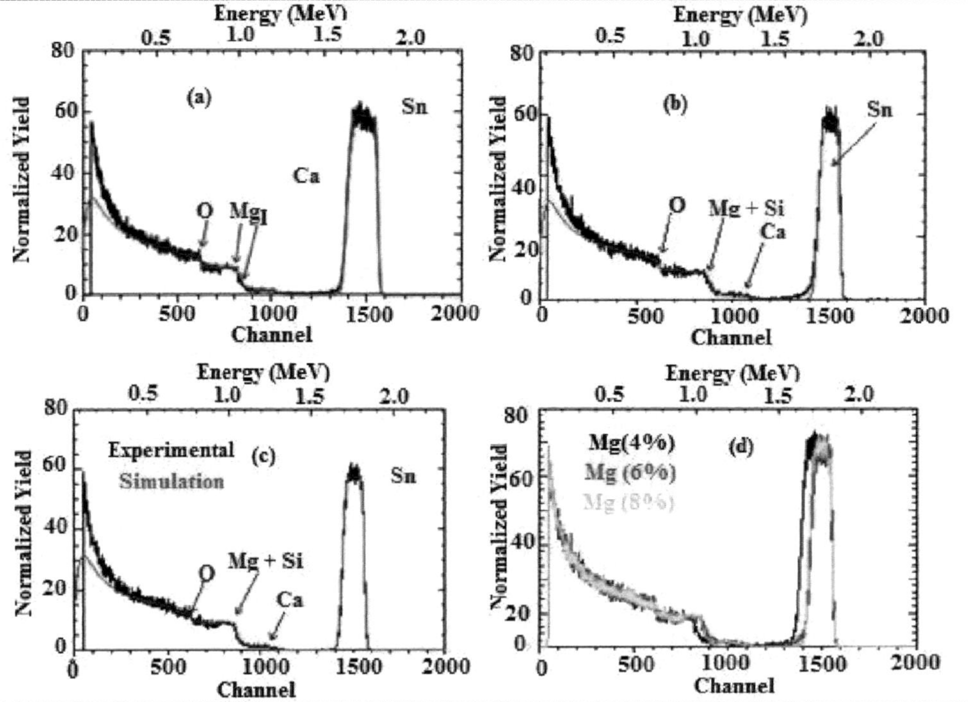

Figure 3. RBS and depth profile spectra of $Mg_xSn_{1-x}O_2$ thin films with increasing concentration of dopant shown in (a), (b), (c) and (d), respectively.

FTIR spectra of $Mg_xSn_{1-x}O_2$ thin films are shown in fig.4. The stretching/vibrational peaks corresponding to 595, 760 and 850 cm^{-1} of deposited pure SnO_2 thin films. The peaks assigned at 595cm^{-1} to O-Sn-O and 760 cm^{-1} to Sn-O-Sn stretching vibration, while the peak at 850 cm^{-1} is due to the phonon vibration. These peaks shifted to 620,775 and 880 cm^{-1}in case of Mg doping. B. Zhang et al [17] reported the peaks at 468 and 619 cm^{-1} which assign to O-Sn-O and Sn-O. K.K. Purushothaman *et. al.* [19] reported the band at 437, 536,586 and 642 cm^{-1} belongs to the vibration of SnO_2. The shape of the FTIR and position of the peaks have been shown to vary with synthesis routes and particle size [20].

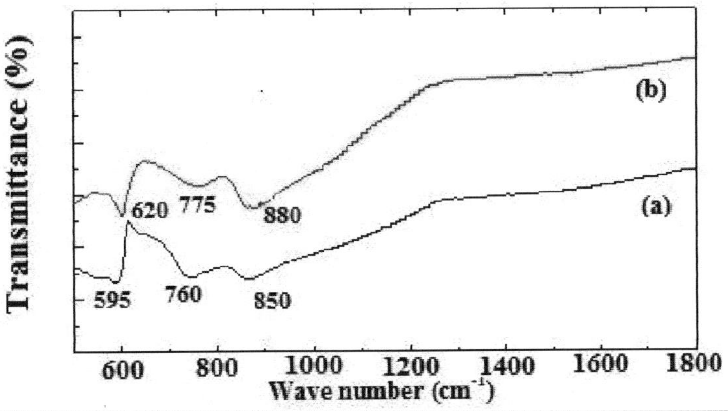

Figure 4. Typical FT-IR spectra of Pure SnO_2 and doped $Mg_xSn_{1-x}O_2$ thin films.

Atomic force microscopic analysis was performed to determine the grain size of the deposited $Mg_xSn_{1-x}O_2$ thin films. For an optical surface, roughness is normally considered as an important parameter. Surface roughness not only the light scattering but also give an idea about the quality of the surface under investigation, in addition to providing some insight on the growth morphology. A systematic description of various analytical method used for roughness characterization can be found in ref [21]. Root mean square roughness (R_{rms}) which is defined as standard deviation of the surface height profile from the mean height, is the most commonly reported measurement of the surface roughness and is given by,

$$R_{rms} = \left(\frac{1}{N} \sum_{i}^{n} \left(h_i - \langle h \rangle \right)^2 \right)$$

Where N is the number of pixels in the image h_i is the height of ith pixel and <h> is the mean height. Two dimension AFM images of the $Mg_xSn_{1-x}O_2$ as shown in the figure 5(a) and 5(b). The value of R_{rms} decreases (54 – 23nm) with increasing the Mg doping.

The transmission spectra of $Mg_xSn_{1-x}O_2$ were recorded in UV-visible near the infrared region (300-850nm). Figure 6 shows the variation of transmittance with the wave length for $Mg_xSn_{1-x}O_2$ thin films. The average transmission in the visible region was found to range of 55 to 78% depend on the Mg concentration with the increase in transmission was observed.

Journal of Physics: Conference Series **439** (2013) 012013 doi:10.1088/1742-6596/439/1/012013

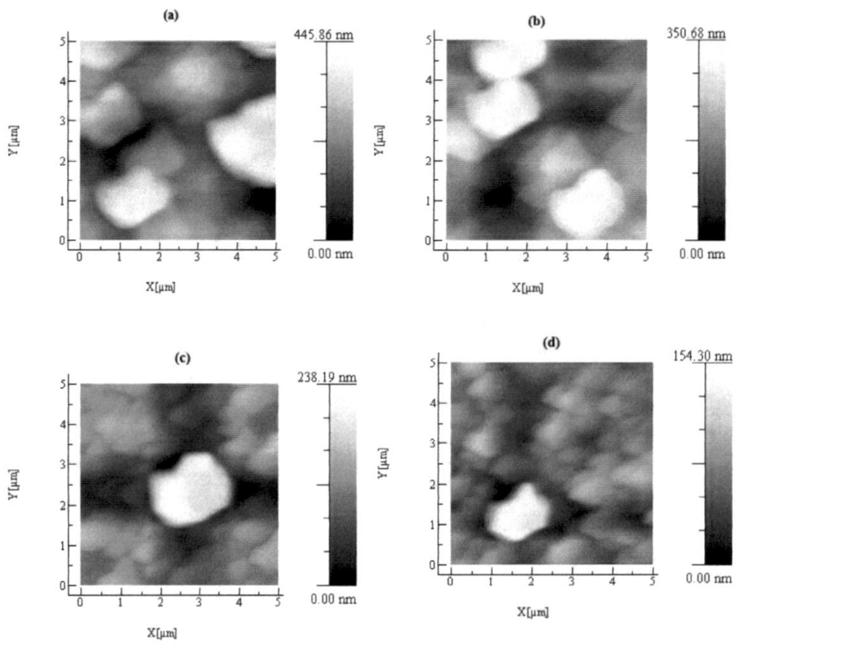

Figure 5. Surface morphology of pure and Mg doped TiO2 thin films. The surface roughness increase with the increase of Mg concentration confirmed by AFM images(5µm x 5µm)

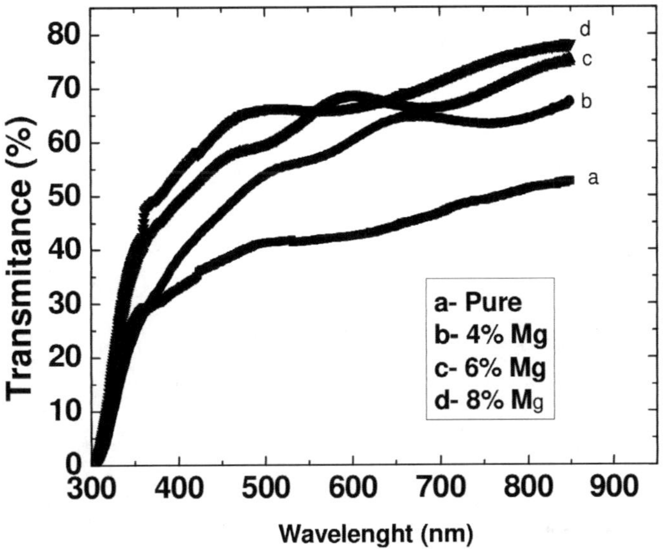

Figure 6. Transmission plot of pure and Mg doped SnO_2 thin films with increasing dopand concentrations.

The absorption coefficient (α) was calculated from the transmission spectra using equation [12],

$$\alpha = \frac{1}{t}\ln\left(\frac{1}{T}\right)$$

Where T is the optical transmission and t is the film thickness. The direct band gap of thin films were calculated from figures 7(a), (b), (c) and (d) respectively using the formula [22],

$$\alpha h\nu = A\left(h\nu - E_g\right)^m$$

Where α is the absorption coefficient, h is the Planck's constant, ν is the frequency of incident light, E_g is the energy band gap of material and m is the factor governing the direct/indirect, etc. transition of electron from the valance band to the conduction band.

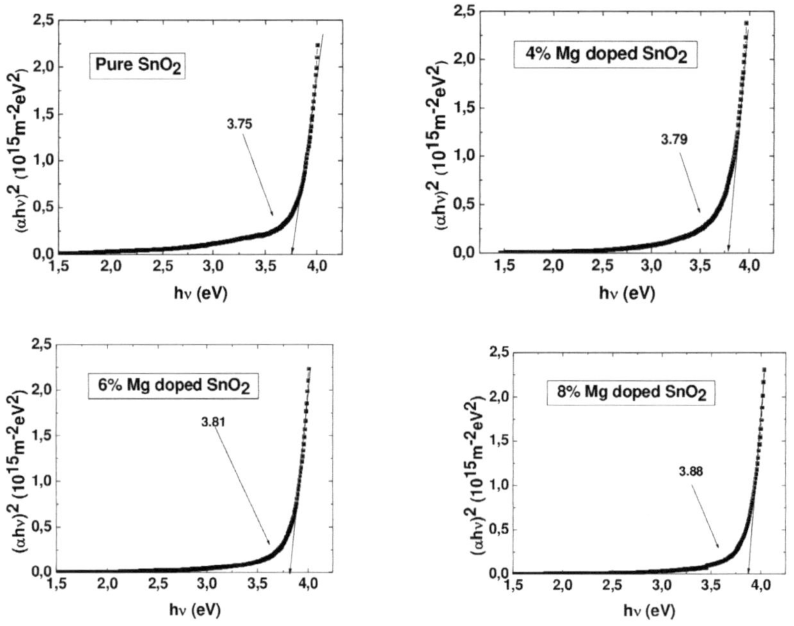

Figure 7. Tauc plot of pure and Mg doped SnO_2 for estimation of optical band gap.

It was found that the band gap of pure SnO_2 that the band gapes of un-doped SnO_2 is 3.76eV which is larger to the typical value quoted for bulk SnO_2 (3.6eV). It is related with the oxygen vacancies and the structural defects. The value of band gape decreases with increase in the Mg doping for low concentration and increase at doping concentration (i.e. 8 wt %) as shown in figure 8.

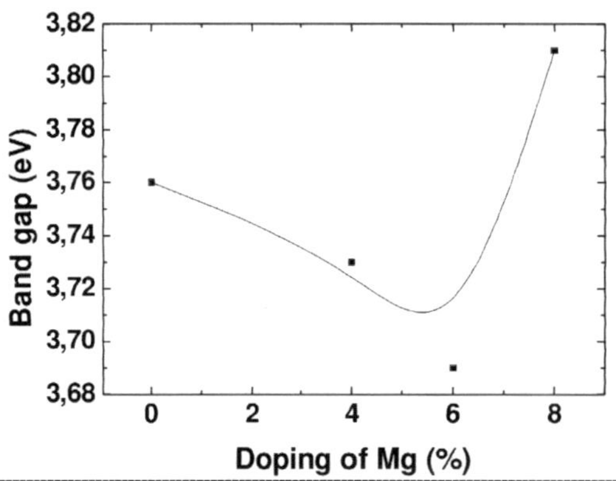

Figure 8. Variation of band gape of thin films with Mg doping.

The increasing of the band gape as usually observed for oxide semiconductor doped with donors, and reduces if doped with accepters. The decreasing in the band gap was attributing to the presence of holes in $Mg_xSn_{1-x}O_2$. Each substitution Mg atom will create two holes in O 2p state because of the lower-valence than Sn^{+4}. They form the impurity band inside the forbidden gap region. The electrons in the valance band are exited, and they can be combining with the holes in the impurity band first. However they are not stable. The trapped electrons will recombine with the hole in the conduction band, resulting the shrinking of the band gap. At the 8% doping of Mg sample has a larger band gap than the 6% Mg doping as shown in Figure 8. It is due to the Mg interstitials because of the limiting solubility of Mg in SnO_2.

Table 1. Relative concentration and thickness of thin films by RBS.

Sample	**Thickness (nm)**	**Stiochiometry**
Pure SnO_2	350	Sn_1O_2
4% Mg/SnO_2	424	$Sn_{1.03}O_2Mg_{0.13}$
6% Mg/SnO_2	300	$Sn_{1.03}O_2Mg_{0.18}$
8% Mg/SnO_2	320	$Sn_{1.08}O_2Mg_{0.23}$

4. Conclusions

$Mg_xSn_{1-x}O_2$ (x = 0-0.08) thin films have been prepared on a glass substrate by aero-sole assisted chemical vapor deposition. The tetragonal structure of pure and Mg doped SnO_2 was conformed and lattice parameters and volume of unit cell were calculated by XRD. The thickness of the deposited thin films are in the range of 300nm, where as the elemental composition of the Mg and Sn in the thin films are in accordance with the proposed composition of the deposited material, confirmed by the RBS and XRD analysis. The transmission was increased from 55 to 78% and energy band gap (3.67 to 3.69eV) was decreased by increasing the Mg (0 to6%) doping. The refractive index (1.61 to 1.53) is increased as the Mg concentration increase. The Mg doped SnO_2 transparent conducting oxide is very useful for window layer of solar cell due to high reflectance of infrared radiation and transparent of visible light.

Acknowledgment

The authors are thankful to the National Center for Physics and The Islamia University of Bahawalpur for the research support.

References

[1] Jin Z C , Hamberg J, and Granqvist. C G 1988 Opticalproerties of sputter-deposition ZnO:Al thin films *J. Appl Phys* **64**, 5117.
[2] Rusu R S and Russia G I 2005 On the electrical and optical characterization of CdO thin films *J.Optoelectron. Adv. Mater* **7**, 823.
[3] Penza M, Cozzi S, Taglient M A, Mirenghi L, Martucci C and Quirini A1999 Characterization of transparent and conductive electrodes of indium tin oxide thin films by sequential reactive evaporation. *Thin Solid Films* 71, 349.
[4] Ishibashi S, Higuchi Y, Ota Y and Nakamura K 1998 Low resistivity indium–tin oxide transparent

conductive films. I. Effect of introducing H_2O gas or H_2 gas during direct current magnetron sputtering *J. Vac. Sci. Techno* **A8**, 1403.

[5] Khan A F , Mehmood M, Rana A M, Bhati M T, Mahmood A 1991 Optical Characterization of rf- Magnetron Sputtered Nanostructured SnO_2 Thin Films *China. Phys. Lett.* **7** 77803 (2009)

[6] Cavicchi R and Semancik S. The growth of thin, epitaxial SnO_2 films for gas sensing application *Thin Solid Films* **81** 206.

[7] Jiang J C, Lian K and Meletis E I 2002 Influence of oxygen plasma treatment on the microstructure of SnOx thin films *Thin Solid Films* **411** 203

[8] Wang S, Huang J, Zhao Y, Wang S, Wu S, Zhang S and Huang W 2006 Nanostructure SnO_2 and supported Au catalysts: Synthesis, characterization, and catalytic oxidation of CO *Mater. Lett.* **60** 1706.

[9] Wan C F, McGrath R D, Keenan W F and Frank S N 1989 LPCVD of Tin Oxide from Tetramethyltin and Oxygen *J. Electrochem. Soc.* **136** 1459

[10] Seigel R W, Hu E H and Roco 1997 M C WTEC panel Report on R & D Status and Trend in Nanoparticales, *Nanostructured Material, Nanodevices*. Workshop

[11] Szezuko D, Werner J, Oswald S, Behr G Wetzing K 2001 XPS investigation of surface segregation of doping in SnO_2 *Appl. Surf Sci* **179** 301.

[12] Roy S S and Podder Gilberto J 2010 Synthesis and optical characterization of pure and Cu doped SnO_2 thin film deposited by spray pyrolysis *J. Optoelectron and Adv. Mater* **12** 1479.

[13] Ricardo H R Castro, Pereira Gilberto J, Douglas Gouvea 2007 Surface modification of SnO_2 nanoparticles containing Mg or Fe:Effects on sintering *Appl. Surf. Sci.* **253**, 4581

[14] Yuhua Xiao, Shihui Ge, Li Xi, Yalu Zhou 2008 Room temperature ferromagnetism of Mn-doped SnO_2 thin films fabricated by sol–gel method *Appl. Surf. Sci.* **254** 7459

[15] Rajpure K Y, Kusumade M N, Neumann-Spallart Michael N, Bhosale C H 2000 Effect of Sb doping on properties of conductive spray deposited SnO_2 thin films *Mat. Chem. And Phy.* **64**, 184

[16] Boben Thomas, Skariah Benoy, Radha K K 2008 Influence of Cs doping in spray deposited SnO_2 thin films for LPG sensors *Sens. Actuators* **133B** 404

[17] Zhang B, Tian Y, Zhang J X and Cai W 2011 Fourier transformation infrared spectrum studies on the role of fluorine in SnO_2:F films *Appl. Physics Lett.* **98** 021906

[18] Mc Carthy, Welton G,1989 "X-ray differaction data for SnO_2. An illustration of new powderdata evaluation method" *J. Powder Differaction* **4** 156

[19] Purushothaman K K, Dhanashankar M, Muralidharan G 2009 Preparation and characterization of F doped SnO2 films and electrochromic properties of FTO/NIO films *Current Appl. Physics* **9** 67

[20] Ristic M, Ivanda M, Popovic S Music S 2002 Dependence of nanocrystalline SnO_2 particle Size on synthesis route. *J. Non-Crystall. Solids* **303** 270

[21] Bennett J M, Mattson L 1989 Introduction to Surface Roughness and Scattering. *Optical Society of America. Washington DC*

[22] Majumder S, Hussain S, Chaudhuri R Pal A K 1991 Sn-doped Zinc Oxide thin films for LPG sensors *Thin films* **204L** 255

Journal of Physics: Conference Series **439** (2013) 012014 doi:10.1088/1742-6596/439/1/012014

Frequency dependent dielectric measurements of Cd^{2+} doped Mn-Zn nano ferrites prepared by sol gel and coprecipitation methods

A Humaira[1] and M Asghari [1*]

[1]Thermal Transport Laboratory, Department of Materials Engineering, School of Chemical and Materials Engineering (SCME), National University of Sciences and Technology, H-12, Islamabad.

[*]*E-mail: tpl.qau@usa.net*

Abstract

Ferrite powders of Mn-Zn doped with Cd^{2+} ($Mn_{0.5}Cd_xZn_{0.5-x}Fe_2O_4$) were prepared by the sol gel auto- combustion and coprecipitation methods. The effect of Cd^{2+} doping on the crystalline phase formation was investigated by X-ray diffraction which confirms the spinel structure of prepared samples. Crystallite size found to be lower for samples prepared by coprecipitation method. The samples prepared using coprecipitation, found to be denser as compared to sol gel method. SEM analysis confirmed the nano structure of the prepared samples. Dielectric properties were studied using impedance analyzer (20 Hz – 5 MHz) and RF material analyzer (1 MHz - 3 GHz) respectively. DC electrical measurements showed highly resistive nature of the prepared samples and values are high for samples prepared from sol gel method. Impedance spectroscopic analysis was carried out in the frequency range from 20 Hz to 5 MHz for the prepared samples, showing resistance of grain boundary is higher than grains. Impedance plane plots gave information about the resistive and reactive parts of impedance. The presence of semicircular arcs at lower frequency side, confirms that the high resistance of samples is due to increase in the number of grain boundaries in the nano synthesised samples.

Key words: Nano structure; Sol gel combustion; Coprecipitation; X-ray diffraction; Dielectric constant

1. Introduction

Nano scale polycrystalline ferrites with unique electrical and magnetic properties are of great interest for scientist and engineers [1-3]. Even after more than half of the century the scientist, researchers, technologist, and engineers are still excited in various types of nanocrystalline as well as bulk ferrite materials. Transition metal oxides (MFe_2O_4) are magnetic materials with cubic spinel structure and have been used in various electronic and electrical applications for the last decade. The high permeability in the radio frequency region, make them suitable for use in electronic devices [4]. The Mn-Zn ferrites are soft ferrite having low magnetic coercivity and high electrical resistivity. The high electrical resistivity and good magnetic properties make it excellent core material for use in electronic and telecommunication applications [5]. The dielectric behavior of polycrystalline soft ferrites is very important to understand the polarization mechanism. The electrical conductivity and dielectric properties greatly depends on the method of preparation, sintering conditions and doping of cations [6]. The change in method of preparation and chemical composition alters the physical properties for use in a particular application. The nano phases obtained using soft chemical methods such as sol gel auto-combustion or coprecipitation methods have an edge over conventional ceramic route with improved properties [6]. Ferrites are very effective as microwave absorbers due to their low cost, low eddy loses and high resistivity. The addition of non magnetic ions (i.e. Zn^{2+}, Cd^{2+}) in small amount results in the modification of saturation magnetization [5]. Batoo *et al.* [7] have investigated the influence of Al doping on the structural and electrical properties of Ni–Cd mixed ferrite. Binu *et al.* [8] have studied structural and magnetic properties of Gd^{3+} doped Ni-Cd mixed ferrites. Nikumbh *et al.* [9] have studied the structural, electrical and magnetic properties and cation distribution of cadmium-substituted nickel ferrite. Very few studies are reported on the doping of Cd^{2+} ions in manganese and zinc ferrites [10]. In the present work comparison of Cd^{2+} doping on the Mn-Zn mixed ferrites has been presented over the wide range of frequency. The change in various physical parameters and their characteristic properties that affect the performance of these materials can be known and explained using complex impedance spectroscopy. In the present paper we have studied the effect of preparation technique on the structural, electrical and impedance properties of $Mn_{0.5}Zn_{0.5-x}Cd_xFe_2O_4$ (x = 0, 0.15, 0.25, 0.35) soft ferrites as a function of composition and frequency at room temperature.

2. Experimental details

2.1. Sample preparation

A series of nano crystalline samples with general formula $Mn_{0.5}Zn_{0.5-x}Cd_xFe_2O_4$ (x = 0, 0.15, 0.25, 0.35) were synthesized using sol gel auto-combustion and coprecipitation methods [5, 8]. The starting materials, for sol-gel auto combustion method, were of analytical grade $Mn(NO_3)_2.4H_2O$, $Zn(NO_3)_2.H_2O$, $Cd(NO_3)_2.4H_2O$, $Fe_2(NO_3)_3.9H_2O$ and $C_6H_8O_7.H_2O$ (99.9%). Metal nitrates and citric acid were dissolved in deionized water in 1: 1.5 molar ratios at room temperature. The pH of solution was adjusted to 7 using ammonia solution. The solution was allowed to evaporate and transform into gel on heating at 363 K. Further heating allows the gel to burn in a self ignition manner and form the fluffy loose powder on complete combustion followed by heating at 473 K for 4 h to remove organic materials. The as-burnt precursor powder was ground well in a pestle mortar and then sintered at 1073 K for 8 h. The sintered powders were then pelletized in circular disks of 10 mm diameter by applying a uniaxial load of 5 kN.

Nanocrystalline $Mn_{0.5}Zn_{0.5-x}Cd_xFe_2O_4$ (x = 0, 0.15, 0.25, 0.35) particles were synthesized by chemical coprecipitation method [11, 12] using AR grade $Mn(NO_3)_2.4H_2O$, $Zn(NO_3)_2.H_2O$, $Cd(NO_3)_2.4H_2O$ and $Fe_2(NO_3)_3.9H_2O$. For synthesis x = 0 composition, 0.05molar solutions of $Mn(NO_3)_2$. $4H_2O$, 0.05 molar $Zn(NO_3)_2.H_2O$ and 0.2 molar $Fe(NO_3)_2.9H_2O$ were mixed in their stoichiometric ratio and homogenized at 358 K. The 2 M solution of NaOH was added to adjust the pH of the solution between 10.5 -11.5. The mixture was then heated at 363 K for about 1h. The precipitates were thoroughly washed with distilled water until the washings were free from sodium and chloride ions. The product was dried in an electric oven at 378 K for overnight to remove water contents. Ferrite powders were heated at 1073 K for 8 h. These were then pelletized in circular disks of 10mm diameter by applying a uniaxial load of 5kN using hydraulic press. The samples prepared from both techniques were then characterized through various techniques

2.2 Characterizations

The crystalline phases of the prepared samples were determined using the X-ray diffraction (XRD) technique. XRD patterns were taken using Cu Kα (λ = 1.5406 Å) radiation at room temperature. Scanning was performed from 20° to 80° at a step size of $0.02^{o}s^{-1}$. The crystallite size (t_{311}), lattice parameter (a), X-ray density (ρ_m) and porosity (P) were calculated using simple formulae [13]. The surface morphology was performed using a scanning electron microscopy (JEOL-instrument JSM-6490A). The DC electrical resistivity was obtained by a simple two-probe method within temperature range 300 – 573 K. The relation between electrical resistivity (ρ) and temperature [13] may be expressed as

$$\rho = \rho_o exp\ (\Delta E/k_BT) \qquad (1)$$

where ΔE is the activation energy in (eV) for conduction, k_B is the Boltzmann constant and T is the temperature in Kelvin (K).

Dielectric and impedance spectroscopic measurements were carried out using WAYNE KERR LCR METER (6440B) in the frequency range from 20 Hz to 5 MHz at room temperature. The samples were pressed into circular disc shaped pellets, and placed in a specially designed sample holder. The real part of the dielectric constant (ε') was calculated by the relation [13, 7]

$$\varepsilon' = C_p d/\varepsilon_o A \qquad (2)$$

where C_p is the parallel plate capacitance in Farad, d the thickness of the pellets in m, A the cross-sectional area of a flat surface of the pellet sample in m^2 and ε_0 is the permittivity of free space, equal to 8.854×10^{-12} F/m. The dielectric loss tangent (tan δ) which is the loss of energy from the applied field into the sample in the form of heat, can be determined in terms of real and imaginary parts of dielectric constant as

$$tan\delta = D = \varepsilon'' / \varepsilon' \quad (3)$$

Impedance measurements were performed in the frequency range from 20 Hz to 5 MHz at room temperature by taking the absolute value of impedance $|Z|$ with varying complex angle θ. The real and imaginary parts of impedance can be written as:

$$Z' = |Z| cos\theta \quad (4)$$
$$Z'' = |Z| sin\theta \quad (5)$$

The variation in dielectric properties in the frequency range of 1 MHz to 1 GHz were also studied using RF material analyzer.

3. Results and discussion

3.1 Structural properties

The X-ray diffraction pattern of nano ferrite samples with composition $Mn_{0.5}Zn_{0.5-x}Cd_xFe_2O_4$ (x = 0, 0.15, 0.25, 0.35) are shown in Figure 1.

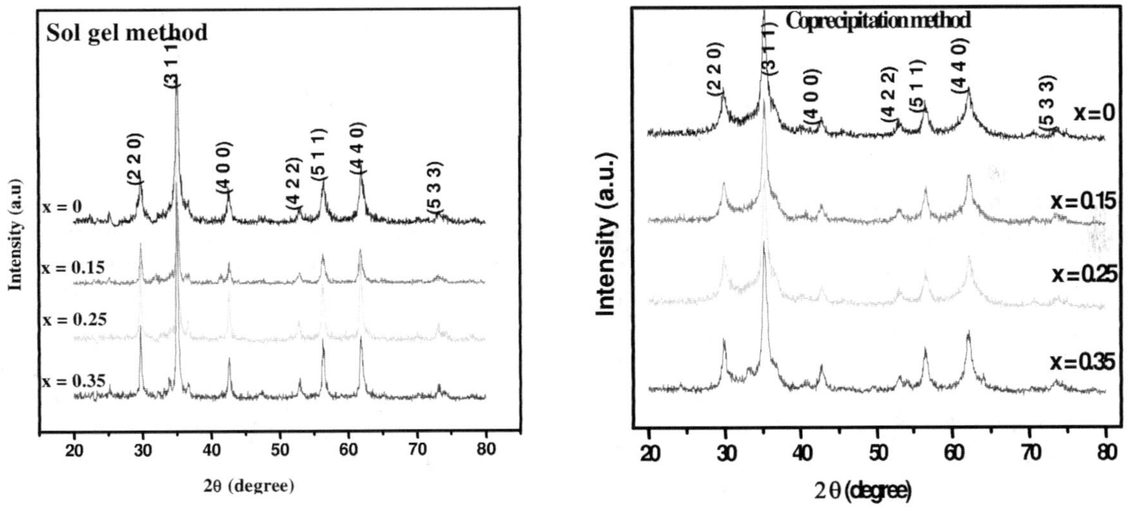

Figure. 1 Indexed X-ray diffraction pattern for the $Mn_{0.5}Zn_{0.5-x}Cd_xFe_2O_4$ nano ferrites.

The XRD data were analyzed by X' Pert High Score software which shows that all the samples exhibit single-phase and polycrystalline FCC spinel structure. Absence of any extra peak implies that Cd^{2+} ions have been dissolved in to complete spinel structure. The lattice constants were determined using data obtained from XRD patterns. The slight increase in lattice constant is observed with increase in Cd^{2+} concentration which is due to the replacement of Zn^{2+} ion (0.83Å) of small ionic radius with Cd^{2+}(0.97 Å) ion of slightly large ionic radius. The results are in good agreement with the reports in the literature [6]. The values of X-ray density (theoretical density) and apparent density (experimental density) of the samples are listed in Table 1.

Table 1 The average crystallite size (t (ave)), lattice constant (a), X-ray density (D_x), physical density (D_m), DC electrical resistivity (ρ_{dc}), and activation energy (ΔE) of the $Mn_{0.5}Zn_{0.5-x}Cd_xFe_2O_4$ nano ferrites with varying Cd^{2+} ions concentrations $0 \leq x \leq 0.35$ prepared by sol gel and coprecipitation methods respectively.

Composition	***x = 0***	***x = 0.15***	***x = 0.25***	***x = 0.35***
t (average) nm	24	31	40	50
a(Å)	8.479	8.488	8.497	8.488
Dx(g-cm^{-3})	5.15	5.28	5.37	5.49
Dm(g-cm^{-3})	3.31	3.35	3.35	3.30
P (fraction)	0.36	0.37	0.38	0.40
ρ(Ω-cm)at 373 K	5.1 x 10^{10}	5.8 x 10^{10}	2.2 x 10^{11}	3.3 x 10^{10}
ΔE (eV)	0.84	0.70	0.98	0.90

Coprecipitation	***x = 0***	***x = 0.15***	***x = 0. 25***	***x = 0.35***
t (average) nm	14	17	19	22
a(Å)	8.415	8.421	8.438	8.437
D_x(g-cm^{-3})	5.26	5.41	5.48	5.59
D_m(g-cm^{-3})	3.28	3.44	3.60	3.50
P (fraction)	0.37	0.36	0.34	0.37
ρ(Ω-cm)at 373 K	7.13 x 10^{8}	5.87 x 10^{9}	1.44 x 10^{10}	3.08 x 10^{9}
ΔE (eV)	0.71	0.55	0.58	0.70

X-ray density increased with addition of Cd^{2+} ions concentration. This may be due to the fact that atomic weight and density of Cd atoms are 112.4 amu and 8.65gm/cm^3 respectively, which are greater than those of Zn atoms (65.4 amu and 7.14 gm/cm^3). The apparent density is less than X-ray density due to presence of pores which depends on the sintering conditions [14]. The porosity increases because X-ray density increased and measured density decreased with the addition of Cd^{2+} ions.

The morphology of the samples prepared by both techniques was examined by scanning electron microscopy (SEM). The microstructural analysis gives information about the type and size of grain growth in the samples, which influence the physical and electrical properties of materials. The SEM images are shown in Fig. 2. The nano size nature of the prepared samples is clear from the micrographs and agglomeration of particles gives the magnetic nature of prepared samples. The comparison of micrographs for x = 0.25 prepared by both methods indicates that samples prepared by coprecipitation are more uniform, homogenous and smaller in size as compared to sol gel method.

Figure 2. SEM micrographs for the $Mn_{0.5}Zn_{0.5-x}Cd_xFe_2O_4$ (x = 0.25) nano ferrites prepared by (a) sol gel (b) coprecipitation method.

3.2. *DC electrical properties*

The DC electrical resistivity of ferrites depends on grain size, chemical composition, density, porosity, sintering conditions etc [15, 16]. The variation in DC electrical resistivity as a function of temperature is shown in Fig. 2. The decrease in resistivity with temperature confirms the semiconducting nature of ferrites. The values of resistivity varies from 3.3 x 10^{10} to 2.2 x 10^{11} (ohm-cm) and 7.13 x 10^{8} to 1.44 x 10^{10} at 373 K for sol gel and coprecipitation method respectively. These high values of resistivity at room temperature show that ferrites are insulators at room temperature. The variation in resistivity with Cd content is consistent with the results reported by Venkataraju *et al* [10]. According to Verwey and de Boer [17, 18] electron exchange take place between the ions of same element present in more than one valence state and distributed randomly over crystallographic lattice sites.

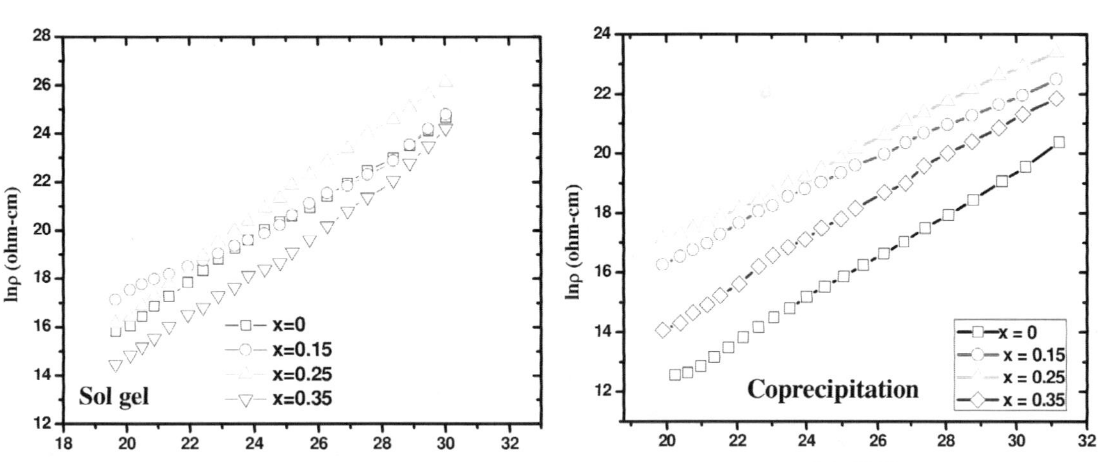

Figure 2. Change in DC electrical resistivity with temperature for $Mn_{0.5}Zn_{0.5-x}Cd_xFe_2O_4$ nano ferrites.

The decrease in DC electrical resistivity with increase in temperature is due to increased drift mobility of the charge carriers. The electrons hopping energy in ferrites is responsible for conduction mechanism depends on the activation energy and is associated with electrical energy barrier during electron hopping [19]. The activation energies are calculated from slope of resistivity plots. The value of activation energy increased with the addition of Cd^{2+} ions for sol gel samples which suggests that the conduction mechanism in the present ferrite system is due to hopping of polarons.

3.3 Dielectric properties

3.3.1 Dielectric constant

The dielectric constant is represented in terms of real and imaginary parts as

$$\varepsilon = \varepsilon' - j\,\varepsilon'' \tag{6}$$

The real part of dielectric constant gives the stored energy and imaginary part gives the dissipation of energy in applied ac field. The variation in dielectric constant (ε') with applied field of 1V/m in the frequency range from 20 Hz to 5 MHz, for all compositions is shown in Fig.
3. The dielectric constant decreases with increase in frequency for all compositions. This decrease is rapid at lower frequency region and slows down at high frequency.

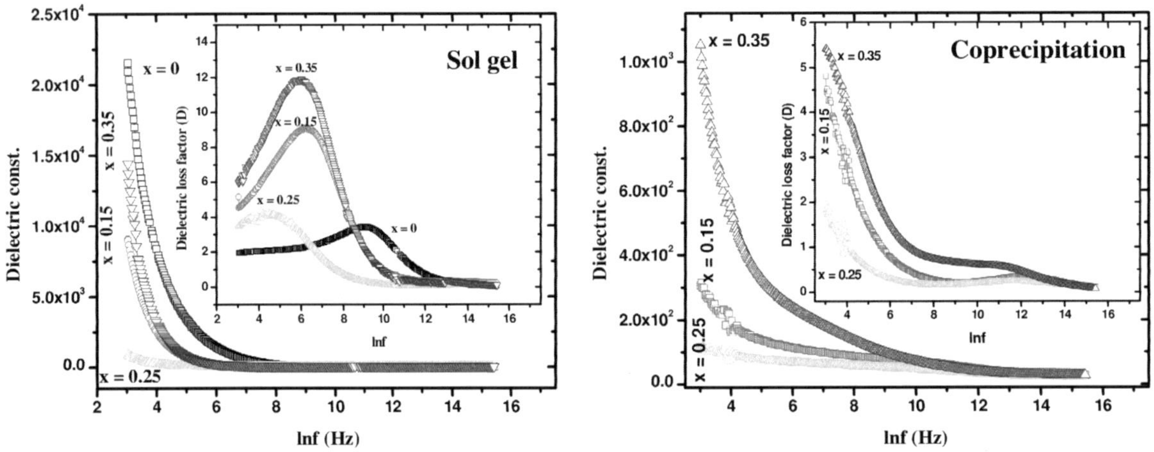

Fig. 3 Variation in (a) dielectric constant and (inset) dielectric loss factor (*D*) with frequency up to 5 MHz for $Mn_{0.5}Zn_{0.5-x}Cd_xFe_2O_4$ nano ferrites.

The dispersion in dielectric constant is due to Maxwell-Wagner [20] type interfacial polarization, which agrees with the Koop's phenomenological theory. The large values of ε' at lower frequency are due to predominance of species like Co^{2+}, Fe^{2+} ions, oxygen vacancies, grain boundary defects, interfacial distribution, etc. [6, 21]. The process of polarization in ferrites is similar to the conduction mechanism [22]. The transfer of electron between Fe^{2+} (ferrous ions) and Fe^{3+} (ferric ions) and holes between Mn^{3+} and Mn^{2+} ions leads the displacement of electrons in the direction of applied field and hence cause the polarization. This type of behavior is reported earlier for Mn-Zn-Cu [23] and Ni-Co [15] systems. At high frequency polarization decreases and reaches a constant value due to the fact that after certain frequency of the applied field the electron hopping does not follow the alternating field. The slight decrease in the dielectric values between 1 MHz to 1 GHz is represented in Fig. 4. Table 2 shows that the value of dielectric constant is minimum for x = 0.25 concentration of Cd^{2+} ions in both types of samples.

3.3.2 Dielectric Loss tangent

The energy dissipation in the dielectric system is represented by the dielectric loss factor (*D*) or loss tangent (*tanδ*). The domain wall resonance is considered to be the cause of this loss. These losses are low at high frequency due to hindrance in the domain wall rotation. The dielectric loss factor is proportional to the imaginary part of dielectric constant according to Eq. (3). The variation in dielectric loss factor (*D*) with applied frequency at room temperature for $Mn_{0.5}Zn_{0.5-x}Cd_xFe_2O_4$ (x = 0, 0.15, 0.25, 0.35) composition is shown in the inset of Fig. 3. The dielectric loss factor increased with frequency up to a maximum value and then starts decreasing with increase in frequency. The appearance of these loss peaks for the sol-gel prepared samples shows the relaxation process occurred with the change in frequency in the polycrystalline Mn-Zn soft ferrites. The samples prepared by coprecipitation method gives no such peak.

Table 2 Dielectric constant (ε') and dielectric loss factor (*D*)of the $Mn_{0.5}Zn_{0.5-x}Cd_xFe_2O_4$ nano ferrites with varying Cd^{2+} ions concentrations $0 \leq x \leq 0.35$ prepared by sol gel and coprecipitation methods.

Composition	*x = 0*	*x = 0.15*	*x = 0.25*	*x = 0.35*
ε′ at 1k Hz	620	98	33	116
ε′ at 1M Hz	20	24	21	32
ε′ at 1G Hz	8.61	11.81	4.06	7.22
tanδ at 1k Hz	2.50	8.38	1.79	9.94
tanδ at 1 MHz	0.17	0.30	0.12	0.26
tanδ at 1 GHz	0.14	0.027	0.022	0.018
Coprecipitation	*x = 0*	*x = 0.15*	*x = 0.25*	*x = 0.35*
ε′ at 1kHz	24962	101	64	188
ε′ at 1MHz	904	33	27	30
ε′ at 1GHz	50	7.68	10.36	8.33
tanδ at 1 kHz	1.53	0.46	0.21	0.98
tanδ at 1 MHz	1.9	0.24	0.17	0.19
tanδ at 1 GHz	0.45	0.01	0.009	0.006

The peak formed when the jumping frequency of charge carriers approximately becomes equal to external applied ac field [14]. The relaxation phenomenon also occurred in 1 MHz to 1 GHz range for sol-gel prepared samples (see inset of Fig. 3). The dielectric loss is also minimum for the concentration of Cd^{2+} at x = 0.25 given in Table 2 (1.79 at 1 kHz and 0.12 at 1MHz). It is known that Zn^{2+} ions occupy the tetrahedral A-site and Mn^{2+} and Fe^{3+} ions have tendency to occupy both A and B-sites [24]. It is reported that Cd^{2+} ions occupy A-site [25]. The substitution on of Cd^{2+} ions increases the resistivity of the Mn-Zn ferrites. The dielectric behavior is directly proportional to the square root of conductivity that is why both dielectric constant and loss tangent decreased with cadmium concentration. Similar results are reported by Hankare *et al.* [26] for Cr substituted Mn-Zn ferrites.

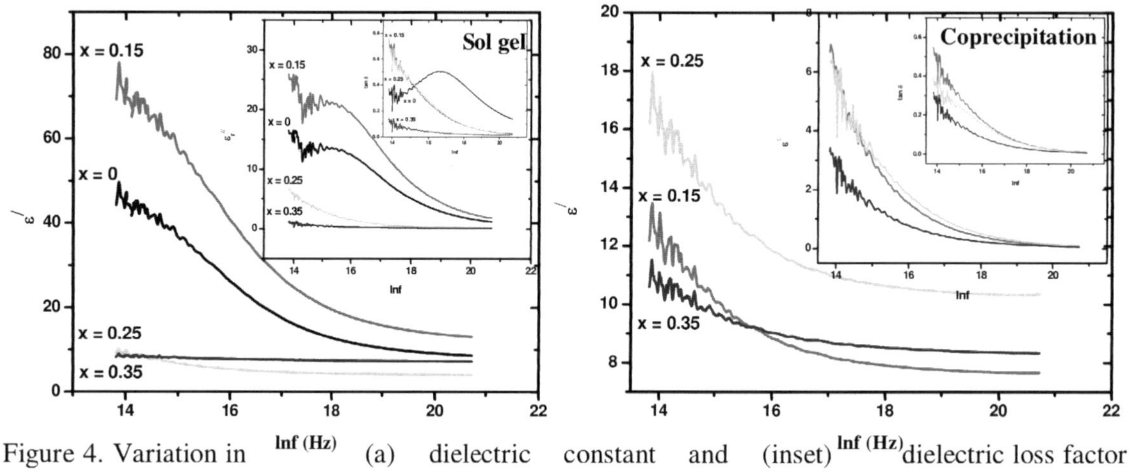

Figure 4. Variation in (a) dielectric constant and (inset) dielectric loss factor (*D*) with frequency up to 1GHz for $Mn_{0.5}Zn_{0.5-x}Cd_xFe_2O_4$ nano ferrites.

3.3.3 Impedance analysis

The impedance spectroscopy is widely used to characterize the electrical properties of materials and interfaces present in these materials. The impedance measurements data gives both resistive (real) and reactive (imaginary) components for a material. To separate the grain and grain boundary contributions of the Cd doped Mn-Zn ferrite samples, complex impedance plane plots (Cole-Cole plots) have been drawn in the frequency range from 20 Hz to 5 MHz at room temperature. It is clear from Fig. 5 that all the samples show single semicircle which is due to the conduction of the grains and one incomplete arc is due to grain boundary conduction, except for x = 0.35

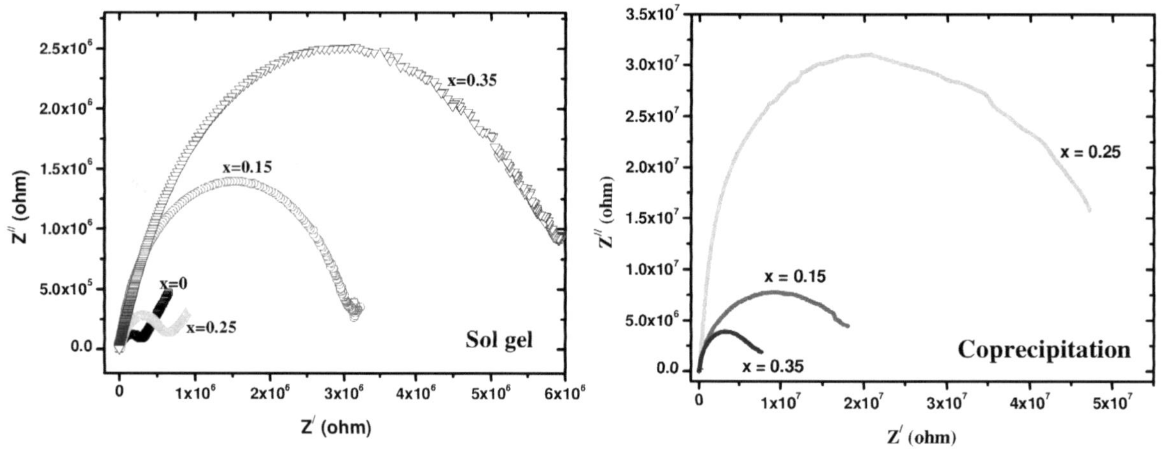

Fig. 5 Cole-Cole plots for $Mn_{0.5}Zn_{0.5-x}Cd_xFe_2O_4$ nano ferrites at room temperature.

The incomplete circular arc in low frequency region shows that grain boundary resistance is out of measurement scale or presence of some additional relaxation phenomena which occur outside the measured frequency range [27]. Grain resistance increases with Cd^{2+} concentration which is in agreement with the electrical measurements.

4. Conclusions

The Cd substituted Mn-Zn nano ferrites with composition $Mn_{0.5}Zn_{0.5-x}Cd_xFe_2O_4$ (x = 0, 0.15, 0.25, 0.35) have been prepared by the sol gel auto-combustion and coprecipitation methods. All the samples showed single phase FCC spinel structure. It is observed that the lattice constant, average crystallite size and XRD density increases with the Cd^{2+} concentration in both types of samples. The DC electrical resistivity increased to maximum value of 2.2 x 10^{11} (Ω-cm) (sol gel) and 1.44 x 10^{10} (Ω-cm) (coprecipitation) for x = 0.25 concentration of cadmium. The dielectric constant found to be minimum for x = 0.25 of Cd^{2+} doping in both types of samples. The presence of Debye peak in loss tangent vs frequency curves confirms the existence of single relaxation phenomena in the given frequency range for the sol gel prepared samples. The complex impedance plane plots show a single semicircle, which indicates the capacitive and resistive properties of the materials are due to contribution of grains and grain boundaries in ferrites. The substitution of Cd^{2+} plays an important role to modify the electrical properties of Mn-Zn ferrites required for power applications and microwave devices.

Acknowledgement
Authors are thankful to Higher Education Commission (HEC) Pakistan for supporting this research work through 5000 Indigenous Scholarships (Batch-V), PSF project no. 147 and HEC project no.1326.

References

[1] Lim I I S *et al* 2008 *Nanotechnology* **19** 305102
[2] Fontijn W F J *et al* 1997 *Phys. Rev. B* **56** 5432
[3] Philip J *et al* 2008 *Appl. Phys. Lett.* **92** 043108
[4] Li Q, Wang Y and Chang C. 2010 *J. Alloys Comp.* **505** 523
[5] Veena E G *et al* 2010 *Appl. Phys. A* **99** 497
[6] Farea A M M *et al* 2008 *J. Alloys Comp.* **464** 361
[7] Batoo K M, Kumar S, Lee C G and Alimudddin. 2009 *Curr.Appl. Phys.* **9** 826
[8] Jacob B P, Thankachan S**,** Xavier S and Mohammed E M. 2011 *Phys. Scr.* **84** 045702
[9] Nikumbh A K *et al* 2002 *J. Mater. Sci.* **37** 634
[10] C. Venkataraju, G.Sathishkumar and K.Sivakumar. 2011 *J. Magn.Magn. Mater.* **323** 1817
[11] Pant R P, Kumar V, Haldar S K, Gupta S K and Sukhvir S. 2007 *Int. J. Nanosci.* **6** 515
[12] Kumar V, Rana A, Yadav M S and Pant R P. 2008 *J. Magn. Magn. Mater.* **320** 1729
[13] Gul I H, Abbasi A Z, Amin F, Rehman M A, and Maqsood A. 2007 *J. Magn.Magn. Mater.* **311** 494
[14] Abdeen A M, Hemeda O M, Assem E E and El-Sehly M M. 2002 *J. Magn. Magn. Mater.* **238** 75
[15] Gul I H, Ahmed W and Maqsood A. 2008 *J. Magn. Magn. Mater.* **320** 270
[16] Magalaraja R V, Ananthakumar S, Manohar P and Gnanam F D. 2002 *J. Magn. Magn. Mater.* **253** 56
[17] Verwey E J W. 1947 *J. Chem. Phys.* **24** 174
[18] Verwey E J W and de Boer J H. 1936 *Recl. Trav. Chim. Phys- Bas* **55** 531
[19] Vijaya P, Bhasker R, Ramesh B and Gopal R C. 2010 *Physica B: Condens. Matter.* **405** 1852
[20] Koop's C G. 1951 *Phys. Rev.* **83** 121
[21] Maxwell J C. 1973 *Electricity and Magnetism,* Vol 1. Oxford University Press, New York,
[22] Rabinikin L T and Novikova Z I. 1960 *Ferrites, Acad. Naok, USSR, Minsk* 146
[23] Hankare P P. *et al* 2011 *J. Magn. Magn. Mater.* **323** 389
[24] Ajmal M and Maqsood A. 2008 *Mater. Lett.* **62** 2077
[25] Ravinder D. 2000 *J. Mater. Lett.* **43** 129
[26] Hankare P P, Patil R P, Garadkar K M, Sasikala R and Chougule B K 2011 *Mater.Res.Bullet.* **46** 447
[27] Dar M A, Batoo K M, Verma V, Siddiqui W A and Kotnala R K. 2010 *J. Alloys. Comp.* **493** 553

Influence of Rare Earth (Gd^{3+}) on Structural, Gigahertz Dielectric and Magnetic Studies of Cobalt ferrite

Erum Pervaiz[1]* and I H Gul[1]
[1]Thermal Transport Laboratory (TTL), School of Chemical and Materials Engineering (SCME), National University of Sciences and Technology (NUST), H-12 Islamabad, PAKISTAN

*E-mail: eramsyam@gmail.com

Abstract. A series of Gd^{3+} doped nanocrystalline Co-ferrites $CoGd_xFe_{2-x}O_4$ (x =0.0 to 0.1) has been prepared by sol-gel auto combustion technique. Structural and morphology studies were performed using X-ray diffraction (XRD) and Scanning Electron Microscopy (SEM). Indexed XRD patterns confirm the formation of pure cubic spinel phase. Average crystallite sizes ranges from 16 nm to 25 nm ±2, were calculated from Sherrer's formula and Williamson Hall plots. Crystal strain increases with increase in doping amount of trivalent Gd ion. Lattice constant (*a*) and crystallite size D (*311*) increases with increase in Gd^{3+} concentration due to large ionic radii (0.94nm) of Gd^{3+} replacing Fe^{3+} (0.64nm). SEM images show the spherical morphology and uniform size distribution. Room temperature DC electrical resistivity decreases (~10^6) for x=0.025 then increases up to x=0.1 ~ ($4.5x10^7$). Dielectric properties have been studied using RF Impedance/material analyzer in the frequency range of 1 MHz to 1GHz. All the studied samples show a semi-conducting behavior as Permittivity and tangent loss (tanδ) decreases with the substitution of Gd^{3+} in parent crystal structure and have values of 4.92and 0.016 at 1 GHz respectively. Complex impedance and Complex electric modulus plots were further studied for complete contribution of grains and grain boundary resistances to conduction and resonance frequencies respectively. Magnetic studies by Vibrating Sample Magnetometer (VSM) shows that magnetization (Ms) decreases with increase in Gd^{3+} concentration from 63 emu/gm to 27.26 emu/gm. Coercivity (Hc) first decreases for x=0.025, after which it increases to 2308 Oe for x=0.1.

1. Introduction

Nano phase ferrites are a class of magnetic materials that have been the most attractive area of the research since the last 5 decades due to its remarkable applications in the high frequency and power devices especially for electromagnetic interference suppression (EMIS) [1]. Spinel ferrites have many versatile electric and magnetic properties owing to its feasibility to make a huge number of solid solutions of different metal cations and a large compositional variability. Cobalt ferrite is a hard spinel with high saturation magnetization, coercive field, mechanical hardness and high magneto crystalline anisotropy [2]. As the crystal structure of spinel ferrite is cubic closed pack (fcc) with anions (O^{2-}) linked with two sub-lattices namely tetrahedral (A) and octahedral (B). Distribution of divalent and trivalent cations on the A and B sites imparts specific characteristics to a spinel structure. $CoFe_2O_4$ is known as partially inverse spinel with ferromagnetic behavior having a high specific resistance and low eddy current losses in high frequency applications [3]. These exceptional physical properties can be achieved by choosing the method of synthesis, doping of cation in the host crystal structure, size and morphology of the particles. Various methods has been adopted for the synthesis of cobalt ferrites which includes chemical co-precipitation [4], sol-gel [5], hydrothermal [6], microemulsion [7], solvothermal [8], reverse micelle [9], citrate precursor [10], and solid state reaction [11]. Nano scaled cobalt ferrites have attracted much more attention due to above mentioned characteristics in the field of biomedicine, ferrofluid technology, advanced microelectronics, high frequency data storage, microwave absorbing materials (circulators, phase shifters, isolators) [12-15]. Rare earth oxides are good electrical insulators with high electrical resistivity. In spinel ferrites the proper choice of rare earth cation can alter the electrical and magnetic properties and have a large influence upon the magnetic anisotropy of the system making the spinel ferrite an excellent option in place of hexa ferrites or garnets.

Doping the parent spinel ferrite with rare earth ions (Y^{3+}, Gd^{3+}, Ho^{3+}, Sm^{3+}, Nd^{3+}) leads to structural disorder and lattice strain, thereby increasing the electrical and magnetic parameters [16]. Cheng. et. al [17] have studied the effect of rare earth ion substitution on Curie temperature, B. Ravi [18] has studied the thermoelectric properties of spinel ferrites doped with Gd^{3+}. K.K. Bharathia [19] has studied the magnetoelectric properties of Gd^{3+} doped ferrites. A. Rana [20] has demonstrated the affect of Gd substitution on dielectric properties of cobalt ferrite. Dielectric parameters of cubic spinel ferrites are important in understanding the electrical conduction, behavior of localized charge carrier's, and polarization phenomenon. Impedance and electric modulus studies are another important way to completely analyze all contributions to resistance in nano materials and to observe the relaxation behavior. According to different studies [21-22] the concentration of rare earth doping in the ferrites is important while describing the electrical and magnetic properties of these ferrites. These unique characteristics of rare earth doped systems are due to 4fn electronic states [23].

In the present work we aimed to study the influence of a rare earth ion (Gd^{3+}) on structural, gigahertz dielectric and magnetic properties of Co-ferrites. We have successfully synthesized spinel type nano ferrites $CoGd_xFe_{2-x}O_4$ (x=0.0 to 0.1) by sol-gel auto combustion followed by heat treatment and characterization is done by XRD, SEM, VSM, LCR , meter bridge and RF impedance/Material Analyzer. We report the complex electric modulus and magneto crystalline anisotropy for gadolinium doped Co-ferrites.

2. Experimental techniques

2.1. Synthesis

A nominal composition of $CoGd_xFe_{2-x}O_4$ with x ranging from 0.0 to 0.1 has been prepared by sol-gel auto combustion technique. All the chemicals purchased were of analytical grade and used as received. Aqueous solutions of Iron nitrate Fe $(NO_3)_3$ $.9H_2O$, Gd $(NO_3)_3.6H_2O$ and Co $(NO_3)_2.6H_2O$ were prepared by dissolving stoichiometric amounts of nitrate salts in de-ionized water. Molar ratio of M^{3+} and M^{2+} were kept 2:1. Aqueous nitrates were then mixed with a magnetic stirrer. Aqueous solution of citric acid were prepared with the ratio of Nitrates/citric acid=1/1.5 and mixed with the nitrate solution. The mixed solution was magnetically stirred for 1 hr. The solution were then neutralized with aqueous ammonia and heated at 100^0C till the liquid turns to a gel. Gel then automatically converted to fluffy powders by self ignition. Prepared powders were then dried further in an electric oven at 100^0C for few hours. As synthesized material were then ground and palletized into disc shaped using a hydraulic press under a load of 6 tonns. Disc shaped pellets and powder were sintered at 600^0C for 4 hrs in a muffle furnace for further characterizations.

2.2. Characterization

The crystallographic information (Crystallite size (*D311*); lattice constant, (a); X-ray density, (D_x); and porosity (P)), were found using indexed XRD patterns of the studied samples. XRD patterns were recorded using X-ray diffraction analysis (STOE-Seifert X'Pert PRO) at room temperature using CuKα (λ =1.5406 Å) radiation with 2-theta in the range of 20^0 to 80^0. The crystallite sizes were calculated and compared using the sherrer's formula and Williamson–Hall plots. W-H plots are usually drawn to see the strain effect in crystal structure. According to Williamson-Hall method [24] the width of individual reflections can be expressed as follows,

$$\beta \cos\theta = \frac{k\lambda}{D} + 4\varepsilon \sin\theta \tag{1}$$

where β is the width of peaks, D is the crystallite size, λ is the wavelength of Cu Kα radiation and ε is the crystal strain effect. The morphology and grain size was analyzed using SEM (JEOL-instrument JSM-6490A). A vibrating sample magnetometer was used to study the magnetic properties. DC electrical resistivity (ρ) were measured using two probe method at room temperature. Pellets (13mm in diameter) of samples were used for measuring the impedance and electric modulus properties using precision impedance analyzer (Wayne Kerr 6500B) at room temperature. The complex relative permittivity (έ, ε″) was measured by RF impedance/Material analyzer (Agilent E4991A). The dielectric tangent loss (tanδ) factor measured directly from the impedance analyzer using pellets of 10mm in diameter.

Complex impedance plane plots were drawn to study the electrical properties of $CoGd_xFe_{2-x}O_4$. The real (Z') and imaginary (Z'') parts of impedance were calculated using the relations [25],

$$Z' = Z\cos\theta \tag{2}$$

$$Z'' = Z\sin\theta \tag{3}$$

where θ is the phase angle measured from LCR meter.

Complex electric modulus was calculated from the standard relations [26].

$$M* = j\omega C_o Z^* \tag{4}$$

Magnetic measurements were done at room temperature using a vibrating sample magnetometer (VSM) under an applird field of 10 kOe.

3. Results and Discussion

3.1. Structural Characterizations

The powder XRD patterns of Gd^{3+} doped Co-ferrites for x=0.0 to x=0.1 have been presented in Fig.1 (a). All the patterns can be easily indexed to cubic spinel crystal structure (Fd3m) of $CoFe_2O_4$ (JCPDS card 22-1086) with (220), (311), (400), (511) and (440) diffraction peaks. Broad diffraction peaks represents the nanoscale crystallite size. It can be seen from the XRD patterns that crystallinity decreases by increase in the concentration of Gd^{3+} attributed to the decreased intensity in the observed peaks.

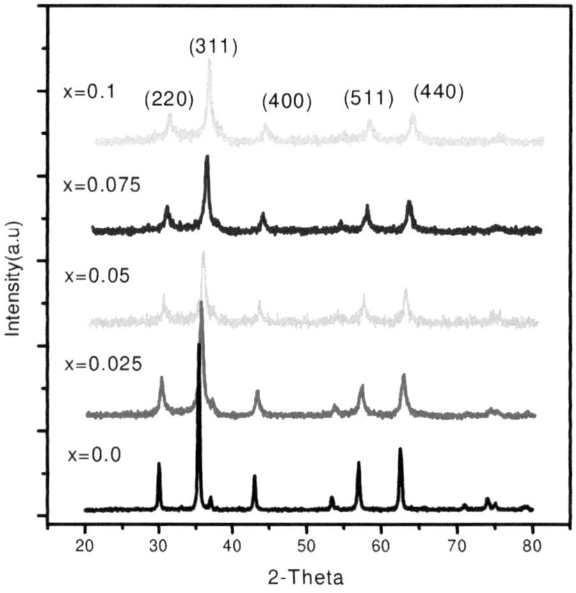

Figure 1. Indexed XRD patterns of $CoGd_xFe_{2-x}O_4$ (x=0.0 to x=0.1) nanoparticles synthesized by sol-gel auto combustion method.

Lattice parameter 'a' for all the samples has been calculated by interplanar spacing (d_{hkl}) and 2-theta values using the standard relation, $a=d_{hkl}(h^2+k^2+l^2)^{1/2}$. Value of lattice constant for x=0.0 comes out to be 8.3865Å, well in agreement with reported value [27]. Lattice constant has increased monotonically with increment in Gd^{3+} concentration as shown in Fig.2. This increase can be easily explained due to substitution of large ionic radii of Gd^{3+} (0.94Å) in place of smaller Fe^{3+} (0.67 Å) ions. Also rare earth ions are usually present at grain boundaries that cause hindrance in the grain growth, therefore crystal size and unit cell parameters increases [6]. Peak broadening (FWHM) has been used to determine the crystallite sizes using the sherrer's formula. Average crystallite size calculated for most intense (311) peak comes out to be 16 nm to 25 nm ±2nm. It is observed that crystallite size also increased by increase in the Gd^{3+} concentration. This increase in lattice parameter and crystallite size induces a lattice strain in the crystal structure. This may be explained on the basis of defects present in the parent $CoFe_2O_4$, which allows the absorption of lattice strain caused by the higher dopant amounts [4]. The value of theoretical density has been calculated by the relation, $D_x = \frac{8M}{Na^3}$ where a^3 is unit cell volume. D_x and measured density (D_m) calculated from weight and dimensions both increases with increase in Gd^{3+} due to increase atomic weight of gadolinium ion. All the values of crystallite sizes, lattice constant and lattice strain have been presented in table 1.

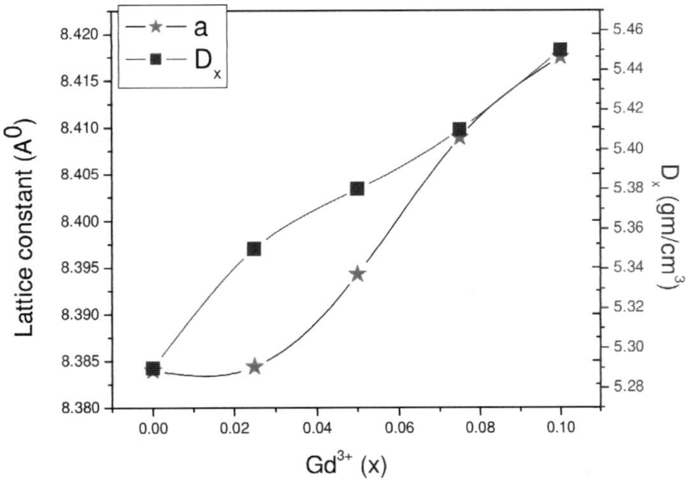

Figure 2. Variation of lattice constant and D_x with composition (x) for $CoGd_xFe_{2-x}O_4$

Fig.3 Shows the W-H plot only for x=0.1. These plots have been drawn to measure the crystallite sizes and lattice strain and compared with the crystallite sizes calculated by sherrer's formula. The positive slopes of the Gaussian fit to the peak broadening shows a tensile strain present in the prepared samples that increases with increases in Gd^{3+} concentration owing to lattice distortion in the host crystal structure by Gd^{3+}. The behavior of the W-H plots also confirm the increase in crystallite size by doping of larger ion Gd^{3+} in Co-ferrites as more the positive slope larger is the crystallite size [24].

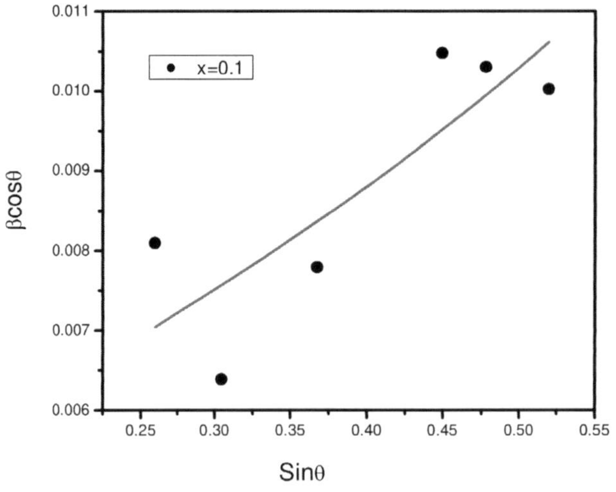

Figure 3. W-H plot for $CoGd_xFe_{2-x}O_4$ nanoparticles synthesized by sol-gel auto combustion method (x=0.1)

3.2. Morphology Analysis

SEM images (Fig.4). Shows the surface morphology and grain size for the prepared sample x=0.0 and x=0.1. It is clear from the image that uniformly distributed; less agglomerated and homogenous spherical particles have been formed in a controlled environment by sol-gel auto combustion technique. Micrographs also confirm the increase in grain size with Gd^{3+} doping in the parent crystal structure. The homogeneity of shape and grain size largely affects the electrical and magnetic properties of ferrites.

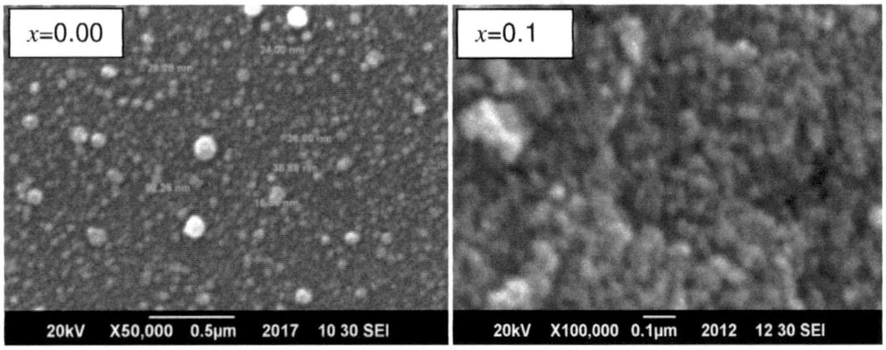

Figure 4. SEM images for $CoGd_xFe_{2-x}O_4$ (x=0.0) and (x=0.1) nanoparticles synthesized by sol-gel auto combustion method

Table 1. Crystallite size (*D_{311}*) sherrer's method, Crystallite size D (W-H) method, lattice parameter (*a*) , X-ray density (D_x), measured density (D_m), Porosity (*P*), dielectric constant (ε'), dielectric loss factor (ε''), and tangent loss (*tanδ),* Relaxation time.

Composition	*x=0.0*	*x=0.025*	*x=0.05*	*x=0.075*	*x=0.1*
Lattice constant 'a'	8.3865	8.3874	8.3947	8.3954	8.4162
D(311) nm	16.92	16.99	21.75	21.93	22.54
D_m (gm/cm^3)	2.47	2.48	2.511	2.523	2.57
D_x(gm/cm^3)	5.29	5.35	5.38	5.41	5.45
P (%)	53.3	53.6	53.3	53.4	52.8
ρ (Ω-cm)	$3.45x10^7$	$1.29x10^6$	$1.03x10^7$	$4.5x10^7$	$9.5x10^7$
ε' *(1MHz)*	15.2	34.9	16.1	15.1	13.6
ε''*(1MHz)*	1.19	11.9	2.81	2.41	0.7
tanδ(1MHz)	0.077	0.3463	0.1750	0.1568	0.043772
ε'*(1 GHz)*	6.20	5.96	5.86	5.22	4.92
ε''*(1 GHz)*	0.473	7.86	0.97	0.842	0.256
tanδ(1 GHz)	0.0324	0.3520	0.0654	0.0598	0.0160
τ_r*(sec)*	$9.88x10^{-7}$	$4.99x10^{-6}$	$8.6x10^{-6}$	$2.54x10^{-5}$	$2.1x10^{-5}$

Table 2. Saturation magnetization (*Ms*), Remanance (*Mr*), Coercivity (*Hc*), Remanance ratio (*Mr/Ms*), Magnetic moments (n_B), Yeffit-Kittel angle ($\alpha_{y\text{-}k}$), Magnetic anisotropy constants (K_1), Anisotropy field (H_k).

Composition	*x=0.0*	*x=0.025*	*x=0.05*	*x=0.075*	*x=0.1*
M_s	63	34.5	28.73	33.32	27.26
H_c	2929	1096	2308	1078	1779
M_r	34.52	14.03	13.49	11.99	10.03
M_r/M_s	0.55	0.40	0.47	0.36	0.37
n_B	1.017	0.59	0.49	0.57	0.46
$\alpha_{Y\text{-}K}$	0	24.74	29.97	31.17	35.52
K_1(erg/cm^3)	$0.35x10^6$	$0.1x10^6$	$0.13\ x10^6$	$0.108\ x10^6$	$0.12\ x10^6$
H_k(*kOe*)	11.55	7.245	12.96	8.067	10.424

3.3. Electrical Characterization

3.3.1. Variation of DC electrical resistivity

Fig.5 shows the DC electrical resistivity of $CoGd_xFe_{2-x}O_4$ with composition (*x*=0.0 to *x*=0.1) of the dopant at room temperature. Co-ferrite is a highly resistive material with high activation energy. Method of synthesis, type and substitution of doped cations at A and B site, particle size and morphology and sintering temperatures largely affects the electrical properties of spinel ferrites [28]. This can be observed from the graph that DC electrical resistivity decreases up to a concentration of *x*=0.025 but then increases with increase in Gd^{3+} concentration. There is a significant increase in the DC electrical resistivity with Gd^{3+} substitution (*x*=0.1). Electrical conduction in spinel ferrites can be

Journal of Physics: Conference Series **439** (2013) 012015 doi:10.1088/1742-6596/439/1/012015

easily explained by Verwey mechanism [29] and is mainly due to electron hopping between Fe^{2+} and Fe^{3+}. The suggested cations distribution for the present samples of $CoGd_xFe_{2-x}O_4$ can be written as: $(Co^{2+}{}_{0.5}Fe^{3+}{}_{0.5})_A\,(Co^{2+}{}_{0.5}Gd^{3+}{}_xFe^{3+}{}_{1.5-x})_BO_4$. Gd^{3+} ion substitution in place of Fe^{3+} at octahedral (B site) decreases the number of Fe^{3+} ions at B sites which also decrease the hopping tendency $Fe^{3+}\leftrightarrow Fe^{2+}$, thereby reducing the conductivity and increasing the DC electrical resistivity. The DC electrical resistivity variations with Gd^{3+} substitution can be explained as for x=0.025, Gd^{3+} ions occupy A site replacing Fe^{3+} due to which number of Fe^{3+} ions at B-site increases and thus Fe^{2+}. This may happen due to lattice distortion caused by doping a concentration of x=0.025. A low concentration of Gd^{3+} causes a severe change in centro-symmetric FCC structure of parent crystal that causes an increase in hopping of electron and a net electric polarization [30]. This reason is reflected in the high dielectric parameters and a sharp decrease in saturation magnetization for x=0.025.

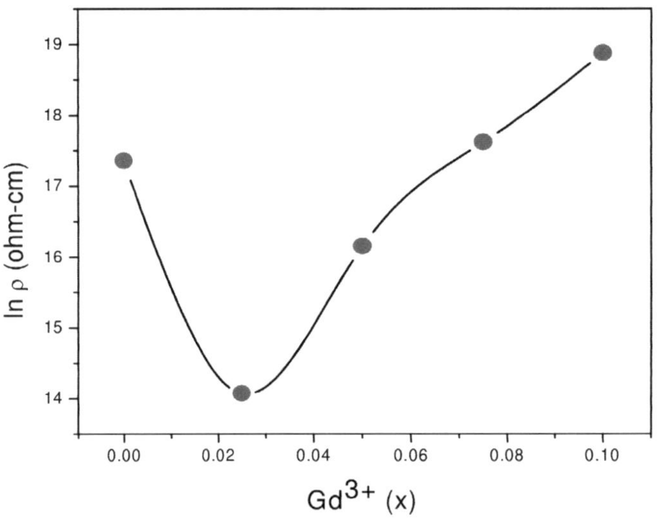

Figure 5. Variation of DC electrical resistivity (*ln ρ*) with composition x (Gd^{3+}) for $CoGd_xFe_{2-x}O_4$ (x=0.0 to x=0.1) nanoparticles synthesized by sol-gel auto combustion method

3.3.2. Dielectric study

Dielectric behavior of nano spinel ferrites mainly depends upon the nature and distribution of metal cations on A-sites and B-sites in the spinel lattice. Spinel Cobalt ferrites are considered good dielectric materials and the high frequency dielectric behavior is mainly dependent upon the particle size and method of synthesis of nano particles [31]. Different studies have been provided relating the dielectric parameters of Gd^{3+} doped ferrites [32-33]. Dielectric parameters (real and imaginary parts of relative permittivity, dielectric loss tangent) for the prepared series of $CoGd_xFe_{2-x}O_4$ (x=0.0 to x=0.1) have been studied in the frequency range 1 MHz to 1GHz at room temperature. Figs.6 (a,b) show the real (ε') and imaginary (ε'') part of relative permittivity with frequency at room temperature. It can be observed from the figure that relative permittivity for all the samples decreases with increase in frequency and ultimately becomes constant at higher frequencies (~GHz). This decrease in permittivity is more rapid in the low frequency region and becomes sluggish as the applied frequency increases. This behaviour is subjected to dielectric polarization under the application of AC field [34].

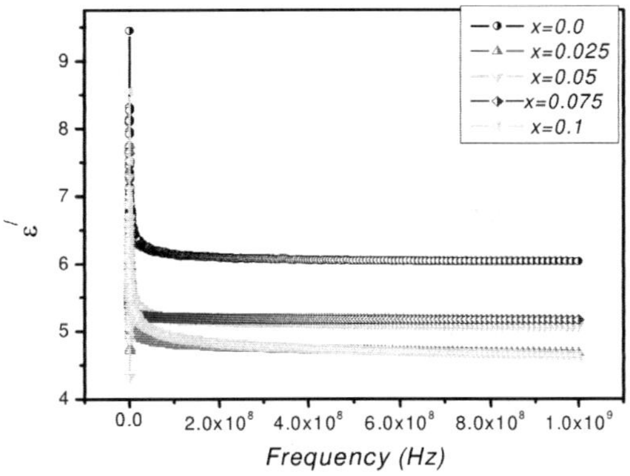

(a)

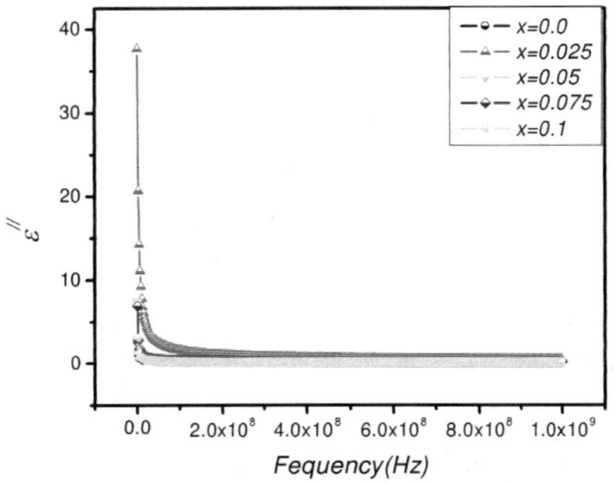

(b)

Figure 6. (a) Variations of Permittivity ε' (real part) and imaginary part ε'' **(b)** with ln f for $CoGd_xFe_{2-x}O_4$ (x=0.0 to x=0.1) nanoparticles synthesized by sol-gel auto combustion method

Permittivity and Resistivity both are electrical properties of ferrites and depends upon the conduction phenomenon. Hopping of electron between Fe^{2+} and Fe^{3+} is responsible for this conduction. This hopping is responsible for polarization at grain boundaries due to local charge displacement. Such kind of dielectric dispersion in ferrites can be explained by Maxwell-Wagner model [35] and koops theory [36] which suggests that ferrite system consist of a combination of highly conducting grains separated by poorly conducting grain boundaries. Electrons by hopping pile up at grain boundaries due to high resistance and polarization takes place there. At high frequencies this hopping frequency does not follow up the field variation thereby making the relative permittivity a constant. Room temperature dielectric parameters have been presented in Table.2. It can be seen from the tabulated data that imaginary relative permittivity increases with increase in Gd^{3+} concentration x=0.025 but after this concentration decreases till x=0.1. As dielectric polarization in ferrites is due to electron exchange $Fe^{2+} \leftrightarrow Fe^{3+}$, magnitude of which depends upon the percentage of Fe^{2+} and Fe^{3+} ion pairs at A and B sites. Fe^{2+} ions concentration largely affects the conduction phenomenon and depends upon type of cation substituting, synthesis route, sintering time and sintering temperatures. In the present case relative permittivity (ε' and ε'') both decreases for Gd^{3+} concentration x=0.1. Which can be explained by low Fe^{2+} ion concentration at B-site causing a low value of resistivity and hence a high value of dielectric parameter. As resistivity and relative permittivity has inverse behavior to each other [37].

Fig.7 shows the variation of *tanδ* with frequency. It can be seen that dielectric loss tangent has the same trend as permittivity losses. It decreases with increase in frequency and becomes constant up to 1GHz due to decreased polarization at high AC fields. This dielectric behavior reversed the trend that has been observed for DC-electrical resistivity. Dielectric losses in ferrite system usually reflect in DC electrical resistivity. Low loss dielectric material will be a highly resistive as well. Also observed that for x=0.025, a slight increase in the loss tangent due to lattice distortion at x=0.025 [38]. Sample with x=0.1 shows a low loss dielectric behavior which allows its use in high frequency data reading/writing in electronic structures.

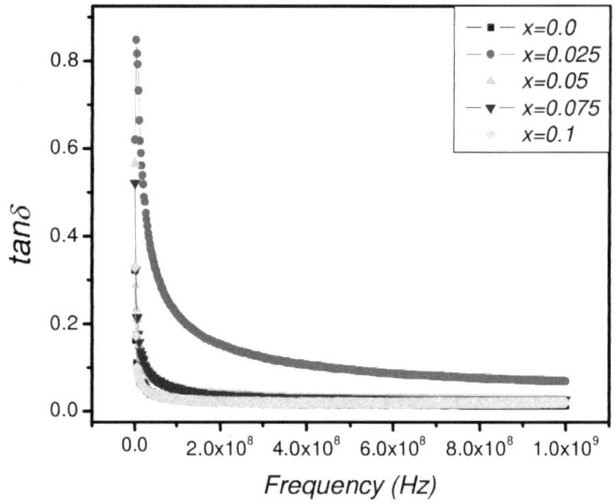

Figure 7. Variations of Dielectric loss tangent (*tanδ*) with ln f for $CoGd_xFe_{2-x}O_4$ (x=0.0 to x=0.1) nanoparticles synthesized by sol-gel auto combustion method

3.3.3. Complex Impedance/Complex electric modulus study

Complex impedance and electric modulus studies were performed at room temperature using the LCR meter in the frequency range of 1KHz to 5 MHz. Impedance analysis is useful to completely understand the electrical properties of spinel type ferrites and give us the data for both resistive (real part) and reactive (imaginary part) contribution to conductivity on the application of AC field [39]. Complex impedance plot also called as Nyquist diagrams gives a complete contribution of microstructure (grains and grains boundary) resistances. Fig.8 shows the complex impedance plane plots (cole-cole) for all the studied samples. These complex plane plots help to distinguish the grain and grain boundary resistances and interfacial resistance of conducting electrodes. It is clear from the Nyquist plot that only one semi circle is present for all the studied samples in the low frequency region. As its shown in different studies [40-41] that there are two consecutive semi circles in complex impedance diagrams, one in low applied fields showing the grain boundary resistances and the other in high field side shows the grains contribution. Only one quarter circle in low field region confirms the predominance of grain boundary resistance for the present studied samples of Gd^{3+} doped Co-ferrite and grain resistance is not well resolved. It can also be seen from complex plane plot that grain boundary resistance increases with increase in Gd^{3+} concentration for x=0.1.

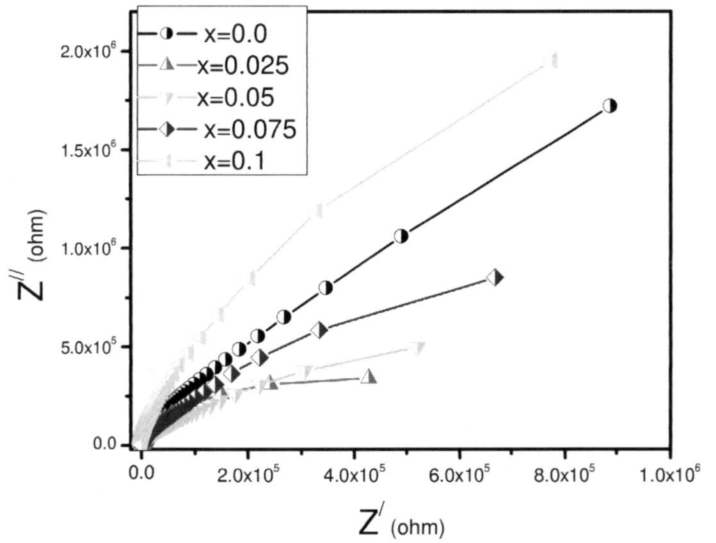

Figure 8. Complex impedance plane plots for $CoGd_xFe_{2-x}O_4$ (x=0.0 to x=0.1) nanoparticles synthesized by sol-gel auto combustion method

Electric modulus were plotted to see the relaxation behavior and resonance frequencies of the prepared samples. To see the influence of Gd^{3+} on the resonance frequencies of cobalt ferrite, cole-cole type diagrams of M* were plotted as the AC data of impedance (Z*) was poorly resolved into semi circle arc in the cole-cole diagram of complex impedance. This possibly could happen for the samples where the change in resistance is dominant as compared to capacitance of the material [42]. Fig.9 (a) shows the imaginary part of the electric modulus variation with applied field. The observed peaks in the plots show that the relaxation phenomenon is occurring in the samples under study. These peaks are due to dielectric relaxation where $\omega\tau=1$, where τ is the relaxation time inverse of relaxation frequency f_r. It can also be observed that the peak shifts to lower frequency side with increase in Gd^{3+} concentration. Showing that relaxation and associated resonance losses are in low frequency region and no resonance is observed for the material at high frequencies. Fig.9 (b) shows the real part of the electric modulus with frequency. Onset frequency also shifts to low frequency region showing the grain boundary resistance is higher for the samples with higher composition of Gd^{3+}. Clearly resolved semi circles have been observed for cole-cole diagram of electric modulus (M' vs M'') fig.9 (c). Sample with x=0.025, show two semi circle arcs one in the high frequency side and other in low frequency region corresponding to grain and grain boundary resistance respectively.

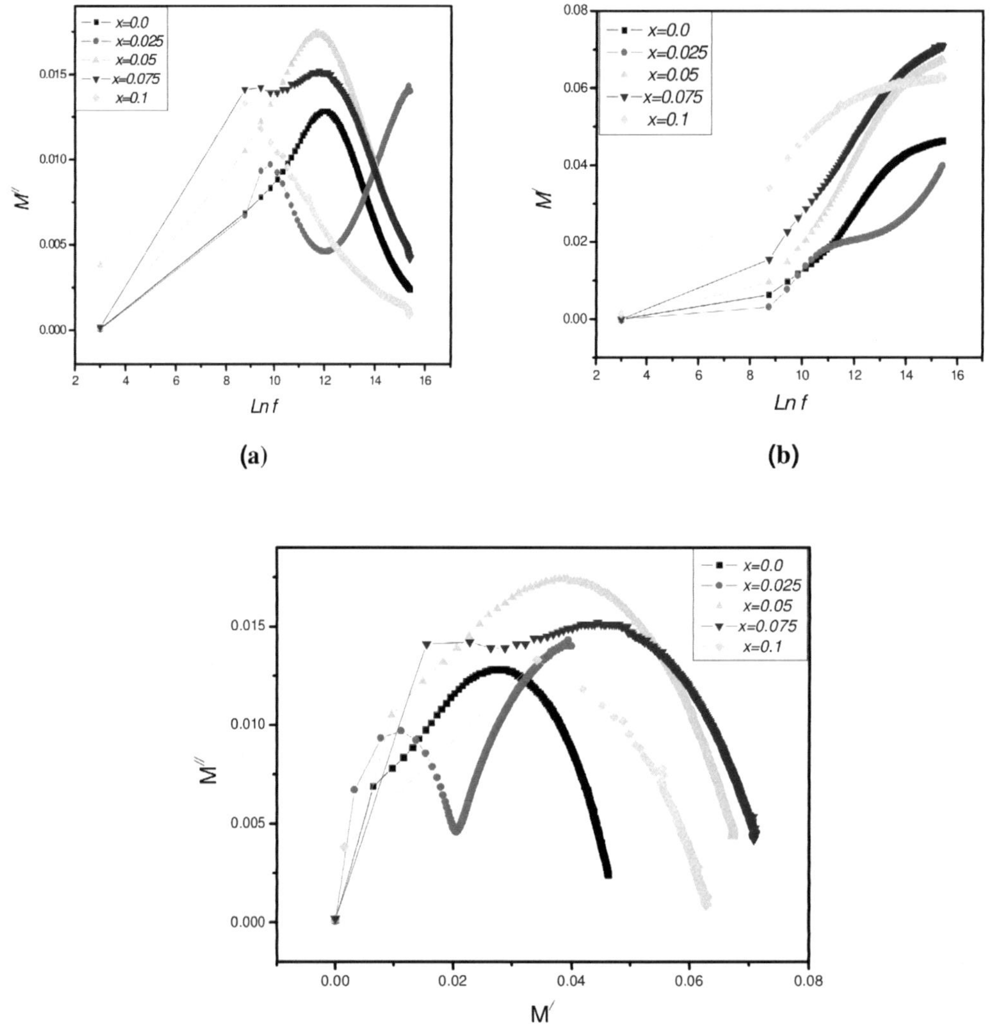

(a) (b)

(c)

Figure 9. **(a)** Variations of complex electrical modulus M'' (imaginary part) with $\ln f$ **(b)** M' (real part) with $\ln f$ **(c)** Complex modulus plots for $CoGd_xFe_{2-x}O_4$ (x=0.0 to x=0.1) nanoparticles synthesized by sol-gel auto combustion method

3.4. Magnetic study

To study the effects of Gd^{3+} doping on magnetization, coercivity, remanent magnetization (*Mr*) and magneto-crystalline anisotropy (H_K) of Co-ferrite, M-H hysteresis loops were recorded using VSM under the applied magnetic field of 10 kOe at room temperature. Fig.10 (a) shows the hysteresis curves for all the samples under investigation. It is clear that all the samples show a fine s shape loops with a decrease in magnetization (*Ms*) and coercivity (*Hc*) with increase in Gd^{3+} concentration. This is due to the substitution of Gd^{3+} in place of Fe^{3+} at B-sites. Gd^{3+} ions have a higher magnetic moment (7 μ_B) in place of Fe^{3+} that has 5 μ_B. The order in magnetic moments of rare earth ions is below room temperature [4], due to this at room temperature Gd ion behaves as non magnetic that cause a decrease in saturation magnetization and coercivity. This substitution causes a lattice distortion that alters the

Journal of Physics: Conference Series **439** (2013) 012015 doi:10.1088/1742-6596/439/1/012015

magnetic characteristics of materials. Magnetic properties of ferrite materials largely based upon the grain size, cation substitution, and A-B exchange interactions [17]. Increase in the grain size and decrease of A-B super exchange interaction causes canting spins at the surface of nano particles that decreases the magnetic characteristics of the present samples. Spin arrangement for the Co-Gd ferrites were analyzed by measuring the value of Bohr's magnetron by the relation,

$$n_B = \frac{\sigma_s}{5585} M_W \qquad (5)$$

using the calculated magnetic moment Yafet-Kittel (Y-K) angles were studied to see the spin arrangement in present ferrite system by the relation [43],

$$n_B = (6 + x) Cos\alpha_{Y-K} - 5(1 - x) \qquad (6)$$

where x represents the composition of doping ion. All the magnetic data including Y-K angles were presented in Table.2. Y-K angle increases with increase in Gd^{3+} concentration owing to triangular spin arrangement of ions and results in decreased A-B interactions [44]. In order to analyze the Gd^{3+} influence on magnetic anisotropy of Co-ferrites, value of anisotropy constant (K_1) is determined using Law of approach (LA) to saturation (Fig.10 (b)). Cubic anisotropy constant (K_1) and anisotropy field (H_k) can be calculated by following equations [45-46]

$$K_1 = \mu_0 M_s \sqrt{105b/8} \qquad (7)$$

$$H_k = \frac{2K_1}{\mu_0 M_s} \qquad (8)$$

where μ_0 is the permeability of the free space, Ms is saturation magnetization from the fit, b is the parameter obtained from the fit. Data for anisotropy constant and anisotropy field (H_k) is mentioned in Table.2. Our results are in comparable to earlier reported values for Co-ferrites [45]. It is clear from the data that cubic anisotropy decreases with increase in gadolinium concentration. Pure cobalt ferrite is has high anisotropy constant and field due to occupation at B-sites. It decreases due to Gd^{3+} occupation at B-sites. Cubic anisotropy constant also decreased doping gadolinium showing a decrease in coercivity for present studied samples. The anisotropy parameters (anisotropy constant and field) are not decreasing monotonically but more abruptly due to differing concentration of doping ion. This behaviour is showing a strong lattice distortion due to Gd^{3+} substitution.

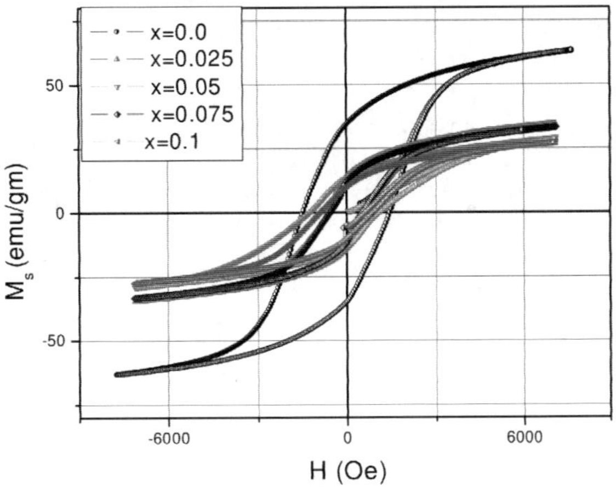

(a)

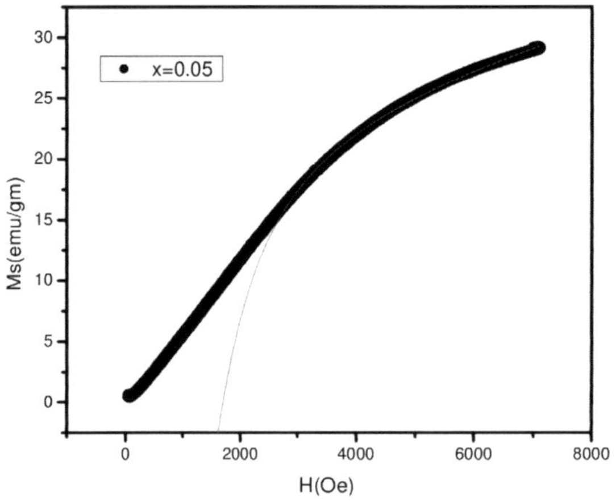

(b)

Figure 10. **(a)** *M-H* loops for $CoGd_xFe_{2-x}O_4$ (x=0.0 to x=0.1) nanoparticles **(b)** Fit to LA for $CoGd_xFe_{2-x}O_4$ (x=0.05) nanoparticles synthesized by sol-gel auto combustion method

4. Conclusions

Nano spinel $CoGd_xFe_{2-x}O_4$ with x in step increment of 0.025 has been synthesized by sol-gel auto combustion. All the studied samples are pure cubic spinel phase ferrites without any impurity metal oxides. Lattice constant and crystallite size increases with increase in Gd^{3+} concentration, owing to increase ionic radii and atomic weight of gadolinium as compared to Fe^{3+}. Substitution of Gd^{3+} ion in parent crystal causes a lattice distortion that can be observed by increased lattice strain in W-H plots. DC electrical resistivity increases to $9.5x10^7$ with Gd^{3+} substitution except for x=0.025 (~ 10^6). Dielectric constant and loss tangent decreases to 4.92 and 0.016 respectively with increase in the dopant concentration showing that the material with x=0.1 is a low loss dielectric. Complex impedance and electric modulus reveals that the grain boundary are offering more resistance to conduction showing that smaller grains with more grain boundaries are formed for present samples. Gd^{3+} substitution have tailored the relaxation behavior making the resonance frequencies lower than the pure cobalt ferrites. Magnetic properties of Gd^{3+} doped Co-ferrites decrease due to influence on A-B exchange interactions and canted spins of electrons on nano particles surface. Saturation magnetization, coercivity and remanance all decreases with increase in dopant concentration. Yeffit-Kittel angles increases with x, showing a triangular spin arrangement for present ferrite system. Magnetic anisotropy of Co-ferrites decreases with increase in Gd^{3+} concentration (x). Thus the rare earth (Gd^{3+}) doped Co-ferrites found an application in high frequency devices and power supply due to high resistivity and low losses.

Acknowledgment

The authors would like to acknowledge TWAS, Italy, Higher Education Commission (HEC) Islamabad project.No. 1326, Pakistan for providing financial support for this work and Pakistan science foundation (PSF) project.No.147.

References

[1] Valenzuela R 2012 Novel application of soft ferrites *Phys. Res. Inter.* **2012** 1-9
[2] Tahar L B, Artus M, Ammar S, Smiri L S, Herbst F, Vaulay M J, Richard V, Grene`che J M, Villain F, Fievet F 2008 Magnetic properties of $CoFe_{1.9}RE_{0.1}O_4$ nanoparticles (RE) prepared in polyol *J. Magn. Magn. Mater.* **320** 3242-3250.
[3] Elkestawy M A, Kader S A, Amer M A 2010 AC conductivity and dielectric properties of Ti-doped $CoCr_{1.2}Fe_{0.8}O_4$ spinel ferrite *Physica. B.* **405** 619-624.

[4] Kim Y I, Kim D, Lee C S 2003 Synthesis and characterization of $CoFe_2O_4$ magnetic nanoparticles prepared by temperature-controlled coprecipitation method *Physica. B.* **337** 42
[5] Kim C S, Yi Y S, Park K T, Namgung H, Lee J G 1999 Chemically prepared magnetic nanoparticles *J. Appl. Phys.* **85** 5223–5225.
[6] Peng J, Hojamberdiev M, Xu Y, Cao B, Wang J, Wu H 2011 Hydrothermal synthesis and magnetic properties of gadolinium doped $CoFe_2O_4$ nanoparticles *J. Magn. Magn. Mater.* **323** 133-138.
[7] Pillai V, Shah D O 1996 Synthesis of high-coercivity cobalt ferrite particles using water-in-oil microemulsions *J. Magn. Magn. Mater.* **163** 243–248.
[8] Cai W, Wan J Q 2007 Facile synthesis of superparamagnetic magnetite nanoparticles in liquid polyols *J. Colloid. Interface. Sci*, **305** 366–370.
[9] Sileo E E, Jacobo S E 2004 Gadolinium-Nickel ferrites prepared from metal citrates precursors *Physica. B.* **354** 241-245
[10] Rashad M M, Mohamed R M, El-Shall H 2008 Magnetic properties of nanocrystalline Sm substituted $CoFe_2O_4$ synthesized by citrate precursor method *J. Mater. Proc. Technol.* **198** 139–146.
[11] Rao K S, Kumara A M, Varmaa M C, Choudary R K, Rao K H 2009 Cation dirtibution of Ti doped cobalt ferrites *J. Alloys. Compd.* **488** 6–9
[12] Goha S C, Chiaa C H, Zakariaa S, Yusoffa M, Hawa C Y, Ahmadia Sh, Huangb N M, Lim H N 2010 Hydrothermal preparation of high saturation magnetization and coercivity cobalt ferrite nano crystals without subsequent calcinations *J. Mater. Chem. Phys.* **120** 31–35.
[13] Sankaranarayana V K, Sreekumar C J 2003 A strategic approach for preparation of oxide Nanomaterial *J. Curr. Appl. Phys.* **3** 205–208.
[14] Suginoto M 1999 The past present and future of ferrites *J. Am. Ceram. Soc.* **82** 269-280.
[15] Bate G Magnetic recording materials since 1975 *J. Magn. Magn. Mater.* **100** 413-424.
[16] Jacobo S E, Uhalde S D, Bertorello H R 2004 Rare Earth influence on the structural and magnetic properties of NiZn ferrites *J. Magn. Magn. Mater.* **272** 2253-2254.
[17] Cheng F X, Jia J T, Xu Z G, Zhou B, Liao C S, Yan C H, Chen L Y, Zhao H B 1999 Microstructure, magnetic, and magneto-optical properties of chemical synthesized Co-RE (RE=Ho, Er, Tm, Yb, Lu) ferrite nanocrystalline films *J. Appl. Phys.* **86** 2727–2732.
[18] Kumar B R, Ravinder D 2002 Electrical Conductivity of Ni-Zn-Gd Ferrites *Mater. Lett.* **53** 441-445.
[19] Bharathi K K, Chelvane J A, Markandeyulu G 2009 Magnetoelectric properties of Gd and Nd-doped nickel ferrite *J. Magn. Magn. Mater.* **32** 3677-3680.
[20] Rana A, Thakur O P, Kumar V 2011 Effect of Gd^{3+} substitution on dielectric properties of nano cobalt ferrite *Mater. Lett.* **65** 3191–3192.
[21] Ravinder D, Kumar K V, Balaya P 2001 High-frequency dielectric behavior of gadolinium substituted Ni–Zn ferrites *Mater. Lett.* **48** 210–214.
[22] Kumar K V, Ravinder D 2002 Electrical conductivity of Ni–Zn–Gd ferrites *Mater. Lett.* **52** 166–**168**.
[23] Maqsood A 2009 Phase transformations in Ho2Si2O7 ceramics *J. Alloys. Compd.* **471** 432-434.
[24] Ahlawat A. et al 2011 *J. Magn. Magn. Mater.* **323** 2049-2054.
[25] Kamble R C, Adhate N R, Chougule B K, Kolekar Y D 2010 Magnetic and dielectric properties of mixed spinel Ni-Zn ferrites synthesized by citrate precursor method *J. Alloys. Compd.* **491** 372-377.
[26] K. M. Batoo K M, S. Kumar S,C. G. Lee C G, Alimuddin 2009 Study of ac impedance spectroscopy of Al doped MnFe2−2*x*Al2*x*O4 *J. Alloys. Compd.* **480** 596-604.
[27] Ceden˜o-Mattei Y, Perales-Pe´rez O 2009 Synthesis of high-coercivity cobalt ferrite nanocrystals *J. Microelectronics.* **40** 673-681.
[28] Mazen S A, Zaki H M, Mansour S F 2007 The infrared absorption and dielectric properties of Li–Ga ferrite, *Int, J. pure. Appl. Phys.* **3** 40-46.
[29] Verwey E J W 1935 Verway transitions in magnetite *Ziest. F. Krist.* **91** 65.
[30] Said M Z 1998 Effect of gadolinium substitution on the structural and electrical conductivity

of Ni-ferrite *Mater. Lett.* **34** 305-307.
[31] Gul I H, Abbasi A Z, Amin F, Anis-ur-Rehman M, Maqsood A 2007 *J. Mag. Magn. Mater.* 311 494
[32] Ahmed M A, Ateia E, El-Dek S I 2003 Rare earth doping effect on the structural and electrical properties of Mg–Ti ferrite *Mater. Lett.* **57** 4256–4266.
[33] Muthafar F, Al–Hilli, Sean Li, Kassim S 2011 Gadolinium substitution and sintering temperature dependent electronic properties of Li–Ni ferrite *J. Mater. Chem. Phys* **128** 127.
[34] Lawrence K, Manoranjan K 2011 Influence of Al^{3+}ion concentration on the crystal structure and magnetic anisotropy of nanocrystalline spinel cobalt ferrite *J. Magn. Magn. Mater.* **323** 2042-2048
[35] Mohamed R M, Rashad M M, Haraz F A, Sigmund W 2010 Structure and magnetic properties of nanocrystalline cobalt ferrite powders synthesized using organic acid precursor method *J .Magn. Magn. Mater.* **322** 2058-2064.
[36] Koop's C G 1951 On the dispersion of resistivity and dielectric constant of some semiconductors at audio frequencies *Phys. Rev.* **83** 121-126.
[37] Gul I H, Maqsood A 2008 Structural, magnetic and electrical properties of cobalt ferrites prepared by the sol–gel route *J. Alloys. Compd.* **465** 227-231.
[38] Verma A, Thakur O P, Hsu J H 2011 Temperature dependence of electrical properties of nickel-zinc ferrites processed by the citrate precursor technique *J. Alloys. Compd.* **509** 5315-5321
[39] Batto K M 2011 Study of dielectric and impedance properties of Mn ferrites *Physica. B.* **406** 382-391.
[40] Chourashiya M G, Patil J Y, Pawar S H, Jadhav L D 2008 Studies on the structural, morphological and electrical properties of $Ce_{1-x}Gd_x O_{2-(x/2)}$ *J. Mater. Chem. Phys.* **109** 39-44.
[41] Farea A M M, Kumar S, Batoo K M, Yousef A, Lee C G, Alimuddin 2008 Structure and electrical properties of $Co_{0.5}Cd_xFe_{2.5-x}O_4$ ferrites *J. Alloys. Compd.* **464** 361–369.
[42] Ya H, Jackman R B, Hing P 2008 Spectroscopic impedance study of nanocrystalline diamond films *J. Appl. Phys.* **94** 7878-7883.
[43] Yafet Y, Kittel C 1952 Antiferromagnetic Arrangements in Ferrites *Phys. Rev.* **87** 290-294.
[44] Chen W, Zhu W, Ke C, Yang Z, Wang L, Chen X F, Tan O K 2010 Impedance spectroscopy and conductivity mechanism of CoFe(2)O(4)-Pb(Zr(0.53)Ti(0.47))O(3) composite thick films *J. Alloys. Compd.* **508** 141-146.
[45] Maaz K, Mumtaz A, Hasanain S K, Ceylan A 2007 Synthesis and magnetic properties of cobalt ferrite ($CoFe_2O_4$) nanoparticles prepared by wet chemical route *J. Magn. Magn. Mater.* **308** 289-296.
[46] Zhao L, Yang H, Yu L, Cui Y, Zhao X, Feng S 2006 Study on magnetic properties of nanocrystalline La-, Nd-, or Gd-substituted Ni–Mn ferrite at low temperatures *J. Magn. Magn. Mater .* **305** 91-94.

Journal of Physics: Conference Series **439** (2013) 012016 doi:10.1088/1742-6596/439/1/012016

Optimization of loading factor of Nanocomposite Coatings deposited by Physical Vapor Deposition

S. Alam[1*], M. Irfan[1], B. D. Soomro[1], M. Shahid, M. Zeeshan[2]

[1]Pakistan Institute of Technology for Minerals and Advanced Engineering Materials (PITTMAEM) PCSIR Laboratories Complex, Lahore, Pakistan.
[2]University of Ulster, Belfast, Northern Ireland, UK

*Email: pitmaem.lhr@gmail.com

Abstract. The Physical Vapor Deposition (PVD) is a technique in which metal or metal alloy vapors are deposited on the surface of the tools in a vacuum chamber. The positioning of the tools i.e. fixture and loading factors (% area of the chamber filled with tools) influence in the flow of the vapors in the chamber. This results in a significant effect on the coating characteristics which affect the film properties. Nano Composite Coatings i.e. $AlTiN/Si_3N_4$ (nACo) or $AlCrN/Si_3N_4$ (nACRo) deposited by the PVD technique show superior hardness, toughness, heat and oxidation resistance. The adhesion and diffusion of these Nano structured films on substrate depend on proper surface cleaning steps and loading factors. In present work, nACo and nACRo coatings were done by changing the loading factors (50% to 75%) for coating. These coated samples were characterized and the tribological, mechanical and wear properties were studied. It was observed that; fixture, surface cleaning steps of substrate and vacuum surface areas (loading factors) have influenced the adhesion and thickness properties of the coated films. The 75% loading factors greatly enhance the efficiency of tools coated with both nACo and nACRo structured films.

1. Introduction

Surface engineering is defined as alteration in the properties of the surface phase in order to reduce the degradation over time. This is done by making the surface robust to the environment in which it will be used. Surface engineering techniques are being used in a number of industries such as; automotive, aerospace, missile, power, electronic, biomedical, textile, petroleum, petrochemical, chemical, steel, power, cement, machine tools, and construction [1]. Titanium Nitride (TiN) and Titanium Aluminum Nitride (TiAlN) have been widely used as protective hard coatings for cutting tools. The demand for the improvement of machining technology requires higher cutting speed, better quality of the machined surface, lower consumption of lubricants and coolants. These requirements can be fulfilled by depositing wear resistant coatings on machine tools [2].
Coated materials are normally used in operations which require high hardness, wear resistance, thermal and chemical stability of the surface. Coatings deposited by physical vapor deposition of the metal vapors or metal on the surface of the stressed materials should satisfy the conditions related to maintaining sufficient quality of the coated parts during their application in practice. Thin coatings applied to the surface of machine parts and tools must show relevant requirements, such as high hardness and modulus, resistance to wear, thermal and chemical resistance, good adhesion and adequate roughness values. Determination of properties of thin PVD coatings mainly depends on the correctly selected methods for their successful application in the practice. [3]. In the present work, nACo and nACRo coatings were deposited by changing the loading factors (50%, and 75%). The tribological, mechanical and wear properties of these coated samples were studied. It was observed that fixture, surface cleaning steps of substrate and vacuum surface areas (loading factors) have influenced the adhesion and thickness properties of the coated films. The 75% loading factors greatly enhances the efficiency of tools coated with both nACo and nACRo structured films.

2. Experimental work

The textile thread guides of SS material were used for present work. In present work, nACo and nACRo coatings were developed by changing the loading factors (50%, and 75%). The samples were coated using the Physical Vapor Deposition System (PLATIT π 80).

2.1 Surface preparation

Textile thread guides were first ultrasonically cleaned by washing them in an ultrasonic bath using Carbon Tetra chloride (CCl_4) for 5 minutes. After washing, the samples were thoroughly treated with distilled water to remove the traces of CCl_4 followed by washing the samples in an ultrasonic bath containing acetone. After the removal of samples from the ultrasonic bath they were carefully dried using an air compressor. Then Samples were placed in a Cleaning solution (Clean-Gel) at 60°C for about 10 minutes.

2.2 Deposition of coating

Two types of coatings, nACo and nACRo with coating thickness of 2 μm and 2.5 μm were done on Textile Thread Guides using Physical Vapor Deposition System (Platit π 80) with two lateral rotating cathodes. The coating deposition was conducted in a flowing nitrogen atmosphere under a working pressure of 1.5 Pa. The pre-cleaned Textile Thread Guides along with test coupons were mounted on carrousel substrate holders. This rotates continuously around the vertical central axis at a speed of 12 rpm. During deposition, a negative DC bias of -70 V was applied to the substrates.

2.3 Characterization of the coating

nACo and nACRo coatings were characterized for their mechanical and tribological analysis, which includes the following techniques:

2.3.1 Coating thickness measurement

The coating thickness was measured with the help of Calo-tester as shown in figure 1. A rotating ball of 20 and 30 mm diameter was pressed on the coating surface with a preselected load. The position of the sphere relative to the sample and the contact load were kept constant. Abrasive slurry was added to the contact zone and a depression with the shape of a spherical cap was abraded into the coating and the substrate. The resulting impression was viewed under the image analyzer system at 100X magnification and values of X and Y was recorded.

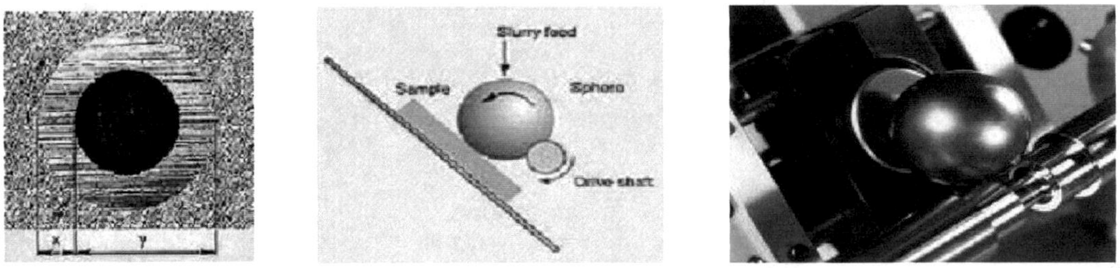

Figure 1. Principle of calo-tester [4]

2.3.2 Cohesion and adhesion behavior

Adhesive behavior of the coating was investigated using Rockwell Hardness Tester (C scale with an applied load of 150 Kg). The resulting impression was investigated under the Image Analyzer System and comparison was done with the chart shown in the figure 2. The acceptable adhesion depends on consumer application:

Normal HSS tools: HF1-HF4 acceptable
Normal HM tools: HF1-HF3 acceptable
High performance tools: HF1-HF2 acceptable

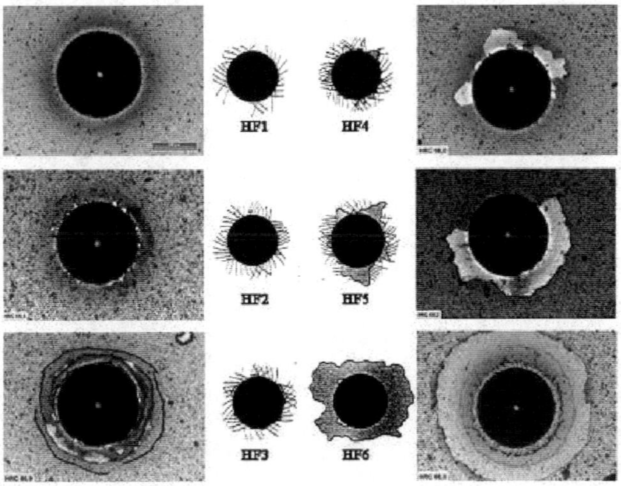

Figure 2. Standard coating adhesion chart [5]

2.3.3 Tribological study

Tribological behavior of the coating was investigated using Atomic Force Microscope (AFM) in the contact mode. Si_3N_4 Tip was used and a 10 μm scan rate was adopted for all the coatings. For each coating, three points were investigated and average values of Roughness (Ra and Rq) were reported.

2.3.4 Mechanical properties

Hardness and Modulus values of the coatings were measured using CSM Nano Indentation Testing equipped with a berkovich diamond indenter (a three-sided pyramid). The Oliver and Phar method was used for hardness calculation [6]. The applied load was 10 mN, while maximum indentation depth was set at 250 nm, which is less than one tenth of the coating thickness. Thus the substrate effect on the measured hardness can be neglected.

3. Results & discussion

3.1 Adhesion testing

Two batches of nACo with coating thickness of 2 and 2.5 micron were deposited. The standard cleaning procedure was adopted on the substrate before coating. Rockwell hardness test was used to access Rockwell adhesion. This test method uses a standard Rockwell hardness tester causing damage adjacent to the boundary of the indentation. The damage to the film was compared with a defined adhesion strength quality. The adhesion values are measured in HF values. HF1-HF4 is defined as good adhesion while HF5-HF6 is defined as poor adhesion. The adhesion values in the range of HF1 were obtained as sown in figure 3 (a and b). These values are showing an excellent adhesion of nACo with coating thickness of 2 and 2.5 microns.

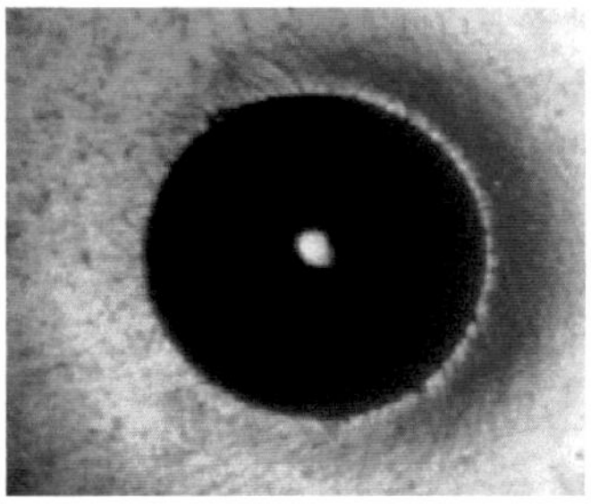

a) nACo coating 2 micron (HF1)

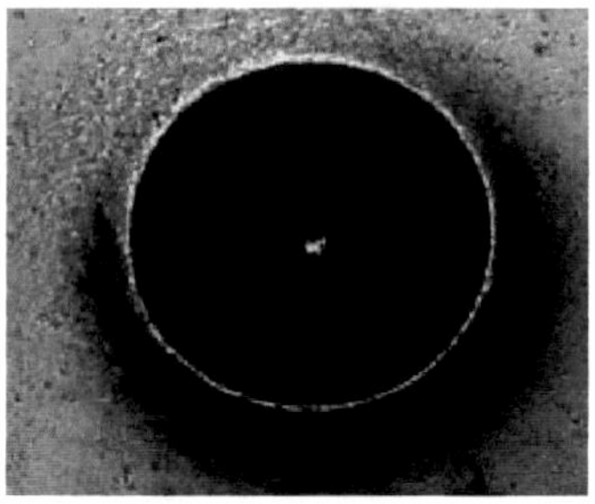

b) nACo coating 2.5 micron (HF1)

Figure 3. Adhesion testing of nACo coatings.

Similarly, two batches of nACRo with coating thickness of 2 and 2.5 micron were deposited. The result shows that the adhesion of nACRo coatings are in the range of HF3 as shown in figure 4 (a & b) which show good adhesion but lower in quality than nACo coatings.

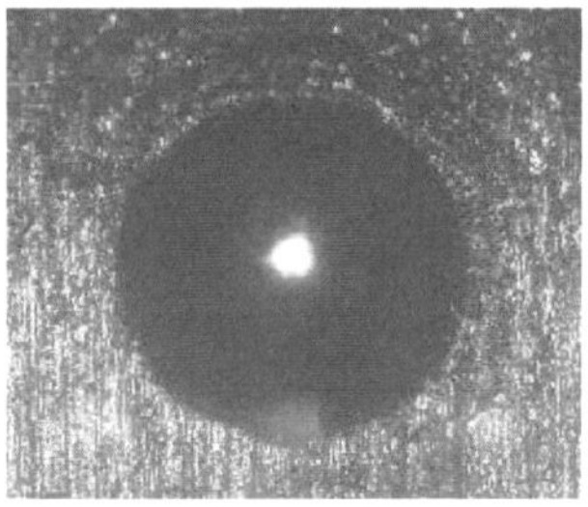

a) nACRo 2 micron coating (HF3)

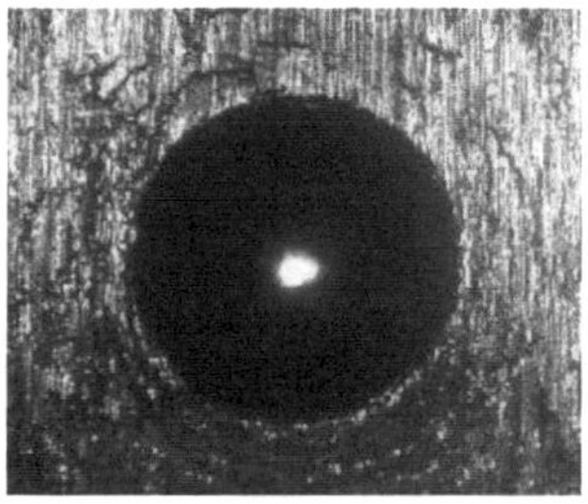

b) nACRo 2.5 micron coating (HF3)

Figure 4. Adhesion Testing of nACRo coatings

3.2 Coating Thickness Measurement

The thickness of coatings deposited on cutting tools is one of the most important characteristic for their practical application as it affects the cutting life of these tools. Rapid and simple determination of coating thickness is carried out by a Calotester. The results of coating thickness values of nACo and nACRo batches are in acceptable range. Table 1 depicts that thickness values are somewhat lower than the values which are set in the recipe during the coating. This loss in thickness can be understood as some of the coating material sticks to the chamber walls and sample holders as well; resulting in slight decrease of the thickness values.

By measuring the parameters X and Y, the thickness of the coating can be calculated by a simple geometrical equation.

$$d = \frac{X^2 - Y^2}{4 \times D \times 1000}$$

Where;

d = Coating thickness (in microns)
X = Diameter of the outer circle
Y = Diameter of the inner circle
D = Diameter of Ball used in the experiment

Table 1. Coating Thickness Values of all the four batches

	X (mm)	*Y (mm)*	*D (mm)*	*d (μm)*
nACo 2μm	563.30	429.50	20	1.66
nACo 2.5μm	474.10	280.90	20	1.82
nACRo 2μm	560.10	430.50	20	1.60
nACRo 2.5μm	860.20	693.50	30	2.15

3.3 Tribological Properties

Tribological behavior of nACo and nACRo batches was studied on AFM. Table 2 shows the roughness values of these coatings.

Table 2. Roughness values of nACo(2 & 2.5 μm) and nACRo coatings (2 & 2.5 μm)

	nACo (2 microns)	**nACo (2.5 microns)**	**nACRo (2 microns)**	**nACRo (2.5 microns)**
Ra	25.22 nm	27.45 nm	29.52 nm	34.06 nm
Rq	32.45 nm	35.37 nm	50.56 nm	58.99 nm

The results show that the average roughness (Ra) values of nACo coatings (both 2 and 2.5 microns) are 25.22 nm and 27.45 nm simultaneously while; the root mean square roughness (Rq) are 32.45 nm and 35.37 nm. The Ra values of nACRo coatings (both 2 and 2.5 microns) are 29.52 nm and 34.06 nm simultaneously while; Rq values are 50.56 nm and 58.99 nm. This comparison shows that both Ra and Rq values of nACo coatings are lower than nACRo Coatings. Figures 5 & 6 explain the 3D analysis of nACo with coating thickness of 2 & 2.5 μm while figures 7 & 8 depict the 3D images of nACRo with coating thickness 2 & 2.5 μm. The 3D images of nACo coatings clearly showing the fineness of surface having lower roughness values than the 3D images of nACRo coatings.

Figure 5. 3D Image of nACo Coating (2 microns)

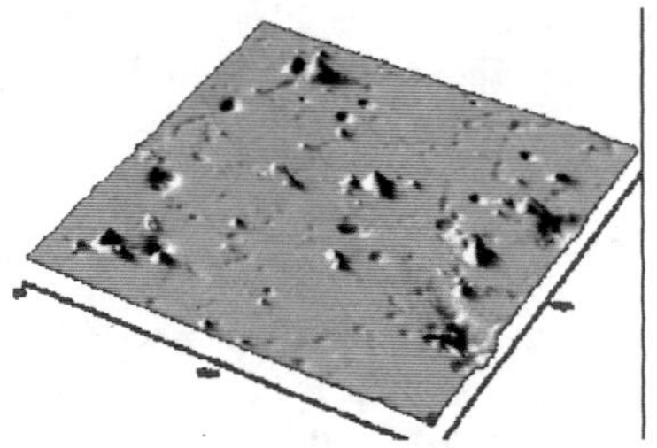

Figure 6. 3D Image of nACo Coating (2.5 microns)

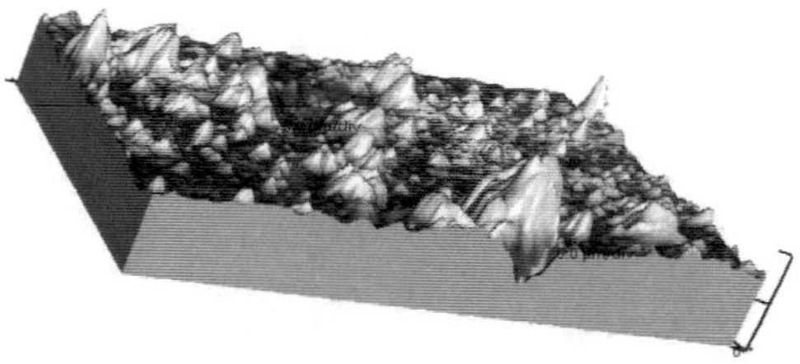

Figure 7. 3D Image of nACRo Coating (2 microns)

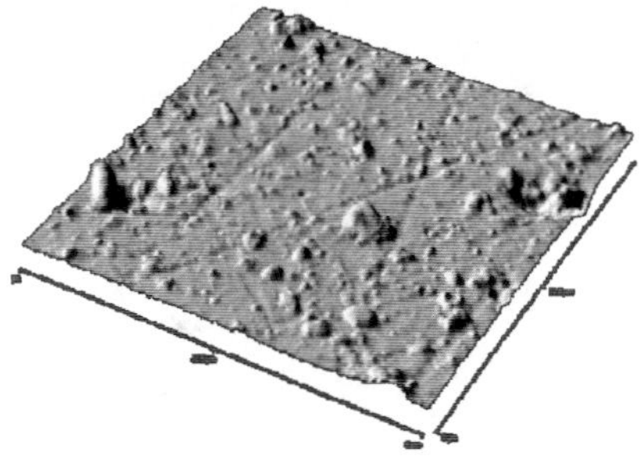

Figure 8. 3D Image of nACRo Coating (2.5 microns)

3.4 Mechanical Properties

Table 3 is showing the hardness (H_{IT}) and modulus of elasticity (E_{IT}) values for nACo and nACRo coatings with 75% loading factor. The applied load was 10mN to ensure that penetration depth should not exceed 200 nm

Table 3. Hardness and Modulus values of nACo (2 & 2.5 μm) and nACRo Coatings (2 & 2.5 μm)

	nACo (2 microns)	**nACo (2.5 microns)**	**nACRo (2 microns)**	**nACRo (2.5 microns)**
H_{IT}	45.10GPa	46.10GPa	48.73 GPa	49.88 GPa
E_{IT}	543.29 GPa	553.42 GPa	466.56 GPa	476.58 GPa

The results show that the hardness (H_{IT}) values of nACo coatings (both 2 and 2.5 microns) are 45.10GPa 46.10GPa simultaneously while; the values of modulus of elasticity (E_{IT}) are 543.29 GPa and 553.42 GPa. The H_{IT} values of nACRo coatings (both 2 and 2.5 microns) are 48.73 GPa and 48.73 GPa simultaneously while; E_{IT} values are 466.56 GPa and 476.58 GPa. This comparison show that the Hardness values increases as the thickness of coating increases whereas the hardness values of nACRo are greater with lower modulus of elasticity than the values of nACo coatings.

4. Conclusions

Nano composite coatings produced a hard layer on the surface of textile guides; certainly these layers will improve the life and other properties of tool which are required in textile industry. Both coatings enhance the efficiency of cutting tool but mechanical evaluation confirms that nACRo coatings are better than the nACo Coatings in performance. In present work, nACo and nACRo coatings were deposited by changing the loading factors (50% to 75%) during vacuum processing. It was observed that fixture; surface cleaning steps of substrate and vacuum surface areas (loading factors) has influenced the adhesion and thickness properties of the coated film. The 75% loading factors greatly enhance the efficiency of tools coated with both nACo and nACRo structured films.

References

[1] Advanced Thermally Assisted Surface Engineering Processes' Kluwer Academic Publishers, MA, USA (now Springer, NY), 2004, ISBN 1-4020-7696-7, E-ISBN 1-4020-7764-5
[2] R Chattopadhyay, "Surface Wear- Analysis, Treatment, & Prevention", ASM-International, Materials Park, OH, USA, 2001, ISBN 0-87170-702-0
[3] Mária HAGAROVÁ, "Experimental Methods of Assessment of PVD Coatings Properties", Journal of Metals, Materials and Minerals. 2007 **17(2)**, pp.29-35
[4] http://www.csm-instruments.com/new/contenus/e/shtml/calotest.shtml
[5] http://platit.com/periphery/quality-control-system?page=0%2C1
[6] W.C. Oliver, and G.M. Pharr, "An improved technique for determining hardness and elastic modulus using load and displacement sensing indentation experiments", *J. Mater. Res.*, vol. **7(6)**, 1992, pp. 1564-1583

Journal of Physics: Conference Series **439** (2013) 012017 doi:10.1088/1742-6596/439/1/012017

Effect of doping of multi-walled carbon nanotubes on phenolic based carbon fiber reinforced nanocomposites

Sadaf Saeed[1,*], Saira Hakeem[1], Muhammad Faheem[1], Rashid Ahmed Alvi[1], Khawar Farooq[1], Syed Tajammul Hussain[2], and Shahid Nisar Ahmad[2]

[1] National Institute of Vacuum Science and Technology, Islamabad, Pakistan

[2] National Center for Physics, Quaid-i- Azam University, Islamabad, Pakistan

[*] *E-mail : sadaf03_2000@yahoo.com*

Abstract. We report on the effect of multi-walled carbon nanotubes (MWCNTs) on different properties of phenolic resin. A low content of MWCNTs (~ 0.05 wt%) was mixed in phenolic resin and a stable dispersion was achieved by ultrasonication, followed by melt mixing. After curing the characterization of these composites was done by using scanning electron microscopy (SEM), thermogravimetric analysis (TGA), and Fourier transform infra-red spectroscopy (FTIR). The thermal and ablative properties of carbon fiber reinforced MWCNTs-phenolic nanocomposites were also studied. The addition of MWCNTs showed improvement in thermal stability and ablation properties.

1. Introduction

Fiber reinforced thermosetting resins have found a widespread use in variety of applications like household items, auto body-panels, aerospace structural components, sports equipments, printed circuit boards (PCBs), high-performance composites, and many more. Phenolic resin, the first synthetic thermoset polymer, is inherently fire resistant. Phenolic resins are commonly based on phenol (carbolic acid) and formaldehyde. High char yield, good thermal stability, low smoke production on burning, and low flammability are the characteristics that make phenolic resins attractive candidate for various applications in the electrical, automotive, construction, and appliance industries. The phenolic resins are widely used in thermal insulation materials, molding compounds, coatings, and composite materials [1-3].

The extraordinary properties of carbon nanotubes (CNTs) have brought new structural concepts in the field of composite materials, and the use of CNTs as nano-fillers for reinforcement of composites has attracted tremendous attention [4-8]. Small addition of CNTs to polymer matrix may drastically change the properties of the polymer in terms of strength, flexibility, thermal conductivity, and electrical conductivity. The doping of carbon nanotubrs in polymer composites as nano-filler has been studied for thermal management [9-12]. Although, the CNTs have been extensively exploited for their use as nano-fillers in various polymers [13-17], but, when it comes to phenolic resins, the field is still open and relatively less attention has been paid to these resins. Mostly pristine CNTs were directly mixed in the phenolic resin for the preparation of CNTs filled nanocomposites [18-20].

Two elements are essential to achieve the desired properties of the nanocomposites; homogeneous dispersion of CNTs in the resin, and degree of interaction between the CNTs with the host matrix [21]. CNTs tend to agglomerate due to strong van der Waal interactions present among them. So they bundle up in the resin, and homogeneous dispersion of CNTs in the resin becomes a difficult task. Even at higher loading of CNTs, poor dispersion may result in a failure to achieve the expected properties of the nanocomposites. The homogeneous dispersion of multiwalled carbon nanotubes (MWCNTs) in poly matrix may stabilize the morphology and improve the properties of the host matrix. This may be attributed to the fact that when CNTs are homogeneously dispersed in the resin matrix, the interactions between MWCNTs and the resin matrix becomes stronger [22, 23], and the load is efficiently transferred from CNTs to the resin. Also, hydrogen bonding between CNTs and polymer chains can enhance the degree of dispersion of CNTs in polymeric matrix. Thus the effective utilization of CNTs in composites strongly depends upon their homogeneous dispersion in the host matrix, which plays a key role to maximize properties of composite system.

We have incorporated MWCNTs into phenolic resin using melt mixing technique. A low loading (0.05 wt%) of MWCNT was dispersed in the phenolic resin. Test samples of CNTs dispersed composites were prepared from prepregs, using compression molding technique, to study the thermal properties of the nanocomposites. The major goal of this study is to evaluate the effect of MWCNT on thermal properties of carbon fiber/phenolic nanocomposites.

2. Experiment

2.1 Materials

Multiwall carbon nanotubes used for the preparation of nanocomposites were obtained from NCP, Islamabad. They were produced by chemical vapor deposition method (purity 95%). Two types of CNTs were present; extra long filaments of carbon with larger diameters, and CNTs in the diameter range of 50-150 nm. The properties of the phenolic resin and carbon fiber used for the preparation of samples are summarized in table 1 and 2, respectively. Ethanol used was 99.9% pure.

Table 1. Specification of phenol formaldehyde resin

Property	Value
Appearance	Dark red
PH	7.4
Dry material content (%)	78-78
Viscosity (cp)	258 at RT
Free phenol content (%)	3.54
Free formaldehyde content (%)	< 0.7
Gel time (sec)	120 at 90°C

Table 2. Specification of carbon fiber

Property	Value
Trade name	T800H
Filament count	12000
Density (gcm^{-3})	1.81
Tex count (g/1000m)	445
Filament diameter (μm)	5
Tensile strength (MPa)	5490
Tensile modulus (GPa)	294

2.2 Nanocomposite preparation

Purification of as received MWCNTs was done by heating at 400°C. To achieve a better state of dispersion, the MWCNTs were first dispersed in ethanol by sonication for 2 hours. This was followed by centrifugation at 6000 rpm to separate the carbon filaments and CNTs. The suspended CNTs were decanted, followed by drying at 60^0C. These CNTs were put in ethanol and sonication was done for 30 minutes. Phenolic resin was added to the CNTs solution and mechanical mixing was done for 2 hours. The apparently homogeneous mixture was then melt-mixed. Melt mixing was done by magnetic stirring of the mixture on a heating plate, and a temperature of 45^0C was maintained for 8 hours. Finally, evacuation was done in a vacuum chamber to remove the remaining solvent. Chopped carbon fiber was impregnated with phenolic resin-CNTs mixture to form a prepreg, using a fiber to resin ratio of 66: 34. The composite samples were prepared form the prepreg by compression molding method at 165°C. A pressure of 5MPa was applied for 30 minutes. The diameter and thickness of the samples were 34.6mm and 5mm, respectively.

2.3 Sample characterization

The characterization of the mixture and the cured composites included the following analysis:

a. Thermal Analyses: Thermogravimetry (TG) was conducted in a Netzsch (STA 449C) equipment. Heating of the samples was done at a heating rate of 10 °C/min from 25°C to 900°C.
b. FTIR Analysis: Fourier transform infrared spectroscopy (FTIR) analysis was done in Perkin-Elmer with a resolution of 4 cm^{-1}, from 4000 to 400 cm^{-1}, in transmission mode.
c. Ablation Test: To evaluate the ablation performance of the composites, the oxyacetylene flame test was carried out according to ASTM-E-285-80.
d. Morphological Characterization: Morphological characterization of the samples was carried out using a scanning electron microscope (Jeol 6480, LV).

3. Results and discussion

The multi-walled carbon nanotubes used in this work were produced by chemical vapor deposition method. The infrared (IR) spectroscopy method has been adopted by many researchers to study the surface properties of CNTs and the bonding between the CNTs and the host matrix.

3.1 FTIR data

The comparison of FTIR data of neat phoenolic resin and CNTs incorporated phenolic resin is shown in figure 1. It suggests that once the nanotubes were incorporated into the Phenolic resin a band at 1699 cm^{-1} was generated due to the formation of linkage between CNTs and the phenolic resin.

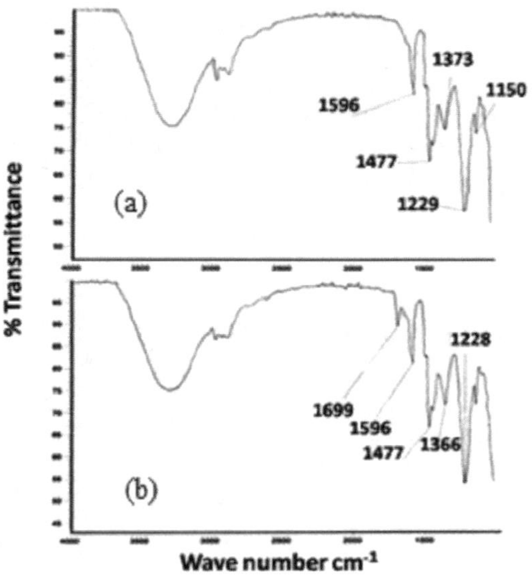

Figure 1. Comparison of FTIR spectrum of phenolic resin. (a) without CNTs, (b) with CNTs

3.2 SEM data

The SEM images of raw CNTs and the processed CNTs (after purification and centrifugation) are shown in figure 2. These images are taken almost at the same magnification. The difference in the diameters between the as received and processed CNTs can clearly be seen. The raw CNTs show a broad size distribution; maximum diameter of ~ 450 nm was observed in this case (shown in part (a) of the figure 2). On the other, the CNTs after centrifugation have a size distribution in the range of 50-150 nm (shown in part (a) of the figure 2).

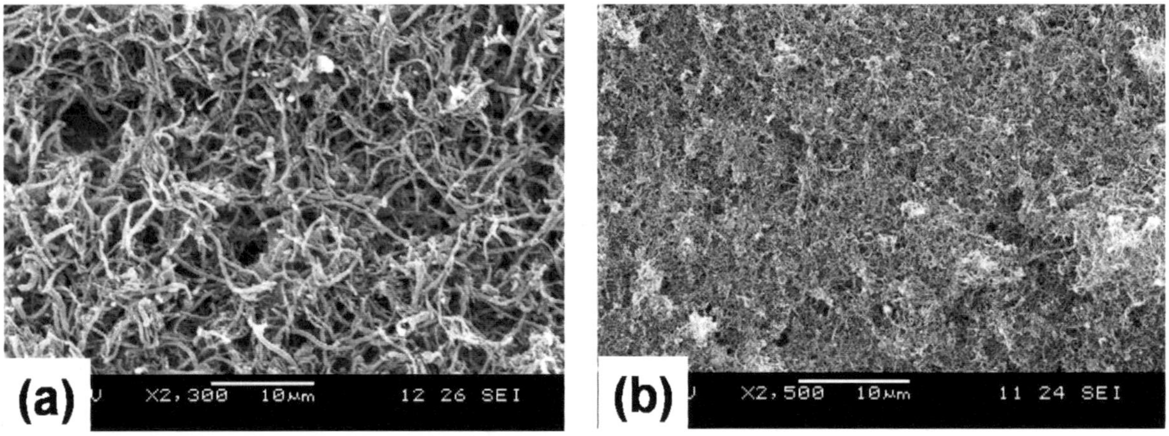

Figure 2. Comparison of SEM images of CNTs. (a) raw CNTs, (b) centrifuged CNTs

3.3. Thermal behavior (TGA data)

The effect of the MWCNTs on the thermal stability of phenolic resin was studied using thermogravimetric analysis (TGA). The TGA curves for phenolic resin and MWCNTs/Phenolic resin nanocomposites, at a heating rate of 20°C/min under an inert atmosphere, are presented in figure 3. The data reveals that thermal degradation of pure phenolic resin and MWCNTs/ phenolic resin nanocomposites exhibits three thermolytic temperature zones. The zone of 100-350°C is the thermal decomposition zone of free or imperfectly cross linked molecules [24]. These molecules include water, unreacted phenol and formaldehyde. The weight loss by the evolution of unreacted monomers contributes up to 5-8%.

The second and third zones at 400-625°C and 625-900°C, respectively, can be attributed to the decomposition of the cross linked nanocomposites with the evolution of carbon mono oxide, carbon dioxide, methane, phenols and cresols. The release of these molecules can be correlated to the post cure reactions and the formation of dihydroxy benzophenone linkages.

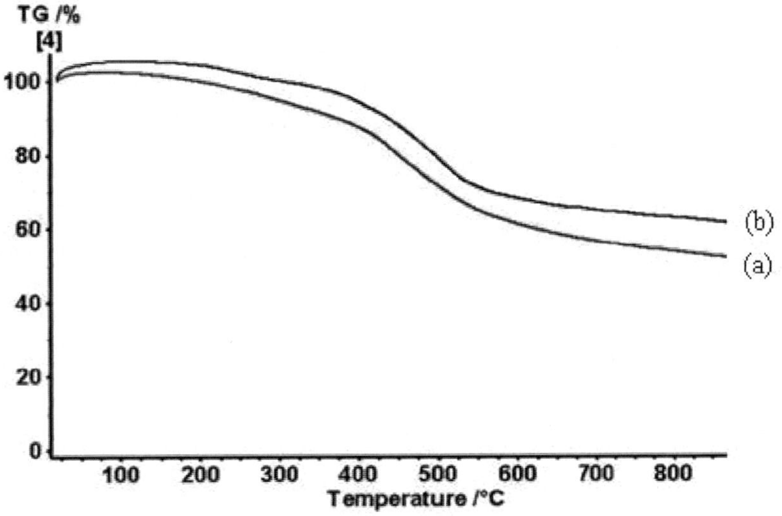

Figure 3. TGA curves of phenolic resin (a) and CNT dispersed phenolic resin (b) nanocomposite

The residual weight (i.e. char yield) is an important performance index for phenolic resin and its composites. Table 3 summarizes the residual weights at 900°C.

Table 3. Residual weight at 900°C

Material	**Residual wt % at 900°C**
0% CNTs-phenolic resin	47
0.05% CNTs-phenolic resin	49

The increased thermal stability can be attributed to better dispersion of CNT in the resin, resin matrix–CNTs interactions, and thermal conductivity of the nanotubes to transmit the heat. Similarly, the char yield of the nanocomposites at 900°C increases with small doping (0.05 wt %) of MWCNTs (the char yield increased from 47% to 49%).

3.3 Ablation Test Results

Ablation test was performed using oxyacetylene flame according to ASTM standard. The samples were exposed to the flame for 30 seconds. A comparison of ablation performance of the phenolic resin-CF composites with and without addition of CNTs is shown in Table 4. The addition of CNTs in the phenolic resin results in the decrease in average weight loss of the Phenolic-carbon fiber composite.

Table 4. Ablative properties of CNTs-PR-CF nanocomposites

Concentration of CNT (%)	Exposure time (Sec)	Average wt. loss (g/Sec)	Fractional wt loss	Comparative decrease (%) in mass ablation
0	30	0.074	--	--
0.05	30	0.057	0.2297	23

4. Conclusion

The primary objective of this paper was to study the effect of MWCNTs on the thermal properties of carbon fiber reinforced phenolic resin composites. The TGA results revealed that MWCNTs incorporated composites had better thermal stability than those without MWCNTs. Decrease in average weight loss was observed on addition of MWCNTs to carbon fiber reinforced phenolic resin composites. The decrease in mass ablation rate of carbon fiber reinforced phenolic resin, on addition of MWCNTs, suggests that MWCNTs may be used in flame retardant materials to enhance their efficiency. The increase in char yield is also observed on addition of MWCNTs which is also a promising result in this regard. CNTs-phenolic interface can play a major role in thermal resistance of the material. The only major issue is a homogenous dispersion of CNTs in the resin matrix which plays a vital role.

5. References

[1] Mathur R B, Singh B P, Dhami T L, Kalra Y and Lal N 2010 *Polym. Compos.* **31** 321
[2] Natali M, Kenny J and Torre L 2010 *Compos. Sci. Technol.* **70** 571
[3] Abdalla M, Dean D, Adibempe D, Nyciro E, Robinson P and Thompson G 2007 *Polymer.* **48** 5662
[4] Grady B P 2010 *Macromol. Rapid. Commun.* **31** 247
[5] Klier J, Tucker C J, Kalantar T H and Green D P 2000 *Adv. Mater.* ***12*** 1751
[6] Chen G X, Kim H S, Park B H and Yoon J S 2006 *Polymer.* **47** 4760
[7] Sengupta R, Ganguly A, Sabharwal S, Chak T K and Bhowmick A K 2007 *J. Mater. Sci.* **42** 923
[8] Kang M, Myung S J and Jin H J 2006 *Polymer.* **47** 3966
[9] Biercuk M J, Llaguno M C, Radosavljevic M, Hyun J K, Johnson A T and Fisher J E 2002 *Appl. Phys. Lett,* **80** 2767
[10] Choi E S, Brooks J S, Eaton D L, Al-Haik M S, Hussaini M Y, Garmestani H, Li D and Dahmen K 2003 *J. Appl. Phys.* **94** 6034
[11] Kim Y A, Hayashi T, Endo M, Gotoh Y, Wada N and Seiyama J 2006 *Scr.Mater.* **54** 31
[12] Knop A and Pilato L A 1985 *Phenolic resins. Chemistry applications performance and future direction. Berlin. Springer.*
[13] Ma C C M, Lee C T and Wu H D 1998 *J. Appl. Polym. Sci.* **69** 1129
[14] Gao J G, Liu Y F and Yang L T 1999 *Polym. Degrad. Stab.* **63** 19
[15] Martin C, Ronda J C and Cadiz V 2006 *J. Polym. Sci. A Polym Chem.* C **44** 1701
[16] Lin J M and Ma C C M 2000 *Polym. Degrad. Stab.* **69** 229
[17] Yeh M K, Tai N H and Liu J H 2006 *Carbon.* **44** 1
[18] Igarashi A, Terasawa T, Kanie M, Yamanobe T and Komoto T 2005 *Polym. J.* **37** 522
[19] Tai N H, Yeh M K and Liu J H 2004 *Carbon.* **42** 2774
[20] Calvert P 1999 *Nature.* **399** 210

[21] Bose S, Rupesh A, Khare R A and Moldenaers P 2010 *Polymer.* **51** 975
[22] Spitalsky Z, Tasis D, Papagelis K and Galiotis C 2010 *Prog. Polym. Sci.* **35** 357
[23] Fyfe C A, McKinnon M S, Rudin A and Tchir W J 1983 *Macromolecules.* **161** 1216

Photocatalytic thin films coupled with polymeric microcapsules for the controlled-release of volatile agents upon solar activation

L F Oliveira[1], J Marques[1], P J G Coutinho[1], P Parpot[2], C J Tavares[1*]
[1]Centre of Physics, University of Minho, 4710-057 Braga, Portugal
[2]Centre of Chemistry, University of Minho, 4710-057 Braga, Portugal

[*]E-mail: ctavares@fisica.uminho.pt

Abstract. This work reportson the application of solar-activated photocatalytic thin films that allow the controlled-release of volatile agents (e.g., insecticides, repellents) from the interior of adsorbedpolymericmicrocapsules. In order to standardize the tests,a quantification of the inherent controlled-release of a particular volatile agent is determined by gas chromatography coupled to mass spectroscopy, so that an application can be offered to a wide range of supports from various industrial sectors, such as in textiles (clothing, curtains, mosquito nets). This technology takes advantage of the established photocatalytic property of titanium dioxide (TiO_2) for the use as an active surface/site to promote the controlled-release of a specific vapor (volatile agentfrom within the aforementioned microcapsules.

1. Introduction

The microencapsulation for the controlled-release of dispersive agents appeared very recently with applications in various areas [1-3]. In one case the authors report an application with emphasis in killing or repelling insectsfrom solar activated materials [4]. In this particular case, this nanotechnology refers to the controlled-release of vapors from within polymeric microcapsules mediated by mechanisms of photocatalysis. Titanium dioxide (TiO_2) is a semiconductor material increasingly reported as photoactive due to the intrinsic photocatalytic ability to dissociate organic pollutants adsorbed on its surface upon direct action of solar light [5-7].Hence, these oxidation-reduction mechanisms can be triggered to dissociate and break-down the polymeric structure of capsules and thus yield a controlled-release of volatile agents (e.g., insecticides, repellents, deodorants, fragrances, etc.). In the case of insecticides and repellents, this technology should be very promising for the prevention of diseases transmitted by airborne insect carriers, such as dengue and malaria; in particular for remote areas in third-world countries without access to electrical supply. The photocatalytic material can be either in the form of a nanostructured thin film or nanoparticles, depending on the final support where they will be coated onto. In Figure1 it is illustrated the layers of the heterostructured material, comprising: a chosen substrate (1) that can be of several types, such as glass, plastic, ceramic, metal, stone, wood, textile, amongst others; a photocatalytic thin film (2) where subsequently the volatile agent-containing microcapsules (3) can be adsorbed onto.

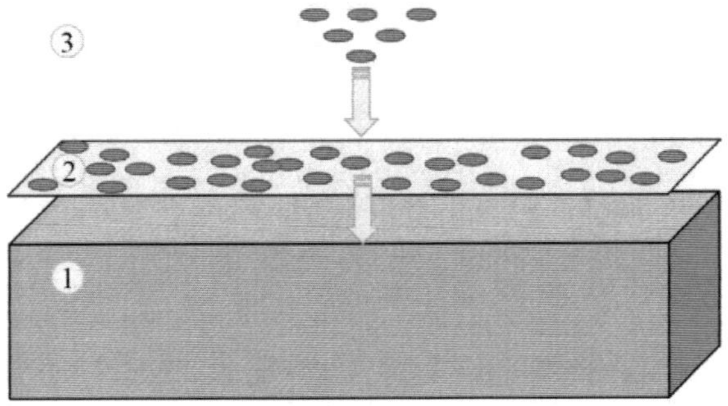

Figure 1. Layers representing the heterostructured material: substrate (1); photocatalytic thin film (2); volatile agent-containing microcapsules (3).

This research focuses on the production of a functional heterostructured material, which may be deposited or attached to various types of surfaces (glass panels, tiles, lamps, textile nets and clothes) in order to induce the controlled-release of a volatile agent. For the particular case of an insecticide or repellent, during daylight the volatile agent is gradually released, being the controlled-release output proportional to the solar radiation projected onto the surface of the photocatalyst-microcapsule system [4]. The solar-activated material has been produced in the form of thin films by reactive sputtering deposition on glass substrates. Several inherent experimental parameters allow the optimization of the resulting thin films. In this work, a specific focus is given to the sputtering gas flow rate. Other experimental parameters such as target power density have been published elsewhere [8,9].Furthermore, the microcapsule synthesis, volatile agent encapsulation and controlled-release monitoring proceduresare also described. Dodecane was selected as a test volatile agent and polyamide/poly(vinyl alcohol) for the formation of the microcapsule. Its output release has been monitored by gas chromatography coupled with mass spectroscopy (GC-MS).

2. Materials and methods

2.1 Materials

All reagents used for preparation of volatile agent and polymeric microcapsules were supplied by Sigma-Aldrich. Deionized water served as the solvent. The microcapsules were synthetized by poly(vinyl alcohol) (PVA – $(C_2H_4O)_x$), ethylenediamine ($C_2H_8N_2$) and sebacoyl chloride ($C_{10}H_{16}Cl_2O_2$) which determines the wall thickness[10], with dodecane ($C_{12}H_{26}$) as encapsulated volatile agent.

2.2 Methods

2.2.1 Synthesis of the TiO_2thin films

The titanium dioxide thin films were synthesized by physical vapor deposition (PVD). In reactive magnetron sputtering mode, impinging argon ions eject titanium atoms from a high purity Ti target, which are subsequently bonded to reactive oxygen molecules when condensing in the glass substrate. Prior to deposition, a base pressure of the order of 10^{-4}Pa was attained in the vacuum chamber. The flow of the oxygen reactive gas was kept constant at 8 sccm for all depositions, while the argon flow rate (working gas) was varied between 35and 275 sccm. A small amount of nitrogen (<30% of total reactive gas) was also introduced in order to dope the TiO_2 lattice with substitutional N anions, which has been proven to reduce the semiconductor band-gap (3.2 eV) and hence enhance the absorption of visible light beyond this absorption threshold [11]. The level of anionic N-doping in the titanium dioxide lattice has been determined elsewhere [12], and is of the order of ~1.2 at. %.The as-deposited coatings are mostly amorphous; in order to induce crystallinity, these coatings were post-annealed in a high-vacuum furnace, at 500 ºC. With this procedure, an improvement of the optical and photocatalytic properties of the thin films was achieved. The average film thickness is ~500 nm.

2.2.2 Synthesis of the formation process of capsules and encapsulation
Polyamide/PVA was chosen for the polymeric microcapsule material that encloses the volatile agent. Under ultrasonic conditions, 265 μL of ethylenediamine was added to the 25 mL of PVAaqueous solution (2%), followed by the addition, drop-by-drop, of a solution of sebacoyl chloride (215 μL) in 2.5mL of dodecane (the volatile agent). Finally, the emulsion was kept for 5 minutes under ultrasonic agitation and an additional 5 minutes under magnetic stirring. The excess of solvent was removed by washing the microcapsules with deionized water.Sebacoyl chloride and ethylenediamine are essential for the encapsulation of dodecane into the microcapsules as an interfacial polymerization process occurs at the water/dodecane interface leading to a polyamide/PVA polymer film that encloses the organic phase.The resulting microcapsules have a liquid doughy aspect.As the physical state of the dodecane-loaded microcapsulesis distinctfrom that of the photocatalytic thin films, it is possible to carry out their bondingby simple physical adsorption without using any chemical reaction; thus, a dispersion of microcapsules was adsorbed onto the photocatalytic thin film samples.

3. Results and discussion

3.1 Evaluation of the photocatalytic activity of the TiO_2thin films
The assessment of the photocatalytic activity is essential to characterize the efficiency of the TiO_2 thin films. For such, the thin films were immersed in an aqueous solution of a dye, methylene blue (MB),acting as an organic pollutant simulator, and subjected to ultraviolet (UV) irradiation for six hours, with an excitation broad-band centered at 380 nm (UV-A). After measuring the dye's absorbance decay withirradiation time (t), and considering its initial concentration (C_o=10^{-5} M), it was then possible to determine the first-order kineticdegradation rate constant, k, from $ln(C/C_o)=-kt$, where C represents the dye's concentration, which is proportional to its optical absorbance.

A correlation of the sputtering working gas (argon) flow rate during deposition with k is illustrated in Figure 2. It was noticed that an increase in argon flow rate up to 250 sccm enhanced the dye degradation efficiency of thin films to a maximum value of 5.3×10^{-3} min^{-1}; above 250 sccm this degradation rate was slightly reduced (5.1×10^{-3} min^{-1}).It should be noted that for the TiO_2:N films deposited with higher Ar flow rates during sputtering, 250 and 275 sccm, the $ln(C/C_o)$evolution with time is biphasic; the k values for these two samples that are presented in Figure 2 are ascribed to the first hour of the experiment; the corresponding values for the subsequent assay are 2.1×10^{-3} min^{-1} and 1.2×10^{-3} min^{-1}, respectively. For optimized conditions, after 6 hours of UV-A irradiation with a relatively low irradiance of ~2 mW/cm^2, the main absorbance peak of the MB dye, and thus its concentration, decreased more than 60%. The variation in k between the minimum and maximum value is approximately one order of magnitude.

Another method to test the photocatalytic effect is to observe the hydrophilicity nature of the surface upon light exposure. In figure 3 it can be seen the evolution of the water contact angle of a 20μL water droplet on the surface of three TiO_2thin films, deposited with varying sputtering Ar flow (in sccm),as a function of irradiation time. The data corroborates with the former data presented in figure 2; as the working gas flow rate increases up to 250 sccm the water contact angle is decreased from ~70º to ~20º; above 250 sccm the water contact angle increases to >40º (not shown). For the hydrophilicity experiments, the samples were pre-irradiated for determined period of time with a 150 W Xe arc lamp, using its full spectrum and with an irradiance level of ~10mW/cm^2; the contact angle measurements were averaged from 3 droplets per sample at the specified time interval.

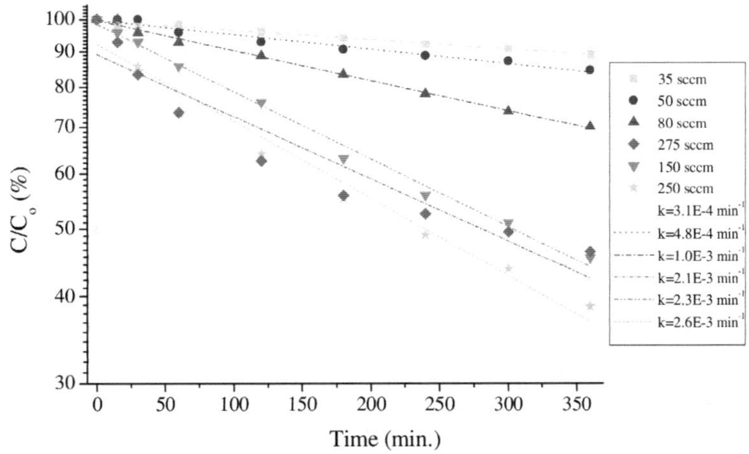

Figure 2.Variation of the 1st-order dye degradation rate as a function of working gas (Ar) flow during the sputtering process.

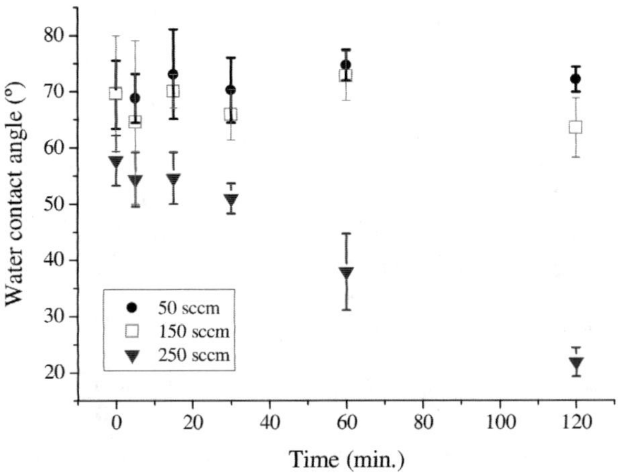

Figure 3.Variation of the water contact angle as a function of working gas (Ar) flow during the sputtering process.

X-ray diffraction (XRD) is essential for the crystallographic analysis of the thin films, allowing, in this particular case, the identification of anatase and rutile as the active crystalline phases of TiO_2, as it can be seen in the diffractograms presented in Figure4, where A and R stand for anatase and rutile, respectively. For lower argon flow rates, and thus lower thin film deposition rates, both anatase and rutile phases are discerned, being the latter diminished for higher Ar flow rates. The average crystallite domain size is ~30 nm. This trend is compatible with the photo degradation results, as anatase is known to be much more active than rutile.

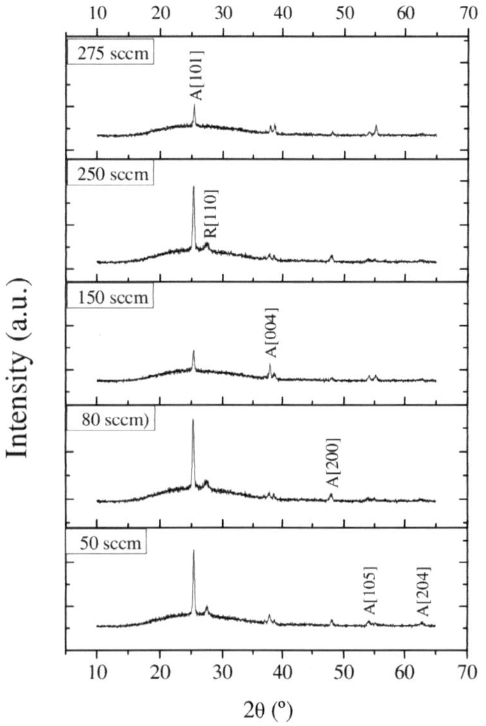

Figure 4. X-ray diffraction patterns for TiO_2 films produced by PVD under varying sputtering gas (Ar) conditions.

3.2 Evaluation the controlled-release by (GC-MS)

The main objective of this study is to upon ultraviolet irradiationevaluate the photocatalysis–induced dissociation process of polymeric microcapsule wall, and subsequent volatile agent release yield, in the presence of TiO_2thin films. The UV-light driven redox mechanisms are responsible for the microcapsule structure rupture witha subsequent release of the volatile agent. For this purpose, gas chromatography coupled to mass spectroscopy allows, besides accurate indexing, the quantification of the released compounds. Different experiments were performed: in the presence and absence of TiO_2 thin films, each with and without UV-A irradiation (6.7mW/cm^2, centered at 365 nm) for two hours. A polydimethylsiloxane (PDMS) fiberwas placed in a vial headspace for 20 minutes, for the adsorption of the released volatile compounds, and then it was proceeded with GC-MS analysis. The experiment was repeated in the absence of UV-A radiation and results were compared in order to verify the difference in intensity.

The GC-MS analysis of the UV-A irradiated samples containing dodecane-loaded microcapsulesadsorbed onto TiO_2 thin films revealedthe presence of dodecane as the principal volatile compound which was released.The chromatograph obtained for the pre-irradiated dodecane-loadedpolyamide/PVA microcapsules in the presence and absence of a TiO_2thin film is given in Figure 4a), while inFigure 4b) a comparison between with and without UV-A irradiation is shown. The chromatographs reveal the peak of dodecane for different experiments. Comparing the UV-A pre-irradiated sample with and without TiO_2(Figure 4a)it is evident that the area of the dodecane peak is much higher when the dodecane-loaded microcapsules are adsorbed onto the photocatalytic thin film. As it can be seen from both Figures5a) and b), even without TiO_2nor UV-A pre-irradiation, some dodecane is released, since the microcapsule hermetical property is still not perfect; this work is in progress and this deficiency has to be attenuated.

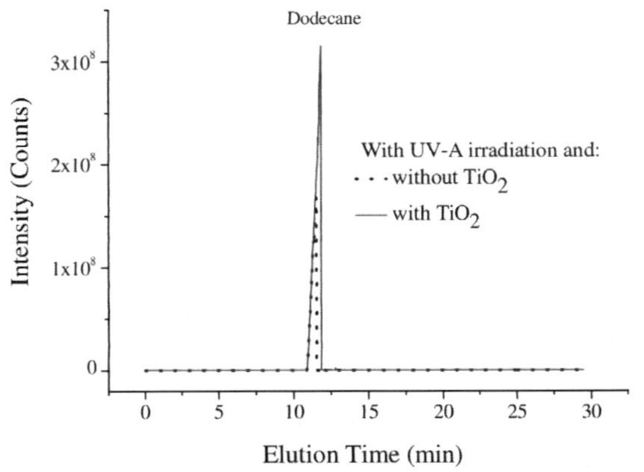

Figure 5a). Chromatograph obtained by GC-MS for dodecane-loaded polyamide/PVA with and without layering onto a TiO_2 thin film, after UV-A irradiation.

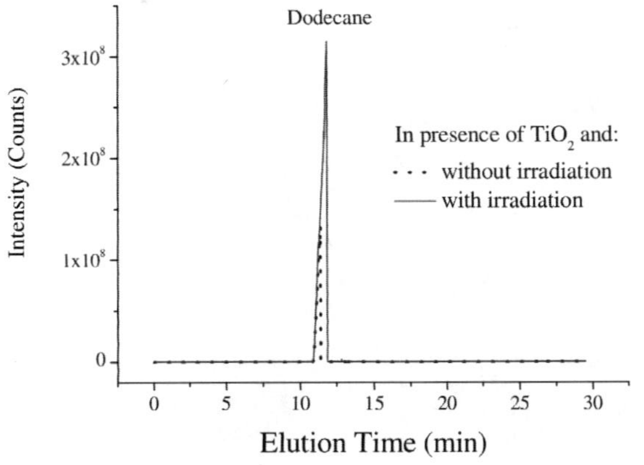

Figure 5b). Chromatograph obtained by GC-MS for dodecane-loaded polyamide/PVA microcapsules that were layered onto a TiO_2 thin film, with and without previous UV-A irradiation.

Figure 6. shows the output concentrations (in ppm) of dodecane obtained from the microcapsules, for different samples and experimental conditions. It can be seen that for this particular case and volatile agent formulation, the maximum output concentration of dodecane was of the order of 160 ppm.

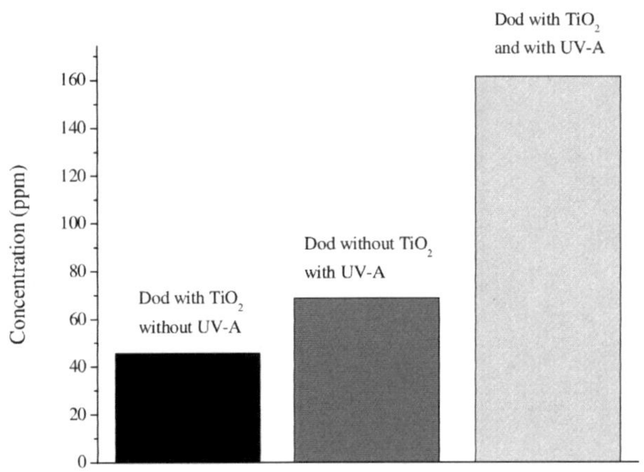

Figure 6. Output release of dodecane (Dod) from the polyamide/PVA microcapsules with and without the presence of the photocatalyst TiO_2 thin film and with and without previous UV-A illumination experiments.

Figure 7. shows a scanning electron microscopy (SEM) micrograph of the dodecane-loaded polyamide/PVA microcapsules, with diameters ranging from 70-100 μm. The spherical nature of these capsules and the porous structure of the outer wallscan be observed from this figure.

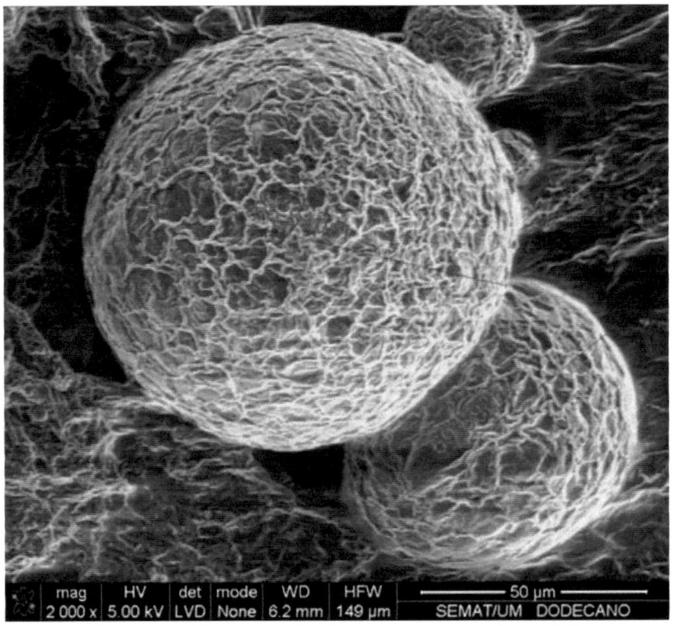

Figure 7. SEM micrograph of dodecane-loaded polyamide/PVA microcapsules

4. Conclusion

The results shown in this work permit to conclude that the overall photocatalytic performance of the TiO_2 thin films can be improved by tailoring specific process parameters, such as the sputtering working gas flow rate; as it was increased from 35 to 250 sccm, the efficiency in degrading a chosen organic dye and the hydrophilic nature of the surfacewere improved; however, above thelattervalue the efficiency was reduced. Furthermore, for higher Ar flow rates the rutile formation is inhibited, leaving anatase as the prominent photocatalytic active polymorph with an average crystallite size of ~30 nm. It was possible to demonstrate by means of GC-MS experiments that for optimized conditions upon UV-

A irradiation of dodecane–loaded polyamide/PVA microcapsules this compound was, in fact, released, identified and quantified. Nevertheless, further work must be endured to improve the polymeric microcapsule synthesis, in order to become more hermetical. Therefore, evidence is provided for the controlled-release of a chosen volatile agent by light activation of a thin film surface.

Acknowledgments
The authors gratefully acknowledge the funding from the Portuguese FCT COMPETE scientific programs, with references NANO/NTec-CA/0046/2007 and PTDC/CTM-NAN/119979/2010.

References

[1] Cohen S, Yoshioka T, Lucarelli M, Hwang L H and Manger R 1991 *Controlled delivery systems for proteins based on poly(lactic/glycolic acid) microspheres*, Pharmaceut. Res. **8 (6)** 713-720.
[2] Sukhorukov G B, Antipov A A, Voigt A, Donath E and Mohwald H 2001 pH-*Controlled macromolecule encapsulation in and release from polyelectrolyte multilayer nanocapsules*, MacromolRapid Comm. **22 (1)** 44-46.
[3] Jeong U, Im S H, Camargo P H C, Kim J H and Xia Y 2007 Microscale Fish Bowls: *A New Class of Latex Particles with Hollow Interiors and Engineered Porous Structures in Their Surfaces*, Langmuir 23 pp 10968-10975.
[4] Tavares C J and Pina F J S February 2, 2011 *Photocatalytic coating for the controlled release of volatile agents* (WO 2011/012935 A2, International Patent (PTC), World Intellectual Property Organization) .
[5] Fujishima A and Honda K 1972 *Electrochemical photolysis of water at a semiconductor electrode, Nature* **238** 37-38.
[6] Fox M A and Dulasy M Ts 1993 *Heterogeneous Photocatalysis, Chem. Rev.* **93** 341-357.
[7] Linsebigler A L, Lu G, Yates J T, Jr., 1995 *Photocatalysis on TiO_2 surfaces – Principles, mechanisms, and selected results, Chem. Rev.***95** 735-758.
[8] Tavares C J,Vieira J, Rebouta L, Hungerford G, Coutinho P, Teixeira V, Carneiro J O, Fernandes A J, 2007 *Reactive sputtering deposition of photocatalytic TiO_2 thin films on glass substrates, Mater. Sci. Eng. B* **138** pp 139–143.
[9] Marques S M, Tavares C J, Oliveira L F, Oliveira-Campos A M F, 2010 Photocatalytic degradation of C.I. Reactive Blue 19 with nitrogen-doped TiO_2 catalyst thin films under UV/visible light, *J.Mol. Struct.* **983** 147–152
[10] Soto-Portas M L, Argillier J F, Méchin F, Zydowicz N 2003, *Preparation of oily core polyamide microcapsules via interfacial polycondensation*, Polym. Int.52 pp 522–527.
[11] Asahi R, Morikawa T, Ohwaki T, Aoki K and Taga Y 2001 Visible-light Photocatalysis in Nitrogen-doped titanium oxides *Science* **293** 269-271.
[12] Tavares C J, Marques S M, Viseu T, Teixeira V, Carneiro J O, Alves E, Barradas N P, Munnik F and Girardeau T. Rivière J P 2009, "Enhancement of the photocatalytic nature of nitrogen-doped PVD-grown titanium dioxide thin films", *J. Appl. Phys.* **106** 113535-1-8

Production and characterization of Ni and Cu composite coatings by electrodeposition reinforced with carbon nanotubes or graphite nanoplatelets

MR Abdul Karim[1,3], M Pavese[1], EP Ambrosio[2], D Ugues[1], M Lombardi[1], S Biamino [1], C Badini[1], P Fino[1,2]
[1] Applied Science and Technology Department, Politecnico di Torino, corso Duca degli Abruzzi 24, 10129, Torino, Italy
[2] Center for Space Human Robotics @ PoliTO, Italian Institute of Technology, Corso Trento 21, 10129 Torino

E-mail: muhammad.abdulkarim@polito.it

Abstract. Electrodeposition is well-known as a versatile and economical processing technique to produce metal coatings on conductive substrates. Recently, it has been gaining increasing interest also for the production of tailored composite coatings, containing for instance floropolymers or silicon carbide. A more novel approach concerns the use of carbon nanotubes or even graphene, in the form of graphite nano-platelets. The production of Ni- and Cu-based nanocomposites containing carbon nanoreinforcements was carried out by using standard electrodeposition conditions, but with a particular attention to the dispersion of the nanotubes. The obtained coatings were strong and well adherent to the steel substrate, and presented rather well dispersed carbon nanotubes or graphite nanoplatelets, even if some agglomerates could be present in samples obtained from highly concentrated suspensions. In the case of nickel-based composite coatings, the size of nickel grains was reduced, and pin-on-disc tests demonstrated a significant increase in the life of the coating. In the case of copper-based composite coatings, thermal diffusivity measurements demonstrated that the carbon nanomaterial does not reduce the conductivity of the pure copper coating.

1. Introduction

Electrodeposition is nowadays a well-known and low cost technique for the preparation of metallic coatings on a conductive substrate [1]. Many metals can be deposited by this technique, with controlled thickness and properties [2]. Recently, composite coatings started to gain interest, due to the possibility to tailor the behaviour of the deposited film by controlling the composition, thus becoming an interesting alternative to standard deposition methods, like PVD [3,4]. In particular nickel coatings are industrially modified with silicon carbide or other ceramics [5-10], for improving hardness, or fluoropolymers [11-12], for reducing attrition and wear, without modifying mechanical properties of core material like it happens when using traditional or novel MMCs [13-15]. Since the discovery of carbon nanotubes [16], carbon nanostructures are gaining significant interest in the field of materials science, as they present very interesting properties, both mechanical and thermal [17-22]. In recent papers, carbon nanotubes or similar materials have also been used to modify the properties of

[3] To whom any correspondence should be addressed.

electrodeposited nickel coatings [23-28]. The composite coating seem to have increased properties with respect to un-modified nickel coatings [27,29,30], and they present interesting hardness [25,31] and wear properties [32]. In particular, both Carpenter and co-workers [31] and Khabazian and Sanjabi [25] demonstrated an increase in hardness with the increase of nanotubes. In both cases however the increase in hardness was possible only with pre-treated nanotubes, while untreated nanotubes hardness was not so different than the pure nickel coating. Functionalization was in both cases obtained by oxidation, in order to increase polarity of the nanotubes so that they could disperse easily in water. Regarding wear properties, Carpenter and co-workers [32] observed a significant reduction of wear rate for carbon-containing composites.

Recently, graphene has also gained much interest as a strong reinforcing carbon nanomaterial [33,34]. However, to the authors' knowledge, the deposition of graphene in electroplated coatings has not been widely studied.

In this work, we deposited nickel and copper coatings containing both carbon nanotubes (CNTs) and –COOH modified graphite nanoplatelets (GNPs). Oxidised graphite nanoplatelets are very similar to graphene since they consist of a very limited number of graphene sheets, oxidised on the surface to make them partially compatible with water-based solutions. The coatings were studied in term of the size of nickel crystallites, hardness, wear/frictionl properties, thermal conductivity.

2. Experimental

Carbon nanotubes with mean diameter of 13 nm and length between 2.5 and 20 μm were obtained from Sigma Aldrich. Pristine and –COOH functionalized Graphene Nano Platelets (GNPs) were obtained from Cheaptubes USA. The GNPs had diameter of 1-2 μm and a surface area of around 700 m^2/g. The –COOH content in the functionalised ones was declared as 7%.

The solutions used for the electrochemical deposition of nickel were Watt's bath ($NiSO_4$ + $NiCl_2$ + H_3BO_3) and Wood's bath ($NiCl_2$ + HCl) purchased from Coventia. The solution used for electrodeposition of copper was a self-prepared sulphate bath, containing 0.85 M of $CuSO_4{\cdot}5H_2O$ (purchased from Sigma Aldrich) and 0.55 M of H_2SO_4. Poly acrylic acid (PAA) with mean molecular weight MW 5100, obtained from Sigma Aldrich, was used as dispersing agent.

The homogenous and stable suspensions for the plating solutions were obtained by ultrasonic agitation for 45 minutes with a probe (Sonics VCX750) and magnetic stirring for 16 hours at 200 rpm. To ensure the homogeneity of the baths, gentle magnetic stirring at 100 rpm was done during the deposition process too. A three electrode cell was used for the deposition, and 99.98% pure Ni plates (10 x 2.5 x 1 cm) and 99.99% pure Cu plates (10 x 3 x 0.1 cm) were used as anodes for nickel and copper depositions respectively. In case of wear test samples, the cathode substrates were 42CrMo4 steel while for thermal analysis samples the used substrates were C20 mild steel. The electrochemical deposition for producing the composites was obtained by means of a switching rectifier AF00 by Giussani S.r.l. The deposition was realised at 30 °C with a controlled pH in the range between 3 and 4.

The compositions of the samples are shown in Table 1. and the aspect of the coating in Figure 1. The N series are the nickel-based coatings, while the C series are the copper-based ones.

Table 1. Composition of coatings.

Sample	**CNTs (mg/l)**	**GNPs (mg/l)**	**PAA (mg/l)**
N0	0	0	0
N1	165	0	140
N2	1000	0	100
N3	0	80	100
C0	0	0	0
C1	1000	0	100
C2	165	0	140
C3	0	65	100
C4	0	190	190

Figure 1. Picture of Ni-coated (left), Cu-coated (middle) and uncoated (right) samples

The morphology of the samples was studied by using Reichert-Jung MeF3 optical microscope (Leica Microsystems S.P.A, Austria) and Hitachi S4000 scanning electron microscope. The microhardness was tested using a LEICA VMHT hardness tester. The wear properties were studied by using a self-made pin on disk apparatus, by using a pin of Co-cemented WC, with 3 mm radius. Surface roughness profiles were also measured by using surface profilometer HOMMEL TESTER T1000. Thermal diffusivity of the samples was measured using a thermal properties analyser FLASH LINE 4010.

3. Results and Discussion

In the first part of this work, the deposition conditions were studied in order to obtain a good and adherent deposit of nickel or copper on the substrate. The chosen substrate was a 42CrMo4 steel for all samples subjected to mechanical testing and a mild steel with 0.2% carbon for thermal tests.

By optical microscopy it was judged the aspect of the deposit, that is shown in Figure 2 for a pure nickel coating (on the left) and a CNT containing one, sample N2 (on the right). A good interface between steel and nickel is observed in both cases and the coating results rather good, without evident defects.

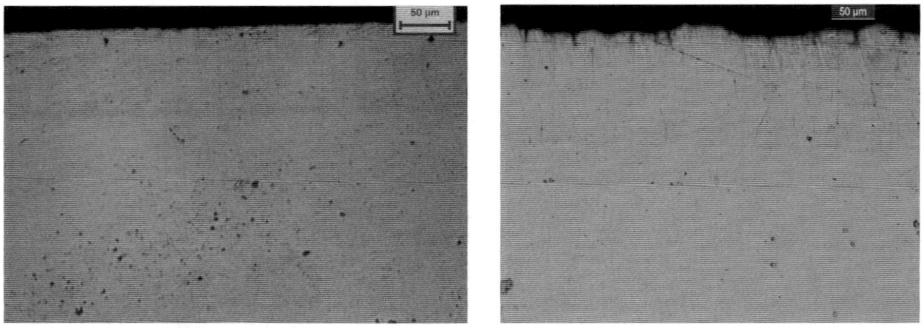

Figure 2. SEM images of pure nickel layer N0 (left) and nickel-CNT composite N2 (right) on steel support.

The realization of CNTs or GNPs containing coating however is much more complicated than the realization of a pure nickel one. In fact, when CNTs or GNPs are added, a significant problem of dispersion of the carbon nanomaterial occurs. While GNPs are functionalized with –COOH groups, that give some polarity to the surface, increasing their dispersibility in water, CNTs are non-polar, and don't disperse easily in water. Since it was thought not to damage the nanotubes with an acid treatment, pristine nanotubes were used, that were dispersed through the addition of polyacrylic acid (PAA), as suggested in the literature [23]. To break the CNTs agglomeration and homogenise the suspension, ultrasonication was applied for 45 minutes, and afterwards the solution was left overnight under magnetic stirring. Magnetic stirring was also continued during the whole course of deposition, in order to avoid re-agglomeration.

In the case of –COOH functionalized GNPs, PAA was also added with the same aim, to improve dispersion and avoid re-agglomeration of the carbon nanomaterial.

Two SEM images with a larger magnification of the nickel coatings containing CNTs and GNPs are shown in Figure 3. The coatings can present zones with good quality and no defects, on the left, or zones with agglomerates of CNTs or GNPs, as shown on the right.

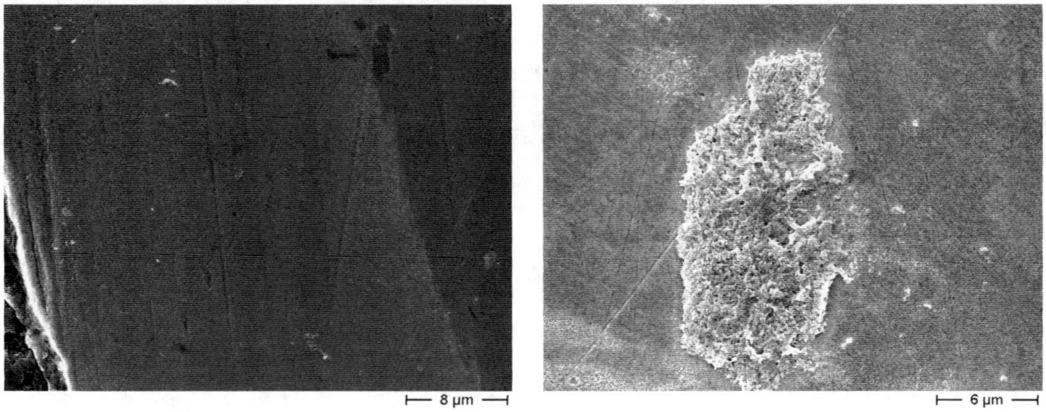

Figure 3. CNTs containing nickel layer with good quality (left); GNPs agglomerate in a nickel layer (right).

In order to measure the size of the nickel or copper crystallites, X-ray diffraction (XRD) was performed and the Scherrer formula was applied,

$$d = K\,\lambda\,/\,b\,\cos\theta$$

where d is the crystallite size, K is a constant depending on the geometry of the grains, λ is the copper K_α wavelength, b is the width of the XRD peaks, and θ is the diffraction angle.

The deposition of pure nickel already gave a rather low grain size, with submicrometric dimension, around 400 nm. The copper grains were a little larger, with a size close to 700 nm. As shown in Table 2, the presence of CNTs or GNPs brought to a significant reduction of nickel grain size, while copper grains were not affected by the presence of carbon nanomaterial.

Table 2. Grain size, calculated by the Scherrer equation, and roughness of nickel and copper-based coatings.

Sample	Grain size (nm)	Roughness Ra (μm)
N0	370	3.43
N1	277	2.27
N2	399	8.96
N3	209	2.08
C0	687	1.58
C1	599	1.99
C2	616	6.46
C3	717	5.52
C4	805	6.12

However, it is important to note that in the case of nickel, samples with too high carbon nanomaterial concentration demonstrated no reduction of grain size. This behaviour suggests that the highest concentrations are not well dispersed and that carbon nanomaterial remains too agglomerated, leaving large bundles of CNTs or GNPs, as the one shown in Figure 2.

With copper deposition instead, carbon nanotubes brings to a slight grain size reduction also in high concentration, while graphite nanoplatelets produce a slight increase of grain size.

The SEM observation of the samples, coupled with XRD, thus suggests that high concentration of nanotubes in nickel brings to a lower quality coating, as also confirmed by roughness measurements, also shown in Table 2. In the case of copper instead, the behaviour is more complicated, probably to interactions between the components of the bath. The most concentrated CNTs containing suspension provide a low roughness coating, while lower concentration of both CNTs and GNPs brings to a much higher roughness of the coating.

Hardness tests, shown in Table 3. were performed on Ni-based coatings, again confirm the previous results. The high concentration CNT-containing suspension brings to a low hardness coating, with HV close to 200. With lower concentration of CNTs or GNPs the hardness is much higher, close to 400 HV, close to the value of the pure nickel hardness. The fact that the composite coating has a lower hardness than the base alloy is due to the presence of low hardness carbon nanomaterial. However, the reduction in grain size of the nickel coating counterbalances the presence of carbon.

Table 3. Hardness values of nickel-based coatings.

Sample	Vickers Microhardness
N0	486
N1	327
N2	212
N3	400

Finally, the wear tests shown in Figure 4. allowed to measure the friction coefficient of the nickel based coating and the rate of removal of the coating. From curve N0 it is possible to observe how the pure nickel behaves. After 200 seconds the coefficient of friction (COF) begins to rise, passing from 0.65-0.7 to around 1.1, where it stabilises, indicating the removal of the coating. The same behaviour, with a faster initial growth, starting immediately, and also with a much higher scatter of the data, is observed for the high concentration CNTs suspension (N2 sample). This is due to the fact that the coating is very rough, and to the presence of many CNTs agglomerates in the coating.

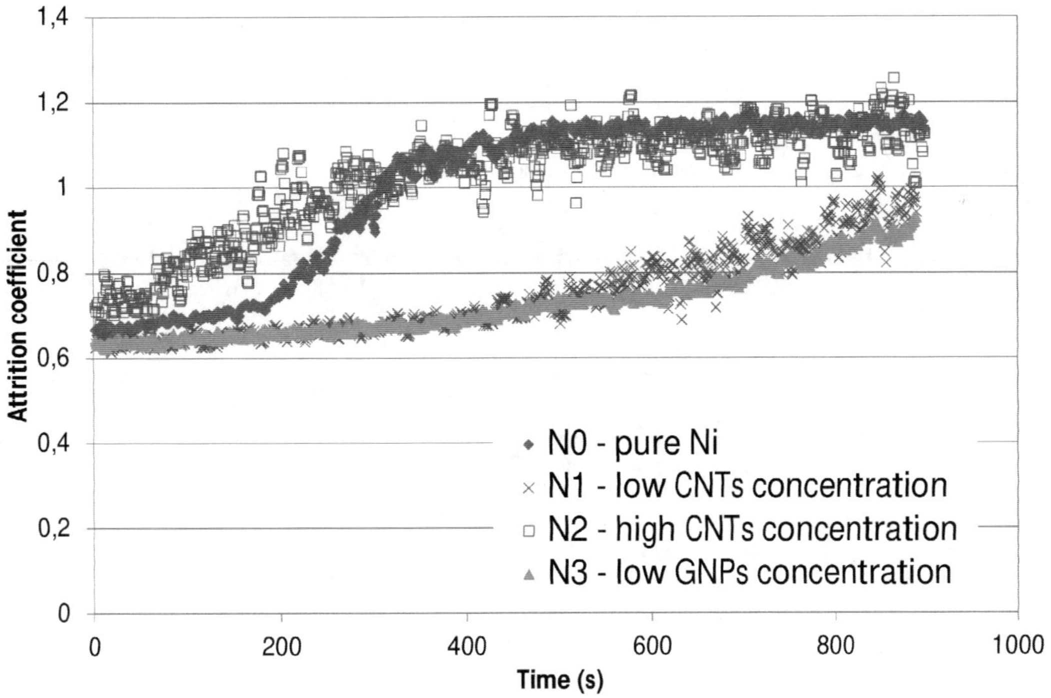

Figure 4. Coefficient of friction behaviour as a function of time for a pin-on-disk test at 5 N weight and 0.1 m/s.

The coatings containing low concentration of CNTs or GNPs (N1 and N3 respectively) instead show a much slower increase of COF. For the first 400 seconds the increase is very marginal, and begins to grow appreciably only after around 600 seconds. After 15 minutes of test, the COF is yet lower than 1 and growing, suggesting that the coating was not yet completely removed by the pin.

Comparing the CNTs and GNPs modified coating, it is possible to note that CNTs containing coatings present more scattered results than GNPs modified ones.

The copper-based samples were submitted to thermal diffusivity measurements, in order to verify if the coating could modify substantially the thermal properties of the support. In Table 4. the results at various temperatures for CNTs and GNPs containing copper coatings are reported.

Table 4. Thermal diffusivity of copper-based coatings.

Sample	**40 °C**	**440 °C**	**790 °C**
support	0.144	0.079	0.048
C0	0.152	0.094	0.06
C2	0.148	0.068	0.044
C3	0.133	0.073	0.044
C4	0.155	0.082	0.047

It is evident from Table 4. that the pure copper coating has a positive effect on the thermal properties of the support at all temperatures. The composite coatings have a more complex behaviour. Close to room temperature the samples obtained with 165 mg/l of CNTs or 190 mg/l of GNPs behave like the pure copper coating, or even slightly better in the case of sample C4. Increasing the temperature the composite coatings show a reduction of thermal diffusivity, probably due to the thermal stresses arising at the interface between the copper matrix and the carbon-based reinforcement. The thermal mismatch is rather high, so that it is possible that some delamination occurs at the interface, increasing the interface thermal resistance.

4. Conclusions

In this work the electrodeposition of composite coatings containing carbon nanotubes or graphite nanoplatelets has been carried out with success using nickel or copper matrices. It was shown that in the case of nickel deposition, the size of the nickel grains is reduced by the presence of the carbon nanoreinforcement, with positive effect on the wear properties, even at the cost of a small reduction of hardness values. In the case of copper deposition, the study shows the feasibility of using these reinforcements to produce well-adherent and solid coatings; regarding thermal properties, probably an increase of dispersion of CNTs or GNPs in the deposition solution is yet needed to fully exploit the exceptional properties of the carbon nanoreinforcement.

5. References

[1] Dini J W 1993 *Electrodeposition - The Materials Science of Coatings and Substrates* (William Andrew Publishing/Noyes)
[2] Gómez E, Pollina R and Vallés E 1995 *J. Electroanal. Chem.* **386** 45
[3] Rosso M, Ugues D, Torres E, Perucca M and Kapranos P 2008 *Int,. J. Mater. Forming* **1** 1259
[4] Lecis N, Ugues D, Previtali B and Demir A G 2012 *Metall. Ital.* **104(6)** 15
[5] Garcia I, Conde A, Langelaan G, Fransaer J and Celis J P 2003 *Corros. Sci.* **45** 1173
[6] Srivastava M, William Grips V K and Rajam K S 2007 *Appl. Surf. Sci.* **253** 3814
[7] Gyftou P, Pavlatou E A and Spyrellis N 2008 *Appl. Surf. Sci.* **254** 591
[8] Robin A, Pinheiro de Santana J C and Sartori A F 2010 *J. Appl. Eletrochem.* **40** 507
[9] Zanella C, Lekka M and Bonora P L 2009 *J. Appl. Eletrochem.* **39** 31
[10] Sun X J and Li J G 2007 *Tribol. Lett.* **28** 223
[11] Ivanov V V, Balakai V I, Kurnakova N Yu, Arzumanova A L and Balakai I V 2008 *Russian J. Appl. Chem.* **81** 2169
[12] Teterina N M and Khaldeev G V 1998 *Prot. Metals* **34** 276

[13] Chawla N and Chawla K K 2006 *Metal Matrix Composites* (Springer)
[14] Manfredi D, Pavese M, Biamino S, Fino P and Badini C 2009 *Comp. Sci. Tech.* **69** 1777
[15] Manfredi D, Pavese M, Biamino S, Antonini A, Fino P and Badini C 2010 *Comp. Part A: Appl. Sci. Manufact.* **41** 639
[16] Iijima S 1991 *Nature* **354** 56
[17] Meyyappan M 2004 *Carbon Nanotubes: Science and Applications* (CRC Press)
[18] Shima H *Materials* **5** 47
[19] Balandin A A 2011 *Nature Materials* **10** 569
[20] Shokrieh M M and Rafiee R 2010 *Mech. Comp. Mater.* **46** 155
[21] Pavese M, Musso S and Pugno N 2010 *J. Nanosci. Nanotech.* **10** 4240
[22] Castellino M, Tortello M, Bianco S, Musso S, Giorcelli M, Pavese M, Gonnelli R S and Tagliaferro A 2010 *J. Nanosci. Nanotech.* **10** 3828
[23] Arai S, Endo M and Kaneko N 2004 *Carbon* **42** 641
[24] An B-G, Li L-X and Li H-X 2008 *Mater. Chem. Phys.* **110** 481
[25] Khabazian S and Sanjabi S 2011 *Appl. Surf. Sci.* **257** 5850
[26] Arai S and Endo M 2005 *Electrochem. Commun.* **7** 19
[27] Guo C, Zuo Y, Zhao X, Zhao J and Xiong J *Surf. Coat. Tech.* **202** 3246
[28] Wang F, Arai S and Endo M *Electrochem. Commun.* **7** 674
[29] Dai P-Q, Xu W-C and Huang Q-Y 2008 *Mater. Sci. Eng.* A **483-484** 172
[30] Jeon Y S, Byun J Y and Oh T S 2008 *J. Phys. Chem. Solids* **69** 1391
[31] Carpenter C R, Shipway P H and Zhu Y 2011 *Surf. Coat. Technol.* **205** 5059
[32] Carpenter C R, Shipway P H and Zhu Y 2011 *Wear* **271** 2100
[33] Choi W, Lahiri I, Seelaboyina R and Kang Y S 2010 *Crit. Rev. Solid State Mater. Sci.* **35** 52
[34] Young R J, Kinloch I A, Gong L and Novoselov K S 2012 *Comp. Sci. Tech.* **72** 1459

Journal of Physics: Conference Series **439** (2013) 012020 doi:10.1088/1742-6596/439/1/012020

A controlled atmosphere tube furnace was designed for thermal CVD

M Rashid[1,3], J A Bhatti[2], F Hussain[1], M Imran[1], I U Khawaja[1], K A Chaudhary[1] and S A Ahmad[1]

[1] Department of Physics, The Islamia University of Bahawalpur Pakistan 63100

[2] Pakistan Vacuum Society, Islamabad, Pakistan

[3] Department of Physics, National University of Singapore, 2 Science Drive 3, Singapore 117542

E-mail: rapakistana@yahoo.com

Abstract: High quality materials were used for the fabrication of hi-tech tube furnace. The furnace was especially suitable for thermal Chemical Vapor Deposition (CVD). High density alumina tube was used for the fabrication of furnace. The tube furnace was found to have three different temperature zones with maximum temperature at central zone was found to be 650°C. The flexible heating tape with capacity of 760°C was wrapped on the tube. To minimize the heat losses, asbestos and glass wool were used on heating tape. The temperature of the tube furnace was controlled by a digital temperature controller had accuracy of ±1°C. Methanol was taken as the representative of hydrocarbon sources, to give thin film of carbon. The a-C: H structure was investigated by conventional techniques using optical microscopy, FT-IR and SEM.

1. Introduction

In the mid seventeenth century the competition to find an ideal and convenient thermometer was under discussion. Galileo began to use and modified the Florentine thermometer. The lower fixed point (ice point) is temperature at which water and ice coexist in thermal equilibrium under a pressure of 1 atm. upper fixed point is the temperature at which water and steam coexist in equilibrium under a pressure of 1 atm [1]. By bonding two metals with different coefficient of thermal expansion, achieved a strip that bends in reaction of temperature change. It is widely used in the home furnaces, being switched ON or OFF by thermostat [2]. Theory of thermocouple based on See beck effect. See beck's finding has resulted in possibly the most general method of temperature measurement. The See beck effect is observable as a voltage potential that occurs when there is a temperature gradient along the length of a conductor. Copper-constantan thermocouples have been used for several years and it is called the T-type. Good quality copper has abnormally attractive properties for use either in thermocouple or as a standard material for thermoelectric assessment purposes. The only disadvantage of copper is its tendency to oxidize quickly at high temperature. Due to this restriction it becomes difficult to use it in air above 400ºC. Because of the oxidation trouble small diameter exposed copper wire has a much shorter useful life than that of larger size. There is a small amount of thermocouples that can be tolerated in reducing atmosphere, so iron-constantan is mainly important in this respect and it is called J- type. A base metal thermocouple with Chromel-Alumel is called K-Type thermocouple that was developed by the Hoskins manufacturing company. In short, a chromel type wire at room temperature that has been recently hard drawn, greatly cold worked, or reduced from high temperature state is in a fully disordered condition [3]. A device which is used to provide controlled temperature for the heat treatment of material is called furnace. It is an enclosed chamber in which heat is produced for heat building, destroying refuse, smelt or refine ores, etc [4].

CVD is a deposition method where chemical precursors are elated in the vapor phase to decay on a heated substrate to form a film [5]. The formation of films depends on the materials and the condition of reactor. These films may be polymers, amorous carbon and glassy carbon. It has become the most important technique of film deposition. It offers the advantage that operates at low temperature. Two types of reactors are widely used in this process. Hot wall reactor and cold wall reactor. In hot wall reactor, a heater surrounds the reaction region. The heater can be heated by electrical resistance [6]. The deposition is not well-organized since the hot walls are coated as well as the substrate. The hot wall reactor is perfect for the situation where the effect is exothermic. The high wall temperature stops the unwanted positions on the reactor walls. Horizontal and vertical tube furnaces are mostly used for CVD. Hot wall reactor is called Tube furnace. CVD synthesis is achieved by putting a carbon source in the gas phase and using an energy source, such as plasma or a resistively heated coil, in order to transfer energy to a gaseous carbon molecule. Commonly used gaseous carbon sources include methane, carbon monoxide and acetylene. The aim of present work is to study carbon film on silicon and glass substrate.

2. Experimental results

The grain size of polycrystalline material is from 1 to 5 microns and its density is 3.95 g $/cm^2$. The water absorption is zero percent and its hardness is about 2000 Vickers. The value of dilation co-efficient at 2000°C is $8.4\times 10^{-6}\,^{\circ}C^{-1}$ and its thermal conductivity is 40 ($w/_{m\text{-}k}$) at 20°C. Alumina (Al_2O_3) has been used as tube reactor and having corundum crystalline structure. It has excellent wear resistance, electrical insulation and good thermal conductivity. It has high hardness compression strength. It is used in ballistic amour, electronic substrate and high voltage insulators in element formers. In thermocouple two different metals having dissimilar number of free electron at same temperature are used. When the hot region is moved along the conductors it passes through area where the metals are connected. There is difference in electron density and therefore difference in charge circulation. Due to this current flow throughout millimeter, as free electrons try to move into equilibrium. This movement will continue until the heat energy is applied. The voltage calculated by this charge distribution is called EMF and proportional to temperature. Asbestos used as a heat resistance in furnaces, insulation on pipes, boilerand hot water tanks. Insulation used to stop heat loss [7]. Glass wool is the superb thermal insulator and the best thermal resistance. Therefore it reduces energy utilization and also reduces variations of temperature in tube furnace [8]. The temperature controller is made with PID (Proportional, Derivative and integral) controller for controlling the temperature. As complete connection and parts that are used in the construction of tube furnace are shown in figure 1.

The tube that is used for the construction of furnace having internal and external diameters of 17 and 24 mm and the length is 500 mm. The ends of the tube are closed by the Teflon covers. The crucible is used for placing the samples in the furnace at different zones. It is divided into the three zones in order to get readings as shown in table 1. Thick insulation divides this trapped air into a greater number of split cells, slowing exchange and the rate of heat flow. The metal sheet is used to pack all these materials and gives the absolute shape to the furnace. The heat loss that controlled by insulation can be seen by the figure 2.Temperature was measured by using the Chromel-Alumel (K type) thermocouple. It gives linear behavior result than copper-constant (T type) and iron-constantan (J type) thermocouples. The comparisons of all these types of thermocouples are shown in the figure 3.The experiment was performed at room temperature and the temperature of the room was 27 °C.

In order to observe the micron size crystals smoothness of the surfaces and crystal boundaries, an optical microscope must be used. A modern Scanning electron microscope (SEM) provides an image resolution typically between 1 nm and 10 nm, not as good as the TEM but much superior to the light microscope. In addition, SEM images have a relatively large depth of focus: specimen features that are displaced from the plane of focus appear almost sharply in focus [9].

Table 1. Distribution of temperature in tube furnace.

Temperature at central Zone (°C)	Temperature at second Zone (°C)	Temperature at corner Zone (°C)
75	65	50
200	170	100
250	215	135
300	265	170
350	300	200
400	335	235
450	370	270
500	400	300
550	440	330
600	470	350

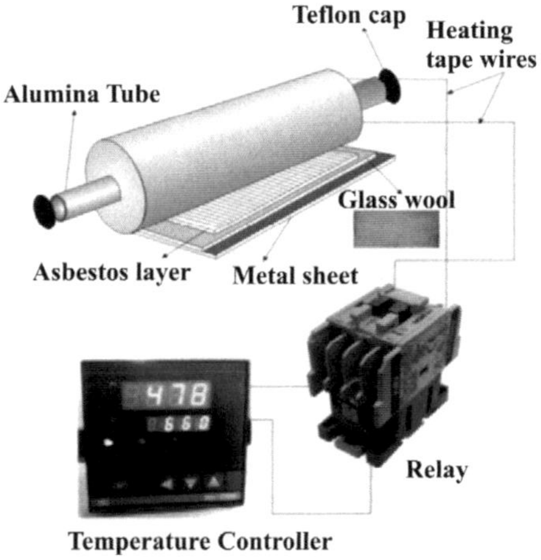

Figure 1. Materials and parts used in the construction of tube furnace

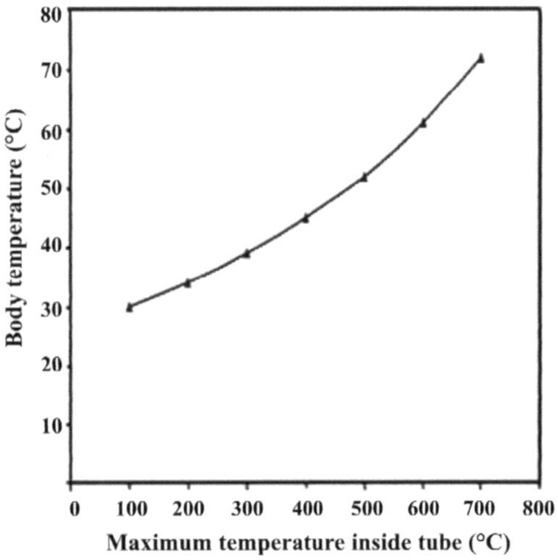

Figure 2. Temperature comparisons between inside and outside tube

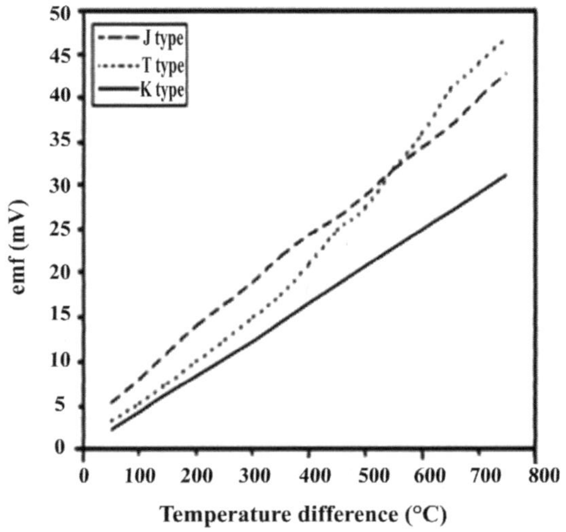

Figure 3. Comparison of three types of thermocouples

Atmospheric pressure thermal CVD (APTCVD) has many possible advantages over other forms of coating technology, which can give important marketable advantages in definite areas of glass coating technology. This is the simplest and widely used method. Glass or silicon is commonly used as a substrate. Heating is normally skilled by use of resistive heating. A hot wall reactor is known as isothermal furnace in which substrate is located. In this reactor, many substrates are deposited at a time. The whole tube is heated and exact temperature control can be achieved with accurate furnace design. A drawback of hot wall design is that deposition occurs on the walls of the tube plus on the substrate. As a result, hot wall reactor must be cleaned to decrease the cracking of particles from the walls which may infect the substrate. The pressure factor of the reactor depends on rate restrictive step of deposition. The tube is radially isothermal and thermal diffusion length is greater than the radius of the tube. Heat can simply pass through by conduction from the wall of tube to central region. The radial temperature remains constant if temperature of the tube wall is constant [10]. Complete experimental arrangement for the inert atmosphere thermal CVD is schismatically shown in figure 4. Air is mainly composed of nitrogen, oxygen, and argon, which together constitute the "major gases" of the atmosphere. The remaining gases are referred as "trace gases, often among which are the greenhouse gases such as water vapor, carbon dioxide, methane, nitrous oxide, and ozone. Thick insulation divides this trapped air into a greater number of split cells, slowing exchange and the rate of heat flow. The metal sheet is used to pack all these materials and gives the absolute shape to the furnace. Typical experimental conditions employed in the present study are summarized in table 2.

Table2. Experimental conditions for the deposition of carbon film.

Sr. No.	Item	Value
1	Source material	CH_3OH
2	Carrier gas	Ar
3	Maximum temperatur	650°C
4	Temperature accuracy	±1°C
5	Crucible material	Alumina
6	Pressure	Atmospheric
7	Substrate	Glass & Silicon
8	Type of Film	Carbon
9	Sintering time	1-30min
10	Ar flow rate	2ml / min
11	Bath temperature	60°C
12	Glass flask temperature limit	500°C

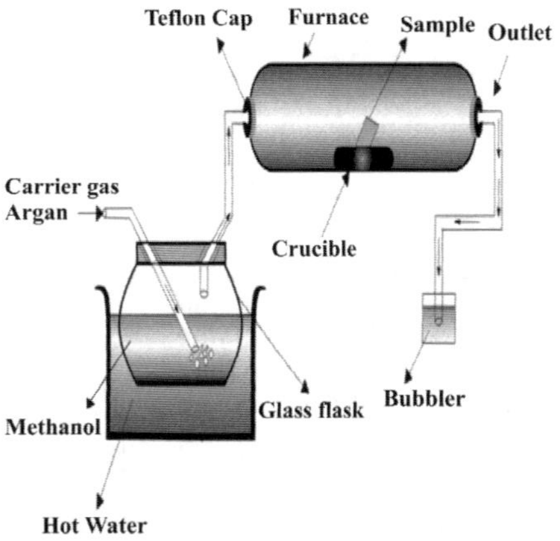

Figure4. Complete experimental setup of atmospheric pressure thermal CVD

3. Results

Some important specifications are given below in table 3.

3.1. Optical microscopy

To analyze the photos captured by the optical microscopy, their structures are explained. At 400°C, almost hexagonal structure was observed; the film deposition was uniform as shown in figure 5. At 550°C, the clear hexagonal structures are shown in figure 6. Hexagonal orientation was observed at the surface and it had poly crystalline structure with a combination of (111) and (100) surfaces. The film distribution was uniform. Film was so thin that its thickness could not be measured by the optical microscope.

Table 3. Specifications of the tube furnace.

Temperature range	0 - 760°C
Thermocouple	K-type (Chromel-Alumel)
No. of Zones	Three
Length of each Zone	200mm, 100mm, 200mm
Internal Diameter of Tube	17mm
External Diameter of Tube	24mm
Length of the Tube	500mm
Controller Diameter	$8100mm^2$
Length of Crucible	80mm
size of Crucible	10mm
Temperature Display	Digital
Accuracy	±1°C
Power	220VAC/50Hz
Both Ends	open or Closed (Teflon Caps)

Figure 5. Optical micrograph of carbon film at 400°C

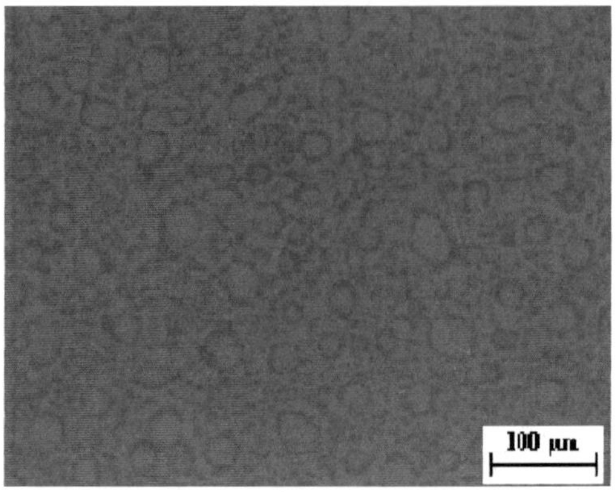

Figure 6. Optical micrograph of carbon film at 550°C

3.2. Inter atomic bonding using methanol as source material

Fourier Transform Infrared (FTIR) is the preferred method of infrared spectroscopy. In infrared spectroscopy, IR radiation is passed through a sample. Some of the infrared radiation is absorbed by the sample and some of it is transmitted. Figure 7 shows the IR spectrum of the sample when the carbon film was deposited on Si (100) substrate at 500ºC for 30 minutes. The upward peaks are due to the loss of reagents while negative peaks reflect the generation of products during the oxidation of methanol within the thin layer transparent cell. The band at 3724 cm^{-1} can be assigned to symmetrical stretching vibration mode of O-H from isolated terminal silanoals (Si-OH) groups. The peak at 1700 cm^{-1} was a resultant of oscillation of C-O double bond in carboxylic group. The bands at 1644 cm^{-1} and 1570 cm^{-1} are due to stretching of carboxyl group C-O.

Figure 8. shows IR spectrum of the sample when the temperature was 600ºC and deposition time was 20 minutes. The Pears of strong IR bands at 2862 cm^{-1} 2964 cm^{-1} due to methyl (CH_3) symmetric and asymmetric stretching demonstrate the presences of methanol and evidence of successful methanol adsorption. The presence of C-H stretching band at 2811 cm^{-1} that associated with methoxy ($O-CH_3$) shows generation of methoxy species due to methanol adsorption. The small upward peaks at 1450 cm^{-1}also reflect the depletion of methanol during oxidation process. In figure 7. and figure 8. the band at 2324 cm^{-1} was due to the atmospheric CO_2.Particularly at location, where large numbers of vacancies are present, methanol can dissociate all its hydrogen resulting in the generation of CO. The samples were prepared at various time duration, only effect on the transmittance of the peaks and the wave number of the peaks almost same. FTIR spectrum shows that there are many hydroxyl groups with oxygen atom distributing on the surface of carbon microspheres.

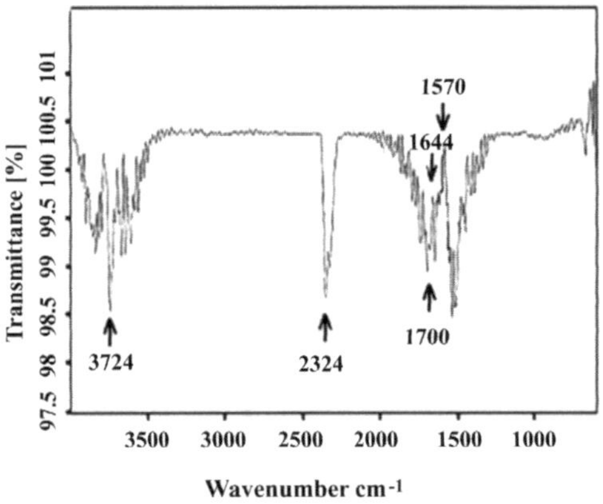

Figure 7. FTIR spectrum of carbon film on Si(100) at 500°C for 30 minutes

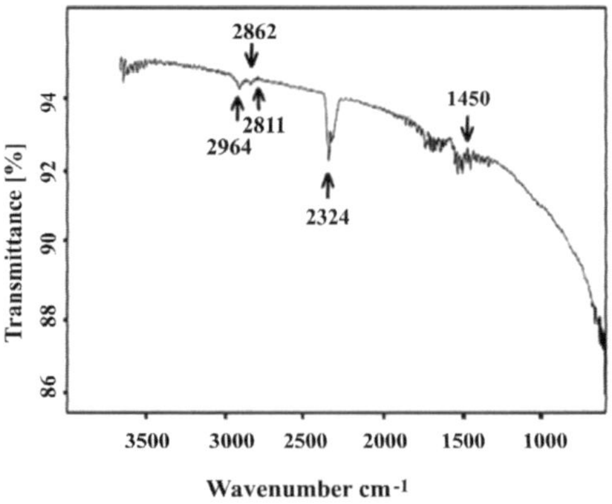

Figure 8. FTIR spectrum of carbon film on Si(100) at 600°C for 20 minutes.

3.3. Surface studies of carbon film

A series of experiment was performed to investigate the deposition of carbon on Si (100) using methanol as a source material and argon as a carrier gas. These experiments were performed at deposition temperatures ranging from 350°C to 600°C. There was no film deposit on the samples that were operated below 350°C. Figure 9(a), (b) and (c) show the results when the film was deposited at 400°C, 500°C and at 600°C respectively.

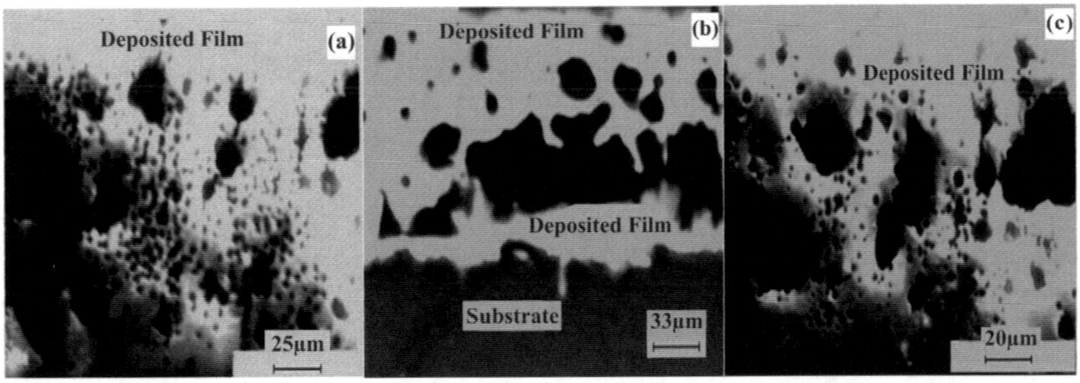

Figure 9. SEM images of deposition of carbon film (a) at 400°C(b) at 500°C and (c) at 600°C.

4. Conclusion

A moderate temperature, in the range, of 0 – 750 °C was produced by a flexible heating tape in a high density alumina tubing. The temperature was controlled by a digital temperature controller with an accuracy of ±1°C up to 650 °C. High performance materials were used for the construction of tube furnace. To illustrate the usefulness of the furnace, a variety of materials were chosen. Specifically Silicon (Si), Barium Titanate ($BaTiO_3$), Methanol (CH3-OH), Yitrium Barium Copper Oxide ($Y_1Ba_2Cu_3O_7$) and Calcium phosphate ($Ca_2P_2O_7$) were selected as the representatives of their respective families Semiconductors, Ferroelectrics, Hydrocarbons, Superconducting and Bio-ceramics respectively. All the above said materials were fabricated in the tube furnace successfully and characterized by standard techniques like OPTICAL MICROSCOPE, XRD, FTIR and SEM but the present research discusses hydrocarbon only.

References

[1] Tippens P E 1978 *Appl. Phys.* **2** 4
[2] Drever R W P 1983 *Appl. Phys. B* **31** 97-105
[3] Kinzie P A 1973 *Thermocouple Temperature Measurements* (R. David Moore and Company) pp1-35
[4] Mullinger P and Jenkins B 2008 *Industrial and Process Furnace* (Butterworth- Heinemann)
[5] Larsson I A S, Elianne M L, Lundstrom T S, Daniel M V and Simon T 2009 Kiln Aerodynamics visualization of merging flow by usage of PIV and CFD *Seventh International Conference on CFD in the Minerals and Process Industries CSIRO (Melbourne, Australia 9-11*
[6] Barron A R 1997 *J. Chem. Mat.* **4** 967
[7] Wilson A 2010 *Avoiding the Global Warming Impact of Insulation* (SAP 2009 England)
[8] Carman P C 1937 Fluid flow through a granular bed *Transactions of the institution of chemical engineers (Electronic Materials* Vol 15) ed E Scholl (London: Chapman and Hall) pp 150- 167
[9] Goodhew P J, Humphreys J and Richard B 2008 J. Materials **29** pp 43-53
[10] HsinTien C, Shiow C, Chung T 1998 *J. Elsevier sci.* **17** 2190

Journal of Physics: Conference Series **439** (2013) 012021 doi:10.1088/1742-6596/439/1/012021

Investigation of failure mechanism of thermal barrier coatings (TBCs) deposited by EB-PVD technique

M R Shahid [1,3], Musharaf Abbas[2]

[1] National University of Science and Technology, H-12, Islamabad 44000, Pakistan
[2] Beijing University of Aeronautics & Astronautics, No. 37 Xueyuan Road, Beijing 100191, China

E-mail: shahidmr786@hotmail.com

Abstract. Failure mechanism of thermal barrier coatings (TBCs) prepared by electron beam physical vapor deposition (EB-PVD) technique owing to formation of micro cracks was investigated. The TBCs were deposited on the Ni-based super alloy IN-100 and the micro cracks were observed within the top ceramic coat of thermally cycled TBCs at 1050°C. It was observed that these cracks propagate in the ceramic coat in the direction normal to interface while no cracks were observed in the bond coat. SEM/EDS studies revealed that some non-uniform oxides were formed on the interface between ceramic top and metallic bond coat just below the cracks. Study proposed that the cracks were initiated due to stress owing to big difference in Pilling-Bed worth ratio of non-uniform oxides as well as thermal stress, which caused the formation of cracks in top ceramic coat leading to failure of TBCs

1. Introduction

Demand for enhanced jet engine efficiencies has led to significant increase in combustion temperatures and operating pressures. These requirements have proved to be a big driving force to improve the super alloy technology, such as alloying, production techniques and different coating techniques. Consequently, an increase in turbine inlet temperature demands materials with higher melting temperatures i.e. intermetallics or ceramics. An intermediate development stage is the use of the super alloys with an oxidation resistant bond coating and a ceramic top coating, so-called thermal barrier coatings (TBCs). TBCs are layer systems deposited on thermally highly loaded metallic components, as for instance in gas turbines [1-4]. The TBC concept is valuable because the mechanical properties of the metallic super alloy are maintained and the alloy is protected against the high temperature due to the low thermal conductivity of the top ceramic coating. It is now well established that yittria partially stabilized zirconia (YSZ) TBCs due to low thermal conductivity can lead to significant improvements of turbine engine efficiency and hot component's life time [5]. The state-of-the-art TBC consists of two layers. Top ceramic coat, 6.5-8 wt.%Y_2O_3 partially stabilized zirconia (YSZ), protects the substrate from high temperature. Metallic bond coat MCrAlY (Where M indicates Ni, Co, Fe or their combination) is deposited next to substrate to protect it from oxidation and corrosion at high temperatures as well as it helps in reducing thermal expansion mismatch between the ceramic coat and the super alloy [6,7]. This bond layer also provides the mechanical

[3] To whom any correspondence should be addressed.

bonding between the substrate and the top ceramic coat. This multilayer system is not thermally stable under long term exposure at elevated temperatures. A thermally grown oxide (TGO) is formed by oxidation at the bond coat/top ceramic coat interface. This scale consists mainly of alumina (Al_2O_3) formed by selective oxidation of aluminum from the bond coat. The thermally grown oxide at the bond coat/ceramic coat interface plays vital role for spallation and lifetime of TBCs.

Previous investigations mainly focused on plasma sprayed TBCs, the most prominent characteristics of which are the loss of adherence and spallation of ZrO_2. The Use of EB-PVD method opened a new era of fabricating TBCs of very high quality. Schematic diagram of EB-PVD is shown in figure 1.

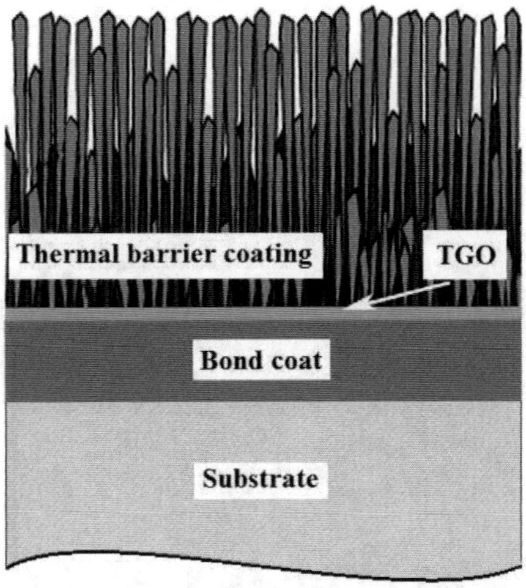

Figure 1. A schematic diagram of an electron beam physical vapor deposited YSZ coating with elongated pores aligned perpendicular to the substrate surface.

EB-PVD TBCs offer significantly extended lifetimes compared with APS TBCs due to its better controlled processing [8,9]. TBCs failure is expected to result either from stresses developed upon heating to elevated temperatures or from stresses developed upon cooling to ambient temperatures. Failure by either mode may be influenced or even controlled by the initial residual stress state of the coating or by the effects of degradation due to the thermally activated processes such as bond coat oxidation, or bond coat plastic deformation and phase transformation [10].

A large number of publications deal with the improvements in lifetime of the TBCs [10-14]. Significant advances in coating durability have been realized through improvements in coating materials and processing conditions for plasma spray and other coating techniques.

Future improvements will be greatly facilitated if coating failure mechanism is more rigorously investigated. Thus the purpose of this study is to elucidate the effects of oxidation of the metallic bond coat under cycling conditions on the failure mechanism of the TBCs.

2. Experimental

The specimens used for experimental purpose were cut from a turbine blade material IN-100 superalloy with nominal composition of 10Cr-60Ni-15Co-3Mo-5.5Al-4.7Ti-1.0V-0.15C-0.06Zr-0.015B. Specimens were ultrasonically cleaned in alcohol bath. At a constant temperature of the substrate and working pressure in the chamber, the microstructure and porosity of ceramic ZrO_2-(6.5-8) wt.%Y_2O_3 condensate were evaluated with respect to the deposition rate (Vd). After deposition of TBC, all the samples were then submitted to cyclic test to evaluate ceramic spallation and crack growth. Each cycle consisted of a 10 min heating up to 1050 °C, an exposure at 1050 °C for 1.5 hour, followed by a 10 min cooling in air. The specimens were then finely polished using standard metallographic methods. The microstructure of the TBCs was investigated with scanning electron microscopy (SEM) equipped with energy dispersive X-ray spectrometry (EDS). Elemental concentration analysis was made by using EPMA to acquire precise, quantitative elemental analysis and elemental variation in the critical regions identified from SEM observations.

3. Results and Discussion

3.1. Initial microstructure

Cross-sectional image of SEM morphologies of the sample under the as-deposited condition with Vd=1.8 are shown in figure 2. Figure shows the uniform flat surface morphology and the overall cross section of the specimen with a 110 μm thick YSZ top coating.

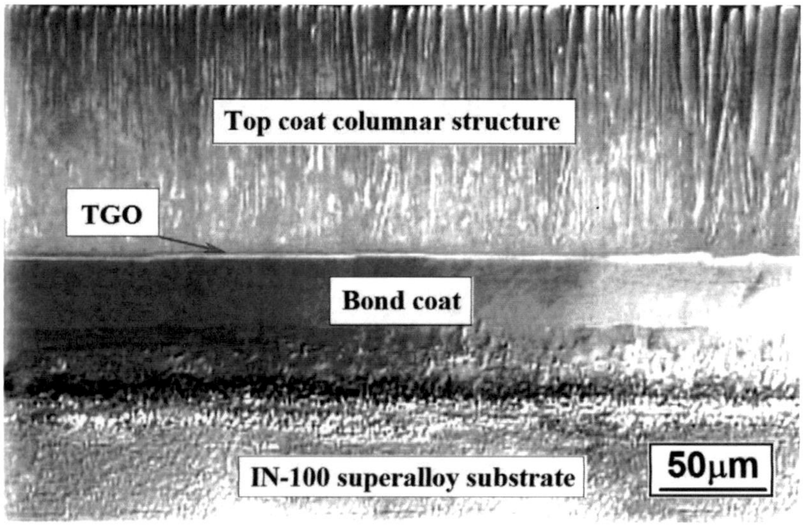

Figure 2. SEM cross-sectional image of IN-100 specimen coated with TBC exhibiting columnar structure.

The total thickness of TBC coating is about 150 μm with a bond coat of about 40 μm and a top ceramic coat of about 110 μm. The top ceramic coat exhibits typical columnar structure which is desirable owing to higher strain tolerance and dense structure.

3.2. Effect of deposition rate

A slight increase in Vd resulted in more dense layer of ZrO_2 formed on the bond coat-ceramic interface in the initial period of deposition, as well as a reduction of the total porosity of the ceramic layer. The formation of a denser ceramic layer with the increase of Vd is accompanied by the growth of the cross section of ZrO_2 columnar crystallites. The average diameter of the single crystallite is increased from 1.5-2 μm at Vd = 1 μm per minute up to 3-4 μm at Vd = 2 μm per minute. The preferred structural growth as observed and shown in Figure 1 with increasing Vd is also substantiated with the finding by previous researches (15-17).

3.3. Crack growth analysis as a function of thermal cycles

The cyclic temperature testing for ceramic spallation and crack growth study shows that the maximum thermal life occurs for ceramic top coat produced in the 1.2-2.0 μm per minute range when the micro porosity of the ceramic layer is 12-16%. Average crack length as a function of thermal cycles for the case of Vd = 2 μm per minute is shown in Figure 3. It shows that the observed crack length increases approximately linearly with the number of thermal cycles, with slight slow rate in the thermal cycles between 30~50 showing crack arrest followed by accelerated crack growth in higher thermal cycles. Such behaviour is in compliance with the findings of fewer past studies (18, 19). Figure 4 shows the cross sectional image of the specimen coated with TBC after maximum cycles (80 cycles) considered in this study at 1050 °C. From the figure, a micro crack can be observed in the top ceramic coat at the interface between the bond coat oxide layer and top ceramic coat. The micro cracks after initiation, propagate through the ceramic top coat perpendicular to the interface.

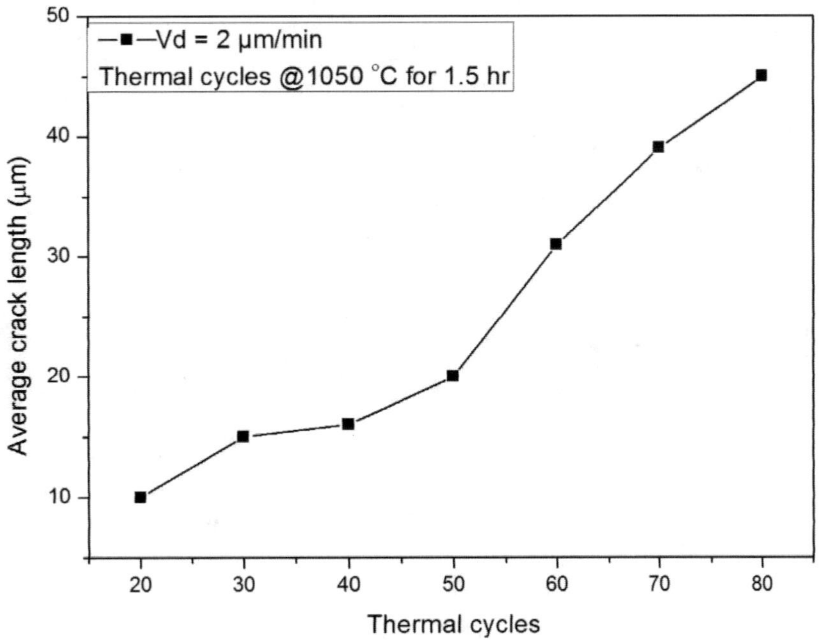

Figure 3. Average crack length as a function of thermal cycles.

Along a line normal to the interface, at about 8 μm from the crack, elemental concentration analysis was made by using EPMA. Figure 5 shows the elemental concentration profiles of Ni, Co, Cr and Al. Further analysis showed that just beneath the crack, Al element was found about 12wt. %, abnormally in higher content than the area nearby.

Figure 4. SEM cross-sectional image of thermally cycled IN-100 specimen coated TBC after 80 cycles at 1050 °C showing the crack.

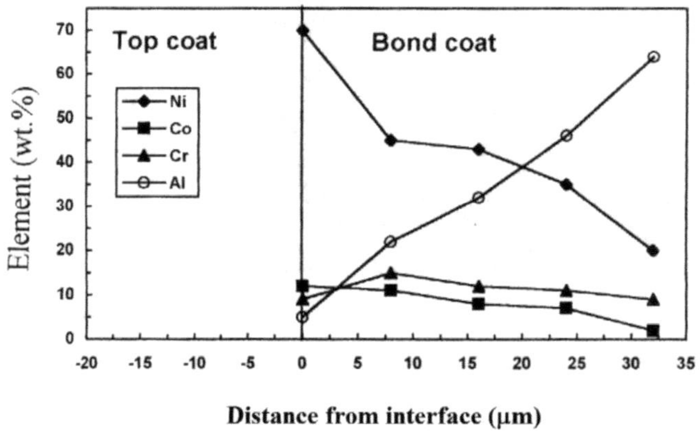

Figure 5. Elemental distribution perpendicular to the interface after thermal cycling of TBC coated specimen from 1050 °C.

3.4 Oxidation kinetics and damage mechanism

During high temperature oxidation, the oxygen from the atmosphere can easily diffuse to the metallic bond coat owing to the high porosity, segmentation and ionic conductivity of the zirconia top coat. The bond coat suffers with oxidation attack. An oxide layer builds up to the interface between the bond coat and the top coat and resulted volume expansion causes internal stress on the interface. Presence of higher Al content just beneath the crack indicates the formation of Al_2O_3 on the interface between the top coat and the bond coat. Formation of Cr_2O_3 and NiO on the interface, along with Al_2O_3 has also been reported elsewhere [13, 14, 20]. It is believed that the growth of Al_2O_3 on the interface is less detrimental to TBCs as compared to that of NiO, CoO or other oxides [14]. The formation of non-uniform oxides on the interface results into high growth stresses. First, the growth rate of Al_2O_3 is relatively slower than that of other oxides, e.g Cr_2O_3, NiO and CoO [20] Secondly, the PBR (Pilling-Bedworth ratio, volume of oxide : volume of metal) of Al_2O_3 is smaller than other oxides present on the interface (1.28 for Al_2O_3 and 1.65, 1.86, 2.07 for NiO, CoO, Cr_2O_3 respectively). Thus the growth stress of Al_2O_3 should be smaller as compared to that of non-uniform oxide scale [21]. Thirdly, growth of NiO or CoO occurs predominantly by cations diffusion through oxide scale and the oxide growth stresses should be generated in the top coat-bond coat outgrown oxide scale interfacial region where spallation of the top coat initiated. However, for the growth of Al_2O_3, oxide growth stresses are generated in the less critical Al_2O_3 bond coat interfacial region by inward diffusion of oxygen through Al_2O_3 scale [22]. Thus it is inferred that a non-uniform oxide formed on the surface of the bond coat leads to the generation of high growth stresses resulting in initiation of cracks in the top ceramic coat causing the failure of TBCs.

Several mechanisms of micro crack formation have been reported in the literature [23-25]. Movchan *et al.* [26] have suggested that in two layer metal / ceramic PVD coatings, the cracks initiate from the base metal and then spread out into the ceramic layer, finally causing the spallation of ceramic coat. Tsai *et al.* [27] found that the spallation of the top coat was initiated and propagated mainly by the cracks within the laser-glazed plasma sprayed top coat near the interface of the top coat /bond coat and same also verified by P. K. Wright. [28] Kh.G. Schmitt-Thomas *et al.* [13] and B.C. Wu *et al.* [21] have used the Al_2O_3 diffusion barrier layer between the bond coat and top ceramic coat to reduce the probabilities of initiation of cracks in the later.

From the observations noticed, cracks are likely first initiated from interface between the ceramic/oxidation layer (Figure 4) and then the non-uniform oxide layer forms between the bond coat and the top coat and the resulted volume expansion causes thermal stress at the interface. When the oxide layer has attained a critical thickness, cracks can first be observed. With increasing oxide thickness, the internal stress increases and the ceramic layer spalls off. According to the calculation, an oxide layer with a thickness of 3μm can produce about 12 MPa of oxidation stress. In case of formation of non-uniform oxides, stresses will be even much higher. Another factor is the thermal stress that can reach about 9.88 MPa. This is caused by the difference between the thermal expansion coefficients of the top and the bond coats when the specimens were cooled to between 700 and 600°C for stress measuring purpose. If the specimens would be cooled to room temperature for stress measuring purpose, a greater thermal stress can be expected.

Combined with the factors discussed above, the cracks in the TBCs produced by EB-PVD technique are first observed on the ceramic/bond coat interface owing to bond coat oxidation. These cracks then propagate along a direction normal to the interface in the ceramic layer. The growth of these cracks finally results in the spallation of top ceramic coat ultimately leading to the failure of the TBCs.

4. Conclusion

In this study the TBCs were prepared by EB-PVD method and the following observations about the formation of initial cracks have been made:

a. In the thermally cycled TBCs, the cracks were observed to initiate from the interface between the ceramic top coat and the MCrAlY bond coat owing to the oxidation of bond coat. Cracks after formation, propagate in the ceramic coat in the direction normal to the interface. No cracks were observed in the bond coat.
b. At a point just below the cracks, abnormal higher Al concentration was observed; it is believed that non-uniformed oxides formed at the interface caused the formation of micro cracks in the top ceramic coat.
c. The techniques which can reduce the oxidation of metallic bond coat and avoid the formation of non-uniform oxides at the interface will be helpful to stop the initiation of cracks in the top ceramic coat and its spallation which ultimately causes the failure of the TBCs.

References

[1] Song M, Ma Y, Gong S 2011 *Prog. Nat. Sci.: Mater. Int.* **21** 262
[2] Abbas M, Guo H B, Shahid M R 2012 *Ceram. Int.http://dx.doi.org/* 10.1016/j.ceramint.2012.06.051
[3] Wu J, Guo H B, Abbas M, Gong S 2012 *Prog. Nat. Sci.: Mater. Int.* **22(1)** 40
[4] Wu Q, Yang R B, Wu Y X, Li S S, Ma Y, Gong S K 2011 *Prog. Nat. Sci.: Mater. Int.* **21** 496
[5] Baufeld B, Biest O van der, R-Scheibe H J 2009 *Ceramic and Engineering Science Proceedings* Vol. **28**, Issue 3
[6] ASM 2000 Specialty Handbook: Nickel, Cobalt, and Their Alloys (America)
[7] Zhao S, Zhang C, Wu N M, Wang H M 2011 *Prog. Nat. Sci.: Mater. Int.* **21** 301
[8] R-Scheibe H J and Schulz U 2007 *Surf. Coat Techol.* **201** 7880
[9] Sudhangshu B, High Temperature Coatings 2007 Elsevier Science & Technology Books (USA)
[10] Miller R A and Lowel Carl. E 1982, *Thin solid Films* **95**, 265
[11] Kokini K, Choules B D and Takeuchi Y R 1997 *Journal of Thermal Spray Technology* Vol. **6** (1) 43
[12] Kokini K, Choules B D and Takeuchi Y R 1996 *Journal of Surface and Coating Technology* Vol. **82**, 77
[13] Schmitt-Thomas Kh G and U Dietl 1994 *Surf. Coat. Techlol.* 68
[14] Wu B C, Chao C H, Chang E and Chang T C 1990 *Mater. Sci. Eng.* Vol. **A124** 215
[15] Scott G. Terry, Jennifer R. Litty and Carlos G. Levi, 1999 *The Minerals, Metals and Materials Society, Warrendale, PA*, 13-26
[16] Gualco G C, Cordano E, Fignino F, Gambaro C, Ahmaniemi S, Tuurna S 2002 *Proceedings of the International Thermal Spray Conference, Essen, Germany*, 196-199
[17] Johnson C A, Ruud J A, Bruce R, Wortman D, 1998 *Surface and Coatings Technology* **108-109**, 80
[18] Thomas S Hille, Thijs J Nijdam, Akke S J Suiker 2009 *Acta Materialia* **57** 2624
[19] He M Y, Evans A G, Hutchinson J W, 1998 *Physica Status Solidi* A **166**, 19
[20] Brik N and Meier G H 1983 Intro to High temperature Oxidation of Metals (London: Edward Arnold) 54
[21] Nakamori M, Harada Y and Hukue I 1983 *The Metallurgical Society of AIME*, Warrendale. PA 175
[22] Kofstad P 1985 *Oxid. Met.* **24 (5-6)** 256
[23] Alperine *et al.* 1997 TBCs - The Thermal Conductivity Challenge *AGARD Report 823* (15-16) 1.1 -1.10
[24] Teixeira V *et al.* 2000 *Journal of Thermal Spray Techol.* Vol. **9(2)** 191
[25] Haynes J A *et al.* 2000 *Journal of Thermal Spray Tech* Vol **(I)** 38
[26] Movchan B A, Malashenko, Yakovchuk K Yu, Rybnikov A I, Tchizhik A A 1994 *Surf Coat. Tech* Vol **67** 55
[27] Tsai H L and Tsai P C 1995 *Surf. Coat. Technol.* Vol.**71** 53
[28] Wright P K, Evans A G 1999 *Current Opinions in Solid State and Materials Science* **4** 255

Journal of Physics: Conference Series **439** (2013) 012022 doi:10.1088/1742-6596/439/1/012022

Enhancement of superconductivity of $TlBa_2Ca_3Cu_4O_{12-\delta}$ superconductors by Be doping

Nawazish A Khan and M Shamraiz[1]

Materials Science Laboratory, Department of Physics, Quaid-i-Azam University, Islamabad, Pakistan, 45320

E-mail: muhammadshamraiz6@gmail.com

Abstract. A reproducible synthesis and characterization of the $TlBa_2(Ca_{3-y}M_y)Cu_4O_{12-\delta}$ (M=0, Be; y=0, 0. 5, 1.0, 1. 5, 2) superconductors have been studied by using x-ray diffraction, resistivity, ac-susceptibility and FTIR absorption measurements. The x-ray diffraction (XRD) studies of these samples have shown a tetragonal structure in which the length of cell parameters and volume of the unit cell have been found to decrease with increased Be-doping. The zero resisitivity critical temperature [T_c(R=0)] have been suppressed with Be doping. A significant increase in the onset temperature of superconductivity have been observed in Be doped O_2 annealed samples. The magnitude of diamagnetism of as prepared samples is higher for particular level of doping. In comparison with as-prepared samples, an improvement in the magnitude of diamagnetism could be witnessed in all oxygen post-annealed samples. It is most likely that oxygen optimization in the unit cell which promotes increase the superconducting properties of $TlBa_2(Ca_{3-y}Be_y)Cu_4O_{12-\delta}$ (y=0, 0.5, 1.0, 1.5, 2) samples in terms of increased magnitude of diamagnetism. The softening of apical oxygen mode of the type Tl-O_A-Cu(2) and CuO_2 planar oxygen mode have been observed providing unequivocal evidence of the enhanced inter-plane coupling. In comparison with Be as-prepared samples, the magnitude of diamagnetism is significantly suppressed in O_2 annealed samples. It is most likely that oxygen optimization in the unit cell suppresses the density of the carriers and hence the superconducting properties of the final compounds. The doped Be atoms suppress the density of anti-ferromagnetic aligned moments existing in the inner CuO_2 planes of $TlBa_2(Ca_{3-y}M_y)Cu_4O_{12-\delta}$ [M=Be] which enhance the magnitude of superconductivity in the final compound.

1. Introduction

Single layer Tl-based $TlBa_2Ca_3Cu_4O_{12-\delta}$ superconductors are important due to their easy method of preparation, less toxicity and higher critical current density [1-2]. The crystal chemistry of this compound tells that it has charge reservoir layer and four conducting CuO_2 planes [3]. The outer two CuO_2 planes (OP) being in the immediate vicinity of charge reservoir layer are over-doped with carriers whereas the inner two CuO_2 planes are under-doped with the carriers [4]. The under-doping of the inner CuO_2 planes promotes anti-ferromagnetic alignment of small Cu spins [5-7]. These d-wave superconductors are an-

[1] To whom any correspondence should be addressed

isotropic and their an-isotropy [$\gamma= \rho_c/\rho_{ab}=H_{c\perp}/H_{c\|} =\lambda_c/\lambda_{ab}=\xi_{ab}/\xi_c$] decreases with the increased coherence length along the c-axis [8]. Since the coherence length through the Fermi-vector K_F depends on the carrier's concentration [9], therefore, the superconducting properties of this compound are marginalized due to decreased carrier's concentration due to the under-doped IP in the unit cell [10]. The coherence length in the ab-plane is almost the same whereas its values along the c-axes vary from material to material depending on the thickness of superconducting layers. A decrease in the coherence length along the c-axis, therefore, would increase the anisotropy of the final compound and hence suppress the superconductivity parameters (such as higher T_c, higher J_c and magnitude of diamagnetism). It is, therefore, advisable to suppress the an-isotropy of $TlBa_2Ca_3Cu_4O_{12-\delta}$ superconductors and enhance the coherence length along the c-axis [11]. One way of decreasing the an-isotropy of these compounds is to decrease the c-axis length, which would develop the better inter-plane coupling and in turn would decrease the thickness of the charge reservoir layer. In the unit cell of $TlBa_2Ca_3Cu_4O_{12-\delta}$ superconductors, Ca atoms separate the CuO_2 planes and develop inter-plane coordination. If doped with smaller sized atoms of the same group, it would help in enhancing the inter-plane coupling. We have chosen Be atoms for this purpose and doped them at the Ca sites in $TlBa_2Ca_3Cu_4O_{12-\delta}$ superconductors [12-13]. These atoms were chosen since the electro-negativity of Be is higher than Ca. It was expected that these atoms would help in developing a better correlation among the carriers existing in various CuO_2 planes. The doping efficiency of $TlBa_2O_{4-\delta}$ charge reservoir layer in $TlBa_2Ca_3Cu_4O_{12-\delta}$ superconductors is studied by carrying post-annealing of the samples in the oxygen atmosphere. The intercalation of O_δ oxygen atoms of the charge reservoir layer controls the carriers doping in the conducting planes [14]. Doped oxygen atoms in the charge reservoir layer control the flow of the carriers to the conducting planes due to their higher electro-negativity; the higher magnitude of it would stop the flow of electron whereas its smaller quantity would help the electron's flow to the conducting planes [14-15].

As mentioned previously that in the low dimensional superconductors like $TlBa_2Ca_3Cu_4O_{12-\delta}$ there are four un-equivalent CuO_2 planes [outer planes (OP) and inner planes (IP)] as for as charge carrier's concentration is concerned [10,16]. The outer CuO_2 planes (OP) which are directly connected to the charge reservoir layer and are over-doped with carriers whereas the inner CuO_2 planes have under-doped carrier's concentration. Due to the smaller density of the carrier in the IP, the small magnetic moments existing at the Cu sites get anti-ferromagnetic aligned spin arrangements [17-18]; these atoms do this to minimize their energy in the under-doped state. The existence of anti-ferromagnetism in the IP suppresses the over-all superconducting properties of the final compound [18-19]. The substitution of Be would help promoting the uniform distribution of the carriers in the IP and OP planes and in turn suppress the anti-ferromagnetism which would help enhancing the superconducting properties of the final compound.

2. Experimental procedure

The $TlBa_2(Ca_{3-y}M_y)Cu_4O_{12-\delta}$ (M=0, Be; y=0, 0. 5, 1.0, 1. 5, 2) samples were prepared by the solid state reaction method accomplished in two stages. At the first stage $Ba_2(Ca_{3-y}M_y)Cu_4O_{12-\delta}$ (M=0, Be; y=0, 0. 5, 1.0, 1. 5, 2) precursor material was prepared, by thoroughly mixing $Ba(NO_3)_2$, $Ca(NO_3)_2$, BeO_2 and Cu(CN) in a quartz mortar and pestle in appropriate ratios. The mixed material was fired twice at 880°C in a quartz boat for 24 hours and furnace cooled to room temperature. At the second stage, the precursor material was ground for about an hour and mixed with Tl_2O_3 to give $TlBa_2(Ca_{3-y}M_y)Cu_4O_{12-\delta}$ (M=0, Be; y=0, 0. 5, 1.0, 1. 5, 2) as final reactants composition. Thallium mixed material was palletized under 3.8 tons/cm^2 and the pellets were enclosed in a gold capsule. Gold capsule containing pellets were annealed for about 10 minutes at 880°C followed by quenching to room temperature. The samples were characterized by resistivity measurements by four probe and ac-susceptibility measurements by following mutual inductance methods. The structure of the material was determined by X-ray diffraction scan from Brouker using a CuKα source of wavelength 1.54056Å and the cell parameters were determined by a cell refinement computer program. The phonon modes related to the vibrations of various oxygen atoms in $TlBa_2(Ca_{3-y}M_y)Cu_3O_{12-\delta}$ (M=0, Be; y=0, 0. 5, 1.0, 1. 5, 2) unit cell were observed by Nicolet 5700 Fourier Transform Infrared Spectrometer (FTIR) in the wave number of 400-630cm^{-1}. The post-annealing of the samples in flowing oxygen atmosphere was carried out in a tubular furnace at 500°C for six hours.

3. Results and discussion

3.1. As-Prepared Be Doped $TlBa_2(Ca_{3-y}Be_y)Cu_4O_{12-\delta}$ (y=0, , 0.5, 1, 1.5, 2.0) samples

The X-ray diffraction scans of Be-doped $TlBa_2(Ca_{3-y}Be_y)Cu_4O_{12-\delta}$ (y=0, 0.5, 1.0, 1.5, 2) samples are shown in Figure 1. The planar reflections are fitted very well to the tetragonal structure following P4/mmm space group. The calculated cell parameters have shown a decrease in the length of cell parameters and volume of the unit cell with increased Be-doping. The decrease in the cell parameters and volume of the unit cell is an evidence of incorporation of the Be into the unit cell. The measurements of resistivity versus temperature of Be-doped as-prepared $TlBa_2(Ca_{3-y}Be_y)Cu_4O_{12-\delta}$ (y=0, 0.5, 1.0, 1.5, 2) samples are shown in Figure 2. The resisitivity variations as a function of temperature down to onset of superconductivity are metallic for all the samples. The variation of room temperature resistivity in these samples is observed between 0.015 and 0.062Ω-cm. The onset of superconductivity in these samples is observed around 112.1, 107.9, 110, 106, 104K and T_c(R=0) around 106, 103, 105, 102, 101.1K, respectively. A maximum enhancement of T_c(onset) is observed in the samples with Be doping y=0.5.

The ac-susceptibility measurements of $TlBa_2(Ca_{3-y}Be_y)Cu_4O_{12-\delta}$ (y=0, 0.5, 1.0, 1.5, 2) samples are shown in Figure 3. The magnitude of the superconductivity increases in all Be-doped samples except the sample with y=0.5. It is to remember that this decrease is observed in such a sample which has shown highest T_c onset.The FTIR absorption measurements of $TlBa_2(Ca_{3-y}Be_y)Cu_4O_{12-\delta}$ (y=0, 0.5, 1.0, 1.5, 2) samples are shown in Figure 4. In Be free $TlBa_2Ca_3Cu_4O_{12-\delta}$ samples the apical oxygen mode of the type Tl-O_A-Cu(2) is expected to be observed and this mode is observed 545.8cm^{-1}. The CuO_2 planar oxygen mode is observed around 601cm^{-1}. In the samples with Be-doping of 0, 0.5, 1.0, 1.5, 2 the apical oxygen mode is observed around 544.4, 553.1, 550.1, 548.6, 548.1. The CuO_2 planar oxygen mode is observed around 606, 598, 602, 599cm^{-1} for the Be doping of y=0.5, 1.0, 1.5, 2 respectively. The softening of the planar and apical oxygen mode oxygen mode is most likely a finger print of enhanced inter-plane coupling. In comparison with Mg-doped samples there is more enhancement in the inter-plane coupling which possibly induces more tilt in the CuO_2 planar oxygen atoms and hence not only the softening of planar oxygen modes but also of apical oxygen modes too. Such tilts in the planar oxygen atoms are reflection of decreased in the magnitude of anti-ferromagnetic aligned spins of the Cu-atoms existing in the inner CuO_2 planes; these anti-ferromagnetic aligned spins of Cu atoms suppress the magnitude of the superconductivity.

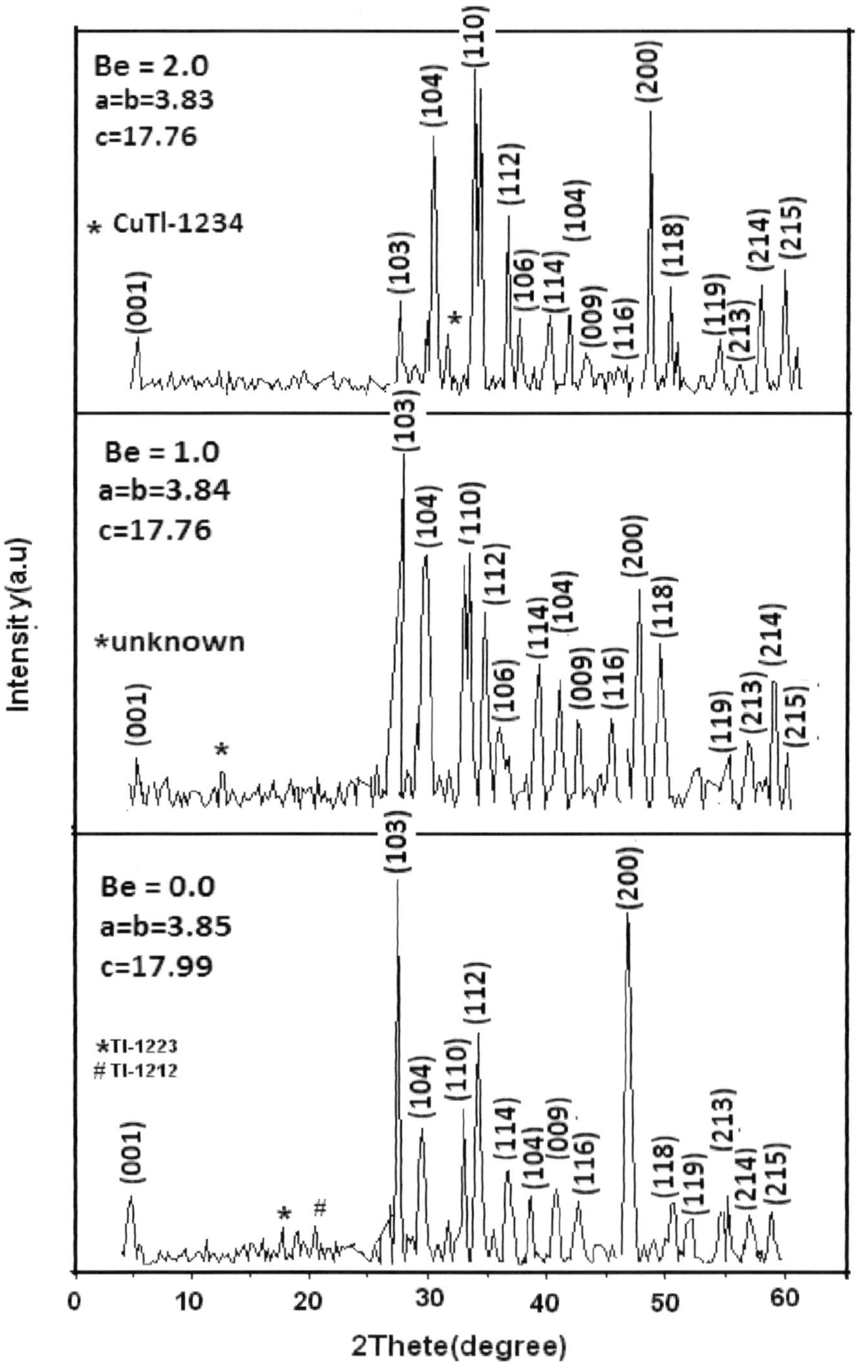

Figure 1. XRD Pattern of $Tl_1Ba_2Ca_{3-y}Be_y Cu_4O_{12-\delta}$ (y=0, 1, 2) superconductor samples.

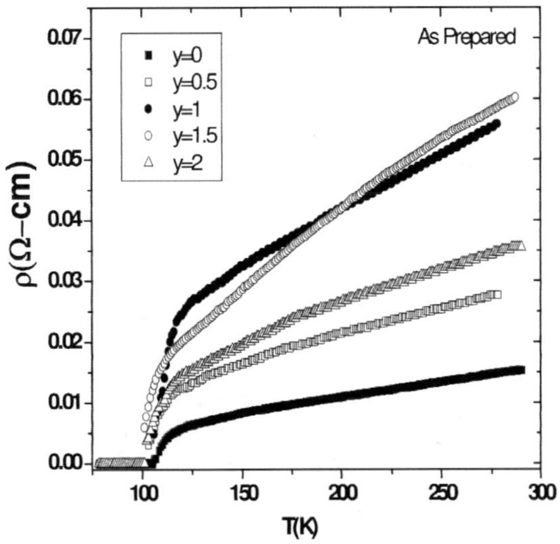

Figure 2. Plot of resistivity versus temperature of $Tl_1Ba_2Ca_{3-y}\,Be_yCu_4O_{12-\delta}$ (y=0, 0.5,1, 1.5, 2) as prepared superconductor samples

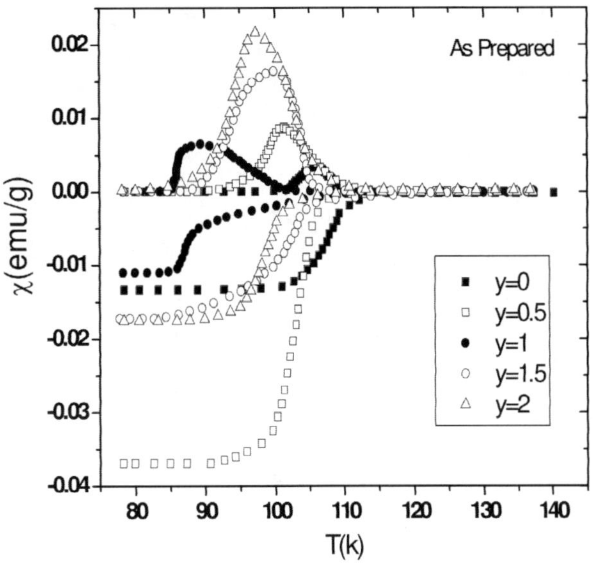

Figure 3. Plot of temperature versus ac-susceptibility of $Tl_1Ba_2Ca_{3-y}\,Be_yCu_4O_{12-\delta}$ (y=0, 0.5, 1, 1.5, 2) as-prepared superconductor samples

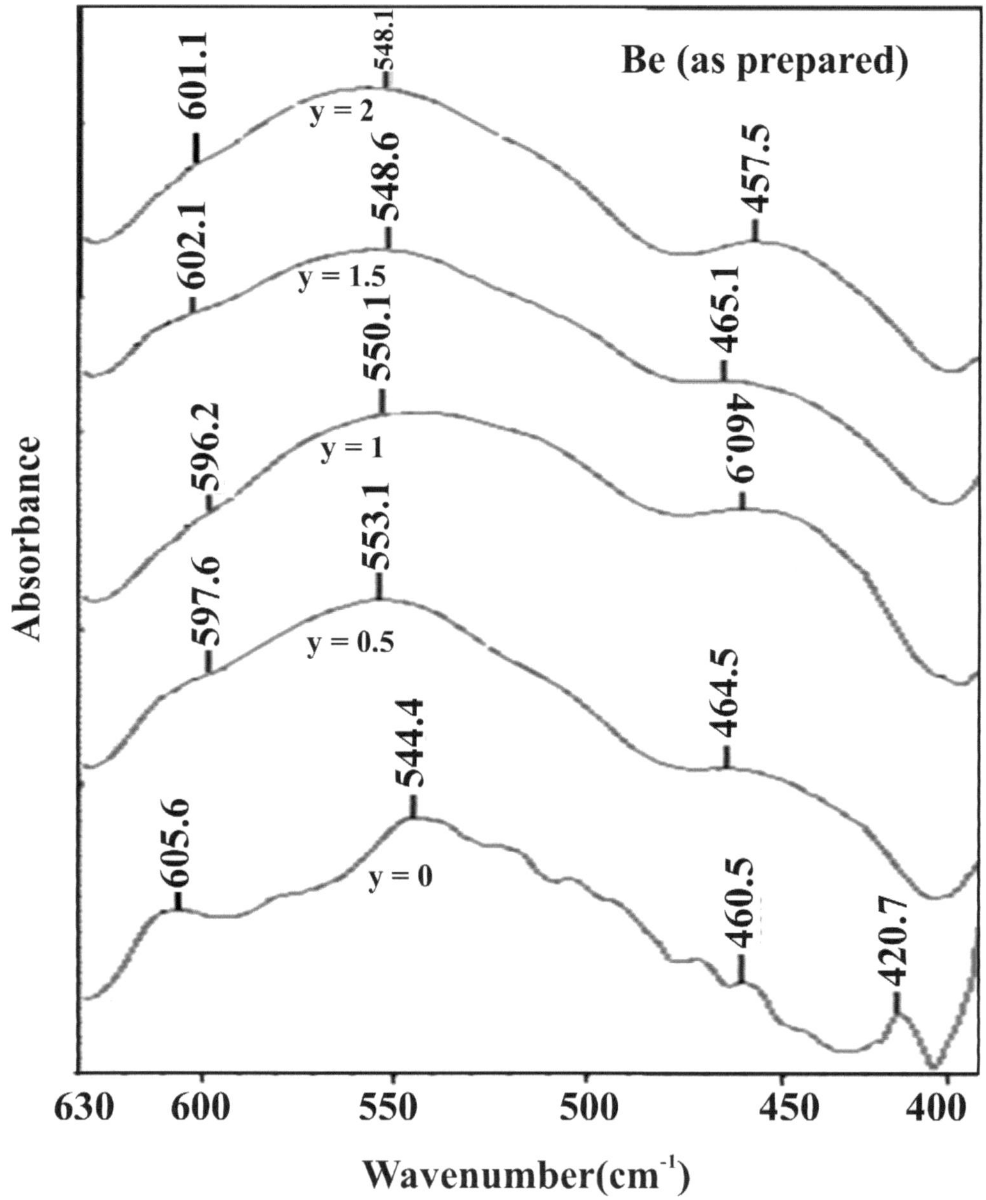

Figure 4. FTIR absorption measurements of $Tl_1Ba_2Ca_{3-y}Be_yCu_4O_{12-\delta}$ (y=0, 0.5, 1, 1.5, 2) as-prepared superconductor samples.

3.2. Be Doped Oxygen Post-annealed $TlBa_2(Ca_{3-y}Be_y)Cu_4O_{12-\delta}$ (y=0, 0.5, 1, 1.5, 2.0) samples.

The resistivity measurements of oxygen post-annealed Be-doped $TlBa_2(Ca_{3-y}Be_y)Cu_4O_{12-\delta}$ (y=0, 0.5, 1.0, 1.5, 2) samples are shown in Figure 5. These samples have also shown metallic behavior of resisitivity variations from room temperature down to onset of superconductivity. The room temperature resistivity in these samples is found to vary from 0.012 to 0.075Ω-cm. A suppression in onset temperature of superconductivity and $T_c(R=0)$ are observed for the Be-doping of y=0.5. The ac-susceptibility measurements of oxygen post-annealed $TlBa_2(Ca_{3-y}Be_y)Cu_4O_{12-\delta}$ (y=0, 0.5, 1.0, 1.5, 2) samples are shown in Figure 6. The magnitude of diamagnetism is significantly increased after oxygen post-annealing. It is most likely that oxygen optimization in the unit cell optimizes the density of the carriers in the conducting planes thereby enhancing the superconducting properties of the final compounds.

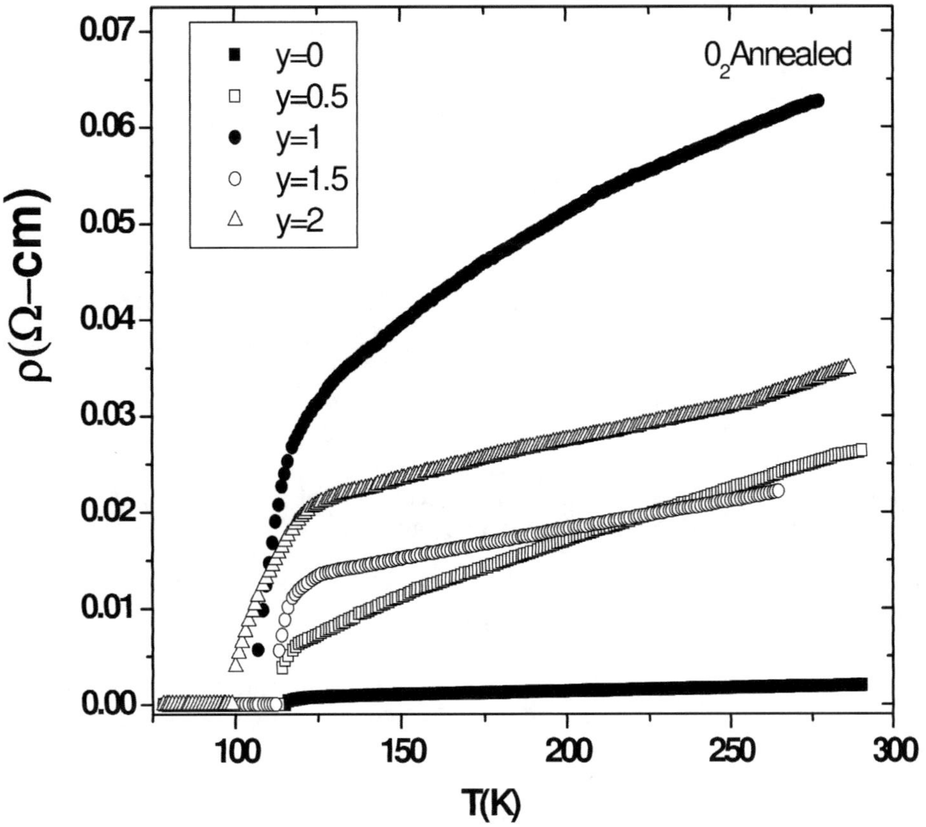

Figure 5. Plot of resistivity versus temperature of $Tl_1Ba_2Ca_{3-y}Be_yCu_4O_{12-\delta}$ (y=0, 0.5, 1, 1.5, 2) O_2 annealed superconductor samples

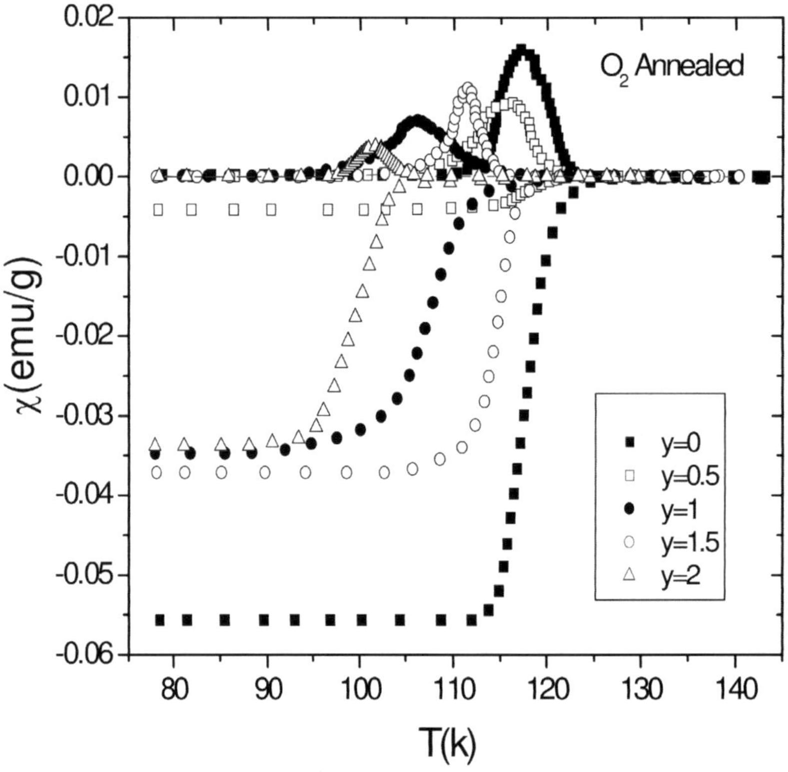

Figure 6. Plot of temperature versus ac-susceptibility of $Tl_1Ba_2Ca_{3-y}Be_yCu_4O_{12-\delta}$ (y=0, 0.5, 1, 1.5, 2) O_2 annealed superconductor samples.

The FTIR absorption measurements of oxygen post-annealed $TlBa_2(Ca_{3-y}Be_y)Cu_4O_{12-\delta}$ (y=0, 0.5, 1.0, 1.5, 2) samples are shown in Figure 7. In Be free $TlBa_2Ca_3Cu_4O_{12-\delta}$ samples the apical oxygen mode of the type Tl-O_A-Cu(2) is observed around 544.6cm^{-1}, the CuO_2 planar oxygen mode around 600cm^{-1}. The apical oxygen mode of the type Tl-O_A-Cu(2) is observed around 548.1, 544.6, 543.8, 542.1 cm^{-1}for the Be-doping of y=0, 0.5, 1, 2, respectively. The CuO_2 planar oxygen mode is observed around 596.2, 590, 576, 580cm^{-1} for Be doping of y=0, 0.5, 1, 2, respectively. It is interesting to note that in comparison with un-annealed samples a further softening of both apical and planar oxygen modes is witnessed, which showed that optimized density of the carriers in the conducting further induces tilts in planar oxygen atoms. However, the tilts in the planar oxygen atoms are not found to enhance the overall magnitude of anti-ferromagnetic aligned spins of the Cu-atoms existing in the inner CuO_2 planes otherwise we would have observe a suppression in the magnitude of superconductivity.

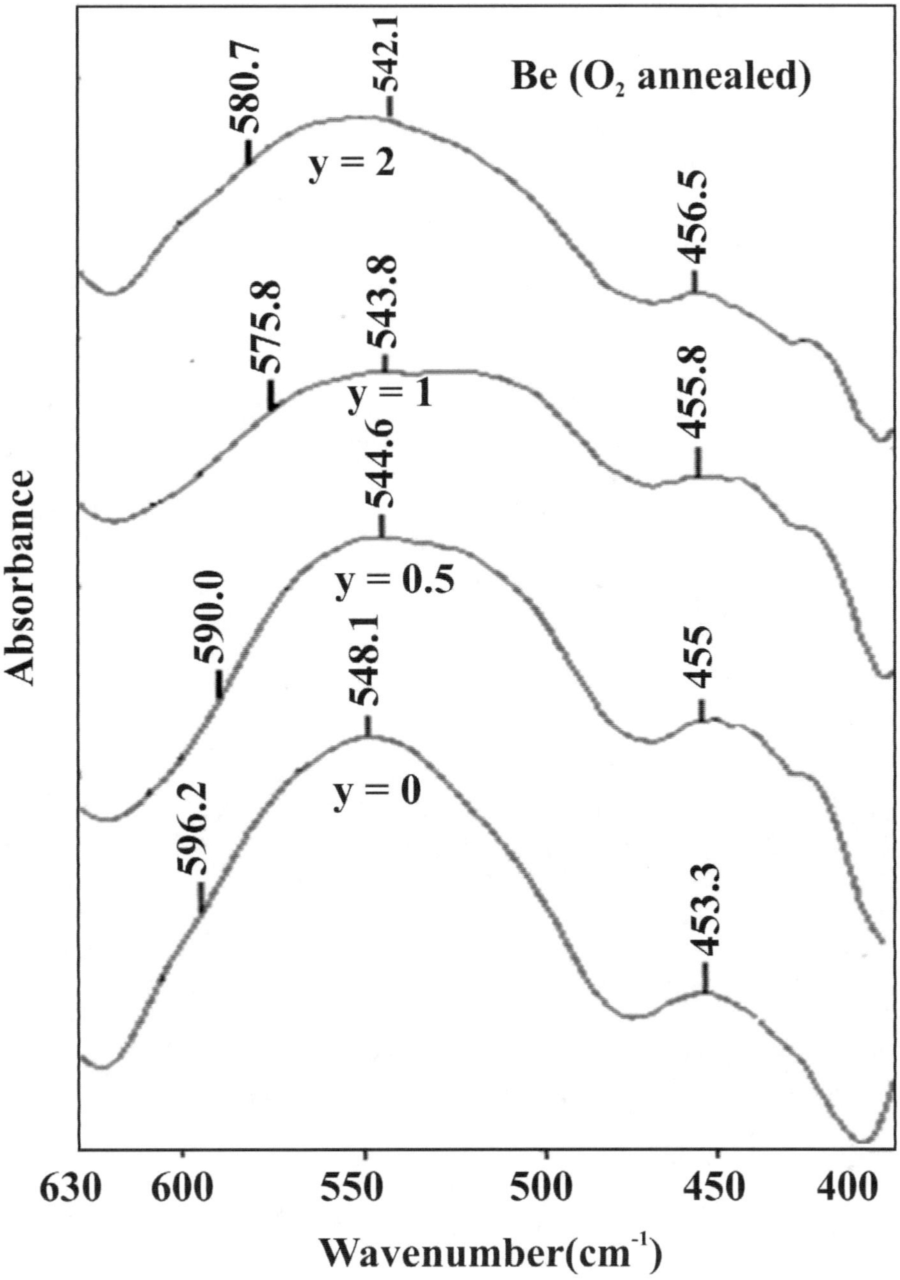

Figure 7. FTIR absorption measurements of $Tl_1Ba_2Ca_{3-y}$ $Be_yCu_4O_{12-\delta}$ (y=0, 0.5, 1, 2) O_2 annealed superconductor samples.

4. Conclusions

We have for first time synthesized Be doped $Tl_1Ba_2(Ca_{3-y}M_y)Cu_4O_{12-\delta}$ (M= Be; y=0, 0.5, 1.0, 1.5, 2.0) samples at normal pressure and studied their superconducting properties. The main objective behind such synthesis was to enhance the inter-plane coupling and reduce the density anti-ferromagnetic aligned spins of copper atoms existing in these compounds. The existence of such anti-ferromagnetic aligned spins of the copper atoms in the inner planes can suppress the over-all superconducting properties of the final compound. These experiments have shown that doped Be atoms promote enhancement of inter-plane coupling on one hand and reduce the population of anti-ferromagnetic aligned small spins of Cu atoms on the other hand. The former objective is most likely accomplished by the higher electro-negativity and the smaller sizes of Be in comparison with Ca atoms, whereas the later objective is achieved by the enhancement of the density of the carriers in the inner planes. The enhanced inter-plane coupling reduces the magnitude of anti-ferromagnetic aligned spins of Cu-atoms existing inner CuO_2 planes. The influence of which could be witnessed in the form of increased magnitude of diamagnetism in Be doped samples. The maximum increased magnitude of diamagnetism, however, depends on the solubility limit of the doped Be atoms at the Ca-sites in the $Tl_1Ba_2(Ca_{3-y}M_y)Cu_4O_{12-\delta}$ (M= Be; y=0, 0.5, 1.0, 1.5, 2.0) superconductors. Due to enhanced inter-plane coupling the an-isotropy of the final compound is suppressed which promote an enhancement in the parameters of superconductivity.

References

[1] Ihara H, Iyo A, Tokiwa K, Tukamato T B J, Terada N,Tokumato M and Umedo M 1996 *Czesh. J Phys.* **46** 3185

[2] Ihara H, Sugise R, Hirabayashi M, Terada N, Jo M, Negish A, Toumoto M, Kimura Y and Shimomura 1985 *Nature* **334** 510

[3] Ihara H, Tanaka K, Tanaka Y, Iyo A, Terada N, Tokumoto M, Tateai F, Kawamura M, Ishida K, Miyashita S and Watanabe T 2000 *Physica B* **284-288** 1085

[4] Kotegawa H, Tokunaga Y, Ishida K, Zheng G.-q, Kitaoka Y, Asayama K, Kito H, Iyo A, Ihara H and Tanaka K 2001 *Journal of Physics and Chemistry of Solids* **62** 171

[5] Ihara H, Iyo A, Tanaka Y, Terada N, Kazuyasue, Watanable T, Tokunaga Yo, Ishida Kenj, Yoshio Kitaoka and Noriaki Hamada 2000 *Physica B: Condensed Matter* **292** 238

[6] Magaishi Ko-Ichi, Kitaoke Y, Zheng G-q, Asayama K, Tokiwa K, Iyo A and Ihara H 1995 *Journal of Physical Society of Japan* **64(12)** 4561

[7] Khan N A and Najam-ul-hassan 2007 *Material Chemisty and Physics* **105** 298

[8] Ihara H, Iyo A, Tanaka K, Ishida K, Terada N, Tokumoto M, Sekita Y, Tsukamoto T, Watanabe T and Umeda M 1997 *Physica C* **282-287** 1973

[9] Ihara H, Iyo A, Tanaka K, Tokiwa K, Tareda N, Tokumoto M, Ishida Ke and Umeda M 1997 *Advance in superconductivity* ĺX 277

[10] Kotegawa H, Tokunaga Y, Ishida K, Zheng G.-q, Kitaoka Y, Asayama K, Kito H, Iyo A, Ihara H, Tanaka K, Tokiwa K and Watanabe T 2001 *Journal of Physics and Chemistry of Solids* **62** 171

[11] Ihara H, Iyo A, Tokiwa K, Tareda N, Tokumoto M and Umeda M 1996 *Advance in Superconductivity* VIII 247

[12] Khan N A and Khurrum A A 2005 *Appl. Phys.Lett.* **86** 152502

[13] Khan N A and Husnain G 2006 *Physica C* **436** 51

[14] Khan N A, Khurrum A A and Mazhar M, 2004 *Physica C* **407** 23
[15] Khan N A, Khurrum A A and Javed Asim 2005 *Physica C* **422** 9
[16] Kotegawa H, Tokunaga Y, Ishida K, Zheng G.-q, Kitaoka Y,Asayama A, Kito H, Iyo A, Ihara H, Tanaka K, Tokiwa K, Watanabe T and Ihara H 2001 *Physical Review B* **64** 064515
[17] Mumtaz M and Khan N A 2010 *Low Temp. Physics* **36** 154
[18] Mumtaz M and Khan N A 2009 *Physica B* **404** 3973 and reference therein
[19] Irfan M and Khan N A 2010 *Cryogenics* **50** 61

Journal of Physics: Conference Series **439** (2013) 012023 doi:10.1088/1742-6596/439/1/012023

Design and development of laser eye protection filter

K Ahmed[1], A N Khan, A Rauf, A Gul and M Aslam
Institute of Industrial Control System, GPO Box-1398, Rawalpindi, Pakistan

E-mail: drkamalahmed@yahoo.com

Abstract. Laser based devices, have been operational for measurement of distances horizontally and vertically in avionics and surveillance industries. These equipments are functional on pulsed Nd:YAG laser at 1064nm, this wavelength elevate the risk of eye exposure to personnel at unexpected levels. In this paper the eye protection filters, for the wavelength 1064nm were developed with soft (ZnS) and hard (TiO2) coating materials by using thin film vacuum coating technique. The damage threshold of the filter is 0.2 J/cm^2. Transmission characteristics are measured and discussed. Optical damage threshold (for eye 5 x 10^{-6} J/cm2) at various distances is also simulated.

1. Introduction

Laser applications have been drastically thrived in the 21st century from enclosed to al fresco environment like remote sensing, surveillance, aviation and forestry. Laser range finder, laser altitude meter and laser target designator which are operational on pulsed Nd:YAG laser wavelength 1064nm raise the threat of eye exposure to unexpected levels. The need of eye protection devices is obligatory for the personnel working above mentioned fields.

The laser light in the visible and near infrared region 400 to 1400 nm can cause of retinal damage and it is known as "retinal hazard region". Unfortunately, the wavelength of Nd:YAG Laser:1064nm lies in this region. The pulsed Nd:YAG laser has been considered one of the most harmful laser because its peak power and functioning wavelength is near infrared and invisible (Sliney 1985). Infrared lasers are particularly perilous for human eye, since the body protective blink reflexes is triggered only by visible light that means the eye does not feel pain or noticeable immediate damage during exposing of laser radiation. The coherence & low divergence of laser radiation and focusing mechanism of the eye create condition that laser light can be concentrated into a small spot on the retina and burns the surface. A transient increase of only 10 ^{o}C can destroy retinal photoreceptor and if the retina heated up to 100 ^{o}C, it creates a permanent blind spot. So that in the light of these facts, eye is extremely sensitive to the Nd:YAG laser radiation and can be permanently or partially injured from direct or reflected beam.

Now days, range finding and altitude measuring laser base devices are frequently employed in the aviation and military industries. These devices deliver the high precision collimated laser beam toward the target (Bryan and Sliney 1998), at a shorter distance from these devices, if eye exposes with laser beam the high optical density (OD) of radiation could damage the eye instantly, as

[1] To whom any correspondence should be addressed.

these radiation approaches at far distance from the source, the diameter of laser beam increases due to this optical density per unit area decreases, eventually this decreases the eye damage probability too. In the above mentioned situation, the solution for bare eye and eye behind the sighting device, it is necessary that an eye protection filter must be placed in front of eye, to protect the eye by the laser radiation.

In this article, eye protection filters were developed with different materials by using physical vapour deposition method and their behaviour and optical response is discussed.

2. Experimental

By using thin film coating techniques, number of thin film coating schemes were employed with soft and hard coating materials. These coating were designed on software OPTALIX.

In first experiment, thin film coating was developed and the coating materials MgF_2 & ZnS were used. The Coating scheme is as follows:

At reference wavelength λ_o: 1064nm
LHLHLHLHLHLHLHLHLH 0.5L

The polished substrate of K9 (Chinese) optical glass (Diameter: 25mm and thickness 25mm) was used and thin film coating was carried out on Coating machine. Coating was performed at substrate temperature at 50°C and pressure of chamber was maintained in the range of 10^{-6} mbar. MgF_2 and ZnS were evaporated through thermal evaporation and electron beam evaporation techniques respectively. It is a soft and fragile coating therefore coating was cemented in between of other K9 substrate for the protection of environmental effects. These coated samples have gone through the environmental test figure 1.

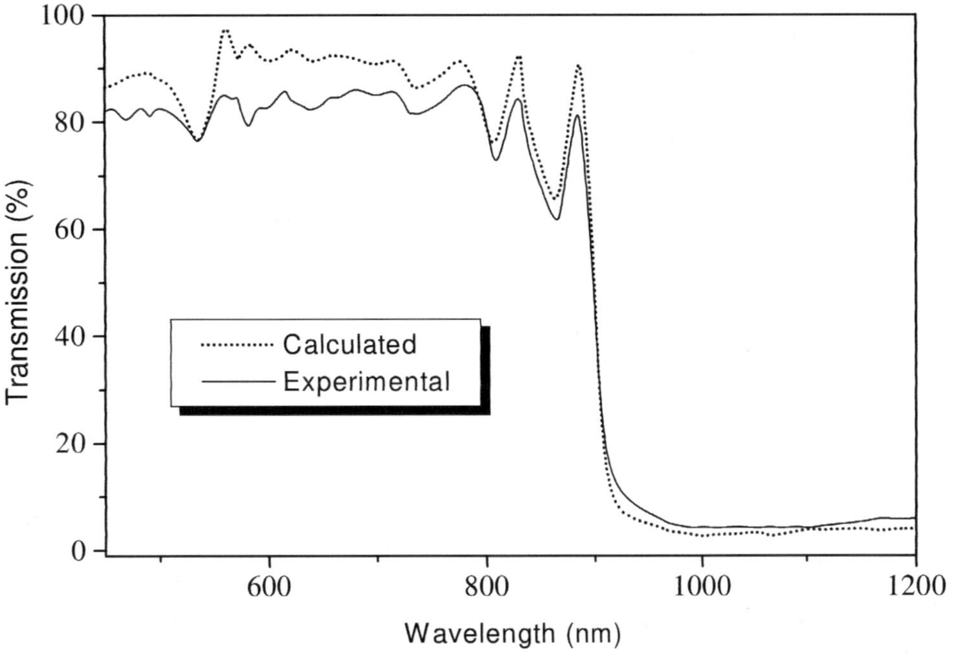

Figure 1. Transmission spectrum response of soft thin film coating (optical glass K9).

In second experiment, thin film coating was designed and developed and the coating materials MgF_2 & TiO_2 were used. The coating scheme is as follows:

At reference wavelength λ_o: 1064nm
0.92LH0.92LH0.92LH0.92LH0.92LH0.92LH0.5L

The polished substrate of K9 optical glass was used and thin film coating carried out on Coating machine. Coating was performed at substrate temperature at 280°C and chamber pressure was maintained in the range of $5x10^{-6}$ mbar. MgF_2 and TiO_2 were evaporated through thermal evaporation and electron beam evaporation respectively.

TiO_2 is a reactive coating material therefore during the coating process a partial pressure of oxygen was maintained from $6.0x10^{-5}$ to $8.0x10^{-5}$mbar range inside the chamber (Chen et al. 1997). It is a hard coating and components were processed from environmental test and qualified. The transmission test was carried out for both soft and hard coating samples on Spectrophotometer shown in figures 1 & 2.

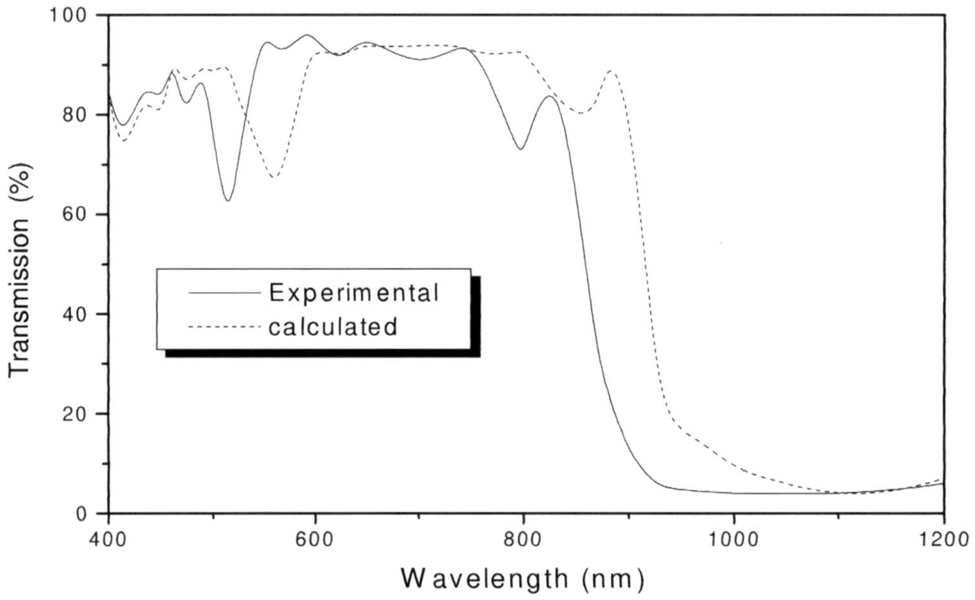

Figure 2. Transmission spectrum response of hard thin film coating (optical glass K9).

3. Results and discussion

The eye protection filters were developed with two types of thin film coating materials soft and hard coating, it is generally known as hot mirror coating.

The colour of coating on polished optical components was greenish blue and spectral response of theoretical simulation and experimental response is shown in figure 1. The theoretical simulation and experimental spectrum is quite similar only in the visible region experimental response is slightly less 5% as compare to theoretical simulation it is due to coating thicknesses and variation in experimental conditions during the coating process. However, in the IR region the theoretical simulation and experimental response is slightly less as compared to experimental curve, it is due to substrate thickness and transmission is nearly 4 percent in the NIR region which is acceptable for eye protection filter for Nd:YAG laser system.

The colour of coating on polished optical components was reddish green and spectral response of theoretical simulation and experimental response is shown in figure 2. The theoretical simulation and experimental spectrum was approximately similar. But experimental spectrum shifted towards shorter wavelength side it was due to the effect of temperature and pressure inside the chamber during thin film coating process (Selhofer et al.). The transmission profile is similar in the visible and NIR region. The cut off started from 950nm in calculated curve while in experimental it started from 875nm. Average transmission in calculated and experimental is approximately is 85% in visible region, however in NIR region the transmission is approximately 4 % which is similar to transmission observed with the soft coating eye protection filter.

Laser damage threshold of these filters is 0.2 J/cm^2 and performance and fitting wise hard coated filter is suitable for binocular. These filters transmit the 1064nm radiation approximately 4%, this yield that these filter capable of reducing optical density $1.5x10^{-2}$ J/cm^2 to safe MPE (Maximum Permissible exposure for Eye) value of $5x10^{-6}$ J/cm^2 (Yeo 1990). These filters are able to protect the eyes of surveillance personnel, viewing the sight by using binocular 8x30 at a distance mention in following Table 1.

Table 1. Performance of Eye Protection Filter to safe the eye with and without filter.

Typical Equipment	Laser	Energy, divergence & pulse width	Safe Viewing distance	
			Without filter	With filter
Laser Finder	Range	10mJ, 1mRad. & 10nsec	500m	25m
Laser Designator	Target	100mJ, 0.3mRad. & 20nsec	07 Km	600m

4. Conclusions

In the light of above observation, reflection at 1064nm should increases up to 1% which enhance the eye protection capability. It is suggested that those personnel responsible of surveillance, their sighting equipment must contain laser eye protection filter or even safe side they should wear laser protection goggle for their eyes. Finally it is suggested that the laser safety training programs are needed to be conducted in work places for surveillance personnel to avoid the laser hazard accident.

References

[1] Sliney D H 1985 *International Ophthalmology Clinics* **25(3)** 151
[2] Bryan P K Sliney D H 1998 *Laser Safety, Eye safe laser system & Laser Eye Protection* (Proceeding Volume)
[3] Chen J, Chao S, Kao J, lai G and Wang W 1997 *Appl. Opt.* **36(19)** 4403
[4] Selhofer H, Ritter E and Limbod R 2002 *Appl. Opt.* **44(4)** 758
[5] Yeo R 1990 *Health & Safety at Work (Eye Safety)* 28

Journal of Physics: Conference Series **439** (2013) 012024 doi:10.1088/1742-6596/439/1/012024

Dielectric, electric and thermal properties of carboxylic functionalized multiwalled carbon nanotubes impregnated polydimethylsiloxane nanocomposite

Sadia Sagar[1,2] , **Nadeem Iqbal**[1], **Asghari Maqsood**[1]
[1]Thermal Transport Lab (TTL), School of Chemical and Materials Engineering (SCME), National University of Sciences & Technology (NUST), H-12, Islamabad, Pakistan

E-mail: sadiasagar@scme.nust.edu.pk

Abstract. The dielectric, electric and thermal properties of carboxylic functionalized multiwalled carbon nanotubes (F-MWCNT) incorporated into the polydimethylsiloxane (PDMS) were evaluated to determine their potential in the field of electronic materials. Carboxylic functionalization of the pristine multi walled carbon tubes (Ps-MWCNT) was confirmed through Fourier transform infrared spectroscopy, X-ray diffraction patterns for both Ps-MWCNTs and F-MWCNTs elaborated that crystalline behavior did not change with carboxylic moieties. Thermogravimetric and differential thermal analyses were performed to elucidate the thermal stability with increasing weight % addition of F-MWCNTs in the polymer matrix. Crystallization/glass transition / melting temperatures were evaluated using differential scanning calorimeter and it was observed that glass transition and crystallization temperatures were diminished while temperatures of first and second melting transitions were progressed with increasing F-MWCNT concentration in the PDMS matrix. Scanning electron microscopy and energy dispersive x-ray spectroscopy were carried out to confirm the morphology, functionalization, and uniform dispersion of F-MWCNTs in the polymer matrix. Electrical resistivity at temperature range (100-300°C), dielectric loss (tanδ) and dielectric parameters (ε', ε'') were measured in the frequency range (1MHz-3GHz). The measured data simulate that the aforementioned properties were influenced by increasing filler contents in the polymer matrix because of the high polarization of conductive F-MWCNTs at the reinforcement/polymer interface.

1. Introduction

Polydimethylsiloxane (PDMS) has shown good flexibility, thermal stability, electrical resistance and dielectric strength due to the cross-linking density and degree of polymerization [1]. PDMS composites are used for the wide range of applications from sensor technology to radar absorbing materials [2, 3]. In addition, PDMS is used for substrate pattern transfer to various substrates due to its soft and flexible nature [4, 5]. During fabrication it leads to residual deformation, the large expansion coefficient of the polymer induces residual stresses in the polymer.

[2] To whom any correspondence should be addressed.

This phenomenon causes misalignment of different structures. The misalignment of structure is usually resolved by incorporating fillers e.g. whiskers, nanoparticles and nano-tubes, etc [6, 7]. The highly conductive nano particles/tubes filled polymer matrix composites are a rapidly developing field in switchable and flexible microelectronics technology [8, 9]. The dielectric property of PDMS is modified by incorporating F-MWCNTs as nano filler in the PDMS matrix. Since carbon nanotubes (CNTs) possess large aspect ratio with extra ordinary thermal, mechanical & electrical properties; CNTs help in enhancing the thermal, mechanical and electromagnetic properties of the PDMS composite [10]. The focus of this research is on the fabrication of carboxylic functionalized multiwalled carbon nanotubes (F-MWCNT) reinforced in PDMS based composite. These PDMS composites are designed for high thermal stability and dielectric strength. When used as interconnects in semi-conducting devices, the conducting multiwalled carbon nanotubes (MWCNTs) can route electrical signals at speeds up to 10 GHz; this is usually attributed due to electron transfer over a wide range of length without intermediate interruption [11]. In this paper, Ps-MWCNTs are functionalized for uniform dispersion in PDMS matrix to study the effect on electric and thermal properties of the composite. The dielectric, electrical, morphological, spectroscopic and thermal analyses are performed and it is observed that all these properties are strongly influenced by increasing F-MWCNTs content in the polymer matrix.

2. Experimental

2.1. Materials

Ps-MWCNTs with 90% purity and average length and diameter were found to be around 30μm and 25nm were purchased from Nanoport. Co. Ltd, China. Room temperature vulcanized - polydimethylsiloxane was purchased from Wacker, Germany and it was used as received. Nitric acid, ammonium hydroxide and toluene were received from Merk. Polypropylene (PP) membranes with 0.2μm pore size were bought from Pall.Co., China.

2.2. Sample preparation

2.2.1. Surface modification of Ps-MWCNTs

To functionalize Ps-MWCNTs, purification was performed at 450^{0}C with a temperature rise rate of 10^{0}C per minute for 6 hours followed by their treatment with hydrochloric acid to remove metal traces, amorphous carbon and other impurities. In the second step, 2g purified MWCNTs were immersed in 68% conc. HNO_3 at room temperature. Then, the solution was sonicated in ultrasonicator bath at 40 KHz, 90^{0}C for 4hours. Subsequently, the solution was washed with ultra-pure water five times and the pH of the solution was neutralized with ammonium hydroxide. The pH value was maintained at 5.5 and the solution was filtered with 0.2μm PP membrane. At the end, F-MWCNTs were collected by drying the nanotubes in the oven at 100^{0}C for 12 hours. Functionalization mechanism is demonstrated in figure 1b.

2.2.2. Fabrication of F-MWCNTs based PDMS composite

The dispersion of Ps-MWCNTs in the polymeric systems is a difficult task. The carboxylic moieties make inter bond between multiwalled carbon nanotubes and polymer matrix. Initially, F-MWCNTs were dispersed in toluene using ultrasonication bath for 2 hours. This was followed by dispersion in PDMS using a mechanical stirrer at 5000rpm for 30 minutes. Furthermore, the F-MW with 90% purity and average length and diameter were found to be around 30μm and 25nm NT/PDMS solution was poured into the 6” x 6” x 2.5” mold. The pre-curing of F-MWNT/PDMS composite was carried out for 1 hour at 120^{0}C and then temperature was raised to 160^{0}C. The composite was kept at 160^{0}C for 30 minutes. Dibenzoyl peroxide (0.2pphr) was used as a crosslinker for PDMS. Flow chart in figure 1b shows the route for the fabrication of the composite. Five different types of F-MWCNT/PDMS composites P0, P1, P2, P3, and P4 were fabricated keeping 0 ,0.1, 0.3 ,0.5, 0.7 wt% ratio of F-MWCNTs in PDMS matrix, respectively.

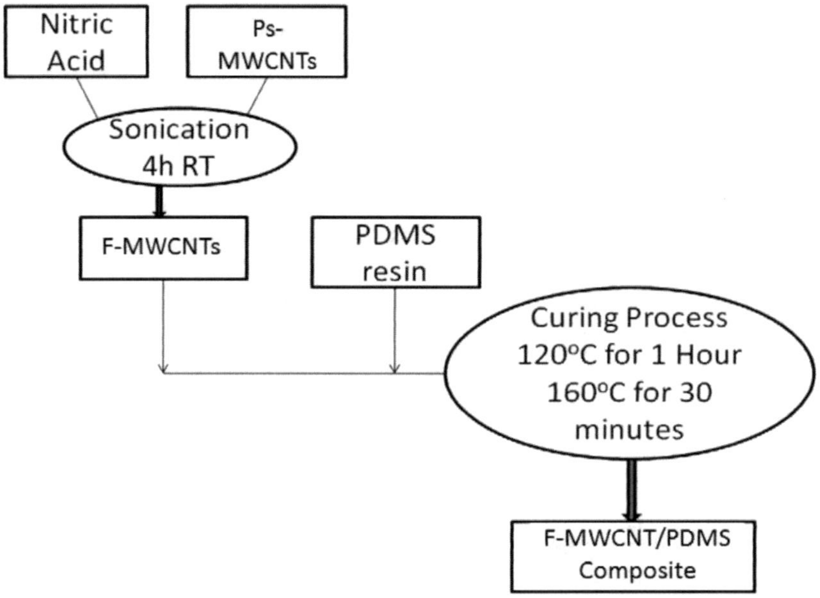

(a)

Figure 1a. Functionalization mechanism and Flow chart for composite preparation.

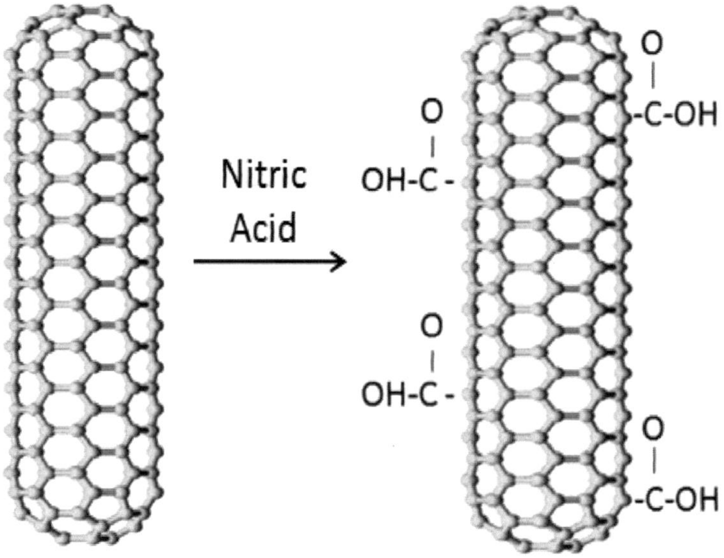

(b)

Figure 1b. Schematic illustration of attached carboxylic moiety with CNTs.

2.3. Characterizations

The spectroscopic characterizations of Ps-MWCNTs /F-MWCNTs were performed using Perkin Elmer Fourier Transformation Infrared (FTIR) spectrometer with KBr disc to study the functionalization and X-

Ray Diffraction (XRD) was done to confirm the crystalline nature of MWCNTs. Thermal properties of pristine/functionalized MWCNTs and PDMS composites were elaborated using Perkin Elmer Diamond thermal gravimetric/differential thermal analysis (TG/DTA). The isothermal crystallization (Tc)/glass transition (Tg) /melting (Tm) temperature responses of PDMS composites were investigated by using Perkin Elmer differential scanning calorimeter (DSC). The surface modification and dispersion of F-MWCNTs in the host polymer matrix were analyzed using scanning electron microscope along with the energy dispersive x-ray spectroscopy (SEM/EDS) Jeol JSM 6490 A. HP 4339B high resistance meter was used to measure the electrical properties with temperature augmentation. The dielectric parameters ($\varepsilon^{/}$, $\varepsilon^{//}$, tanδ) were measured using the bridge LCR meter (Model HP 4284 A) and Agilent network analyzer.

3. Results and discussion

3.1. FTIR & XRD analysis of MWCNTs

Figure 2a represents FTIR spectra of Ps-MWCNTs and F-MWCNTs. The presence of the broad band in F-MWCNTs spectra in the range 2400-3500cm^{-1} confirms the presence of hydroxyl group and carboxylic group in KBr disc. The appearance of a peak approximately at 1450-1650 cm^{-1} corresponds to the C–O stretching which indicates the introduction of carboxylic groups due to surface oxidation of nanotubes. The peak at 1100-1120 cm^{-1} indicates a C–O vibration. The fingerprint region at 550-600 cm $^{-1}$ may indicate a bending vibration of C-H. Therefore, the spectrum clears the evidence for covalent attachment of organic moieties on the surface of F-MWCNTs.

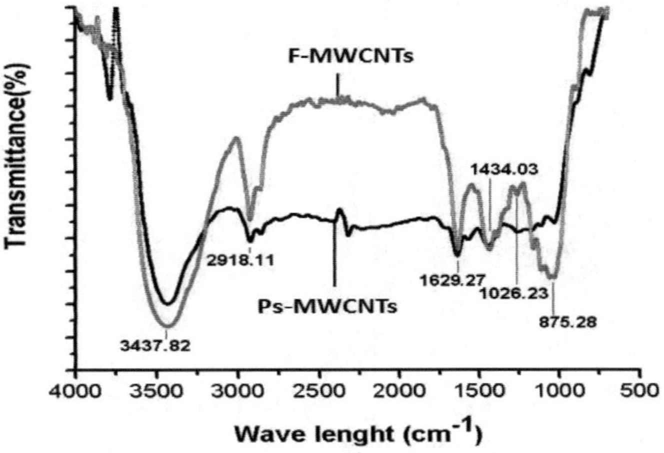

Figure 2a. FTIR spectra for MWCNTs

Crystalline nature of F-MWCNTs is also confirmed from (10° to 80° (2θ)) XRD pattern with a scanning step of 0.02^{o}/s; figure 2b shows the diffraction peaks at 26.71 and 44.91 corresponding to graphite structure derived from MWCNTs. Furthermore, the intensity of (002) and (100) peaks of F-MWCNTs are much closer to Ps-MWCNTs, provides the evidence for the fact that the treatment does not damage the graphene layer organization.

3.2. Scanning electron microscopy (SEM) and EDS analysis

SEM coupled with EDS was performed to investigate the morphology/compositional analysis of Ps-MWCNTs/F-MWCNTs. CNTs and composite samples were coated with gold to make them conductive for SEM analysis. The comparative analysis of Ps-MWCNTs and F-MWCNTs morphology was carried out as obvious in figure 3 at 50Kx magnification along with their EDS. Figure 3b confirms the oxidation of Ps-MWCNTs in EDS analysis. To analyze the dispersion of F-MWCNTs in the polymer matrix,

Journal of Physics: Conference Series **439** (2013) 012024 doi:10.1088/1742-6596/439/1/012024

samples were broken in liquid nitrogen and then the cross-section of P3 (0.5wt% F-MWCNT/PDMS composite) was observed in SEM.

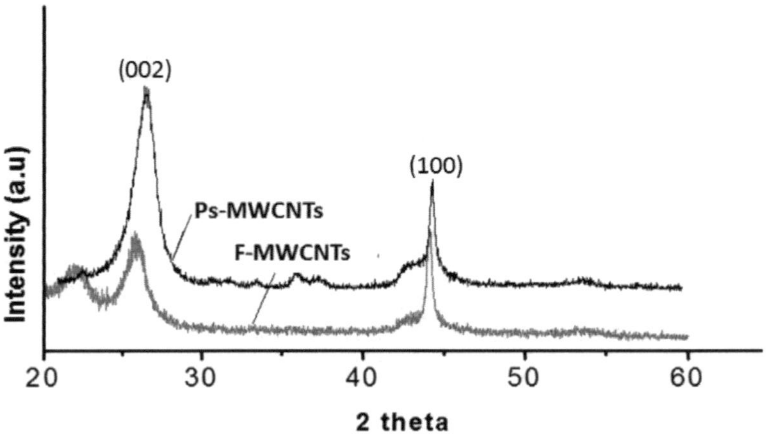

Figure 2b. XRD pattern for MWCNTs

Figure 3a. SEM images with EDS of Ps-MWCNTs

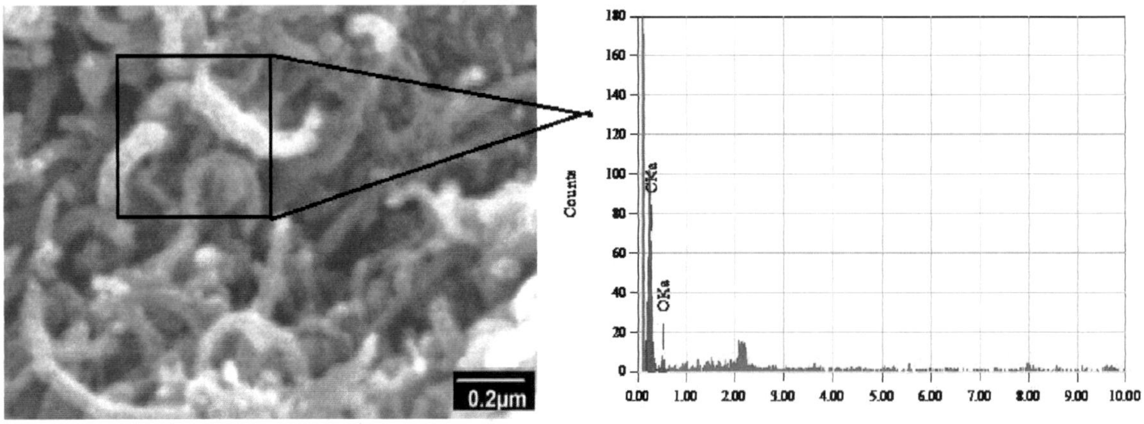

Figure 3b. SEM images with EDS of F-MWCNTs

Figure 4(a-b) shows the SEM micrograph of P3 (0.5wt% F-MWCNT / PDMS composite) at 50Kx and 90Kx magnification confirming the uniform dispersion of CNTs in the host polymer matrix. The uniform dispersion of F-MWCNTs in PDMS strongly influences the dielectric, electrical, structural and thermal properties of the polymer due to the strong interfacial interaction between polymer chains and nanotubes surface.

Figure 4. SEM of F-MWCNTs reinforced polymeric composite at 50Kx (a) & at 90Kx (b)

3.3. Thermogravimetric/differential thermal analysis (TG/DTA) of PDMS composites

TGA was used to determine the thermal degradation behavior of Ps-MWCNTs/F-MWCNTs and relative effect of different wt% loadings of F-MWCNTs in PDMS host matrix. Figure 5(a, b) represents the TGA and DTA traces between 25-800^{0}C under the oxygen environment. The thermal properties of Ps-MWCNTs/F-MWCNTs and composites are also given in table 1. Thermal gravimetric analysis of Ps-MWCNTs (P6) and F-MWCNTs (P5) in figure 5a elucidate that P5 thermally degrade 13% in excess than P6 due to –COOH debonding from F-MWCNTs surface in the form of water vapors and CO_2 exhaust. The weight loss behavior of F-MWCNTs incorporated polymer composite is ranked as P4<P3<P2<P1<P0, as can be seen in figure 5a & table 1.

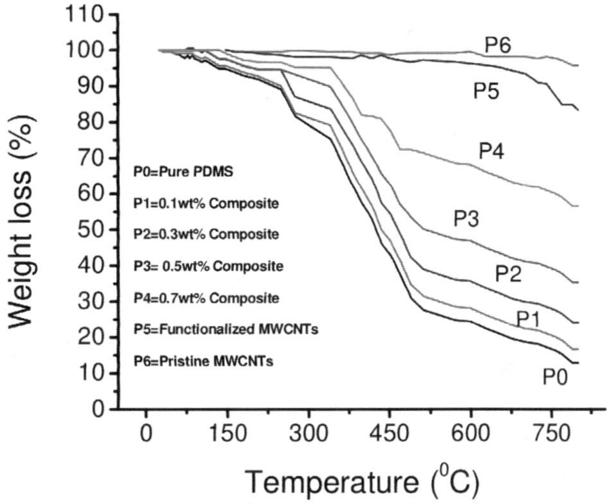

Figure 5a. TGA of MWCNTs and PDMS nanocomposite

This behavior defines the efficacy of increment of F-MWCNTs in PDMS polymer matrix. The nano scale interaction of F-MWCNTs with polymeric chains of host matrix causes the maximum thermal stability of the F-MWCNT/PDMS composites.

Table 1. %Weight loss at different temperature.

Sample ID	%Wt Loss at 250^{0}C	%Wt Loss at 375^{0}C	%Wt Loss at 525^{0}C	%Wt Loss at 800^{0}C
P0	7.76	35.90	72.86	87.09
P1	7.19	31.93	69.17	83.39
P2	5.01	24.84	61.79	75.70
P3	4.91	18.86	50.41	64.63
P4	3.22	12.95	29.36	43.04
P5	0.66	2.71	3.10	13.30
P6	0.38	0.67	0.94	5.20

In conclusion, thermograms confirm that F-MWCNT/PDMS composites exhibit more thermal stability than pristine PDMS due to the heat quenching ability of F-MWCNTs as obvious in DTA thermograms. In DTA curves, it is clear that F-MWCNTs absorb more heat than Ps-MWCNTs due to bond cleavage of covalently attached moieties on the surface of MWCNTs. The DTA behavior of both polymer matrix and CNTs dispersed composites is completely endothermic but the polymer absorbed additional heat due to the presence of F-MWCNTs. Figure 5b illustrates the upward shift in thermograms which portrays that addition of F-MWCNTs enhance the thermal stability of the host polymer.

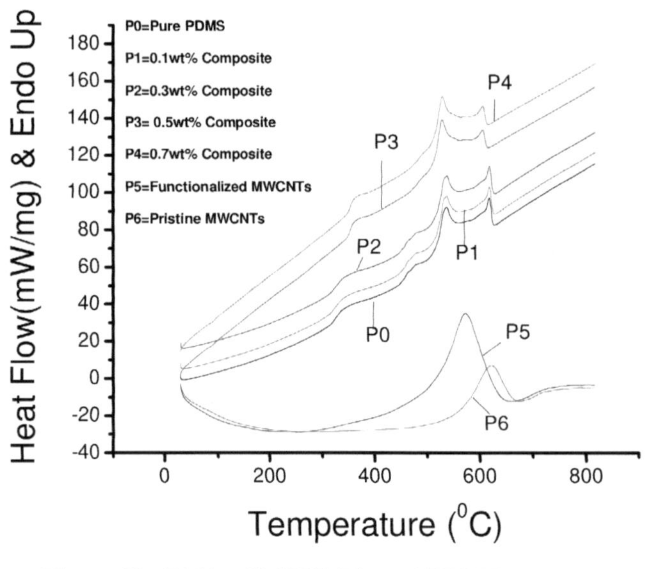

Figure 5b. DTA of MWCNTs and PDMS nanocomposite.

3.4. Differential scanning calorimeter (DSC) analysis

Figure 6 elaborates the DSC scrutiny obtained from pure PDMS and different wt% reinforced F-MWCNTs in PDMS after heating at a rate of 10^{0}C/min from -165^{0}C to 250^{0}C. Crystallization temperature (Tc, Tc*) gradually decreases with the wt% increment of F-MWCNTs from 0.1% to 0.7% due to the reduction of spherulite sites in the polymeric composite [11]. The same behavior is revealed in the melting temperature (Tm, Tm_1) response of the composite specimens.

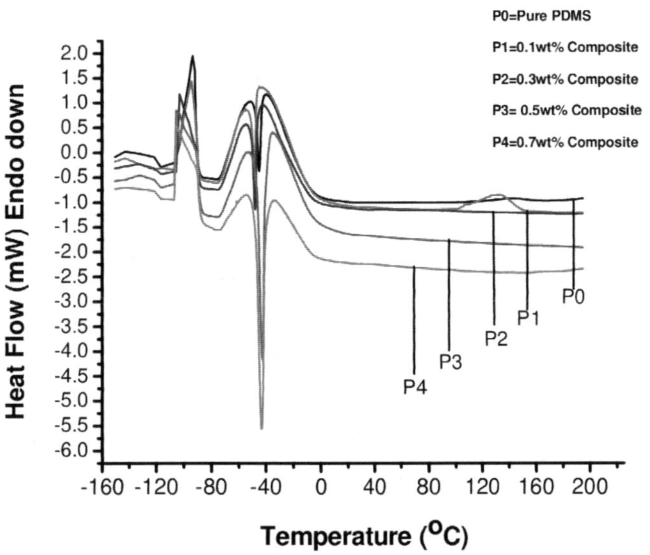

Figure 6. DSC profiles displaying Tg, Tc, Tm of the PDMS nanocomposites

Figure 6 also simulates that glass transition temperature (Tg) of F-MWCNT/PDMS composite decreases with percent enhancement of F-MWCNTs because of the interaction of polymer chains with nanotubes through carboxylic moieties. The nano reinforcement improves the ductile/rubbery behavior of polymer matrix up to ~13^{0}C. The reduction of Tc, Tg & and enhancement of Tm confirm the proficiency of functionalized nanotubes due to the intermolecular interaction between PDMS molecular chains and F-MWCNTs as evident from table 2.

Table 2. Table of DSC parameters.

Sample ID	Tg (^{0}C)	Tc (^{0}C)	T_{m1} (^{0}C)	T_{m2} (^{0}C)	Tc*(^{0}C)
P0	-121.37	-93.23	-74.84	-45.25	-51.90
P1	-122.19	-93.23	-74.93	-47.56	-54.34
P2	-125.56	-96.60	-73.93	-47.80	-55.72
P3	-129.66	-98.24	-73.11	-42.79	-55.99
P4	-134.67	-99.88	-72.66	-43.61	-56.90

3.5. Electrical, dielectric and microwave permittivity measurements

The dc electrical resistivity was measured in the temperature range 100-300^{0}C using two probe resistivity apparatus. Standard Arrhenius equation [12, 13] was used to calculate dc-electrical resistivity (ρ_{dc}) for all fabricated composites. The dc electric resistivity with respect to temperature is represented by figure 7(a). It is observed that dc resistivity decreases with the increase in temperature which confirms the semiconducting behavior of fabricated composites. It is also observed that dc electric resistivity decreases with the increase in F-MWCNTs contents and minimum value is observed for 0.7 Wt% F-MWCNTs filled composite.

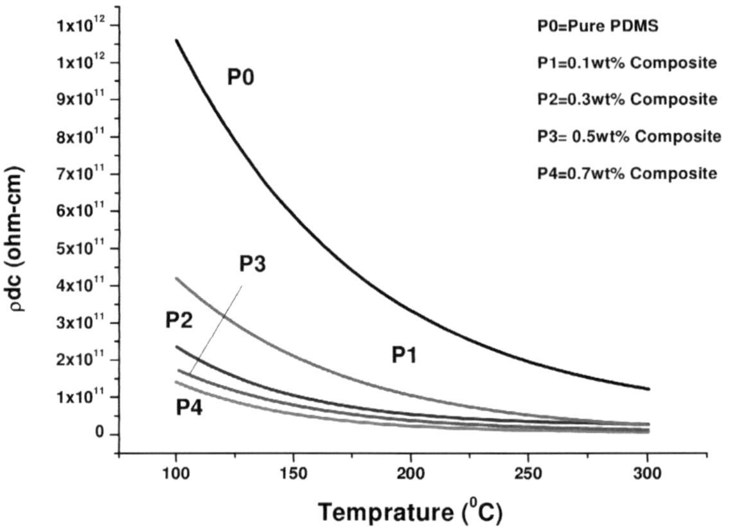

Figure 7a. The dc electrical resistivity with temperature (ρdc).

The dc electrical resistivity at 100^{0}C was found in the range 0.1-1.25x10^{12}Ω-cm. The electrical resistivity data confirms that as F-MWCNTs wt % increases, the electric conductivity of fabricated composite increases. The accumulation of electrostatic charging, usually observed on insulating matrix surface leads to a serious problem and high electric conductivity above $\sigma=10^{-6}$ Sm^{-1} is required. F-MWCNTs (highly electrical conducting nanofiller) were used to get a maximum value of electrical conductivity in the polymeric matrix. The dielectric constants ($\varepsilon^{/}$) and dielectric loss (tanδ) of various filler concentrations with frequency, represented in figure 7 (b, c), and were calculated using standard relations [14].

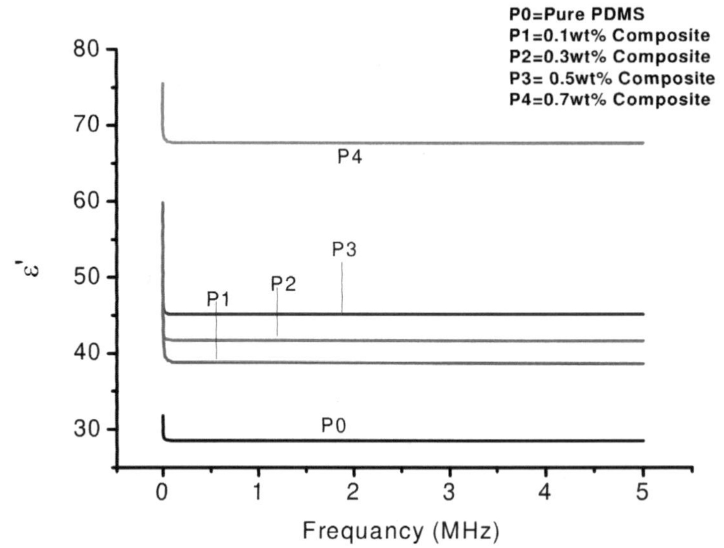

Figure 7b. Dielectric parameters as a function of frequency (ϵ/) of nanocomposites

The value of both these of parameters was found in ranges 31-75 and 0.092-0.31 at 100Hz. This indicates that with the increase in nanotubes concentration, both of these parameters are influenced. The observed dielectric behavior in fabricated composites is due to the shift in electric charges from their mean or equilibrium position between the PDMS/F-MWCNT interfaces. Furthermore, the wt% increase in F-MWCNTs results in heterogeneous phases having different polarizability, which in turn leads to accumulation of different charges at interfaces. As a result, these parameters decrease with the increase in frequency [15, 16]. The effect of microwave frequency on dielectric parameters was measured in the frequency range (1MHz-3GHz).

The behavior of dielectric parameters at frequency range 0 to 3GHz for P0-P4 is represented in figure 8 and 9. The effective decrease in measured values is observed for both of these parameters at 1GHz. The permittivity originates at the interface of composites due to the orientation polarization, space charge induced polarization, electrostatic and electronic polarization. The polarization is influenced by resonance generated at the expanse of higher frequencies, especially at microwave frequency range leads to electronic polarization at interfaces [17]. The negative values of permittivity for composites are attributed by quantum dot excitation at interfaces explained theoretically [18].

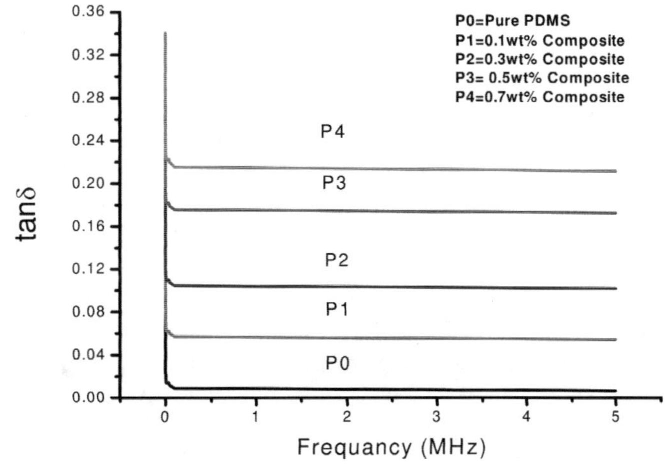

Figure 7c. tanδ as a function of frequency of nanocomposites.

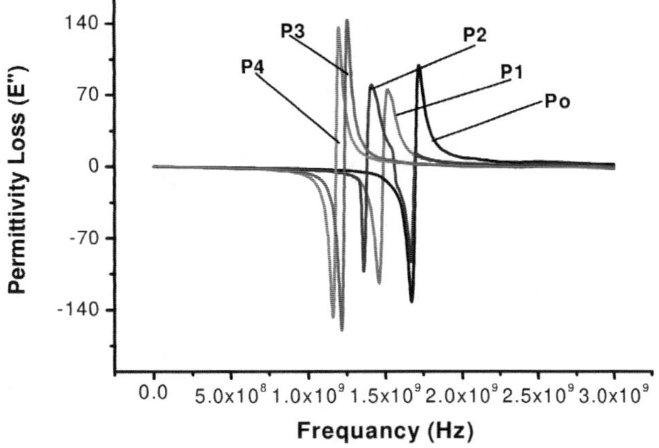

Figure 8. Dielectric loss at microwave frequency range (1MHz-3GHz) of fabricated F-MWCNT/PDMS

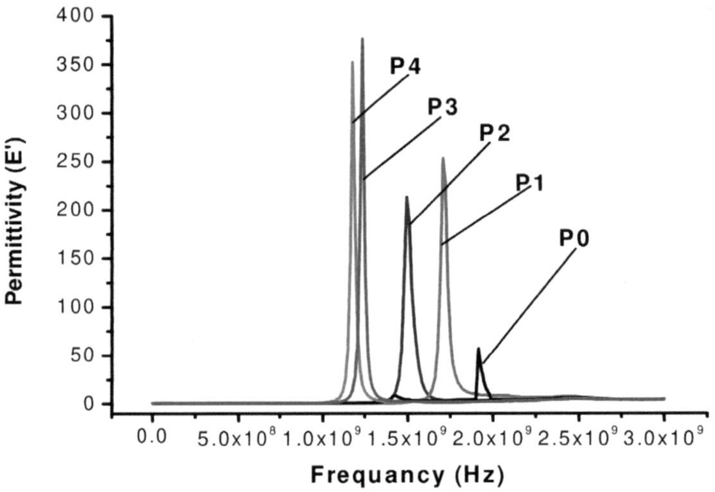

Figure 9. Permittivity of all F-MWCNT/PDMS composite at microwave frequency range (1MHz-3GHz)

4. Conclusion

The wt% increment of the nanotubes contents in PDMS matrix influences and improves dielectric/electrical & thermal behavior of fabricated PDMS/F-MWCNT composites. The remarkable enhancement in thermal stability and heat quenching capability of PDMS with the incorporation of F-MWCNTs are clearly observed in TG/DTA. The reduction of Tc & Tg, and improvement in Tm are augmented by DSC analyses. The uniform dispersion of F-MWCNTs in PDMS is obvious from the SEM images. XRD results corroborate that crystalline nature does not change with the carboxylic functionalization of Ps-MWCNTs. The effective increase is observed in dc electric conductivity with% increase in nanofiller content. The decrease in electric resistivity by temperature and dielectric parameters with frequency confirmed the semiconducting behavior of fabricated composite. The dielectric parameters are influenced with increasing F-MWCNTs incorporation in PDMS due to the interface polarization between the filler and matrix. These polymeric composites are resonated due to atomic, electrostatic and electronic polarization at 3GHz.

Acknowledgements

The Sadia Sagar would thankfully acknowledge her uncle Mr. Hamid Hayat. The authors would gratefully acknowledge HEC PhD scholarship for this research work and Dr. M. Nasir (PIEAS, Islamabad) for their kind help to perform DSC analysis of fabricated samples.

References

[1] Mei J J, Zhi M D, Sheng H Y, and Jinbo B 2008 *J. Chem. Phys. Lett.* **457** 352
[2] Dollase T, Spiess H W, Gottlieb M, and Yerushalmi R R 2002 *Euro. Phys. Lett.* **60** 390
[3] Pergal M V, Dzunuziv J V, Kicanovic M, Vodnik V, Pergal M M, and Jovanovic 2011 *Russ. J. Phys. Chem. A.* **85** 2251
[4] Ren P, Guan J, and Cheng X 2006 *J. Mater. Chem. Phys.* **98** 90
[5] Chen N, Yang K, and Gu M 2010 *J. Alloy. Compd.* **490** 609
[6] Nataraj S K, Kim B H, Cruz J F, Aminabhavi T M, and Yang K S 2009 *Mater. Lett.* **63** 218
[7] Huang C L and Chen Y B 2006 *Mater. Lett.* **60** 198
[8] Mei G, Gang D, Shuping G, and Dongxiang Z 2012 *J. Eur. Ceram. Soc.* **32** 883

[9] Penn S J, Alford N M, Templeton A, Wang X, Xu M, and Reece 1997 *J. Amer. Ceram. Soc.* **80** 1885
[10] Yoon S H, Kim D W, Cho SY, and Hong K S 2006 *J. Eur. Ceram. Soc.* **26** 2051
[11] Pratibha P, Chauhan R S, Semwal R P, Banerjee S, and Rajeev J 2004 *J. Appl. Poly. Sci.* **91** 2278
[12] Kalaivani S, Amit D, Dagmar S, Manfred K, and Gert H 2011 *Eur. Pol. J.* **47** 2234
[13] Yanke Z, Jian Y, Hailong T, Jiachun Z, Rui Z, and Xiaoboo L 2012 *J. Mater. Sci.* **23** 206
[14] Mohammad H, Alimuddin A , Shalendra K, Koo B H, Sagar E S, and Mohammad E M 2012 *J. Alloy. Compd.* **518** 11
[15] Liu L D, Duan Y P, Ma L X, Liu S H, and Yu Z 2012 *J. Appl. Surf. Sci.* **257** 842
[16] Fan Y Z, Yang H B, Li M H, and Zou G T 2012 *J. Mater. Chem. Phys.* **115** 696
[17] Tong G X, Wu W H, Miao Y Q, Guan J G, and Qain H S 2011 *J. Alloy. Compd.* **509** 451
[18] Jun L, Yongbao F, and Tai Q 2011 *J Magn. Magn. Mater.* **323** 3071

Journal of Physics: Conference Series **439** (2013) 012025 doi:10.1088/1742-6596/439/1/012025

Piezoelectric and pyroelectric properties of Sr-doped PZT (PSZT) with minor manganese additions

M A Aleem*, H Nawaz, M Shuaib, S Qaisar and M S Akbar

Institute of Industrial Control Systems (IICS), Rawalpindi, Pakistan

*E-mail: adeel713@gmail.com

Abstract. A systematic study was performed to see the effect of Manganese addition and temperature gradient on the electrical properties of PSZT. $Pb_{0.96}$ $Sr_{0.04}$ $(Zr_{0.52}Ti_{0.48})$ O_3 (PSZT) containing 0.3%, 0.5%, and 1% Mn was prepared by the sol gel method in order to ensure good stoichiometry and enhanced purity. The powders were calcined at 550 °C and sintered at 1200 °C to achieve 98% of the theoretical density. High field ac study was performed by (P-E) hysteresis measurements at different temperatures (RT, 60, 90, 120 and 150 °C) using an electric field up to 3 kV/mm. It was observed that for a lower Mn concentration P-E loops are pinched at the center while this constriction is found to decrease for greater concentrations. The optimized results were obtained for the 1 mol% of Mn content with 4 mol% of Sr. The values of Q_m, k, d_{33} and tanδ were measured as 756, 0.38, 257 and 0.002 respectively. Higher temperatures coupled with a gradual increase of the electric field resulted in a shift of the hysteresis loops along electric field axis, indicating the presence of an internal bias field. Dependence of pyroelectric properties on applied electric field was also investigated. The value of pyroelectric coefficient was found maximum 6.25 $\times 10^{-4}$ (C/m^2K) at 3 kV/mm.

1. Introduction

Lead Zirconate Titanate (PZT) ceramics are famous for their enormous applications in ultrasonic transducers, actuators, vibration sensors, IR (pyroelectric) sensors, dynamic RAM's, electro-caloric coolers for computers and capacitor array [1-3]. Main peculiarity of PZT materials is hysteresis dependence of permittivity, polarization, and pyroelectric behavior on doping and applied electric field [4]. Electrical properties of PZT can be tailored by incorporating appropriate dopants to meet technical demands of user based applications. Donor dopants (La^{3+}, Bi^{3+} (for A-site) and Sb^{5+} and Nb^{5+} (for B-site)) impart soft characteristics to the host material, while, acceptor(K^+, Na^+ (for site A) and Fe^{3+}, Al^{3+}, Mn^{3+} (for site B)) dopants are responsible for hardening effect [5,6,7]. The deficiency of positive charge occurred as a result of acceptor doping is compensated by oxygen vacancies and excess positive charge is responsible of lead vacancies [8].

Along with doping electrical after effects are also responsible for change in properties. Volume effect, domain effect, and domain boundary effect are three famous mechanisms used to explain the effect of added impurities and ageing related phenomena [9]. Z. Cao [10] observed improved pyroelectric property along with double hysteresis loop for Mn-doped PLZT system. The possibility of presence of antiferroelectric phase was rejected on the basis of high piezoelectric and pyroelectric coefficients. For practical purpose it is normally preferred to use more than one dopant to maneuver a broad range of properties. Strontium (Sr^{+2}) is a famous isovalent dopant and it is found to have profound effect on dielectric constant but has relatively little effect on other parameters i.e. Q_m, k (piezoelectric coupling), resistivity and temperature dependence. However on the other hand, acceptor dopants are reputed for raising the Q_m, amplitude and lower resistivity [7].

In this study, Sr and Mn were co-doped at A and B sites in PZT (52/48), respectively to improve its electrical properties. Effect of complex doping on piezoelectric, pyroelectric and hysteresis properties was studied. The impact of electric field and temperature was also investigated to understand the resultant effect.

2. Experimental

$Pb_{0.96}Sr_{0.04}(Zr_{0.53}Ti_{0.47})_{1-x}Mn_xO_3$ for x = 0.03, 0.05 and 0.1 was prepared using sol-gel method. For this purpose; Lead acetate was dissolved in appropriate quantity of Isopropanol with small additions of acetic acid at 80 ˚C. The solution was distilled to remove water and extra solvent. Strontium acetate and manganese acetate were added to this solution. Separately, titanium isopropoxide and zirconium-n-propoxide were refluxed in isopropanol at 100 ˚C and stabilized the solution with acetyl acetone. The two solutions were mixed together drop wise and stirred for an hour. Hydrolysis was carried out by adding water in 1:2 to the metal mixture. The sol was then converted to a viscous gel in a rotary evaporator and was further dried in vacuum oven at 120 ˚C. The dried gel was then calcined at 550 ˚C for 4 hours. Calcined powder was milled in high energy ball mill with Zirconia balls in water for six hours. Phase confirmation was made by XRD [11] PVA was added as binder in resulting slurry and dried at 50 °C. The powder was pressed in the form of small pallets in a die. Binder was then removed at 270 °C for an hour and sintered at 1150 °C for two hours. Sintered pallets were electroded with silver paste and fired at 550 °C for 30 min. Density greater than 99% of the theoretical density was achieved for all compositions.

Polarization electric field (P-E) hysteresis measurements were carried out using a computer-controlled function generator (Agilent 33220A) in conjunction with a high-voltage (×1000) amplifier (Trek 10/10B). A two probe sample holder was immersed in a silicone oil bath placed at a heating stage. The temperature was controlled by the temperature controller (Eurotherm 818P) with accuracy of ±0.5C. AC pulses of varying magnitude with frequency 2 Hz were applied to the sample and the induced current was integrated to calculate the polarization in order to obtain the P-E characteristics of material. Schematic diagram of the measurement setup is shown in Figure 1. After P-E measurements, samples were heated at 450 C for 30 min to nullify the ageing effects. Poling was carried out at 120 °C with a dc field of 2-3KV/mm in silicone oil bath. Piezoelectric charge coefficient d_{33} was measured using Piezometer (piezometer Systems, UK). Dielectric constant, resistance and loss factors were measured using a precision LCR meter (7600 Quad Tech, USA). Admittance-susceptance loop was taken using (Transducer analyzer TA-2000, Lecore systems, UK) to obtain Q_m, k and G. Pyroelectric coefficient was derived from temperature dependence of remnant polarization (P_r).

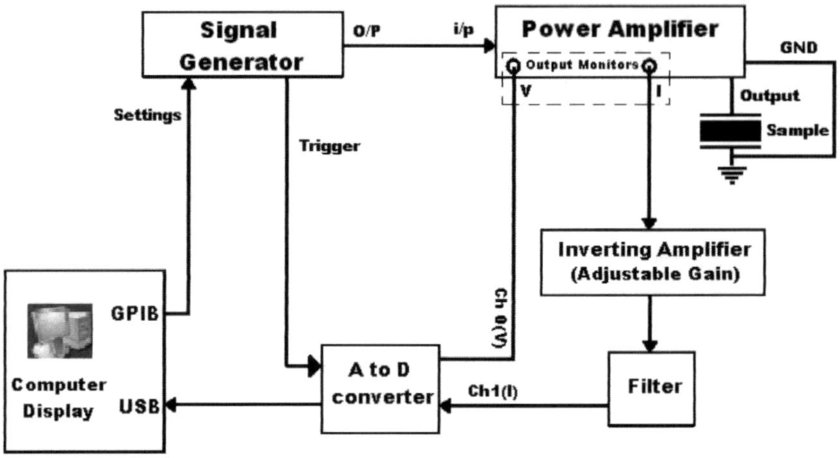

Figure 1. Schematic Diagram of P-E system.

3. Results and discussion

Figure 2 shows P-E hysteresis loops of (Sr, Mn) doped PZT for varying Mn content at room temperature, 90 °C and 150 °C. At room temperature; variation in Mn content doesn't seem to have considerable effect on polarization, but this effect is pronounced at higher temperatures. At 90 °C the values of remnant polarization (Pr) are 0.022, 0.0295 and 0.048 C/m^2 for 0.3, 0.5 and 1.0 mol% of Mn respectively, which shows an increasing trend with concentration. These values almost doubled for 150 °C as 0.029, 0.041and 0.093 C/m^2 respectively, following the same trend. It is evident from Figure. 2 that the hysteresis loops are thin and constricted/pinched at the center which confers to the hard nature of material. Unease of switching at such high field of 3 kV/mm can also be regarded as the side of the constriction. It is observed that the switching is improved with rise in temperature for 1.0 mol% Mn only while for lower concentrations of Mn constriction is still present.

As strontium ion (Sr^{2+}) replaces lead ion (Pb^{2+}) at A-site as an isovalent dopant; it doesn't disturb the electro neutrality of the crystal. It also improves the dielectric constant and ease of switching in result [7,10, 12]. However, Manganese as an acceptor dopant at B-site creates positive charge deficiency which is compensated by oxygen vacancies. These oxygen vacancies form defect dipoles and result in the pinning of domain rotation. The field which opposes the change in domain orientation resulting in the resistance to domain switching is called internal bias field [7, 10, 13, 14]. Internal bias can also be regarded as a strong reason for constriction of hysteresis loops. Internal bias; if present in a material causes a shift of P-E hysteresis loop from its centre along the electric field axis [15]. The dependence of internal bias field on temperature and applied electric field is shown in figure 3 for different manganese concentrations.

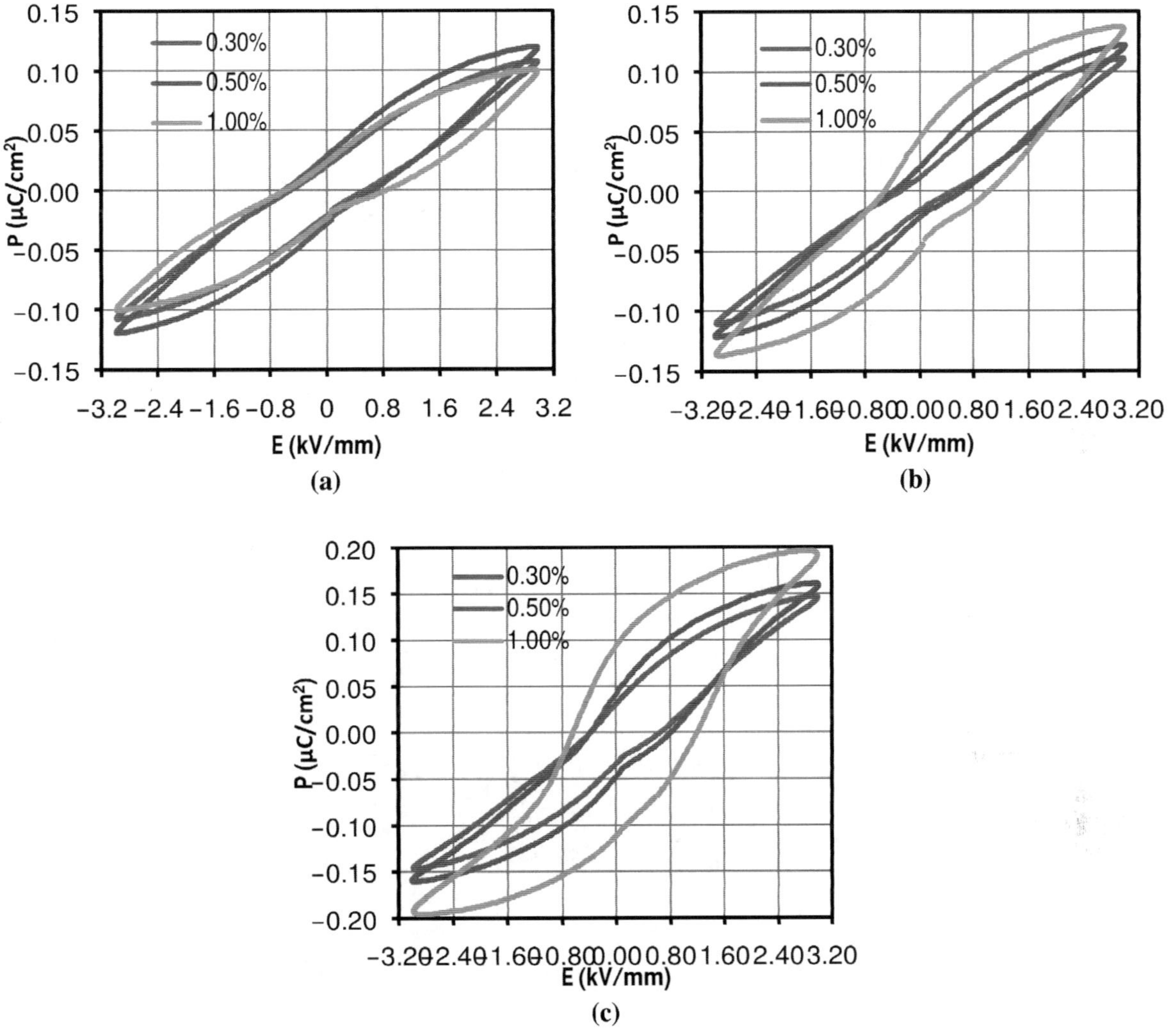

Figure 2. P-E hysteresis loops at (a) room temperature (b) 90 °C (c) 150 °C.

It can be seen that at low field strength the bias field is approximately negligible but it is significant at higher fields especially for the values where switching starts to appear: 1.6 kV and above. It is apparent that with the ease of hysteresis switching the opposing force is also rising to keep the system in low free energy state. Same like hysteresis saturation trend, the bias field is again higher, 0.15 kV/mm for 1.0 mol% and decreasing gradually for 0.5 and 0.3 mol% Mn as 0.005 and 0.0025 (kV/mm) at room temperature. From figure 2 we deduce that the temperature also plays a significant role in the rise of internal bias which can be explained in terms of the relationship of temperature and broadening of hysteresis loop resulting in an increase in bias field too.

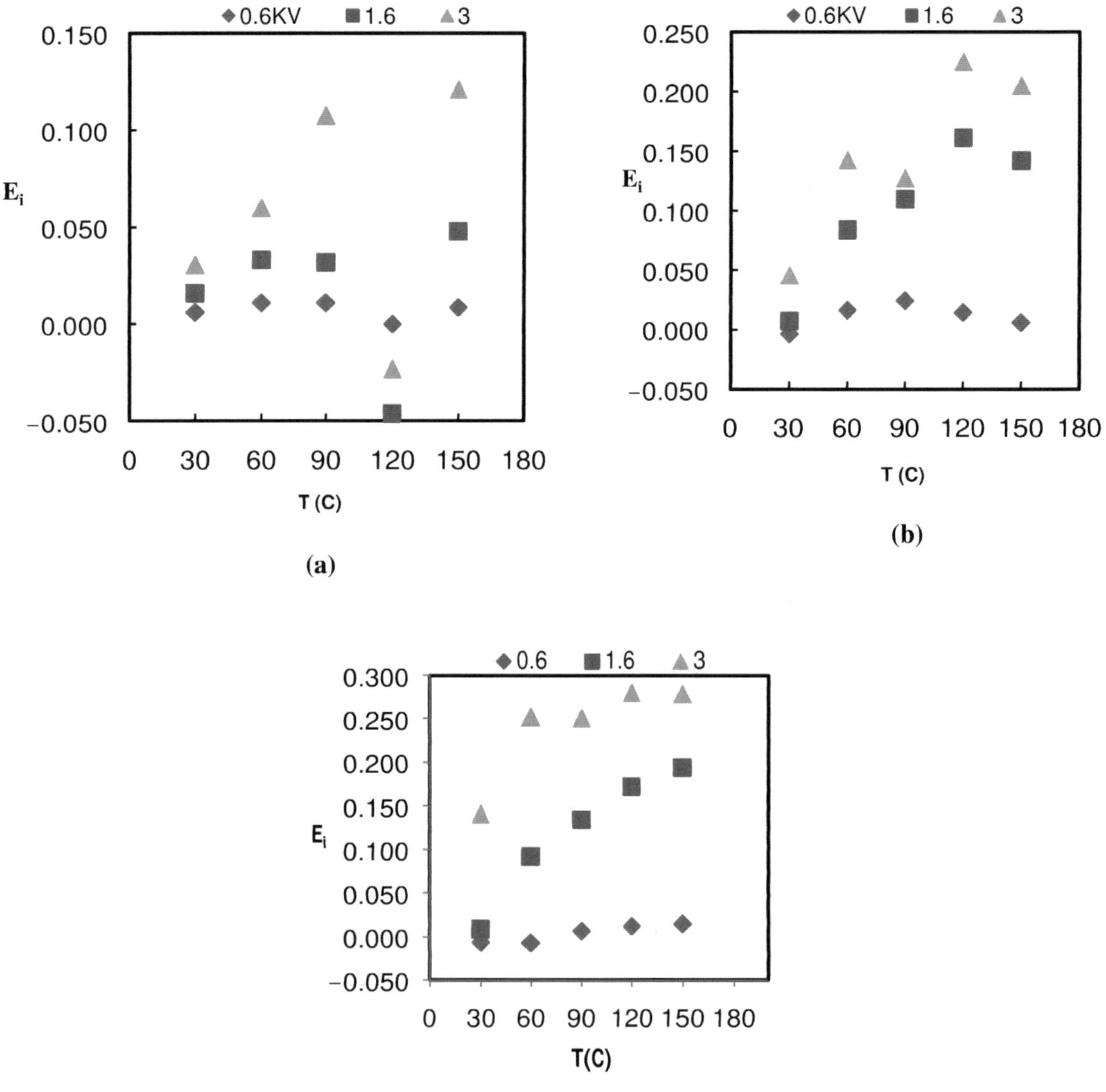

Figure 3. Dependence of internal bias field on temperature and applied field for different mol% of manganese (a) 0.3, (b) 0.5 and (c) 1.0.

Figure 4 shows plots of remnant polarization versus temperature of 1.0 mol% Mn-doped PZT for different applied electric fields. The overall trend of remnant polarization is increasing with increasing temperature. For fields up to 1 kV/mm this dependence is linear however beyond this field non-linearity subjected to hysteresis switching increases, resulting a sharp increase in remnant polarization with temperature, which is maximum for 3 kV/mm.

The temperature dependent polarization of PZT makes it important for pyroelectric applications. The pyroelectric coefficient (p) can be determined from the temperature dependence of remnant polarization using the following relation [16].

$$p = \frac{dP_r}{dT} \tag{1}$$

There are two modes of operation of pyroelectric IR detectors. The first one operates without external electric field at temperatures considerably lower than Curie point while in second mode an optimal dc field is applied at temperatures near phase transition to exploit the peaking of polarization near curie point. The later mode permits higher sensitivities and lower [17]. Pyroelectric coefficient was determined from the slope of P_r versus temperature plots at different bias fields from figure 4. The field dependent values of pyroelectric coefficients are 5.17×10^{-5}, 1.31×10^{-4}, 3.22×10^{-4}, 4.13×10^{-4}, 5.05×10^{-4} and 6.25×10^{-4} C/m^{2}K for 0.5, 1.0, 1.5, 2.0, 2.5 and 3.0 kV/mm, respectively.

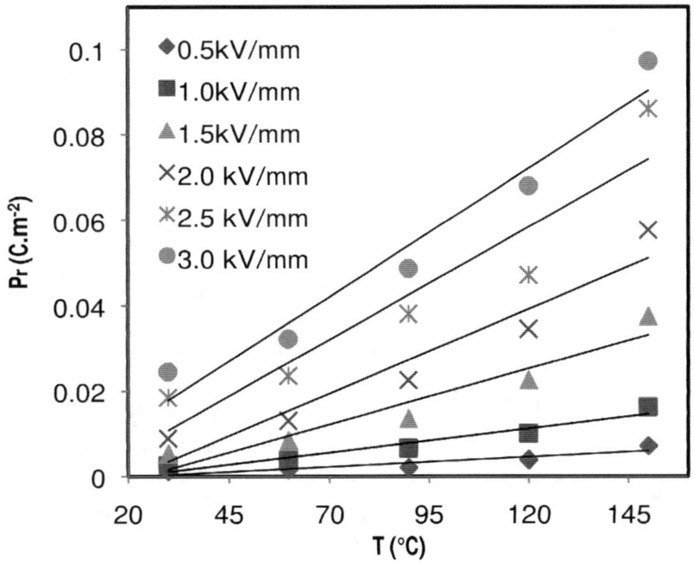

Figure 4. Variation of remnant polarization with temperature at different applied fields.

Piezoelectric properties of the poled samples are shown in Table 1. It is evident that the relative dielectric constant is decreasing with increasing Mn content. There is a little effect on piezoelectric coefficient d_{33} and coupling coefficient k, however the former shows a decreasing and later presents an increasing trend. Manganese addition has lifted the mechanical quality factor Q_m to an appreciable. Because the Ionic radii of Mn^{3+} and Mn^{4+} are 0.62 Å and 0.52 Å respectively, as compared to the host ions Zr^{4+} (0.79) and Ti^{4+} (0.68), substitution of Mn at B-site leads to the shrinkage of crystal cell [10] which may be the cause of improvement in quality factor.

Table 1. Electrical properties of manganese doped PSZT

Composition	ε_r	d_{33}(pC/N)	Q	R (kΩ)	G (mS)	K	tanδ
MPSZT 0.3	1137	273	342	1.80	20.8	0.3	0.003
MPSZT 0.5	1044	270	506	2.60	35.6	0.32	0.008
MPSZT 1.0	974	257	756	1.26	64.5	0.38	0.002

4. Conclusions

PZT system was co doped with Mn (acceptor dopant) and Sr (isovalent dopant) to enhance the electrical properties of the material. Sr was added to facilitate the switching process and Mn was impregnated for the improvement of mechanical quality factor. The pyroelectric property was also investigated from the temperature dependent behavior of remnant polarization. The optimized results were obtained for the 1 mol% of Mn content with 4 mol% of Sr. The values of Q_m, k, d_{33} and tanδ were measured as 756, 0.38, 257 and 0.002. Higher values of Q_m with low loss factor makes this material more suitable for application purposes. Maximum value of pyroelectric coefficient 6.25×10^{-4} was obtained at 3 kV/mm.

References

[1] Scott J F 2007 Science **315** 954
[2] JaffeB William R C Jaffe H 1971 *Piezoelectric ceramics* (London: Academic Press)
[3] Hall D A 2001 *J. Mat. Sci* **36** 4575
[4] Morozovska A N 2007 *Cond. Matter. Phy* **10** 85
[5] Hizebry A Attaoui H E l Sa^Adaoui M Chevalier J and Fantozzi G J 2007 *Euro. Ceram. Soc* **27** 557
[6] Boucher E Guiffard B Lebrun L and Guyomar D 2006 *Ceram. Inter* **32** 479
[7] Berlincourt D *J. Acoust. Soc. Am* **91** 3034
[8] Eyraud L Guiffard B Lebrin L and Guyomar D 2006 Ferroelectrics **330** 51
[9] Carl K Hardtl K H 1978 *Ferroelectrics* **17** 473
[10] Cao Z Ding A Zhang Y Qiu P Zhang X 2004 *Sol. Stat. commun* **131** 57
[11] Nawaz H Aftab S Shoaib M and Siddiq M 2012 *Key. Eng. Mat* **510-511** 520
[12] Shao Q Li A Xia Y Wu D Liu Z and Ming N 2006 *J. App. Phy* **100**, 036102
[13] Lee B W Lee E J 2006 *J Electroceram* **17** 597
[14] Barranco A P 2001 *J. Eur. Cer.* Soc **21** 523
[15] Hall D A Ben-Omran M M and Stevenson P J 1998 *J. Phys: Cond. Mat* **10** 461
[16] Ok K M Chiand E O Halasyamani P S 2006 *Chem. Soc. Rev* **35** 2710
[17] Muralt P 2001 *Rep. Prog. Phys* **64** 1339

Journal of Physics: Conference Series **439** (2013) 012026 doi:10.1088/1742-6596/439/1/012026

Electrochemical impedance spectroscopy based MEMS sensors for phthalates detection in water and juices

Asif I Zia*[1,4], A R Mohd Syaifudin[1], S C Mukhopadhyay[1], P L Yu[1], I H Al-Bahadly[1], Chinthaka P Gooneratne[2], Jűrgen Kosel[2] and Tai-Shan Liao[3]

[1] School of Engineering and Advanced Technology, Massey University, Palmerston North, New Zealand.
[2] Sensing, Magnetism and Microsystems Group, King Abdullah University of Science and Technology Saudi Arabia.
[3] Instrument Technology Research Centre, National Applied Research Laboratories, Hsinchu, Taiwan.
[4] Department of Physics, COMSATS Institute of Information Technology, Islamabad, Pakistan.

E-mail: a.i.zia@massey.ac.nz

Abstract. Phthalate esters are ubiquitous environmental and food pollutants well known as endocrine disrupting compounds (EDCs). These developmental and reproductive toxicants pose a grave risk to the human health due to their unlimited use in consumer plastic industry. Detection of phthalates is strictly laboratory based time consuming and expensive process and requires expertise of highly qualified and skilled professionals. We present a real time, non-invasive, label free rapid detection technique to quantify phthalates' presence in deionized water and fruit juices. Electrochemical impedance spectroscopy (EIS) technique applied to a novel planar inter-digital (ID) capacitive sensor plays a vital role to explore the presence of phthalate esters in bulk fluid media. The ID sensor with multiple sensing gold electrodes was fabricated on silicon substrate using micro-electromechanical system (MEMS) device fabrication technology. A thin film of parylene C polymer was coated as a passivation layer to enhance the capacitive sensing capabilities of the sensor and to reduce the magnitude of Faradic current flowing through the sensor. Various concentrations, 0.002ppm through to 2ppm of di (2-ethylhexyl) phthalate (DEHP) in deionized water, were exposed to the sensing system by dip testing method. Impedance spectra obtained was analysed to determine sample conductance which led to consequent evaluation of its dielectric properties. Electro-chemical impedance spectrum analyser algorithm was employed to model the experimentally obtained impedance spectra. Curve fitting technique was applied to deduce constant phase element (CPE) equivalent circuit based on Randle's equivalent circuit model. The sensing system was tested to detect different concentrations of DEHP in orange juice as a real world application. The result analysis indicated that our rapid testing technique is able to detect the presence of DEHP in all test samples distinctively. [1]

*[1] To whom any correspondence should be addressed

1. Introduction

Esters of 1-2-Benzenedicarboxylic acid, commonly called phthalates are the most abundantly synthesized key industrial chemicals due to a number of their commercial uses. Owing to their low cost, these ubiquitous chemicals are most commonly used as plasticizers, solvents and additives in different products manufactured by industry [1]. For the reason of their use as plasticizer in almost every plastic product manufactured around the globe, phthalates with higher molecular weight i.e. DINP (diisononyl phthalates) and DEHP makeup nearly 80% of the total phthalates produced in the world. These include food and beverage packaging and most importantly, medical applications [2, 3]. Addition of plasticizer induces flexibility and mechanical strength to the Polyvinyl chloride (PVC) products but, unfortunately, do not bond to PVC molecules in the lattice structure. On the other hand, lower molecular weight phthalates, such as diethyl phthalate (DEP) and dimethyl phthalate (DMP) are used in consumer cosmetics, house hold insecticides, paints and pharmaceutical applications [3]. Being an odourless, colourless high density oily compound, phthalates do not get covalently bound to the molecular structure of the polyvinyl chloride (PVC) product, therefore, it has a large potential for non-occupational leaching and ingestion into packed food and beverages, gas out in the atmosphere or directly into human body fluids through medical products [1-3]. Phthalates' carcinogenicity and teratogenicity result in reproductive and developmental toxicities in rodents, and their potential for human exposure declares it an area of concern in environmental public health [2-6]. Figure 1 shows molecular structures of most common phthalates used in cosmetic, toys, PVC and food packaging industry.

The possible routes of human exposure to phthalates are ingestion (dietary), inhalation, dermal and parenteral exposure from plastic medical devices [6, 7]. Dietary exposures occur by accumulating phthalates either during the food processing and packaging from processing equipment or leaching into fatty foods, including dairy products and from packaging during storage. Dairy products consumption by children per kilogram body weight is higher as compared to the adults, posing greater risk of reproductive and developmental toxicities for young children [7]. Polymer toys softened with DINP were estimated exposures ranged from 5-44μg/kg body weight/day with 99th percentiles up to 183μg/kg body weight/day for infants via mouthing activities [8]. World Health Organization (WHO) and European Union(EU) has published a list of priority compounds posing endocrine disrupting hazards to humans and have incorporated DEHP in the field of water policy by setting the guideline value at 8.0 μg/L in fresh and drinking waters [9,10]. Table 1 shows risk assessment of phthalates by agencies in EU, US and Canada.

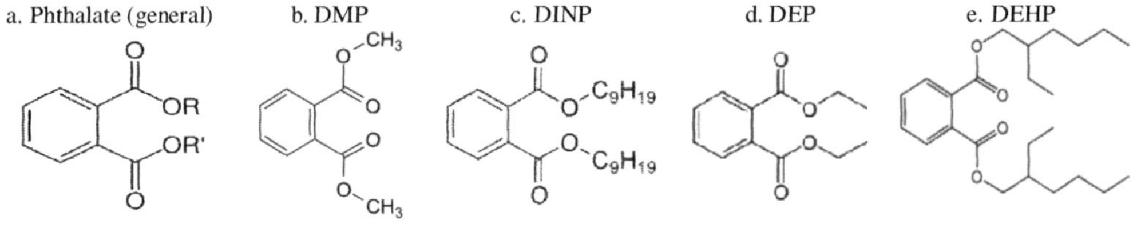

Figure 1. Molecular structures of commonly used phthalates; (a) Phthalate (general) (b) DMP (c) DINP, (d) DEP, (e) DEHP.

Several researchers have given an account for the evaluation of the migration of phthalates from food packaging materials, PVC gloves used in cooking and handling food, paper and cardboard [4,6,11], PET (PETE, polyethylene terephthalate) bottled beverages and mineral water [12,13], corks for glass bottles [6]. In light of the research studies carried out on phthalates, United States, Canada and

European Union have phased out use of phthalates in many consumer products, especially young children's products such as toys, lotions, sprays etc.

In May 2011, Taiwan had to face the worst business set back when its ministry of health reported illegitimate addition of di(2-ethyl-hexyl) phthalate (DEHP), to certain medicines, foods and beverages manufactured in Taiwan. It had been legitimate to use DEHP as plasticizers in production of PVC food contact packaging material, but using as food additive, had never been allowed in food industry. In compliance to the limit set by WHO, TDI for DEHP in Taiwan and Hong Kong is 1.5 ppm. Long-time ingestion of DEHP at levels above the TDI can create hormonal imbalance in the human body, which may result in the decrease of male reproductive ability, and female precocious puberty, breast cancer and loss of gender uniqueness [14].

Table 1. Risk assessment by ASTDR, US-EPA, EU-CSTEE, Health Canada [1]

Phthalate type	**Risk assessment of DEHP and DINP**			
	Country, region	***Committee/year***	***mg/kg bodyweight/day***	***MRL¹/TDI²/RfD³***
DEHP	USA	US-EPA, 1993b	0.02	RfD
DEHP	USA	ATSDR, 2002	0.1	MRL
DEHP	Canada	Health Canada, 1994	0.044	TDI
DEHP	EU	CSTEE, 1998a,b	0.050	TDI
DINP	EU	CSTEE, 1998a,b	0.25	TDI
DEP	USA	ATSDR, 1995	7	MRL
DBP	USA	ATSDR, 2001	0.5	MRL
DBP	USA	US-EPA, 1990	0.1	RfD

[1] minimal risk level; [2] tolerable daily intake; [3] reference dose levels

Detection and quantification of very low concentration levels of DEHP and DINP are purely laboratory based and are seriously limited by the ubiquitous presence of this compound as a contaminant in almost all laboratory equipment and reagents and hence cannot, generally, be accurately quantified below about 2-ppb even in the most controlled laboratory setup [5, 15]. Most commonly gas chromatography (GC) is used to measure the phthalate metabolites in biological matrix [16]. In addition to GC a more sophisticated technique called high performance liquid chromatography (HPLC) is also applied to measure phthalate concentrations in blood plasma and urine samples with a detection limit of 0.345μg/mL and 1.2ng/mL respectively [17]. The chromatography technique separates complex mixtures of organic compounds to detect, identify and measure the concentration of each individual compound by using a specific detector [18]. Detectors specific to identification and measurement of DEHP in samples are electron capture detector (ECD) [16] and flame ionization detector (FID) [19]. Methods based on HPLC coupled with mass spectrometry (MS) or ultra violet (UV) detection are also used for detection of phthalates [5, 20].

Contemporary analytical techniques used to measure phthalate concentrations in food products and beverages, such as gas chromatography/mass spectroscopy (GC/MS), require expensive and complicated instruments, trained professionals, and laboratory environment with stringent conditions over sampling procedures. It involves high testing cost and most importantly long testing time per sample which does not match the time constraint posed by industrial production. There is no low cost technique so far reported for on-site and real-time phthalate detection which may be deemed suitable for industrial use where batches of thousands of products are manufactured on daily basis. Requirement to develop a reliable and highly sensitive quantification system that may provide on-site real-time detection and measurement of phthalates in beverages, food products and medicines dictates paramount importance and urgency at present.

2. Planar interdigital sensors

The term 'planar interdigital' (ID) refers to finger-like, parallel in-plane electrodes built in a periodic pattern on a rigid substrate. Conventional ID sensors are based on repeated patterns of one sensing (working) and one excitation (reference) electrode configuration on a solid substrate. This geometry exploits the capacitive effect produced as a result of applied alternating electric field that fringes through the material sample. The most important benefit of planar geometry for this type of sensor is its single side access to the material under test which is an advantage for in-situ measurements. It also allows non-destructive testing which is mandatory for a sensing system operating in an industrial setup. Since the penetration depth of the electric field depends on the spatial wavelength (distance between two electrode with same polarity) of the ID structures, this configuration is mostly used for bio sensing applications where the sample is pipetted on the sensor surface in a volume range of a few µL [21, 22]. Dielectric profiling and conduction properties of semi insulating sample materials are hence obtained using EIS tools. The dielectric properties measured using this type of sensor, can directly or indirectly be related to other physical characteristics like density, structural integrity, porosity and biological or chemical content in the material under test, hence providing effective means to evaluate parameters vital for scientific and industrial requirements [23].

2.1. New sensor design

To achieve the objective of phthalate detection in bulk sample, a new type of sensor was designed with enhanced penetration depth of the fringing electric field. The higher penetration depth was achieved by increasing the number of sensing (working) electrodes between the reference electrodes. Finite element modeling software COMSOL Multiphysics® was used to simulate the new design for the sensor and evaluate its performance on the basis of theoretical model. Two sensors with different geometries and number of working electrodes were modeled keeping the effective sensing area (2.5mm x 2.5 mm), substrate material (Si, SiO_2), electrode thickness (520nm) and electrode area (25 µm x 2.425 mm) constant. To obtain higher penetration depth and uniform electric field distribution throughout, the sensor's geometry was designed with two configurations of five (1-5-50) and eleven (1-11-50) working electrodes with spatial wavelength of 50 µm. The spatial wavelength for reference electrodes was kept to 425 µm and 875 µm for 1-5-50 and 1-11-50 respectively. Figure 2(a) shows the design parameters of the proposed sensor 5 working electrodes between 2 reference electrodes.

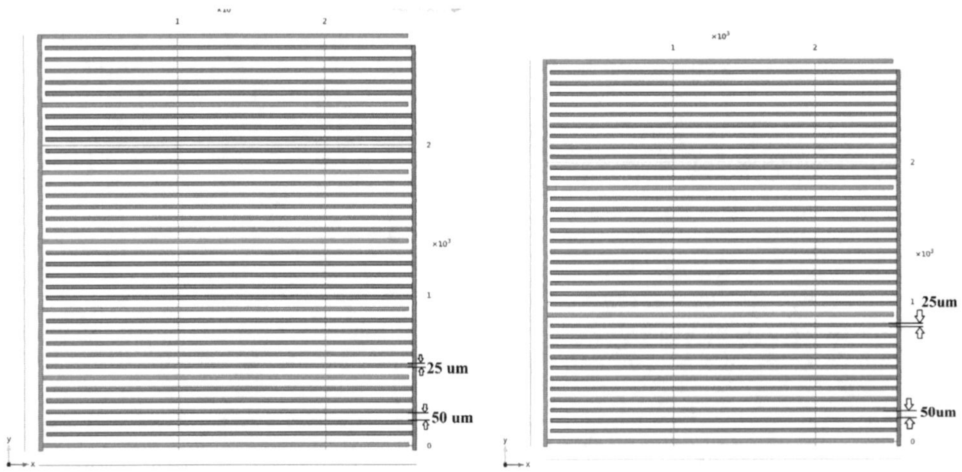

Figure 2. (a) Design parameters of (1-5-50) 5 working electrodes between two reference electrodes (b) Design parameter of (1-11-50) 11 working between two reference electrodes.

Figure 2(b) shows 11 working electrodes between two reference electrodes. Single crystal silicon substrate with thickness 525μm and relative permittivity r) 4.2 was used in the models so that the optimized model design may be fabricated practically on commercially available 4 inch silicon wafer. In order to calculate capacitance, stored electrical energy, penetration depth of electric field lines and field intensity the models were sequentially exposed to a dielectric medium on the sensing area with a dielectric constant of 80.1 and 5.1 simulated at 20°C. These are the dielectric constant values corresponding to pure MilliQ water and DEHP, respectively, which were selected to observe the model response of each sensor towards the material under test. The calculated values of capacitance and total electrical energy stored for the two simulated sensors for MilliQ and DEHP is given in table 2 below.

Table 2: Evaluated capacitance and total electrical energy stored using COMSOL Multiphysics®.

Sensor type	Milli Q. r = 80.1		DEHP r = 5.1	
	Capacitance. (pF)	**Energy density (J)**	**Capacitance. (pF)**	**Energy density (J)**
1-11-50	13.94	6.9e-12	1.52	7.6e-13
1-5-50	8.46	4.2e-12	0.92	4.6e-13

Figure 3 is the plot for the calculated values of capacitance for modeled sensors simulated for different media.

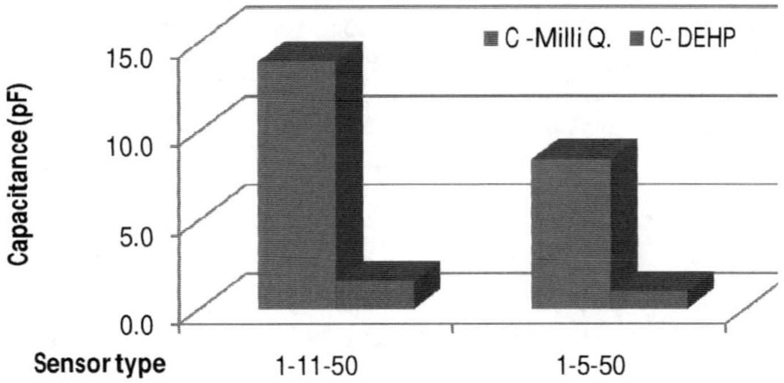

Figure 3. COMSOL simulated sensors' capacitance with MilliQ and DEHP.

2.2. Sensor fabrication

The analysis results using COMSOL simulation displayed better capacitive response for the sensor design with 11working electrodes as compared to the design with 5 working electrodes; therefore 1-11-50 was selected for fabrication. Single crystal Si wafer (4 inch diameter, 525μm thickness) with native SiO_2 layer was used to fabricate the new types of planar interdigital (ID) structures. 36 workable sensors with total area of 100 mm^2 per sensor and a sensing area of 6.25mm^2 per sensor were patterned on 4 inch single crystal Silicon wafer for the fabrication. The patterns were written for periodic interdigital structure with 11 sensing between 2 excitation electrodes for spatial wavelength of 50 μm with the width of electrodes and the sensing area set to values of 25 μm and 2.5 mm × 2.5 mm respectively. The sensors were fabricated using MEMS technology used in semiconductor fabrication industry. It involves the steps of photoresist coating, UV exposed ID pattern transfer, plasma etching, and metal deposition by DC magnetron sputtering and lift off. Owing the properties of inertness, resistive to corrosion and flexible in the methods available for deposition as thin film, Gold was selected for electrodes' material. DC magnetron was used to sputter 500 nm of Gold (Au) on top of 20 nm thin film of Chromium (Cr) which provides strong adhesion of the Gold ID structures on the

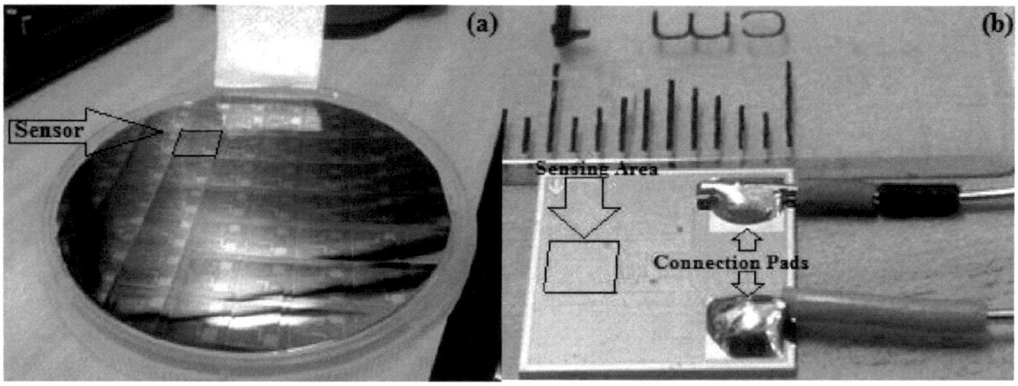

Figure 4. (a) Sensors in wafer form (b) individual sensor.

substrate surface. After lift-off, the wafer was coated with a 1 μm layer of Parylene C by a conformal coater. The Parylene C is an industrial polymer used to coat fabricated circuits to avoid environmental effects on circuit operation. Parylene C coating is uniform and pinhole free. It does not edge, bridge or gap and can withstand continuous exposure to air at high temperatures for long periods of time. It helps avoid corrosion of electrodes by chemical reaction in addition to its inherent dielectric properties which makes it a better choice for capacitive sensing systems. Photolithographic steps were repeated and plasma etching was used to open the area with bonding pads to facilitate the connections to the reference and working electrodes on the sensor. Detail of sensors' fabrication has been explained elsewhere [24]. Figure 4 below shows sensors in wafer form and individually against scale.

3. Methods and materials

3.1. Electrochemical impedance spectroscopy

Electrochemical Impedance Spectroscopy (EIS) is a popular technique used in conjunction with ID sensors that describes the response of an electrochemical cell to a low amplitude sinusoidal perturbation as a function of frequency. This technique has the inherent potential of non-destructive testing, single side access and label-free detection for biological and chemical analysis and has been reported for estimation of fat contents in meat [25], to detect bio toxins in shell fish [26], evaluate electrical impedance of food products [27] and many other applications. The impedance Z of a system is determined by applying a small signal perturbation as a function of frequency and analysis of the resultant current that flows through the system in terms of amplitude and phase shift in comparison to voltage-time function. The complex value of impedance is interpreted in terms of its real and imaginary values at different frequencies. In absence of a redox probe the Z_{real} refers to the ionic molecular concentration present in the material under test, whereas, the Z_{imag} translates capacitive nature of the bulk medium. The results of an impedance measurement can be graphically demonstrated using bode and Nyquist (Cole-Cole) plot for all applied frequencies with real part of impedance Z plotted along X-axis and Imaginary part plotted along Y-axis in the later. The impedance spectrum obtained, allows the characterization of surface, layers and concentrations in addition to the interpretation of exchange and diffusion processes used to describe the system kinetics. This analysis is achieved by translating the impedance spectrum into an equivalent series/parallel circuit, commonly consisting of resistances and capacitances representing the different electrochemical and physiochemical properties of the system under analysis [28].

The most frequently cited equivalent circuit used to interpret EIS experimental spectrum in electrical model form is Randle's equivalent circuit as shown in figure 5(a). It shows uncompensated solution resistance R_s in series to a parallel combination of double layer capacitance C_{dl} to the charge transfer resistance R_{ct} in series with Warburg impedance Z_w [29]. The Nyquist plot (Cole-Cole plot)

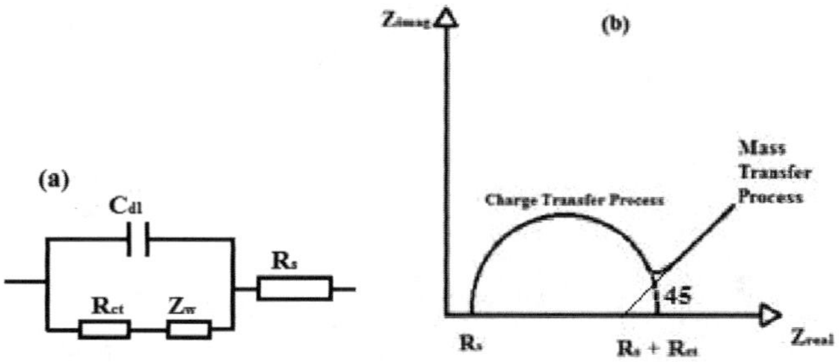

Figure 5. (a) Randle's Equivalent Circuit. (b) Nyquist plot.

for the equivalent circuit shows a semicircular region followed by a 45° straight line as shown in figure 5(b). The straight line interprets a faster mass-transfer limited process due to electron flow from electrode surface in the bulk solution at lower frequencies, whereas, the semicircular portion describes a relatively slower charge-transfer limited process at higher frequencies. C_{dl} can be calculated from the frequency at the maximum of semicircular region in the Nyquist plot using $\omega = 2\pi f = 1/R_{ct}C_{dl}$, whereas, R_{ct} is calculated by extrapolating the semicircle to Z_{real} axis as shown in figure 5(b). The double layer capacitance C_{dl} and charge-transfer resistance R_{ct} are the key electrical parameters to determine the impedance change for analysis of the system kinetics required to observe the change in dielectric properties of the medium due to the presence of phthalates. Macdonald discusses the in-depth analysis of electrochemical impedance spectroscopy technique in [23].

3.2. Sample preparation

Gravimetrically prepared 99.5 % pure di(2-ethylhexyl) phthalate (DEHP) solution with a concentration of 100µg/mL (100 ppm) in ethanol was procured from ChemService® USA. Two set of working solutions were prepared by serial dilution method in deionized water MilliQ® (MILLIPORE USA) at 0.002 through 2 ppm concentration levels of DEHP. Pure deionized water with 2 ppm 99.7% pure ethanol was used as the control solution to create reference standard for impedance change observation. pH of the control and working solutions were tested using IQ Scientific Instrument Inc. USA after calibration with buffer solutions. Commercially available bottled orange juice was spiked at a concentration of 5, 10 and 20 ppm with DEHP for testing the sensor in acidic medium (pH 3.35 at 20°C). These concentrations of DEHP in juice were selected below and above TDI calculated over an average person of 60 Kg mass as given in table 1. Pure juice was used as reference standard for test results' comparison with the spiked samples.

3.3. Experimental setup

EIS experiments were carried out using Hioki 3522-50 LCR Hi precision Tester (Japan) at laboratory temperature (23°C). All experiments were performed using slow mode of testing equipment to achieve an error rate of less than 0.05%. A constant voltage 1V_{rms} sinusoidal signal was applied to the sensor with a frequency sweep of 1 Hz to 10 kHz with 20 data points per decade on log scale. LabView® data acquisition program was used to write the measured values in Microsoft excel® file for post processing and analysis on the attached desktop computer. Automatic data acquisition system was linked using RS232C hardware interface with the Hi precision tester. After setting up the test apparatus and circuitry, the tester was calibrated using the built-in function of self-calibration in open and short circuit mode. This calibration ensures to offset any stray capacitance appearing in the testing data due to the length of testing lead in addition to suppressing any noise in the observed signal, hence improving signal to noise ratio. The sensor was cleaned with acetone, rinsed with deionized water and dried with nitrogen before testing the next concentration. An average of three set of experiments were used to ensure reproducibility and reliability of the results. Figure 6 shows the test bench setup and LabView based front panel of automatic data acquisition system developed for the experiments.

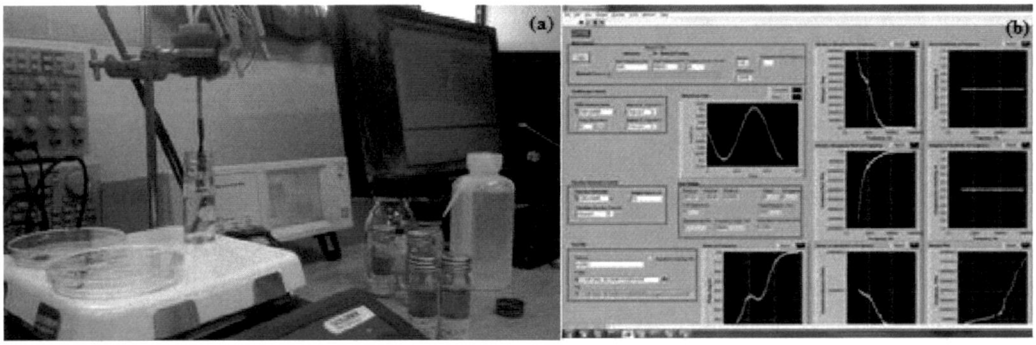

Figure 6. (a)Test bench setup (b) LabView based data acquisition system front panel.

4. Results and discussions

4.1. DEHP detection in MilliQ

MilliQ water (18MΩ cm) with 2ppm ethanol was used as control in the EIS experiments. pH of the control solution at 23°C was measured as 7.3. Gravimetrically prepared 100 ppm DEHP in ethanol was added to MilliQ to obtain a stock solution of 2ppm DEHP in water. Ethanol was used as a polar solvent to dissolve DEHP in deionized water. The pH of the stock solution was measured as 6.98 at 23°C which showed that the addition of phthalate had very less impact on the pH of the system. 0.2ppm, 0.02ppm and 0.002ppm concentrations of DEHP were obtained from the stock solution by using serial dilution method. Sample testing by proposed sensing system was carried out immediately after sample preparation. All the experiments were performed in laboratory environment at 23°C with 47% humidity level. Figure 7 below shows the plot for imaginary and real part of impedance vs. frequency for all four DEHP concentrations. The capacitive reactance (Z_{imag}) shows a good variation with changing concentrations of DEHP especially at lower frequency range (10 Hz-400Hz), whereas, the corresponding change in Z_{real} is not dominant which could be explained on the basis of ionic concentration in the test samples. The pH value observed for all concentrations of DEHP does not vary and is approximately the same for each concentration, therefore, Z_{real} which depends on the Faradic current flowing through the bulk medium is same for all concentrations as shown in the plot in figure 7. A drastic change in $Z_{imaginary}$ (reactance) with variation of DEHP concentrations in samples is however observed in the said plot which is attributed to the dielectric properties of the DEHP molecules. The Nyquist plot shown in figure 8 demonstrates: (1) the fast mass transfer process in form of a straight line which corresponds to the Warburg resistance in the equivalent circuit suggested in the next section and (2) the slow charge transfer process is translated by semi-circular region of different diameters corresponding to the DEHP concentrations in the test solutions. This slow charge transfer process occurs due to the accumulation of DEHP molecules on the electrode solution interface. This happens due to the adsorption of the DEHP molecules on the electrode surface as a result of electron transfer from the electrodes into the solution [23].

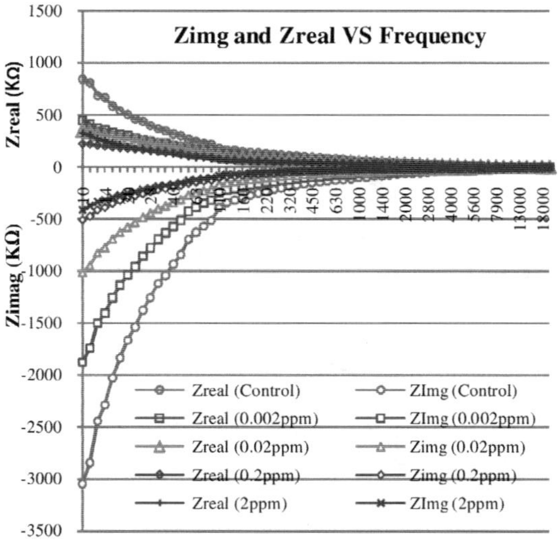

Figure 7. Imaginary and real part of Impedance vs. Frequency.

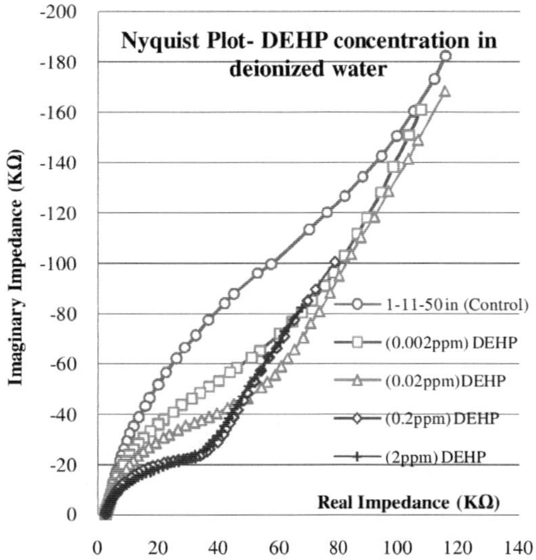

Figure 8. Nyquist Plot for different test concentrations of DEHP.

4.1.1. Data analysis using non-linear least square curve fitting.

The diameter of the semi-circular region in the Nyquist plot dictates the value of charge transfer resistance R_{ct} which in turn changes the double layer capacitance C_{dl} at the sensor-solution interface. The variation in the diameter of the semi-circular trace with changing concentrations of DEHP molecules in the bulk medium shows the capability of the sensor to monitor the variation of DEHP as low as 0.002ppm in deionized water.

Using Randle's cell model, the real and imaginary impedance at the electrode solution interface could be derived as shown by the equation (1)

$$Z(\omega) = R_s + \frac{R_{ct}}{1+\omega^2 R^2{}_{ct} C^2{}_{dl}} - \frac{j\omega R^2{}_{ct} C_{dl}}{1+\omega^2 R^2{}_{ct} C^2{}_{dl}} \tag{1}$$

The equivalent circuit was estimated using theoretical calculations by electrochemical spectrum analyser algorithm which suggests an equivalent circuit by fitting the experimental values over theoretical response of the most probable circuit by performing iterations over the experimental data. The circuit parameters were estimated by non-linear least square fitting technique which fits the measured impedance data on theoretically estimated values. The algorithm performs statistical analysis to calculate the residual mean square for experimentally observed values in the measured spectra against the calculated values based on the theoretical response of suggested equivalent circuit. The optimization of the calculated data is achieved by the number of iterations it takes to solve the mathematical model for the proposed equivalent circuit. Figure 9 shows the electrical circuit equivalent to the kinetic processes taking place in the electrochemical cell for a single concentration of DEHP in bulk solution. This equivalent circuit was proposed at the most optimal curve fitting with an error rate of less than 5% in 300 iterations. The fitted Nyquist plot is shown in figure 10 and the least square curve fitting plot for absolute values of impedance is traced in figure 11. The bold line curve displays the modelled behaviour of the equivalent circuit whereas; the markers represent the experimentally observed response from the electrochemical cell in both plots.

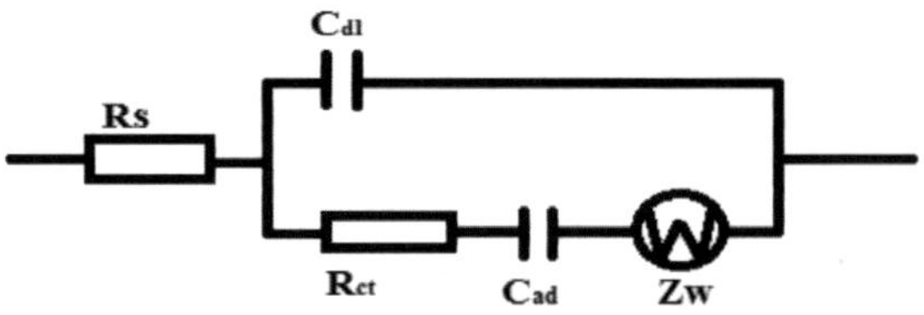

Figure 9. Equivalent circuit proposed by the least square curve fitting algorithm.

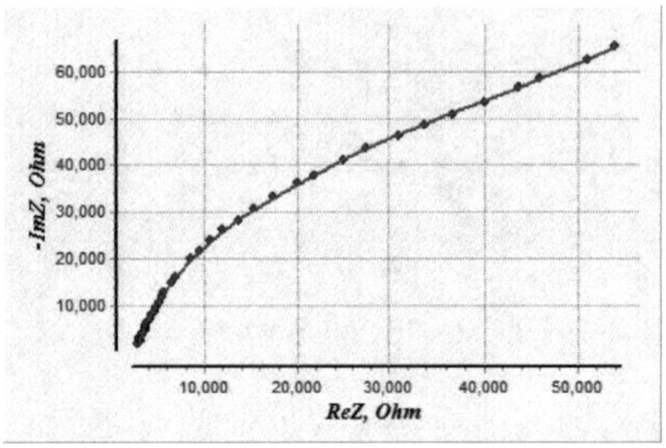

Figure 10. Least square curve fitting for Nyquist plot. Solid line shows calculated curve, markers show experimental data points.

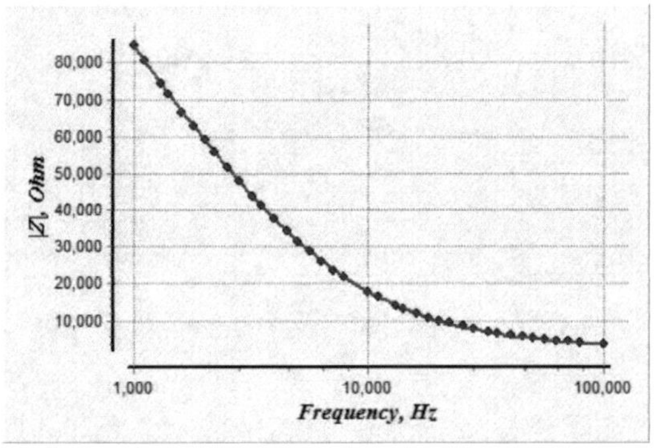

Figure 11. Least square curve fitting for absolute value of Impedance. Solid line shows calculated curve, markers show experimental data points.

The electrical parameters for the components estimated in the constant phase equivalent circuit shown in figure 9 are calculated by electrochemical spectrum analysis algorithm for all concentrations of DEHP in aqueous medium. Table 3 provides the values of double layer capacitance (C_{dl}), charge transfer resistance (R_{ct}), Warburg resistance (Z_w) and solution resistance (R_s). It is observed that an additional parameter adsorption capacitance (C_{ad}) appears in the equivalent circuit shown in figure 9 consequent to electrochemical spectrum analysis for the sensing system. This parameter interprets the process of ion diffusion from the electrode surface into the bulk solution. These ions are produced as a result of the electron transfer from the solution into the electrode surface especially at low frequencies [30]. This parameter tends to increase with an increase in the concentration of phthalates in the samples indicating more number of phthalate molecules adsorbing to the electrodes' surface.

Table 3: Evaluated electrical parameters obtained by using electrochemical spectrum analyzer for 300 iterations.

Eq. Circuit	Electrochemical spectral analysis for DEHP concentrations in MilliQ							
Parameters	2 ppm		0.2 ppm		0.02 ppm		0.002 ppm	
(units)	Value	Error (%)	Value	Error (%)	Value	Error (%)	Value	Error (%)
R_s (Ω)	2021.1	3.10	2681.6	1.985	2845.7	4.47	2890	2.82
Z_w (KΩ)	291.5	4.47	513.1	2.6064	1702.4	3.411	2187.2	1.6126
R_{ct} (KΩ)	48.8	0.68	42.3	0.416	64.5	1.29	91.1	0.57
C_{ad} (F)	1.14E-8	4.03	1.28E-8	2.452	1.32E-8	4.05	5.889E-9	2.72
C_{dl} (F)	1.003E-8	0.52	2.313E-9	0.4254	1.7E-9	0.931	1.584E-9	0.322

4.2. DEHP detection in orange juice

Due to strict health policy observed by MoH New Zealand, the beverages available in the local markets are assumed to be DEHP free. The sensor's response was tested for locally available orange juice available in a glass bottle. Glass-bottled juice was selected to avoid the potential chances of DEHP leaching into the content from the PET packaging of the juice [6]. The juice was spiked with 5, 10 and 20 ppm concentrations of DEHP, prepared at 23°C temperature and was tested immediately after sample preparation. These tested concentrations were selected to observe sensing system's response above and below the minimum risk level (MRL) intake limit for DEHP calculated for an average individual of 60 kg body weight given in table 1. Un-spiked original drink was used as control solution to create standard reference curve. Figure 12 shows the plot for real and imaginary part of the impedance spectra for the spiked juice compared to the standard reference for the original juice. The sensor provides change of impedance in the capacitive reactance (Z_{imag}) dominating the corresponding real impedance values in a frequency range of 10 Hz to 1.3 kHz. The change in Z_{imag} is higher in magnitude at lower frequencies especially below 100 Hz range and the plot shows a sequential change in the capacitive reactance with increasing concentration of phthalate molecules. Figure 13 displays the bode plot for the same frequency range. As observed, the change in absolute impedance (Z_{abs}) is visible only at lower frequencies whereas the phase shift is distinctive for pure and spiked samples especially for a DEHP concentration as high as 20ppm. It is observed that the real impedance of the juice samples has drastically reduced in comparison to the DEHP testing in pure deionized water as shown in figure 7. It is due to the ionic presence in the acidic medium (juice) of pH 3.04 at 23 °C, causing an appreciable change in the solution resistance variable R_s.

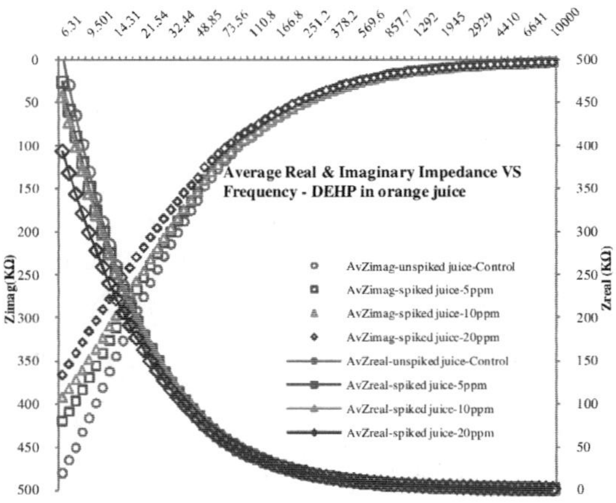

Figure 12. Real and imaginary part of the impedance spectra for different concentrations of DEHP in orange drink plotted against frequency

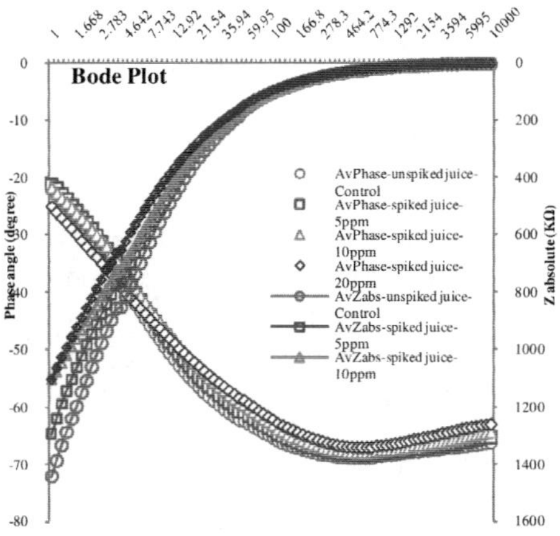

Figure 13. Bode plot for the measured impedance spectra for different concentrations of DEHP in energy drinks.

5. Conclusions

Non-invasive, real time detection technique for phthalates in water and juice has been reported. Sensitivity optimization for designed MEMS based interdigital sensors was achieved on the basis of mathematical modelling simulation software, COMSOL and was discussed to analyse the sensitivity concerns. Optimized design was fabricated using MEMS based semiconductor device fabrication techniques. Sensor was able to detect phthalate ester, DEHP, in deionized water and orange juice as a practical application of the sensing system. Analyses of the results have been carried using non-linear least square curve fitting statistical technique to discuss the kinetic processes taking place inside the electrochemical cell on basis of Randle's cell equivalent circuit model. Sensors performance was evaluated for commercially available orange juice and the experimental results proved its detection ability in such products. Selectivity of the sensing system is the major challenge being worked upon in the on-going research in order to render the testing system usable for industrial applications.

References

[1] J Heudorf U, Mersch-Sundermann V, Angerer J 2007 *Int. J. Hyg. Environ. Health* **210** 623-634
[2] T E D Schettler 2006 *International Journal of Andrology* **29** 134
[3] Hauser R, Duty S, Godfrey-Bailey L, Calafat AM 2004 *Environ. health perspect.* **112** 751
[4] USCPSC 2010 *Toxicity Review of Di(2-ethylhexyl) Phthalate (DEHP).* (Bethesda: MD)
[5] K Fiselier, E Rutschmann, G McCombie *et al* 2010 *European Food Res. and Tech.* **230** 619
[6] R Sendón, A Sanches-Silva, J Bustos *et al* 2012., *Journal of Separation Science* **35** 1319
[7] R Kavlock, K Boekelheide, R Chapin *et al* 2002 *Reproductive toxicology* **16** 655
[8] ATSDR 2002 *Toxicological Profile for Di(2-ethylhexyl) Phthalate.* (Atlanta: GA)
[9] WHO (World Health Organization) 2003 *Guidelines for drinking-water quality* **180** Chap8 3 ed.
[10] European Union Council 2001 *J. Eur. Commun. L* Decision No. 2455/2001/EC. Directive 2000/60/EC Off. p331
[11] M Biedermann, K Fiselier, G Marmiroli *et al* 2008 *European Food Res. and Tech.* 226 1399
[12] X L Cao 2008 *Journal of Chromatography A* **1178** 231
[13] M Wagner and J Oehlmann 2009 *Environmental Science and Pollution Research* **16** 278
[14] L López-Carrillo, R U Hernández-Ramírez, A M Calafat *et al* 2010 *Environ. health perspect.* **118** 539
[15] C Giam, H Chan, and G Neff 1975 *Analytical chemistry* **47** 2319
[16] M R Lee, F Y Lai, J Dou *et al* 2011 *Analytical Letters* **44** 676
[17] K Kambia, T Dine, R Azar *et al* 2001 *International journal of pharmaceutics* **229** 139
[18] K Shintani, M Taki, H Ootera *et al.* 2000 *Plasma processing apparatus using a partition panel* Google Patents
[19] V Triantafyllou, K Akrida-Demertzi, and P Demertzis 2007 *Food Chemistry* **101** 1759
[20] A K Malik, C Blasco, and Y Picó 2010 *Journal of Chromatography A* **1217** 4018
[21] Berggren C, Bjarnason B, Johansson G, 2001 *Electroanalysis* **13** 173
[22] Sadik OA, Aluoch AO, Zhou A. 2009 *Biosensors and Bioelectronics* **24** 2749
[23] E. Barsoukov and J. R. Macdonald, 2005 *Impedance spectroscopy: theory, experiment, and applications*: John (Wiley and Sons)
[24] A I Zia, A Mohd Syaifudin, S C Mukhopadhyay *et al.*, 2012 *IEEE Transactions on Instrumentation and Measurements* 1220
[25] S C Mukhopadhyay, and C P Gooneratne 2007 *IEEE Sensors Journal* **9** 1340
[26] A R M Syaifudin, K P Jayasundera, and S C Mukhopadhyay 2009 *Sensors and Actuators B-Chemical* **137** 67
[27] X Li, K Toyoda, and I Ihara 2011 *Journal of Food Engineering* **105** 563
[28] F Lisdat, and D Schäfer 2008 *Analytical and bioanalytical chemistry* **391** 1555
[29] M I Prodromidis 2010 *Electrochimica Acta* **55** 4227
[30] Kerner, Z Pajkossy 2002 *Electrochimica Acta* **47** 2055

Journal of Physics: Conference Series **439** (2013) 012027 doi:10.1088/1742-6596/439/1/012027

Fabrication of Ni/Ti/Al Schottky contact to n-type 4H-SiC under various annealing conditions

M Yousuf Zaman[*,1,3], S Ferrero[1,3], D Perrone[2,3], L Scaltrito[1,3], N Shahzad[1,2,3] and D Pugliese[1,2,3]

[1]Department of Applied Science and Technology, Politecnico di Torino, 10129 TO, Italy
[2]Italian Institute of Technology, Center for Human Robotics, C.so Trento 21 Torino, Italy
[3]X-Lab Materials and Microsystems Laboratory, via Lungo Piazza d'Armi 6, Chivasso, Italy

E-mail: [*]muhammad.zaman@polito.it

Abstract: Forward I-V characteristics of a silicon carbide Schottky diode, with triple layer metallization Ni/Ti/Al as Schottky contact, are presented. Effects of different annealing conditions on the Schottky barrier height and ideality factor are discussed. The diodes were annealed in inert Ar atmosphere for 30 minutes at temperatures ranging from 600 °C to 800 °C. The ideality factors of the four diodes, chosen out of 20 diodes, range from 1.02 to 1.13 and the Schottky barrier heights range from 1.47 eV to 3.17 eV.

1. Introduction

Silicon carbide (SiC) is favored as a semiconductor for the fabrication of power devices because of its wide energy band gap, high hardness, chemical inertness, high electric field breakdown strength and high thermal conductivity [1-5]. SiC based devices need Schottky as well as low resistance ohmic contacts to operate at high power and high frequencies [6-9]. Among a number of materials that can form low resistance ohmic contacts to SiC one is Ni/Ti/Al. This triple layer metallization after annealing can form low resistance ohmic contacts on both p- and n-type SiC; a very important property with potential application in the fabrication of metal-oxide field effect transistors (MOSFETs) [1]. MOSFET fabrication also requires low annealing temperatures so as to reduce the chances of any chemical reactions between the gate metal and the oxide layer. Ni/Ti/Al forms low resistance ohmic contacts with annealing temperature as low as 800 °C and is considered a suitable candidate for MOSFET fabrication [1].

The focus of this paper, however, will be on the study of Schottky behavior of Ni/Ti/Al/SiC contact which can be achieved by altering the relative thicknesses of the three constituent metals used for the contact deposition i.e. nickel, titanium and aluminum—with the aim to fabricate both Schottky and ohmic contacts simultaneously to make possible the fabrication of devices that require both Schottky and ohmic contacts at low annealing temperature (such as MESFETs). In the following sections we will elaborate the experimental setup and discuss some of the electrical parameters in view of thermionic emission theory of four Ni/Ti/Al Schottky diodes annealed at different temperatures.

* To whom any correspondence should be addressed.

2. Device fabrication

The Schottky barrier diodes characterized in this work were fabricated using 3 inch diameter 4H-SiC wafers. The substrates were purchased from Cree Research Inc. (Durham, NC) and they were constituted by n-type bulk substrate (with a doping concentration N_d of 10^{18} cm^{-3} and a thickness of 360 μm) and an epitaxial layer (also n-doped). The thickness of the epitaxial layer was 5.5 μm, and the doping concentration N_d was 10^{15} cm^{-3}.

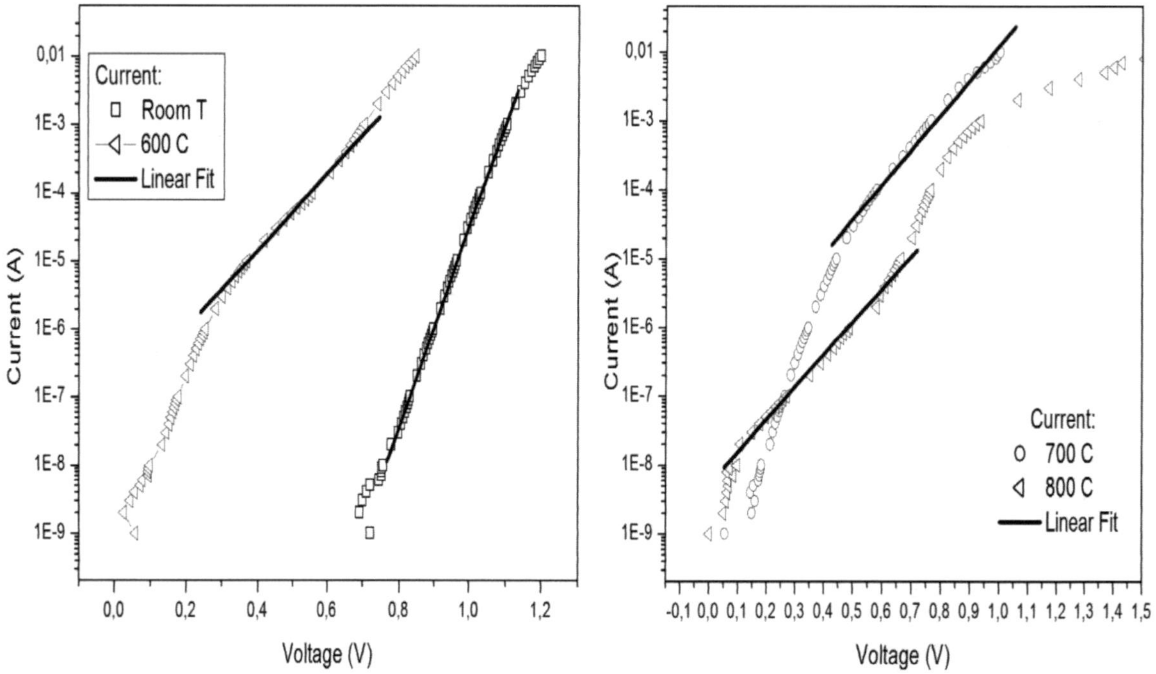

(a) (b)

Figure 1. I-V characteristics of Ni/Ti/Al/4H-SiC Schottky diodes annealed at various temperatures.

Schottky barrier diodes were designed with a p^+-type guard ring and a mesa-type structure with an additional silicon dioxide (SiO_2) passivation layer. Boron ion implantation was used for the formation of the p^+-type guard ring. This step was followed by a Rapid Thermal Process (RTP) at temperatures exceeding 1500 °C, in order to promote lattice damage recovery and to achieve a reasonable electrical activation of the implanted B^+ ions.

The silicon dioxide passivation layer was grown by means of a Low Temperature Oxidation (LTO) process, using TetraEthylOxySilane (TEOS) as precursor. The SiO_2 layer was patterned using standard UV photolithography, and etched by a buffered HF-based solution. The Schottky contact was then realized by lift-off patterning of Ni/Ti/Al triple layer (with respective thicknesses of 20, 50 and 150 nm) on the 4H-SiC epilayer surface. The Ni/Ti/Al deposition was followed by an annealing treatment performed at 600, 700 and 800 °C in inert atmosphere. The ohmic contact was obtained on the wafer's backside by means of a triple layer of Ti, Ni and Ag, without the need of a post-deposition annealing. The deposition of both the Schottky and the ohmic contacts was performed using an electron beam evaporator (ULVAC model EBX-14D), at a pressure of about 10^{-5} Pa. The electrical characterization of the SBDs was performed by acquiring the current-voltage (I-V) curves using Keithley 237 and 238 SMUs. The capacitance-voltage (C-V) characteristics were acquired by means of a HP4192A impedance analyzer.

3. Discussion

Forward bias current-voltage (I-V) plots of four diodes characterized at room temperature and annealed at 600, 700 and 800 °C were applied linear fit. The forward I-V characteristics are shown in figure 1. With the help of thermionic emission theory electrical parameters such as Schottky barrier height SBH (denoted by Φ_B) and ideality factor n, were extracted. The thermionic emission theory relates the diode current to the applied voltage V_a as follows [10]:

$$I = I_s [\exp (qV_a / nK_BT) - 1] \quad (3.1)$$

$$I_s = AA^* T^2 \exp (- \Phi_B / K_BT) \quad (3.2)$$

In (3.1) I_s represents saturation current K_B, Boltzmann's constant and T applied temperature. Similarly in (3.2) A represents area of the diode (1.1 mm^2 in our case), A* represents Richardson's constant (A*=146 $A/cm^2.K^2$) and Φ_B represents Schottky barrier height. Figure 1 shows I-V characteristics of Ni/Ti/Al/4H-SiC Schottky diodes annealed at 600, 700 and 800 °C and one without annealing. Schottky barrier height (extracted from the intercept) is 1.47 eV for the diode without annealing and 2.37, 2.62 and 3.17 eV for the diodes annealed at 600,700 and 800 °C. For the diode without annealing ideality factor (extracted from the slope) is 1.13 and for those annealed at 600, 700 and 800 °C ideality factors are 1.02, 1.03 and 1.08 respectively. These four diodes were chosen out of a group of diodes which also contained a number of diodes with small linear region and poor rectifying behavior with average Schottky barrier height of 1.03 eV and an average ideality factor of 1.46 (extracted from forward I-V plots). Similarly an average Schottky barrier height of 1.31 eV extracted from capacitance voltage (C-V) characteristics.

Figure 2 shows both Φ_B and n in relation to the annealing temperature, general trend shows an increase in the SBH with annealing temperature. For diode that was not annealed the SBH values are close to the commonly accepted values for Ni/SiC Schottky barrier but diodes which were annealed at 600 °C temperature or more show bad rectifying behavior and high Schottky barrier heights as reported also by Kestle et al. in the case of Ni/SiC Schottky diodes [11]. The ideality factor shows a decline with increasing annealing temperatures which indicates an improvement in the quality of the diodes at high annealing temperatures. The diode which was not annealed shows best Schottky behavior out of all the diodes characterized. The diode annealed at 800 °C shows very poor Schottky behavior with ideality factor of 1.08 and SBH of 3.17. The

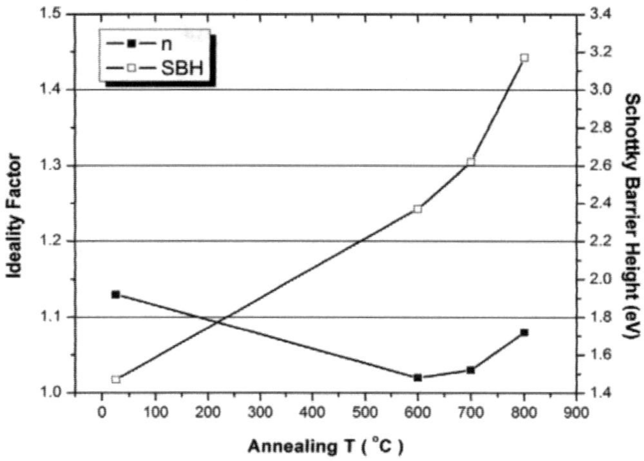

Figure 2. Schottky barrier height and ideality factor versus annealing temperature.

increasing trend in SBH accounts for the varying quality of Schottky contact which is deteriorating at high annealing temperature this behavior is contrary to the normal where diodes tend to perform better after annealing at high temperatures [12]. This could be due to a number of different reasons such as structural defects, laboratory conditions or chemical reactions taking place at high annealing temperatures such as Ni_2Si which itself forms a Schottky contact on n-type SiC [13].

4. Conclusion

Schottky behavior of four ternary Ni/Ti/Al contacts to 4H-SiC was compared for different annealing temperatures. The diode which was not annealed showed better behavior as compared to the ones annealed at 600,700 and 800 °C, with ideality factor as low as 1.13 and Schottky barrier height of 1.47 eV. All the diodes which were annealed at more than 600 °C exhibit excess forward current at low applied voltage which indicates inhomogeneous Schottky barriers. There is dire need to focus on the inhomogeneity of Schottky barrier in order to take full advantage of this ternary Schottky contact material.

References

[1] Tsukimoto S, Sakai T, Onishi T, Ito K and Murakami M 2005 *J. Electon.Mater.* **34** 10
[2] Nikitina I P, Vassilevski K V, Wright N G, Horsfall A B and O'Neill A G 2005 *J. Appl. Phys.* **97** 083709
[3] Haberstroh C, Helbig R and Stein R A 1996 *J. Appl. Phys.* **76** 509
[4] Muench W V and Pfaffeneder I 1977 *J. Appl. Phys.* **48** 4831
[5] Slack G A J. 1964 *Appl. Phys.* **35** 3460
[6] Konishi R, Yasukochi R, Nakatsuka O, Koide Y, Moriyama M and Murakami M 2003 *Mater. Sci. Eng.* B **98** 286
[7] Trew R J, Yan J B and Mock P M 1991 *Proc. IEEE* **79** 598
[8] Davis R F, Kelner G, Shur M, Palmour J W and Edmond J A 1991 *Proc. IEEE* **79** 677
[9] Vasilevskii K V 1992 *Sov. Phys. Semicond.* **26** 994
[10] Tung R T 2001 *Mater. Sci. Eng. R* **35** 1
[11] Kestle A, Wilks S P, Dunstan P R, Prichard M and Mawby P A 2000 *Electron. Lett.* **36** 3
[12] Zhang Q and Sudarshan T S 2001 *J. Electron. Mater.* **30** 11
[13] Roccaforte F, La Via F and Raineri V 2003 *J. Appl. Phys.* **93** 11

Three phase six-switch PWM buck rectifier with power factor improvement

M Zafarullah Khan[1], M Mohsin Naveed[1] and D M Akbar Hussain[2]
[1]Institute of Industrial Control Systems, P.O Box, 1398, Rawalpindi, Pakistan
[2]Department of Energy Technology Aalborg University, Denmark

E-mail: mohsinnaveed30@gmail.com

Abstract. Conventional Phase Controlled Rectifier injects low order current harmonics into the AC mains. Large size filtering components are required to attenuate these harmonics. In this paper, three phase six-switch PWM buck rectifier is presented which operates at nearly unity power factor and provides variable output voltage. Small size energy storing components are required depending upon switching frequency. MATLAB simulation is performed and modified Sinusoidal Pulse Width Modulation (SPWM) switching technique is used in 3kW prototype converter to demonstrate low input current THD, nearly unity displacement factor, well regulated output voltage and reduced switching losses compared to conventional SPWM.

1. Introduction

In most of the power electronic applications, AC/DC converter is widely used as a first stage rectifier. DC obtained from this rectifier is further converted into desired amplitude and frequency. Most common topology of rectifier is three phase controlled rectifier (PCR). PCR is capable of delivering controlled DC voltage but it has a very low input power factor at low conduction angle and introduces unwanted distortion in AC mains. It does not comply with IEEE-519 standard of low input current harmonics. Large size low frequency input filters are required in PCR to improve its power factor. This results in both increased cost and large size of the PCR. PWM rectifiers have replaced the conventional PCR due to their high efficiency, good voltage regulation, nearly unity power factor and small input/output filter size depending upon switching frequency. PWM rectifiers are normally operated at high switching frequency where low frequency harmonics are easily suppressed. PWM rectifiers are increasingly becoming popular due to the availability of fast switching, high voltage and high current IGBT's[1].

Three phase PWM regenerative boost rectifier [2] is shown in figure 1. Output voltage is greater than the peak of the line to line RMS voltage in this rectifier. This is a four quadrant rectifier capable of bidirectional power flow. Operation of this rectifier is similar to boost SMPS. When SW_{R2} is turned on, voltage is applied across the inductor and it gets energized. When SW_{R2} is turned off, inductor de-energizes through the diode of SW_{R1} charging the capacitor. Unity power factor is achieved by controlling the current in the inductor.

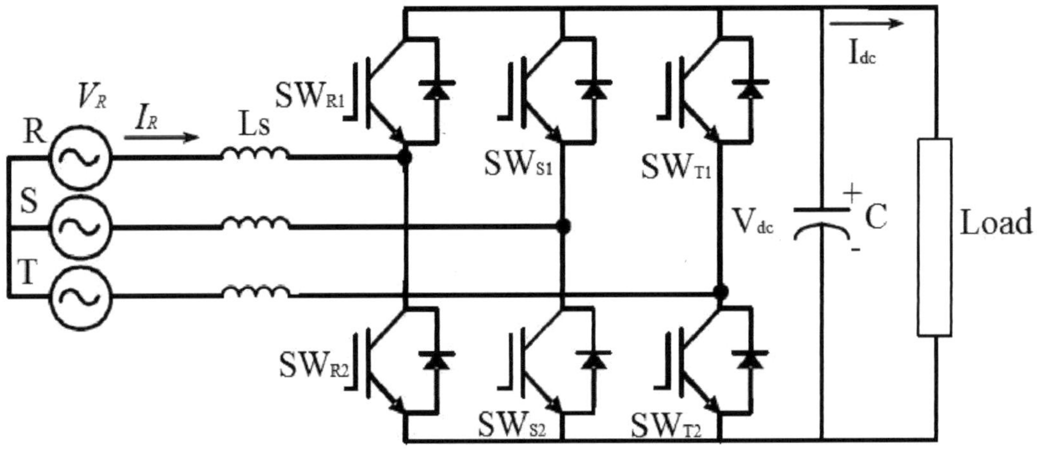

Figure 1. Three phase PWM regenerative boost rectifier.

Three phase six-switch PWM buck rectifier is shown in figure 2. Output voltage of this rectifier is less than 1.26 times line to line RMS voltage [3]. It is a unidirectional converter and power flow is controlled by controlling the depth of modulation and delay angle. Modulation index and delay angle provide a flexible control over output voltage as well as input displacement factor [3]. Change in delay angle produces lagging power factor which compensates the leading power factor of input capacitor resulting in unity displacement factor [3]. However unity displacement factor is not achieved if capacitor current is greater than rectifier current for low loads. Some applications of this rectifier are UPS battery chargers, DC motor speed control, front end converter in wind farms and front end converter in variable speed ac drives.

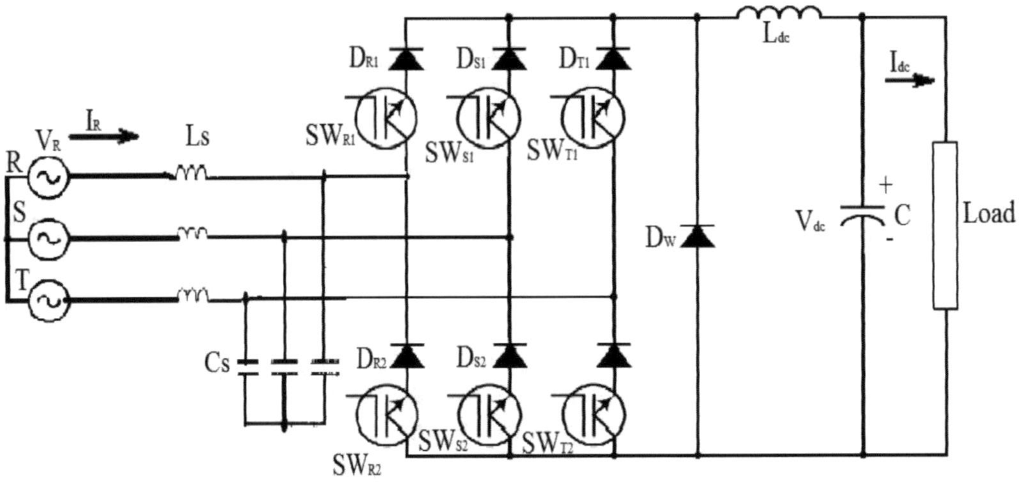

Figure 2. Three phase six-switch PWM buck rectifier.

The rest of this paper discusses the design, simulation and implementation of three phase six-switch PWM buck rectifier optimized to provide variable output voltage with low ripple contents and nearly unity power factor using small filters.

2. Design of three phase six-switch PWM buck rectifier

2.1 PWM technique

Assuming balanced set of three-phase input voltages, the objective of this rectifier is to draw sinusoidal input currents in phase with input voltages. Only active power flows through the bridge. The ripple content is very low in the output voltage and it is well regulated since the devices are being switched at high frequency. As the input voltages are balanced, a three phase modified Sinusoidal PWM (SPWM) is applied to the switches and switching waveforms are shown in figure 3. Switching instants of all the switches are clearly mentioned in figure 3. Each switch is continuously conducting for 60^0 in the center of switching pattern. There are no overlapping pulses in the switching waveforms instead freewheeling diode is used for overlapping instants resulting in the reduction of switching losses up-to 30%. Switching pattern of 2kHz SPWM is generated using analogue discrete IC's and also both the modulation index and the delay angle is controlled.

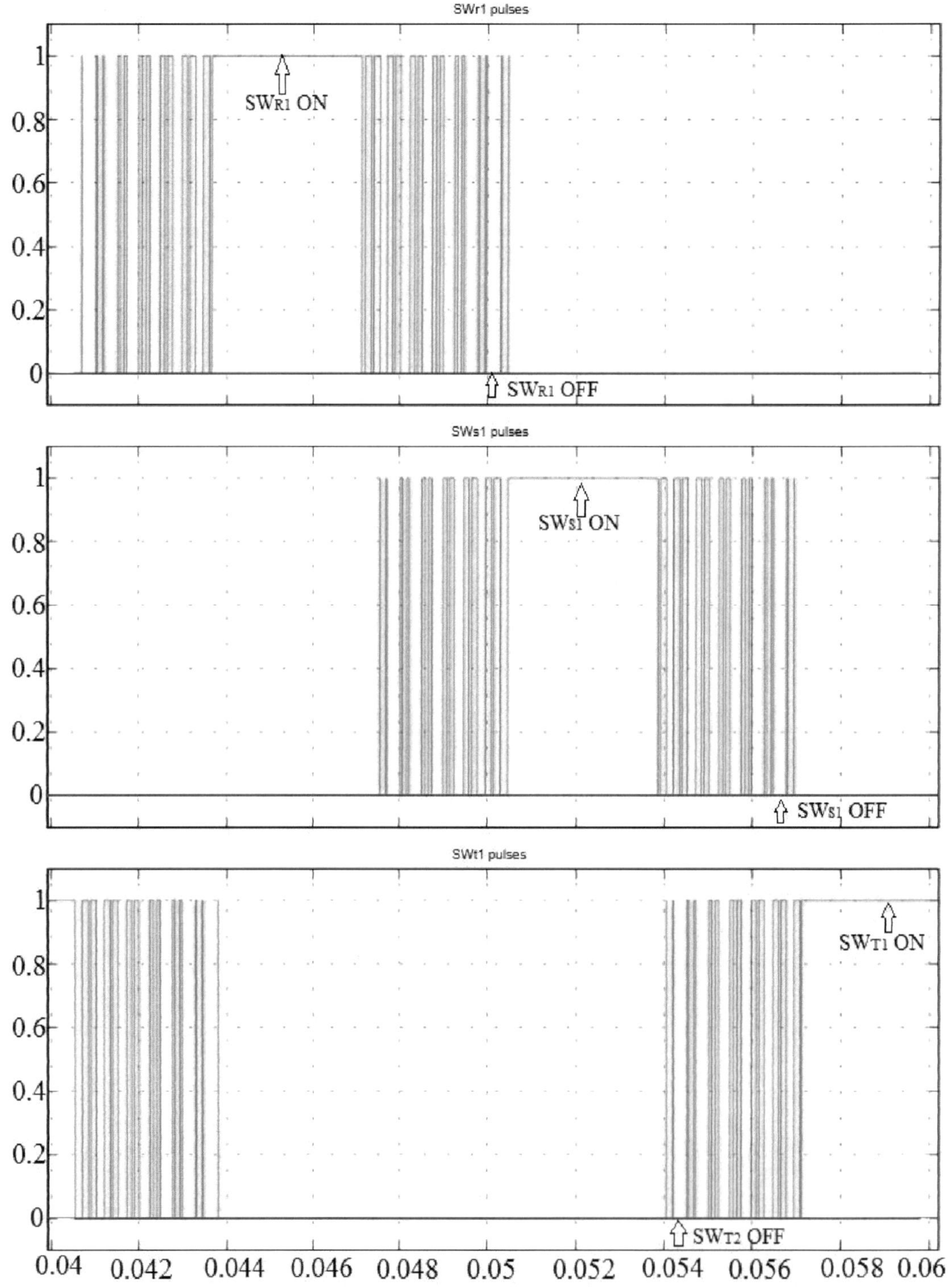

Figure 3. Modified Sinusoidal Pulse Width Modulation.

2.2 Power topology

Power topology consists of power semiconductors devices, AC-side filter [4] and DC-side filter. IGBT's are used with series fast recovery diode to block the reverse voltages across the IGBT. The reverse blocking capability of IGBT is poor and series diode is required. AC-side filter is implemented using gapped core inductor of 3mH and paper capacitor of 10uF. It results in a small size filter with cut-off frequency of 920Hz. DC-side filter consists of an inductor and capacitor. Inductor is designed to maintain low ripple current and continuous flow of load current. Small capacitor is required to filter out high frequency ripple components from the load current.15mH inductor and 1000uF capacitor are used in the prototype and simulation. RC snubber of 10Ω and 0.1uF is used across each switch. Resistive load of 39Ω is used.

2.3 MATLAB Simulink simulation

Three phase six-switch PWM buck rectifier circuit has been simulated with Simulink using the values described in the last section. Input voltage is $400V_{p\text{-}p,\ rms}$ and output voltage is variable from 80V to 342V. Figure 4(a) shows the simulation results of R phase voltage and current. It can be seen that voltage and current are in phase, displacement factor is nearly unity and THD is also low resulting in nearly unity power factor. Figure 4(b) shows the simulation results of output voltage and current. Voltage across D_W is shown in figure 5.

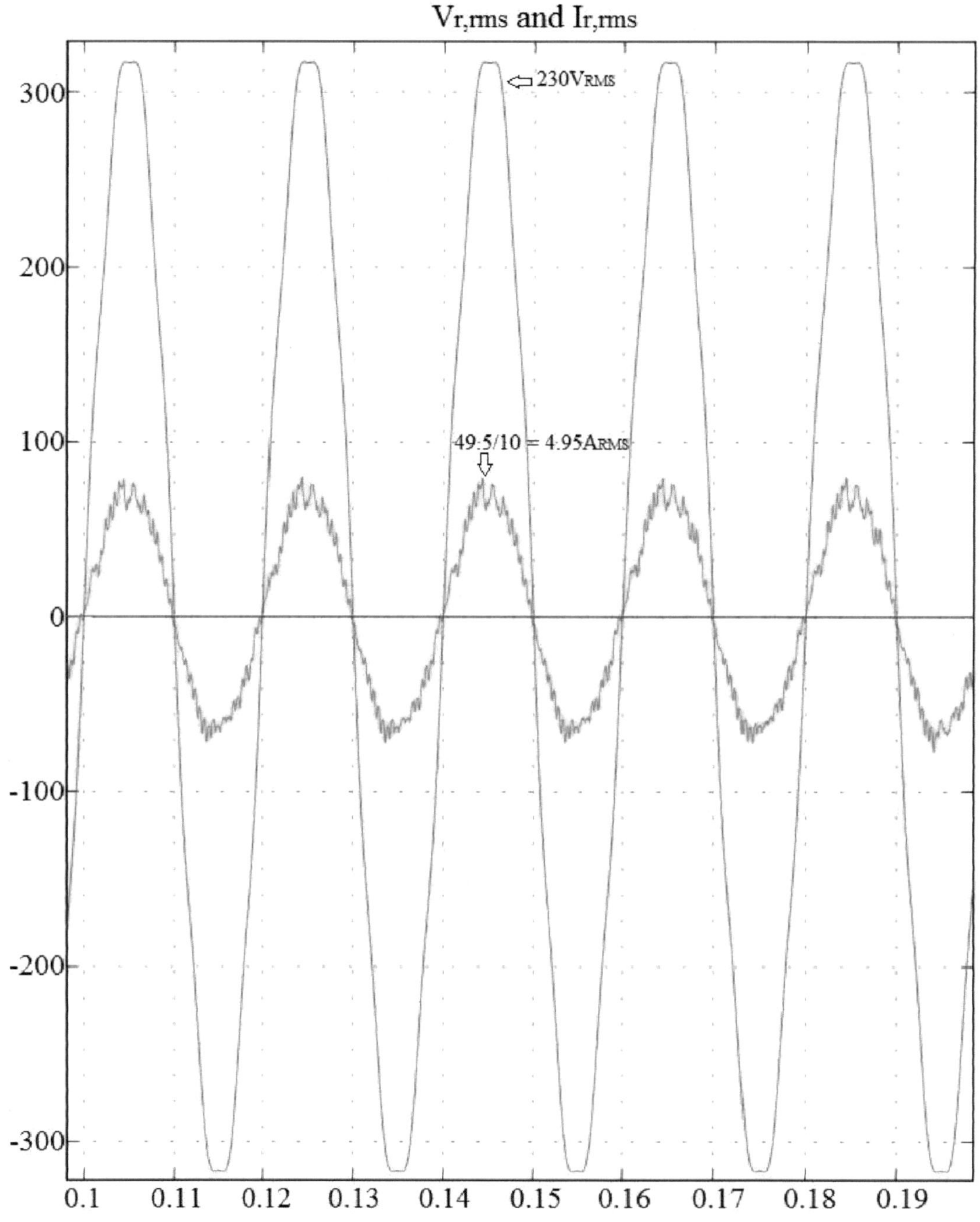

Figure 4(a). $V_{R,rms}$ and $I_{R,rms}$.

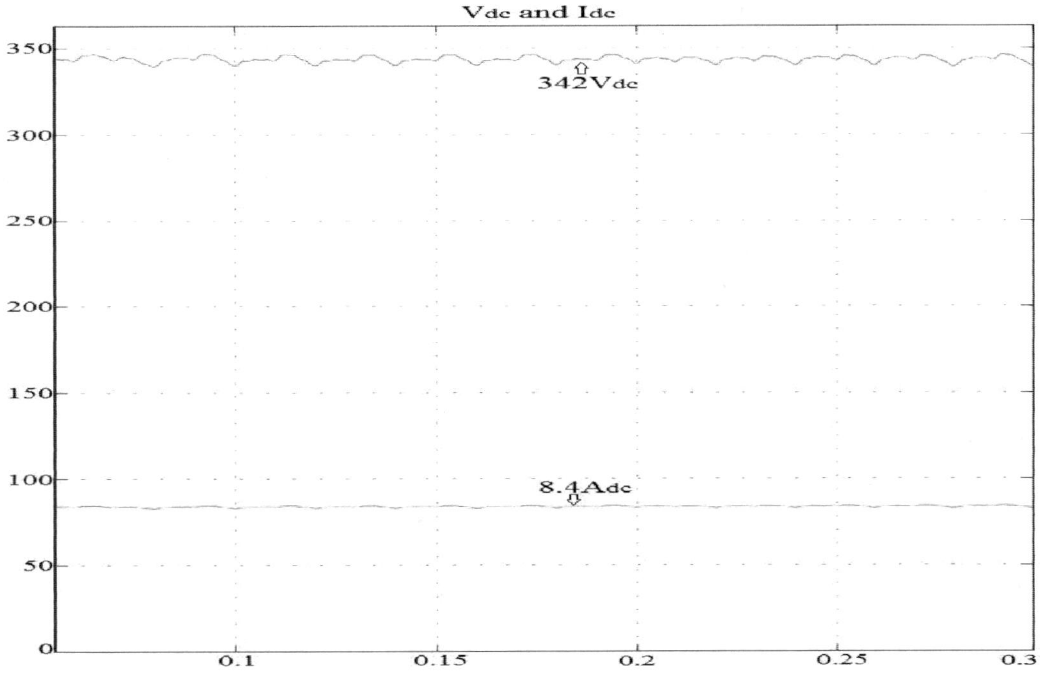

Figure 4(b). V_{dc} and I_{dc}.

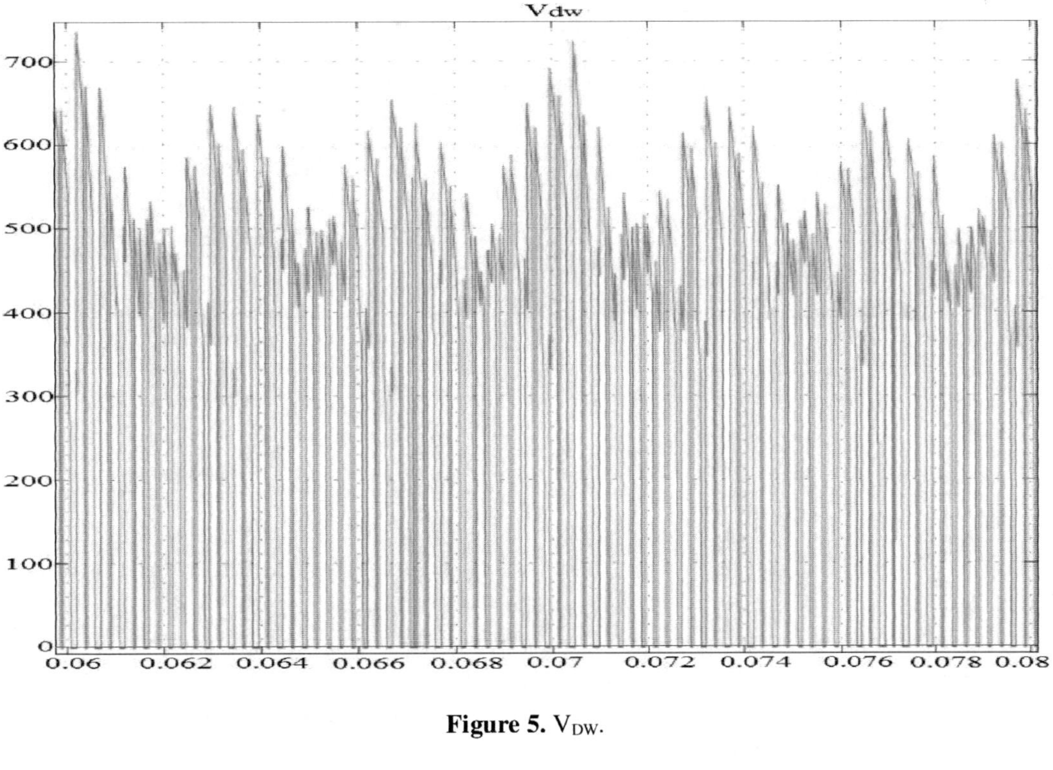

Figure 5. V_{DW}.

2.4 Experimental verification

3kW prototype of three phase six-switch PWM buck rectifier has been implemented using components described in previous section. Input voltage is $400V_{p\text{-}p,\ rms}$ and output voltage is variable from 80V to 342V. Figure 6(a) shows the experimental results of R phase voltage and current. Experimental results clearly verify the simulation results. Nearly unity power factor is achieved. Figure 6(b) shows the experimental output voltage and current. It can be seen that ripple content is very low in output voltage and current. Figure 7 shows the SW_{R1} pulses. It can be seen that there are no overlapping instants in the SW_{R1} pulses resulting in low switching losses. Figure 8 shows the experimental voltages across D_w.

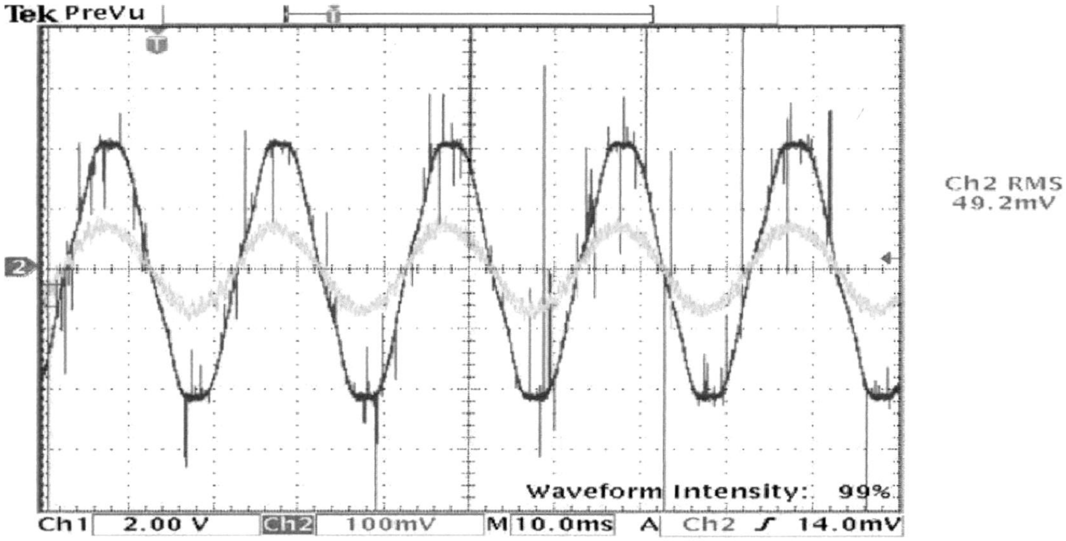

Figure 6(a). $V_{R,rms}$ in ch1 and $I_{R,rms}$ in ch2.

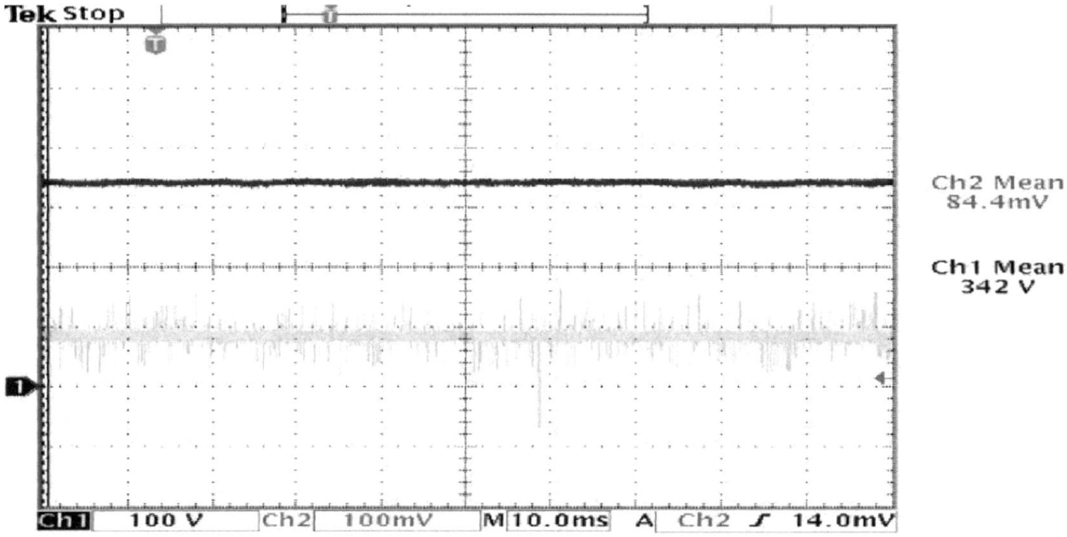

Figure 6(b). V_{dc} in ch1 and I_{dc} in ch2.

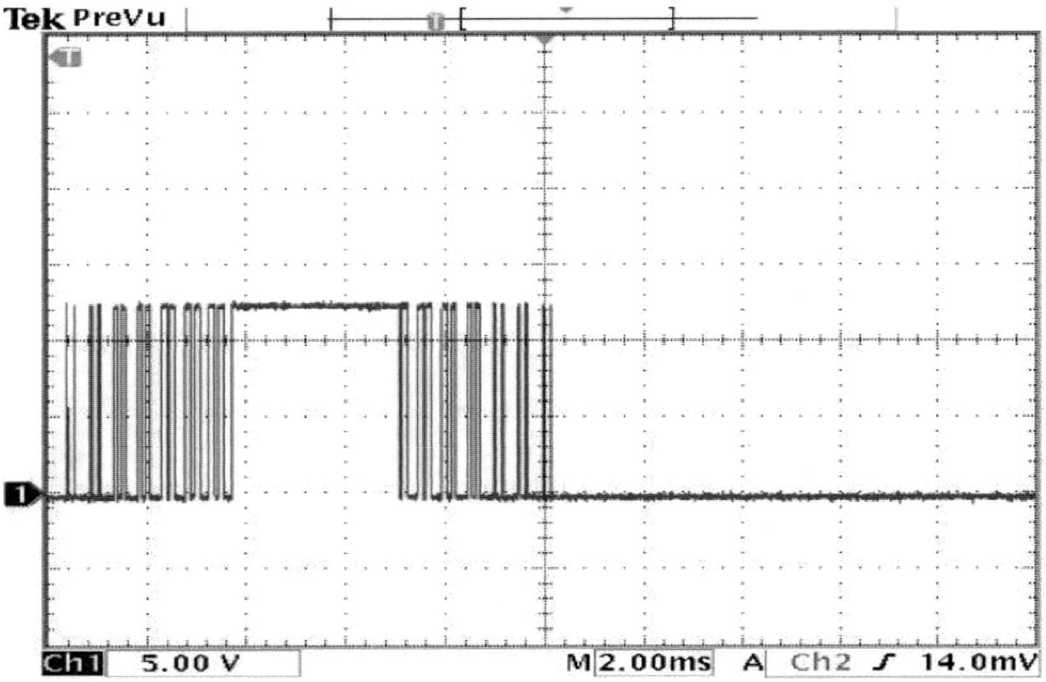

Figure 7. SW_{R1} pulses.

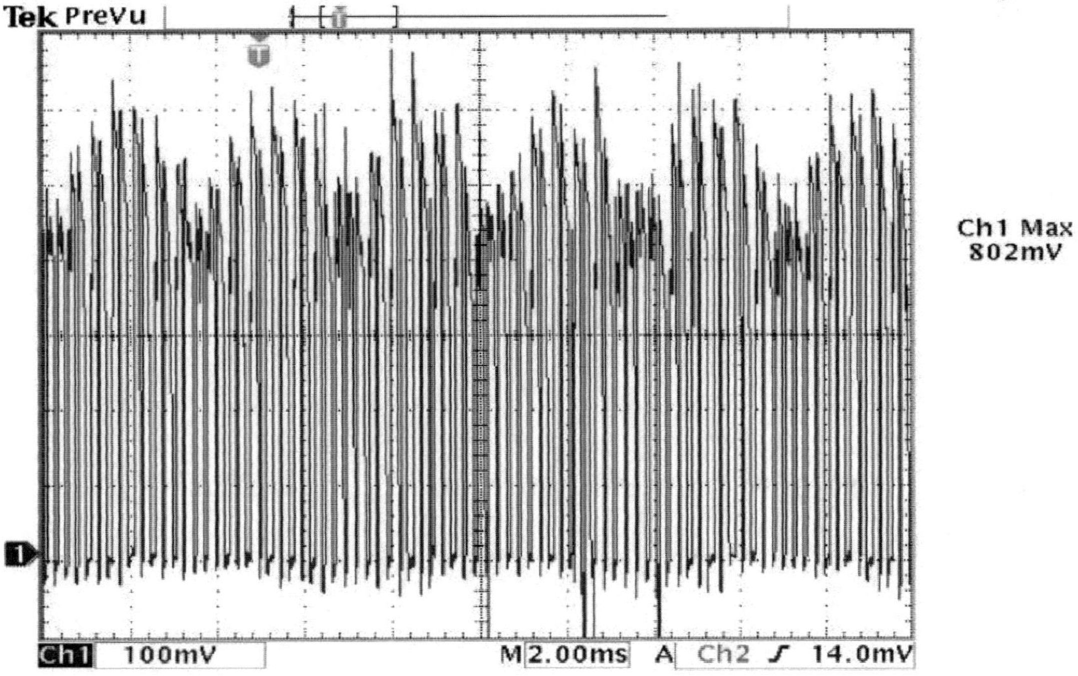

Figure 8. V_{DW}.

3. Conclusions

It has been shown that three phase six-switch PWM buck rectifier is capable of drawing input current with low THD and nearly unity displacement factor using the modified SPWM. Switching losses may be reduced up-to 30% using the suggested SPWM scheme. Output voltage is variable over a wide range. Energy storing components are small in size compared to PCR due to high switching frequency. It is highly suitable for replacing conventional PCR at medium power level.

References

[1] Conde-Enriques J, Benites-Read J S, Duran-Gomez J L and Pachecho-Sotelo J O 1999 *IEE Proceedings-Electric Power Applications* Vol. **146** issue (6) 637

[2] Ming-Tsung, Tsai and Tsai W I 1999 Analysis *IEEE Transactions on Industrial Electronics* Vol. **46** issue (3) 535

[3] Bin Wu High-power converters and Ac drives (John Wiley & Sons, Inc., Hoboken, New Jersey, 2006) p.219

[4] Navid R Zargari, Geza Joos, Phoivos D Ziogas 1994 *IEEE Transactions on Industrial Electronics* Vol. **30** issue (6) 235

Journal of Physics: Conference Series **439** (2013) 012029 doi:10.1088/1742-6596/439/1/012029

Bulk etching of silicon wafer and development of a polyimide membrane

M Aslam
Head, Electronic Materials and Processing Division
Pakistan Vacuum Society, P.O. Box 1880 Islamabad, Pakistan

E-mail; m_aslammalik@yahoo.com

Abstract. Progress in the field of engineering is always dependent on new materials. For quite some time, smart and state of the art systems are seen demanding from a modern point of view. The gas sensing devices are turned into an array of sensors than a single sensor to materialize the concept of electronic nose. The metal oxide gas sensors are actively sensitive to gas species at an elevated temperature of more than 250°C. To achieve this minimum temperature requirement, micro-heaters are developed on a certain substrate which ensures the utilization of heat energy for sensing layer of metal oxide such as tin oxide (SnO_2). In this paper, a substrate having good thermal insulation has been discussed.

1. Introduction

One of the main objectives of this research work is to produce a low heat capacity, rugged membrane support as a thin-film gas sensor substrate. An obvious candidate that is both rugged and will withstand the high temperature needed to sensitise the sensor layer is polyimide (PI) membrane. To realise this type of membrane, the supporting structure is a silicon wafer, which is bulk etched from the area of interest for membrane fabrication. A general membrane structure is as shown in figure 1.

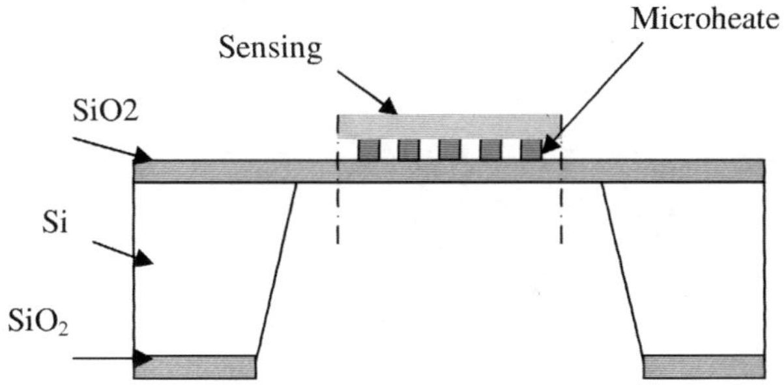

Figure 1. Basic bulk etched membrane.

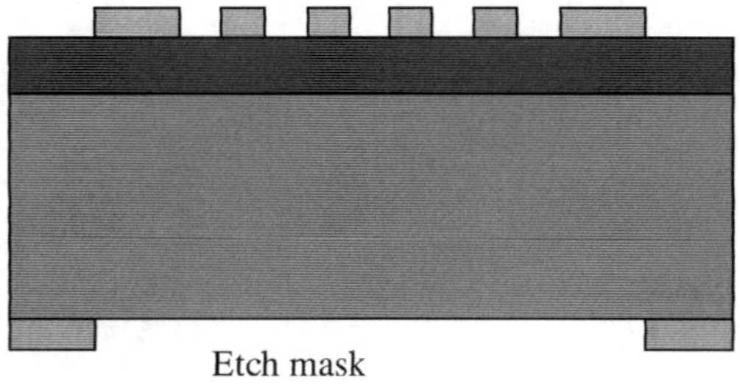

Figure 2. Photolithography of Cr/Au.

Silicon is an attractive material for micromachining for a number of reasons such as: it has useful, well-recognised mechanical properties and well defined processing techniques in the semiconductor industry[1, 2]. Membranes on silicon supporting structures are mostly insulating films based on oxides or nitrides (such as SiO_2 or Si_3N_4) but these membranes tend to be fragile and prone to breakage.

Several other materials apart from silicon can be used for infrastructure support such as Gallium Arsenide (GaAs) and crystalline quartz. Shearwood *et al* [3] and Petrini *et al* [4] have reported PI membrane on a GaAs substrate. According to Shearwood *et al,* a single layer of Probimide 7020 polyimide was spin coated onto the GaAs wafer to achieve a 7 micron thick membrane. The GaAs was then isotropically etched in NH_3: H_2O_2 mixture. This membrane was used to realise a micro speaker for a hearing aid application.

Petrini *et al* manufactured a pressure sensing structure on a PI membrane using PI2555 polyimide. The fabrication approach of Petrini *et al* for instance, starts with a GaAs wafer which is coated with polyimide at a spin speed of 6000 rpm for 60 seconds. After spinning, the polyimide was cured at 300 °C for 30 minutes. The membrane thickness was 2.2 microns. The schematic diagram of the salient features of the manufacturing process is as shown in figure 2. A double alignment was used to align the backside to the upper structure of the wafer for etching. The membrane realisation on the silicon wafer was carried out using a 525 micron thick Si <100> wafer. It was spin coated using PI2575 polyimide at 1000 rpm for about 5 seconds and then at 2000 rpm for 30 seconds to achieve a thickness of 5 microns. After baking at 90 °C for 5 minutes, a subsequent polyimide layer was spin coated again at 2000 rpm for 30 seconds to achieve a thickness of 10 microns. The polyimide was cured at 360 °C for 30 minutes and then a photoresist mask was applied and patterned on the backside of the silicon wafer and etched in silicon etchant, generally called CP4 (3HF: $5HNO_3$: $3CH_3COOH$). The etch rate was about 25-30 microns/ minute. When the sample was immersed in this etchant, it was observed after 10 minutes that the PI coating had come off the surface of silicon wafer.

Next, a CP4 etchant with different composition (1HF: $6HNO_3$: $1CH_3COOH$) was used. The etch rate of this etchant was about 5 microns/ minute. A new sample was immersed in it and the same effect was observed after about 90 minutes. The PI membrane peeled off and the etchant attacked the wafer as shown in figure 3.

It was observed that the PI masks will not withstand the long etch times needed for bulk etching. The polyimide etch mask may be good for GaAs, as claimed by Shearwood and Petrini, but cannot be used for silicon. For this reason another approach was adopted, using tetra methyl ammonium hydroxide (TMAH) anisotropic etchant with silicon dioxide as an etch mask. The etching results were favourable but the membrane structure was very fragile and few membranes were intact after completion of the etching process. The author then developed another mask of Cr/Au on silicon wafer and bulk etched the wafer as shown in figure 5.

Figure 3. Wafer etching in CP4 with PI membrane for longer time resulted in peel off the membrane.

2. Etching of silicon wafer

The wet etching of silicon plays an important role in various techniques in sensor fabrication[5]. Silicon micromachining falls into two categories; surface micromachining[1, 6] and bulk micromachining [1, 7, 8]. The purpose of bulk micromachining is to selectively remove a significant amount of silicon from a wafer in the fabrication of micromachined sensors and actuators, whereas surface micromachining is involved with the wafer surface only and usually involves sacrificial layers. Etching techniques that can be applied to the micromachining of silicon are isotropic wet etching and anisotropic wet etching, (although dry reactive ion, RI, etching is possible). In the case of isotropic wet etching, the most common etchant for silicon is 'HNA', a mixture of Hydrofluoric acid (HF), Nitric acid (HNO_3) and Acetic acid (CH_3COOH) [9]. The overall reaction is:

$$Si + HNO_3 + 6HF \longrightarrow H_2SiF_6 + HNO_2 + H_2O + H_2 \qquad (1)$$

The HNO_3 drives the oxidation of the silicon, while fluoride ions from HF form the soluble silicon compound, H_2SiF_6. The acetic acid has a lower dielectric constant (6.15 for CH_3COOH versus 81 for H_2O) than water, which helps to prevent the dissociation of HNO_3 into NO^{-}_{3} or NO^{-}_{2}, and thereby yields a higher oxidation power for the etch [10]. The acetic acid is also less polar than water and can help in achieving proper wetting of the slightly hydrophobic Si wafers. The etching chemistry is complex and the etch rate depends on chemical mixture and silicon doping.

A useful formulation for HNA is 1:3:8 (1HF: $3HNO_3$: $8CH_3COOH$) concentration [11]. The author used this etchant at the final stage of membrane fabrication to remove the final layer of silicon beneath the polyimide membrane. This etchant was used because it does not attack the polyimide membrane at this strength over a short time period.

The anisotropic wet etching of single crystal silicon in alkaline solutions has become a key technology for micromachining. The definition of anisotropic etching is that the etch rate of silicon is highly dependent on crystal orientation. This means that the anisotropic etchant works much faster in one direction i.e., Si<100> than in another i.e., Si<111>, exposing the slower etching crystal planes over time. According to the crystalline orientation chosen, the silicon <111> planes are generally either at an angle of 54.74° to the wafer surface for silicon<100>, or perpendicular to it for silicon <110>[10].

A number of etchants, including potassium hydroxide (KOH) [6, 12-14], Ethylene-Diamine (EDP)[15, 16] and Tetra methyl ammonium hydroxide (TMAH) [17] are useful for anisotropic etching of bulk silicon. The author used TMAH to bulk etch the silicon wafer.

2.1 Silicon structure and TMAH Etching

In the anisotropic or wet etching process, it was observed that the silicon <111> is least prone to etching, and the silicon<100> and silicon<110> planes are etched with a reasonable etching rate. The etching rate depends on the etchant composition and processing temperature. The lower etch rate for silicon<111> may be due to tight packing of atoms. There may also be a reason that the energy required to remove atoms from the surface is much higher in silicon<111> as compared to silicon<100> planes.

The etching of the wafer depends both on the mask pattern and the crystal structure. The sidewalls of the developing structure are inclined at 54.74° to the surface. Figure 4 shows the structure of anisotropic etching.

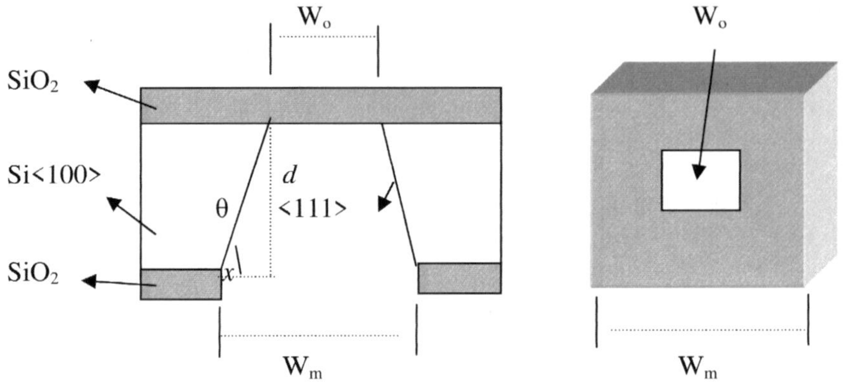

Figure 4. Etched structure of silicon<100>.

From figure 4 above,

$$Tan\ \theta = \frac{d}{x} \tag{2}$$

Then

$$x = \frac{d}{\tan\theta} \tag{3}$$

where x is the horizontal component and d is the wafer thickness. As $\theta = 54.74°$, x can be evaluated to find the dimensions for the etching window. When designing the etch masks for backside etching, the backside window dimensions should be defined according to the following expression.

$$W_0 = W_m - 2x \tag{4}$$

or

$$W_0 = W_m - \frac{2d}{\tan 54.74} \tag{5}$$

or

$$W_0 = W_m - \frac{2d}{\sqrt{2}} \tag{6}$$

or

$$W_0 = W_m - \sqrt{2} \times d \tag{7}$$

or

$$W_m = W_0 + 1.41 \times d \tag{8}$$

where d is the wafer thickness or required etch depth. Hence, the mask window should be in accordance to equation (8). The detailed experimental work carried out during this research on wet anisotropic etching of silicon wafer and membrane fabrication is described in the next section.

3. Tetra Methyl Ammonium Hydroxide (TMAH)

This is a wet etchant based on an alkaline solution. The hydroxide etchants are generally safer and more predictable relative to other types of etchants, but they usually involve alkali metals (like Li, Na, K etc), which makes them incompatible with MOS or CMOS processing. Ammonium hydroxide (NH_4OH) is one hydroxide that is free from alkali metal, but it has a problem of ammonia, which is dissolved into water and evaporates at 90 °C from solution.

The TMAH etchant is an anisotropic etchant and it relies on the fact that alkaline solutions etch at widely different rates in various crystalline structures of silicon wafer. The anisotropic property of TMAH has been investigated by many researchers [25-28]. The first extensive investigation of TMAH was presented by O. Tabata [29] and U. Schnakenberg [30] in 1991. Highly diluted TMAH has been used in the microelectronics field as a developer for positive photoresist.
The main advantage of TMAH compared with other etchants is its relatively easy handling and CMOS compatibility. In addition, the etch rate of silicon dioxide is much slower than in KOH, making SiO_2 a suitable etch mask for TMAH. There is no doubt that TMAH is a very suitable etchant for anisotropic etching of silicon <100>.

It has properties somewhere between organic alkalis like ethylene diamine pyrocatecol (EDP) and metallic hydroxides such as KOH. The most negative aspect of TMAH is its price, which is much higher than KOH. However, due to its CMOS compatibility and lower toxicity, it is becoming more popular in the microelectronics field. The fabrication process is shown in figure 5.

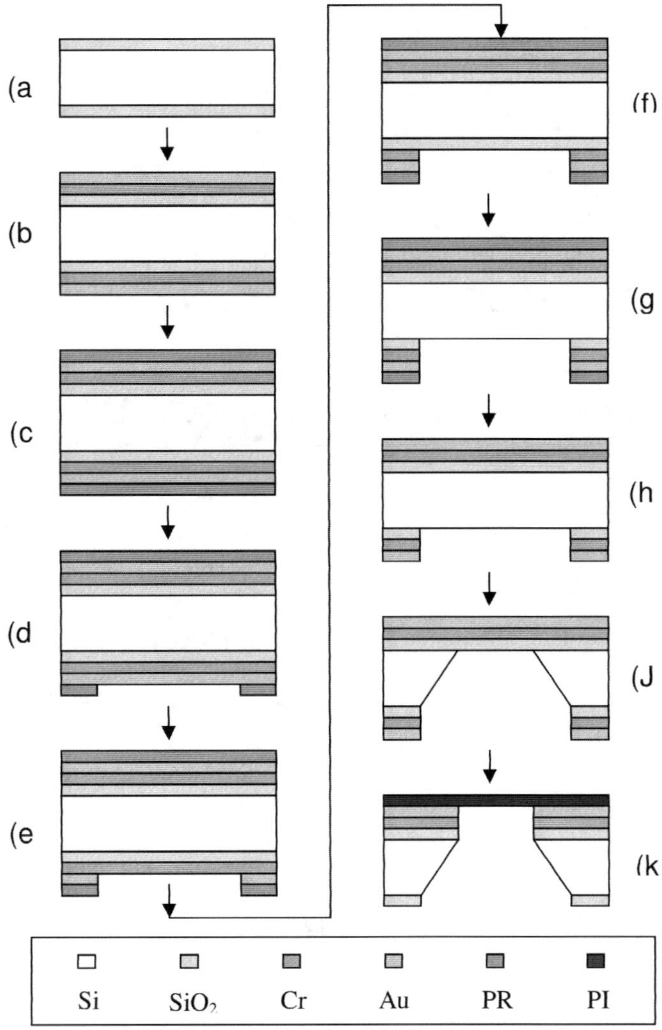

Figure 5. Process Flow Diagram (Micromachining of silicon wafer and realisation of PI membrane).

This sample was then used for bulk etching of silicon in TMAH etchant. The experimental set up is as shown in figure 6.

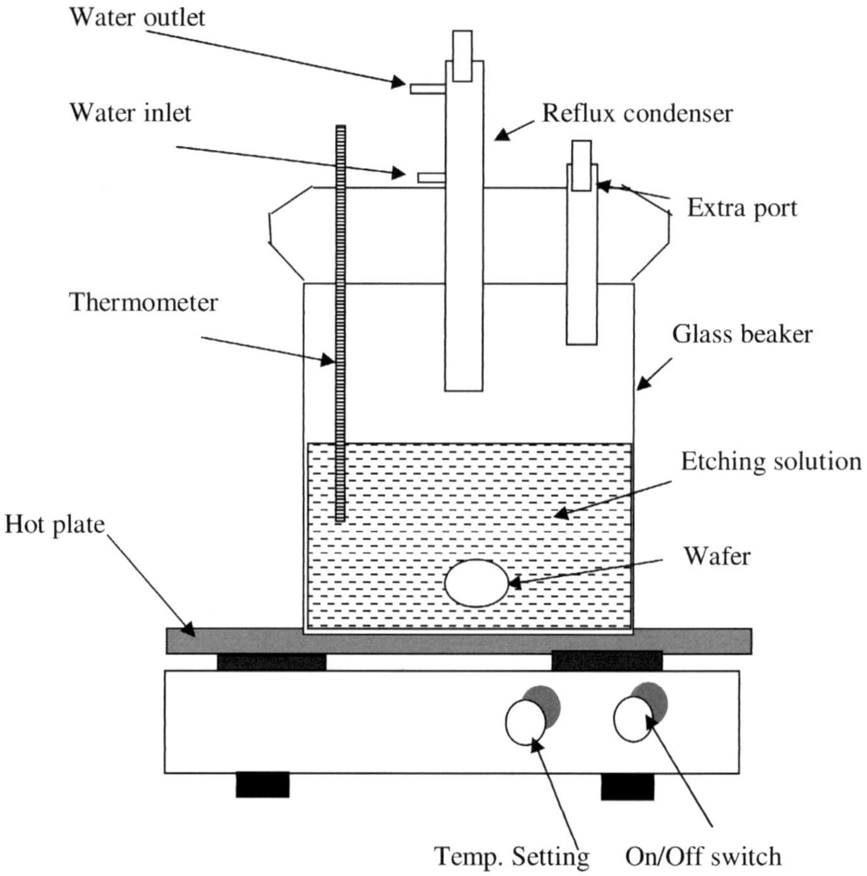

Figure 6. TMAH Etching apparatus.

The hotplate consists of an on/off and temperature control switches. The temperature of the solution can be read from a thermometer mounted inside the container. The lid of the container is equipped with a reflux condenser, which maintains the solution concentration during the long etching process. The etching rate of TMAH depends upon temperature and chemical concentration. The etch rate of TMAH at 90 °C was calculated as about 1 micron/minute. Two types of silicon wafers of different thicknesses (525 and 250 microns) were used for etching. To etch 525 microns thick wafer, it took about 8 hours and for 250 microns wafer, it was approximately four hours.

4. Polyimide membrane fabrication

Polyimides are an interesting group of strong, heat and chemical resistant polymers. Their strength, heat and chemical resistance is such that these materials are replacing metals and glass in many demanding industrial applications. The author has used polyimide as a substrate for a metal oxide gas sensor array due to its good thermal stability, resistance to chemicals like grease, oil and acids, and its physical strength. HDMicroSystems offers a complete product line of polyimide coatings for microelectronics, automotive and aerospace industry [35]. These coatings are formulated as a polyimide precursor in an

NMP- solvent based carrier. The curing process of polyimide precursor into a polyimide film is called imidization. An imide is a group in a molecule that has a general structure as shown in figure 7.

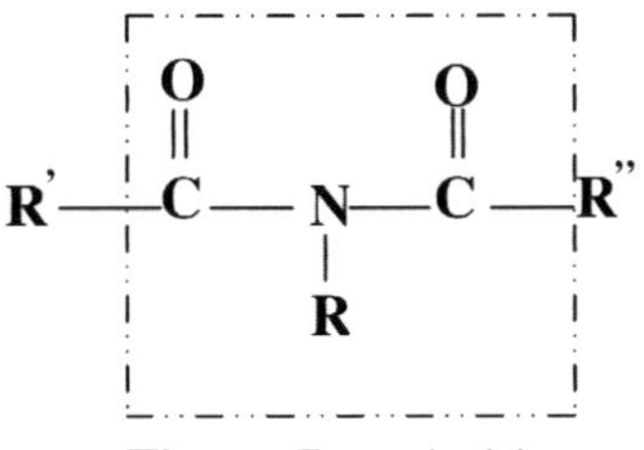

Figure 7. An imide.

When the molecule shown in figure 7 is polymerised, it becomes polyimide. Polyimides usually take one of the two forms: the first of these is a linear structure (figure 8a) where the atoms of the imide group are

Aromatic heterocyclic polyimide

(a) (b)

Figure 8. Chemical structures of polyimides.

part of a linear chain and the second one is a heterocyclic structure where the imide group is part of a cyclic unit in a polymer chain (figure 8b) [36, 37]. Aromatic heterocyclic polyimides, shown in figure 8, are typical of the polyimides from HD MicroSystems . The detailed chemistry of polyimide materials can be found in the literature [36, 38-40]. A polymer which contains a charge transfer complex consists of two different types of monomers, a donor and an acceptor.

The donor has more electrons than the acceptor, so it gives some of its electrons to the acceptor, holding them tightly together. Due to this charge transfer, the chains are stacked together with donors and acceptors very tightly, not allowing them to move around very much, thereby making polyimide so strong. The charge transfer not only occurs with the adjacent units in the polymer chain but also between the chains, making it stronger. Figure 9 shows an aromatic polyimides structure

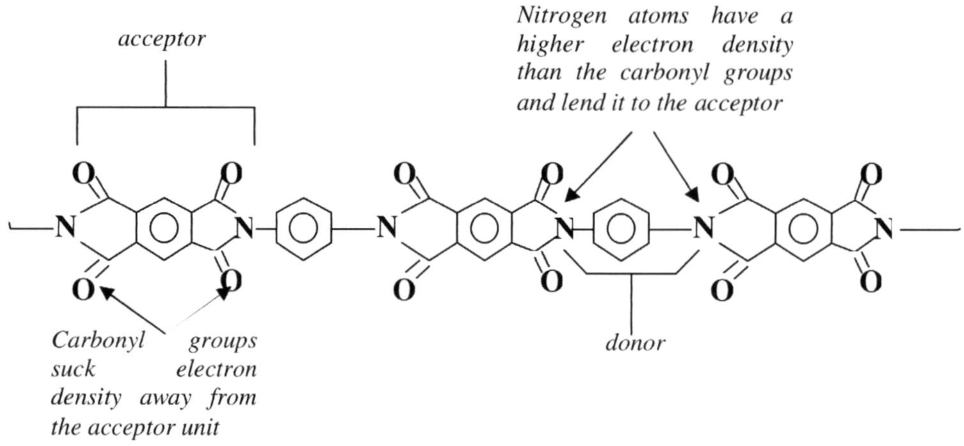

Figure 9. Aromatic heterocyclic polyimides structure[36]

4.1. Experimental Work (PI Membrane)

After etching and suitable washing in DI water, the sample covered with Cr/Au on one side and etched windows on the other side, is ready for realisation of PI membrane. The process flow diagram for the membrane fabrication is as shown in figure 10. The polyimide PI2575 from HD Microsystems is supplied in a form suitable for spin coating applications. The material is a self-priming and does not require any adhesion promoter. The solution is highly viscous and it is necessary to have a short delay prior to spin to allow the polyimide to flow as far as possible and relax. This delay in spin avoids bubbles or comets in coating. The spin speed and time is determined by the film thickness required. It is observed that longer spin time improves the coating uniformity, but reduces the film thickness. The spin speed curve for PI2575 polyimide is as shown in figure 11.

- The polyimide was dispensed on the static etched-back substrate.
- The spinner started at 1000 rpm for 5 seconds. It was then accelerated to the spin speed to 2000 rpm for 30 seconds.
- The wafer was cleaned of the leached polyimide on the backside of the wafer with NMP (N-methyl-pyrrolidone) solution.
- The coated wafer was then pre-baked in oven at 90 ºC for 5 minutes.
- The wafer was again spin coated with polyimide at 2000 rpm for 30 seconds. This step was repeated to increase the thickness of polyimide layer to approximately 10 microns.
- Now, the sample was baked at 120 ºC for 30 minutes before final thermal curing process.

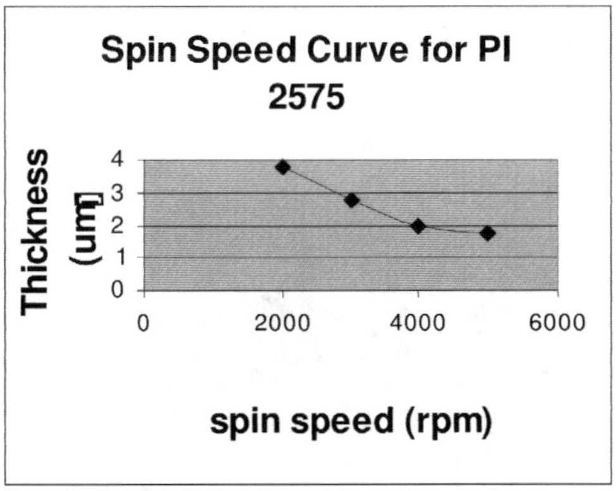

Figure 11. Spin-Speed Curve for PI 2575.

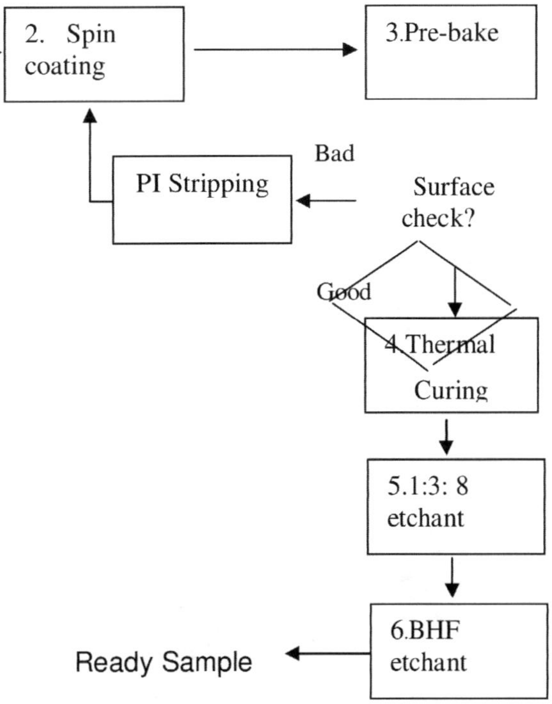

Figure 10. Polyimide Process Sequence.

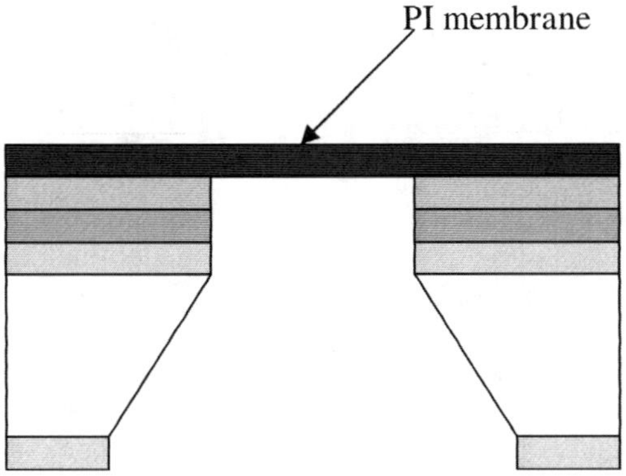

Figure 12. Final membrane structure.

Figure 13. Polyimide membrane on bulk etched Si<100> (each membrane measures 3.5 x 3.5 mm).

After application of the polyimide, a high temperature thermal curing process is required which converts the polyamic acid to the insoluble imide form and drives out the remaining solvent. The curing process requires elevated temperatures and a controlled environment to achieve the best results. A temperature-controlled hotplate was used for curing. According to the curing profile, the sample was heated from room temperature to 360°C, with a ramp rate of 10°C/minute in nitrogen atmosphere. The sample was kept at this temperature for 30 minutes and then cooled down from 360°C to room temperature, with the ramp rate of 10°C/minute in the same environment. The colour of cured polyimide became dark brown. The sample with the cured polyimide membrane was now further processed in a 1: 3: 8 etchant (1HF(49%) : $3HNO_3$ (70%) : $8CH_3COOH$) until the remaining silicon layer was etched away. The sample was then immersed in buffered HF solution to etch away the silicon dioxide layer for approximately 7.5 minutes. With this etching, the chromium film was exposed through the selective windows. The polyimide membrane was released by etching the exposed chromium and gold layers with their respective etchants. The samples were rinsed in de- ionised water and dried with slow blow dry nitrogen. The final membrane structure is as shown in figure 12. The samples with the polyimide membrane on silicon were then used to develop the micro-heater and a sensing array of four sensor elements. The samples with polyimide membranes are shown in figure 13.

4.2. SEM and AFM images

Scanning electron microscopy (SEM) and Atomic Force Microscopy (AFM) are the techniques used to study the material morphology and surface flatness. The micromachined silicon samples with PI membrane were studied using SEM images as shown in figure 14 and figure 15.

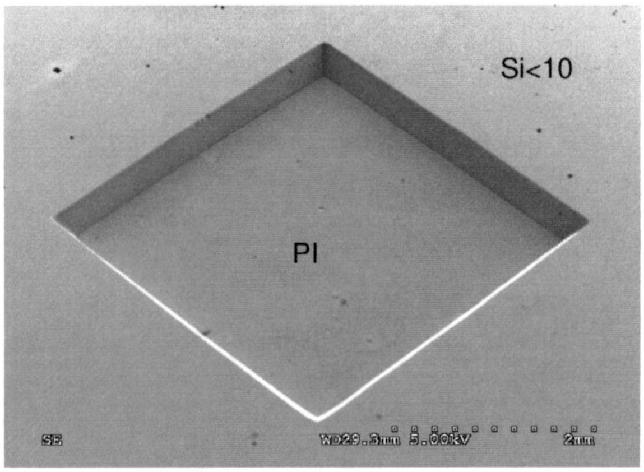

Figure 14. SEM view of bulk etched Si<100> windows *x*25mag.

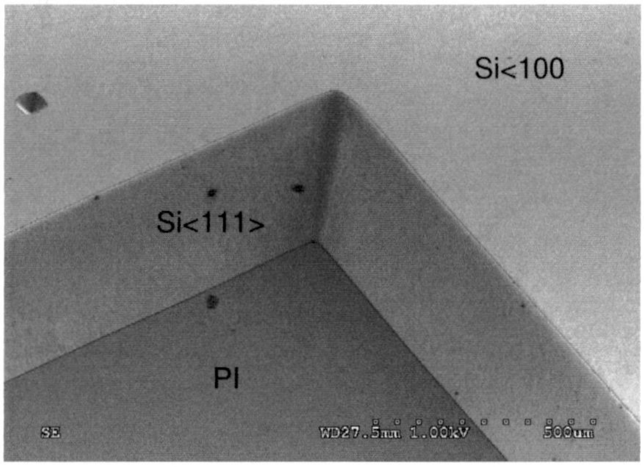

Figure 15. SEM cross-section of bulk etched cell *x*100mag.

From the SEM images, it is apparent that the etching is very smooth and very few pits can be seen on the surface of silicon wafer. The SEM cross-section shown in figure 15 is 100 times magnified image of bulk etched sample and it shows a smooth definition of etching profile and polyimide surface.

Surface irregularities were studied by using the atomic force microscopic (AFM) images. The AFM data and 3D image of the membrane surface are shown in figure 16 and figure 17 respectively.

The PI surface was scanned at three different points with a 10nm diameter stylus. The membrane surface flatness was 15nm for data line1 whereas figure 16 shows peaks of 28nm and 120nm respectively at a certain point. This was probably due to debris or particulate contamination of the surface.

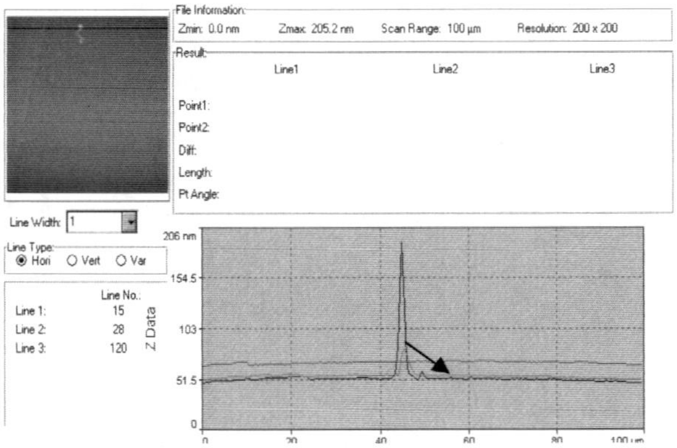

Figure 16. AFM data of PI membrane.

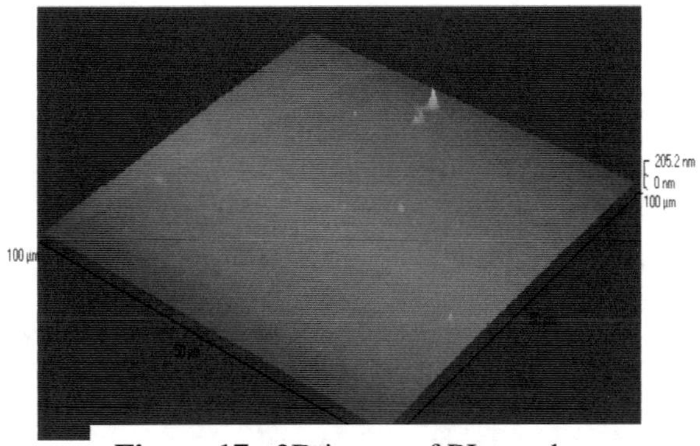

Figure 17. 3D image of PI membrane.

5. Conclusion

In this paper, a detailed description of the fabrication of polyimide membranes is given. The paper is actually in two parts; one describing the micromachining of the silicon wafer to use it as frame for the membrane and second describing the fabrication of the polyimide membrane on top of this frame. The etching was carried out using TMAH solution at 90 ºC for 4 hours for 250 micron thick silicon wafer.

The measurement of the surface flatness of the membrane was also carried out using SEM and AFM techniques. Towards the end of this chapter, a detailed description of some problems encountered, especially during the wet etching process, is discussed. The membranes fabricated by the technique described in this chapter have proved to be more robust and rugged than silicon dioxide or silicon nitride and reproducibility and repeatability is good.

References

[1] Sze S M, 1994 *Semiconductor Sensors* (John Wiley & Sons)

[2] Ristic L 1994 *Sensor Technology and Devices* (Artech House Boston-London) 65

[3] Shearwood, Birch T S and Stevens J C 1996 *Microelectronics Engineering* **30** 547

[4] Ioana Petrini, Avramesu V, Simion G, Nitescu N, Vasilache D, Dascalu D, Knnstantinidis G and Giacomozzi F 2000 *Journal of Micormechanical, Micorengineering* **10** 218

[5] Gregory T. A. 1998 *Proceedings of the IEEE* **86** (8)

[6] Seidal H, Heugerger A and Baumgartel H 1990 *Journal of electromechanical Society* **137** 3621

[7] Shikida M *et al.* 2000 *Sensors and Actuators* **A 80**(2) 179

[8] Rodica Iosub, 2002 *Sensors & Actuators* **A** (3277) 1

[9] Bogenchutz A F, Lochrer K H and Modreanu W 1967 *Journal of Electrochemical Society: Solid State* **114** (9) 970

[10] Madou M J 2002 *Fundamentals of Microfabrication: The Science of Miniaturization*. 2nd ed (Florida 33431: CRC Press LLC 2000 N W Corporate Blvd, Boca Raton)

[11] Peterson K E 1982 *Proceedings of IEEE* **70** 420

[12] Zubel I 1998 *Sensors & Actuators* **A 70** 260

[13] Palik E D, Heard Jr I, Bruno P S and Tenerz L 1991 *Journal of Applied Physics* **70** 3291

[14] Quoc-Bao Vu 1996 *Journal of Electrochemical Society* **143** 1372

[15] Klein, 1967 *Journal Electrochemical Society* **114** 956

[16] Reisman A, M B, Chan S A, Kaujman F B and Green D C 1979 *Journal of Electrochemical Society* **126** 1406

[17] Tabata O *et al.* 1992 *Sensors and Actuators* **A**: *Physical* **34** (1) 51
[18] Elwenspoek M 1993 *Journal of Electrochemical Society* **140** 2075
[19] Elwenspoek M 1996 *Journal of Micromechanics and Microengineering* **6** 405
[20] Hesketh P J, Gowda S, Zanoria E and Danyluk S 1993 *Journal of electrochemical Society* **140** 1080
[21] Allongue P 1993 *Journal of The Electrochemical Society* **140** 1009
[22] Glembocki O. J, 1985 *Journal of the Electrochemical Society* **132** 145
[23] Palik E D, 1983 *Jouranl of Electrochemical Society* **130** 956
[24] Babic D, 1998 *Journal of Electrochemical Society* **145** 4040
[25] Bassous E 1978 *IEEE Transacations on Electron Devices* **ED-25** (10) 1178
[26] Kim C, http://www.ee.eng.hawaii.edu/mems/Han/mnc98_final.htm.
[27] Merlos A *et al.* 1993 *Sensors and Actuators* **A**: Physical 37-38: 737
[28] Yan G *et al.* 2001 *Sensors and Actuators* **A**: Physical **89** (1-2) 135
[29] Tabata O, Funbashi H and Sugiyama S 1991 *Technical digest of* 6th *conference on Solid State Sensors and Actuators, Transducer's* **91** 811
[30] Schnakenberg U, Lochel B, Ullerich S and Lange P 1992 *Sensors & Actuators* **A** 25-27: 1
[31] Kloeck B, de Rooj N F and Smith R I 1998 *IEEE Electron Devices* **36** 663-669
[32] French, Nagao M, and Esashi M 1996 Sensors and Actuators A: Physical **56** (3) 279
[33] Ashruf C M A, Sarro P M, Kazinczi R, Xia X H and Kelly J J 2000 *Journal of Micromechanical Micorengineering* **10** 505
[34] Ashruf C M A *et al.* 1998 *Sensors and Actuators* **A**: Physical **66** (1-3): 284
[35] Dupont, www.hdmicrosystems.com.
[36] Mittal, 1996 *Polyimide: Fundamentals and Applications* (New York: Marcel Dekker)
[37] Bessonov M I, Kudryavtsev V V and Laius L A 1987 *Polyimides: Threnmally Stable Polymers* (New York: Consultants Bureau)
[38] Kricheldorf H R1999 *Progress in Polyimide Chemistry* Vol. **I** (Berlin, London: Springer)
[39] Mittal K L 1984 *Polyimide: Synthesis, Characterisation and Applications* Vol. **1,2** (New York)
[40] Adrora N A 1969 *Polyimide: a new class of heat-resistant polymers* (Jerusalam: I.P.S.T)

Journal of Physics: Conference Series **439** (2013) 012030 doi:10.1088/1742-6596/439/1/012030

Electrical characterization of Au/ZnO/Si Schottky contact

M Asghar[1*], K Mahmood[1], M Faisal[1] and M A Hasan[2]
[1*]Department of Physics, The Islamia University of Bahawalpur, Pakistan
[2]Department of Electrical and Computer Engineering, UNC-Charlotte, NC USA

E-mail: mhashmi@iub.edu.pk

Abstract. In this study, temperature dependent current-voltage (I-V) and capacitance-voltage (C-V) measurements have been performed on Au/ZnO/Si Schottky barrier diode in the range 150 – 400K. The room temperature values for ideality factor and barrier height found to be 2.68 and 0.68 eV respectively. From the temperature dependence of I-V, the ideality factor was observed to decrease with increasing temperature and barrier height increased with increasing temperature. The observed barrier height trend was disagreeing with the negative temperature coefficient for semiconductor. A deep defect with activation energy 0.57 eV below the conduction band was observed using the saturation current plot and deep level transient spectroscopy.

1. Introduction

ZnO is very important material for short wavelengths electronic and optoelectronic devices due to its amazing properties of wide band gap of 3.37 eV and exciton binding energy of 60 meV at room temperature [1]. ZnO finds applications in ultraviolet light emitting devices (UVLED), photo-detectors, sensors, thin film transistors and solar cells [2-4]. Most of these ZnO based devices employed Schottky contacts to use them in practical applications. But good rectifying contacts have been difficult to achieve due to the presence interface states [5] between film and metal. Furthermore room temperature I-V and C-V measurements could not explain the transport phenomena. Therefore study on the quality of metal semiconductor junction is very important and needs to be understood thoroughly.

In this study, the transport properties of Au Schottky contacts fabricated on ZnO have been investigated with the help of temperature dependent I-V and C-V characteristics. The measured values of ideality factor and barrier height for Au/ZnO/Si diode were found to be 2.68 and 0.68 eV respectively. The saturation plot was drawn to calculate the activation energy of deep level defects with energy 0.57 eV below the conduction band. Deep Level Transient Spectroscopy (DLTS) measurements were also performed to confirm deep level defect.

2. Experimental

ZnO thin film is grown on p-type silicon wafer by MBE. The detail of growth can be seen elsewhere [6]. After growth Au metal was evaporated by e-beam evaporation to form metal contacts of diameter 1 mm.

1* To whom any correspondence should be addressed.

3. Results and discussion

Current in the Schottky barrier diode due to the thermionic emission, neglecting series resistance is given as [7],

$$I = I_S\left[\exp\left(\frac{qv}{nkT}\right) - 1\right] \tag{1}$$

Where n is the ideality factor and I_S is the reverse saturation current given by the relation,

$$I_S = AA^*T^2 \exp\left[\frac{-q\phi_B}{kT}\right] \tag{2}$$

Where A is the contact area, A^* is the Richardson constant and its value is 32 AK^{-2} cm^{-2} for ZnO [8]. T is temperature in Kelvin, k is the Boltzmann's constant ($k=1.38\times10^{-23}$ J/K or 8.617×10^{-5}eV/K), q is the electric charge and ϕ_B is the barrier height.

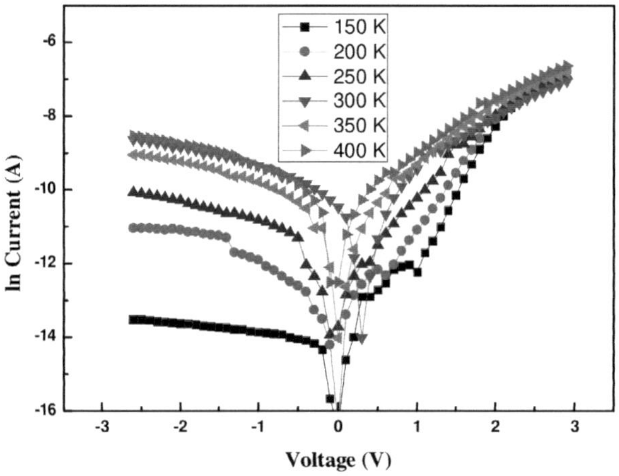

Figure 1. Current-voltage characteristics of Au/ZnO/Si Schottky diode at different temperatures.

Figure 1 shows the semilogarithemic plot of Au/ZnO Schottky contacts at room temperature. By fitting the linear part of this plot and using intercept and slope, we obtained the value of ideality factor and barrier height using following equations

$$n = \frac{q}{kT \times slope} \tag{3}$$

$$\phi_{B(I-v)} = \frac{kT}{q}\ln\left(\frac{AA^*T^2}{I_S}\right) \tag{4}$$

Where k =8.617×10^{-5}eV/K, T is the temperature in Kelvin, A is the contact area =0.0314 cm^2 and A^* is the Richardson constant having value 32 AK^{-2} cm^{-2} for ZnO [8, 9].

The measured room temperature value of ideality factor 2.68 and the corresponding barrier height value was 0.68 eV respectively. The value of the ideality factor is greater than unity. High values of n can be attributed to the presence of the interfacial thin native oxide layer at Au and ZnO interface, a wide distribution of low-Schottky barrier height (SBH) patches (or barrier inhomogeneities), to the series resistance effect and to the voltage dependence of the SBH [10-12].

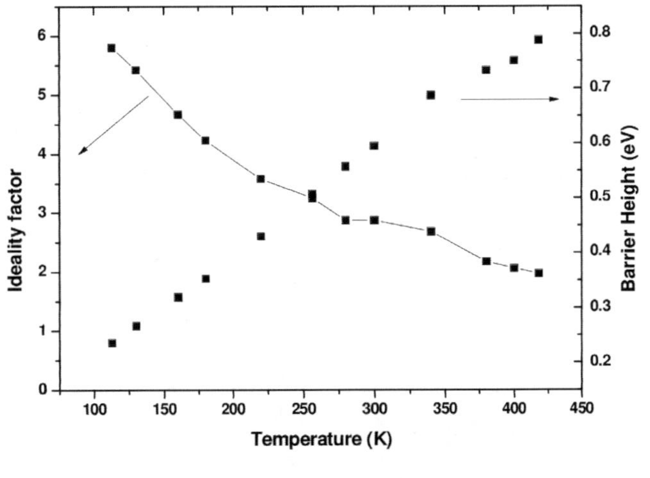

Figure 2. Graph between ideality factor verses temperature and barrier height verses temperature of Au/ZnO/Si Schottky diodes.

Figure 2 shows that ideality factor decreased with increasing temperature for both diodes. These decreasing values of n with increasing temperature, shows that the thermionic transport mechanism is strongly affected in low temperature region or trapped assisted tunneling is responsible for transport mechanism [13, 14].

The plot of temperature versus barrier height $\phi_{B(I-v)}$ at the temperature 150 to 400K is shown in figure 2. Like the ideality factor, $\phi_{B(I-v)}$ is also seemed to be a strong function of temperature. This is attributed to the fact that ZnO readily forms interfacial layer of thickness of one to two monolayer on its surface which is resulted in the existence of abundant surface states on ZnO [15].

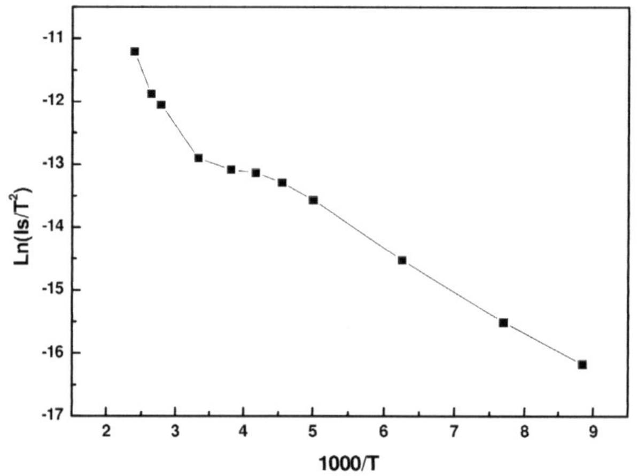

Figure 3. Richardson plot for Au ZnO/Si Scottky diode in the temperature range 150-400K.

Figure 3 show the Richardson plot for the temperature range 150-400K. The value of Richardson constant was found to be 10 $AK^{-2}cm^{-2}$. The deviation of Richardson constant from the theoretical value, 32 $AK^{-2}cm^{-2}$ is due to the barrier height inhomogenity effect. The barrier height consists of low and high barrier area at interface. Due to these potential fluctuations at the interface, the current of diode will flow preferentially through the lower barriers in the potential distribution [16].

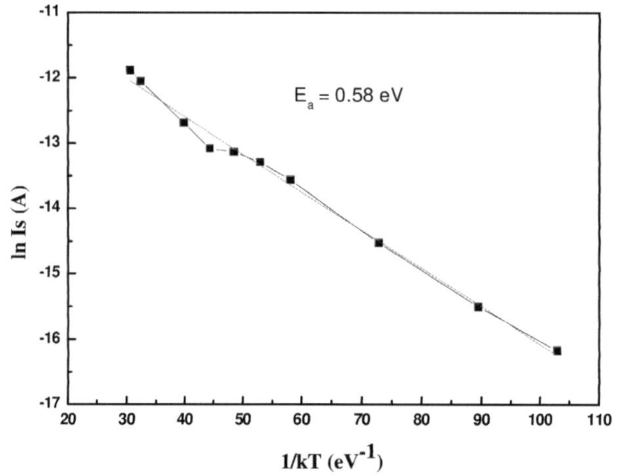

Figure 4. Saturation plot for Au/ZnO/Si Scottky diode in the temperature range 150-400K.

Figure 4 is the saturation plot in the temperature range 145 to 400K. The activation energy of deep level defect that traps the carrier and can be found by the gradient of linear fit to following equation

$$Is = I_o exp(-Ea/kT) \quad (5)$$

Where Is is saturation current measured at different temperatures, Ea is the activation energy of the defect. The activation energy was found to be 0.58eV below the conduction band. To confirm this, we have performed DLTS measurements. Fig. 5 is the DLTS spectra of Au/ZnO/Si Schottky diode. The spectra show one peak with activation energy 0.57 eV below the conduction band.

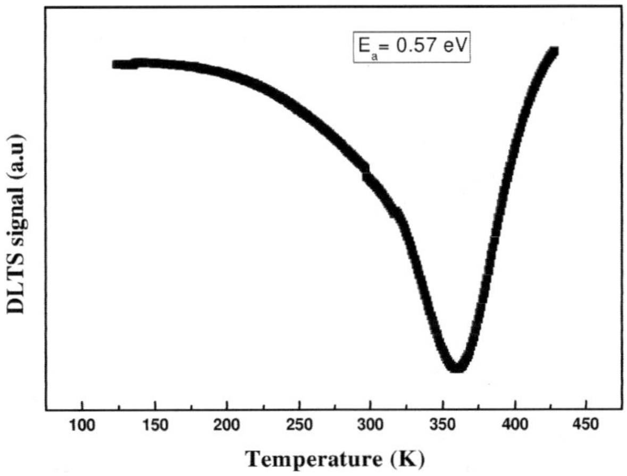

Figure 5. DLTS spectra for Au/ZnO/Si Schottky diode.

4. Conclusion

In this paper, we have investigated the transport properties of Au/ZnO/Si Schottky diode with the help of current-voltage characteristics in the temperature range of 150-400K. From the temperature dependence of I-V, the ideality factor was observed to decrease with increasing temperature and barrier height increased with increasing temperature. The observed barrier height trend was disagreeing with the negative temperature coefficient for semiconductor material that C–V barrier height decreases with temperature. A deep level defect with activation energy 0.57 eV below conduction band was observed from saturation plot and deep level transient spectroscopy.

Acknowledgement

Authors are thankful to Higher Education Commission Pakistan for financial support under project No. 624/2007. The authors are also thankful to UNC-Charlotte, USA for technical support.

References

[1] Klingshirn C, Fallert J, Zhou H, Sartor J, Thiele C, Maier-Flaig F, Schneider D and Kalt H 2010 *Phys. Status. Solidi B* **247** 1424

[2] Hadia N, Kalason P, Nur O, Wahab Q, Asghar and M, Willander M 2009 *J. Appl. Phys.* **105** 123510

[3] Qiang M, Tegush Endah S, Akihisa O and Masaaki N 2011 *Appl. Phys. Lett.* **98** 051908

[4] Samarendra P, Zi-En O, Serene Ng Lay G, Gregory K. Goh L and Ananth D 2011 *Appl. Phys. Lett.* **98** 073302

[5] Mtangi W, Auret F D, Nyamhere C, Janse van Rensburg P J, Diale M and Chawanda A 2009 *Physica B* **404** 1092

[6] Asghar M, Mahmood K, Adnan A and Hasan M A 2012 *Key Eng. Mat.* **511-512** 265

[7] Asghar M, Mahmood K, and Hasan M A 2012 *Key Eng. Mat.* **511-512** 132

[8] Wenksten H V, Kaidashev E M, Lorentz M, Hochmuth H, Biehne G, Lenzer J, Gottscalch V, Pickenhain R and Grundmann M 2004 *Appl. Phys.Lett.* **84** 79

[9] Shawn C, Bernard J R and Kristin M P 2008 *Appl. Phys. Lett.* **92** 012103

[10] Aydo gan S, Cinar K, Asıl H, Coskun C and Türüt A 2009 *J. Alloy. Comp.* **476** 913

[11] Coskun C, Gedik N and Balc E 2006 *Semicond. Sci. Technol.* **21** 1656

[12] Dhananjaya, Nagarajua J and Krupanidhi S B 2007 *Physics B* **391** 344

[13] Padovani F A and Summer G 1965 *Appl. Phys. A* **36** 3744

[14] Altindal S, Karadeniz S, Tugluogiu N and Tataroglu A 2003 *Solid-state Electronics* **471847** 1854

[15] Copa B J, Pulton C C, Hortlieb P J, Shields B J, Nemanich R J and Davis R F 2004 *J. Appl. Phys.* **92** 5886

[16] Dokmi. I, Altindal. S and Bulbul. M.M 2006 *Appl. Surf. Sci.* **252** 7749

[17] Schroder D K 2006 *semiconductor material and device characterization* John Willy &sons Inc, Hosten New jersey Canada

Growth and interface properties of Au Schottky contact on ZnO grown by molecular beam epitaxy

M Asghar[1, 3], K Mahmood[1], Faisal Malik[1] and M A Hasan[2]
[1]Department of Physics, The Islamia University of Bahawalpur, 63100, Pakistan
[2]Department of Electrical and Computer Engineering, UNC-Charlotte, NC 28223, USA

E-mail: mhashmi@iub.edu.pk

Abstract. In this paper, we have discussed the growth of ZnO by molecular beam epitaxy (MBE) and interface properties of Au Schottky contacts on grown sample. After the verification of structure and surface properties by X-Ray Diffraction (XRD) and Scanning Electron Microscope (SEM), respectively, Au metal contact was fabricated by e-beam evaporation to study contact properties. The high value of ideality factor (2.15) and barrier height (0.61 eV) at room temperature obtained by current-voltage (I-V) characteristics suggested the presence of interface states between metal and semiconductor. To confirm this observation we carried out frequency dependent capacitance-voltage (C-V) and conductance-voltage (G-V) demonstrated that the capacitance of diode decreased with increasing frequency. The reason of this behavior is related with density of interface states, series resistance and image force lowering. The C^{-2}-V plot drawn to calculate the carrier concentration and barrier height with values 1.4×10^{16} cm^{-3} and 0.92 eV respectively. Again, high value of barrier height obtained from C-V as compared to the value obtained from I-V measurements revealed the presence of interface states. The density of these interface states (D_{it}) was calculated by well known Hill-Coleman method. The calculated value of D_{it} at 1 MHz frequency was 2×10^{12} eV^{-1} cm^{-2}. The plot between interface states and frequency was also drawn which demonstrated that density of interface states had inverse proportion with measuring frequency.

1. Introduction

ZnO is a material of current research due to its superior optoelectronic and nano-electronic properties which makes it ideal candidate for short wavelength ultraviolet light emitting diodes and photo-detectors [1-2]. To realize these devices for particle applications, high quality Ohmic and Schottky metal contacts are indispensible. But good Schottky contacts are difficult to fabricate due to the presence of interface states between film and substrate which can lead to a large deviation from the ideal behavior. Furthermore different metal such as Al, Au, Ag, Pt and

[3] To whom any correspondence should be addressed.

Pd had been used by researchers to achieve reliable Schottky contacts. These Schottky contacts have barrier heights in the range of 0.6 eV to 0.8 eV indicating the non-negligible impact of interface defect states [3-5]. Moreover these interface states strongly depend upon the defect density of grown film and consequently on growth technique used [6]. Therefore the characterization of these interfaces states between semiconductor and metal contact is highly desirable.

In this paper, we have investigated the effect of interface states on Au Schottky diode contact fabricated on MBE grown ZnO. I-V measurements yielded ideality factor and barrier height to be 2.15 and 0.6 eV, respectively. C-V and G-V measurements performed for a wide frequency range (10 kHz – 2 MHz) at room temperature. The lower value of capacitance at higher frequencies and higher value of barrier height obtained from data C-V as compared to the one from I-V, confirmed the presence of interface states. X-ray diffraction and SEM measurements additionally performed to access the structural and surface properties of grown sample.

2. Experimental

ZnO thin film is grown on p-type silicon wafer by MBE. The detail of growth can be seen elsewhere [7]. After growth Au metal was evaporated by e-beam evaporation to form metal contacts of diameter 1 mm.

3. Results and discussion

Figure 1 is typical XRD pattern of grown ZnO film. It shows three peaks at 2θ angles 34°, 36° and 72° corresponding to ZnO (002), (101) and (004) planes, respectively. The full width half maximum of ZnO (002) peak is 0.3°. This small value of FWHM demonstrates the good quality of film.

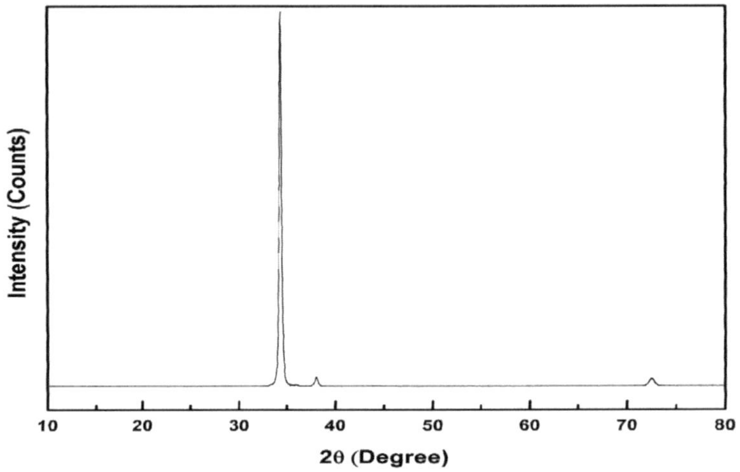

Figure 1. XRD pattern of as-grown ZnO thin film on silicon substrate indicating that preferred growth orientation is along C-axis.

The thermionic emission theory for a bias voltage, V, is valid for the analysis of electrical characteristics of the Schottky barrier structures of Au/ ZnO/p-Si, and hence junction current can be expressed by [8]

$$I = I_s\left[\exp\left(\frac{qv}{nkT}\right) - 1\right] \tag{1}$$

where n is the ideality factor and I_S is the reverse saturation current given by,

$$I_s = AA^*T^2\exp\left[\frac{-q\phi}{kT}\right] \tag{2}$$

where A is diode area, A^* is Richardson's constant, q is charge, Φ_{ap} is apparent barrier height, k is Boltzmann's constant and T is temperature.

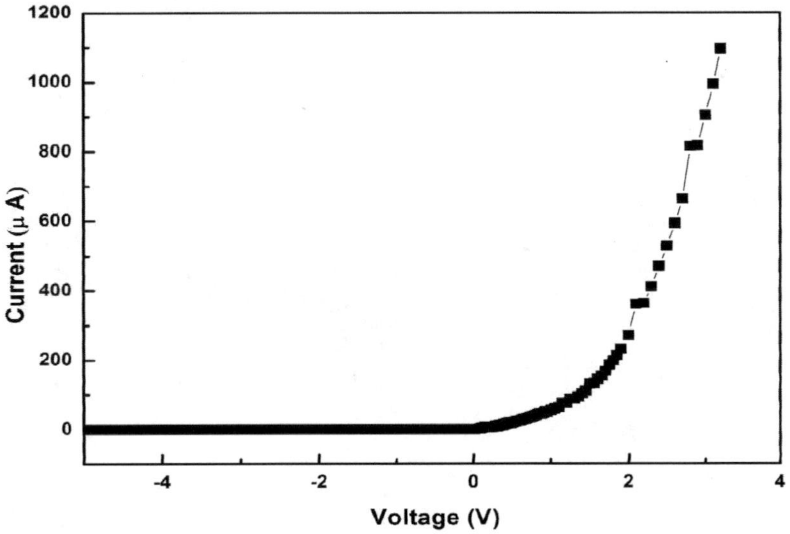

Figure 2. Current-Voltage characteristics of Ni/ZnO/Si Schottky diode at room temperature.

The ideality factor (n) of Schottky barrier is an important parameter that explains the perfection of metal-semiconductor junction. It is determined from the slope of semilog I-V (not shown here) characteristics in the exponential region by the relation;

$$n = \frac{q}{kT \times slope} \tag{3}$$

where q is the charge (C), k is Boltzmann's constant (J/K), T is Temperature (K) [9]. The measured value of ideality factor at room temperature for Au/ZnO Schottky diode is 2.15. The high values of n can be attributed to effects of the bias voltage drop across the interfacial layer and series resistance.

The barrier height is an electrostatic barrier to the charge transfer across the metal/semiconductor interface. The $\emptyset_B$ can be written as

$$\phi_B = \frac{kt}{q \ln\left(\frac{AA^*T^2}{Is}\right)} \quad (4)$$

where k is Boltzmann's constant (J/K), T is temperature (K), q is charge of electron (C), A is Schottky contact area (cm^2), A* is Richardson constant (J/cm^{2}K) and I_s is reverse saturation current (A). The barrier height can be determined by the intercept of the linear part of the forward bias voltage at exponential (region not shown here). The measured value of barrier height for Au/ZnO/Si Schottky contact is 0.6 eV and it is well agreed with the reported value.

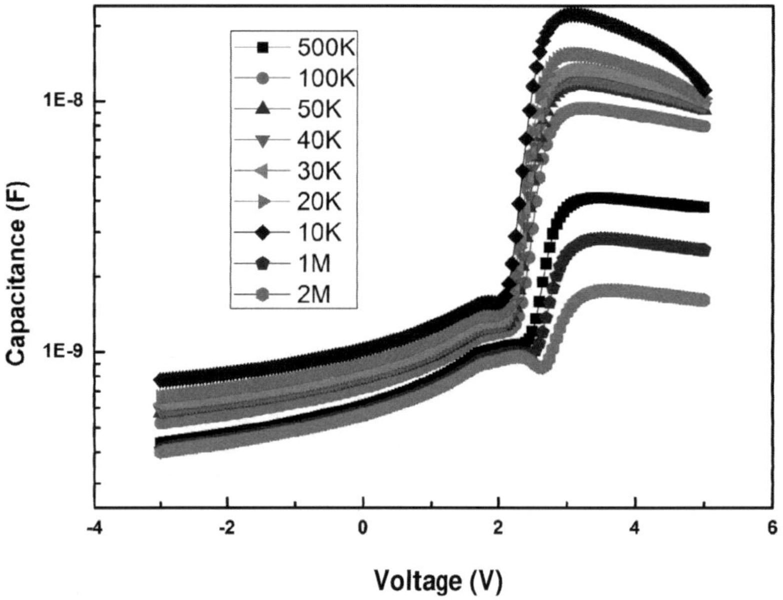

Figure 3. Plot of C-V of Ni/ZnO/Si Schottky contact at different frequencies.

Figure 3 is typical C-V characteristics of Au/ZnO/Si Schottky diode between voltage range of -4 V to +5 V at room temperature and at different frequencies ranging from 10 kHz to 2 MHz. The graph shows that the value of capacitance is decreasing with increasing frequency and plot exhibits a peak in positive voltage. This peak is gradually disappeared as frequency increased from 10 kHz to 2 MHz. The possible reasons of this peak may the presence of deep states in the band gap, series resistance and interface density states [10].

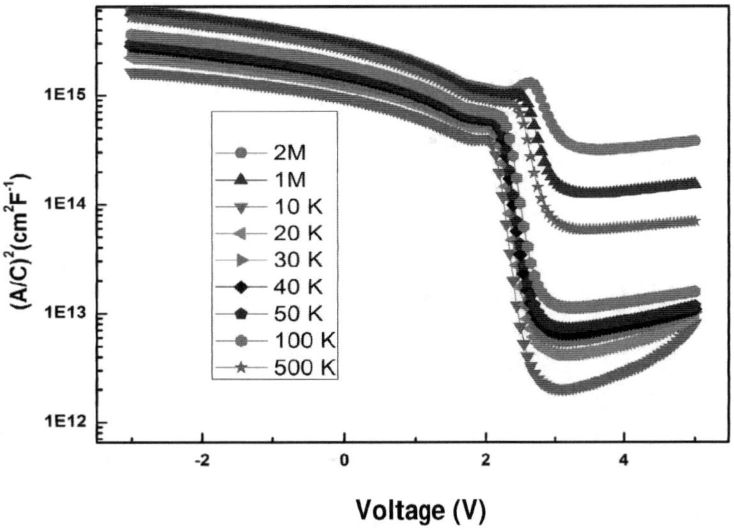

Figure 4. C^{-2} - V characteristics of Au / ZnO / Si Schottky diode at different frequencies.

C - V characteristics can be analyzed by plotting $1/C^2$ vs V and deducing N_d from the slope and V_{bi} from the intercept of reversed biased capacitance-voltage (C-V) characteristics using following equation.

$$N_d = \frac{-2}{q\varepsilon_0\varepsilon_r \frac{d\left(A/C\right)^2}{dv}} \tag{5}$$

Where q (1.6×10^{-19} C) is the charge on an electron, ε_o (8.85×10^{-14}) is the permittivity of the free space, ε_r (=9.66) is the relative permittivity of the medium, A is the area of the Schottky contact ($0.034cm^2$) and C is the capacitance in pF.

The measured value of doping concentration at 2 MHz frequency is 1.4×10^{16} cm^{-3}. The barrier height value can also be deduced from this graph. The barrier height obtained from reversed biased $(A/C)^2$ graph is 0.91 at 1 MHz frequency. The measured value of barrier height from C-V characteristics is high as compared to barrier height extracted from I-V data. This difference is due to the presence of interface states at interface and other possible reasons may be barrier height inhomogenity and image force lowering.

Journal of Physics: Conference Series **439** (2013) 012031 doi:10.1088/1742-6596/439/1/012031

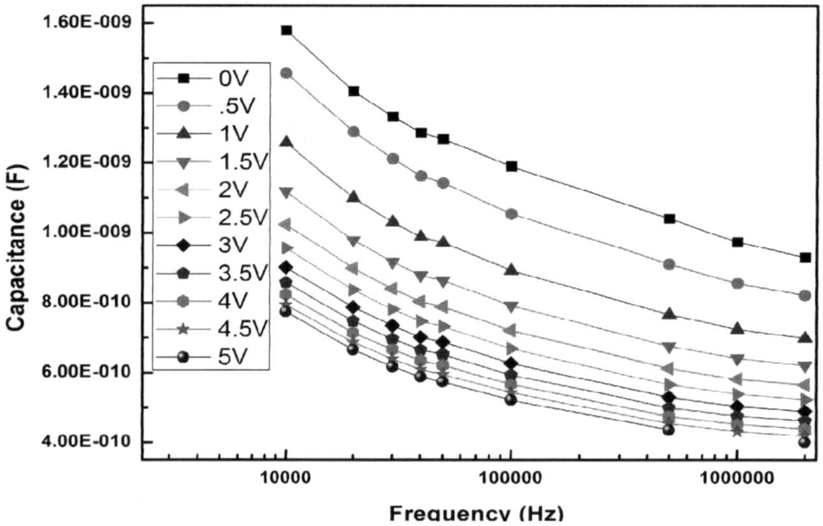

Figure 5. Frequency dependent plot of capacitance of Au / ZnO / Si Schottky diode at different biases.

Generally the capacitance measured for the rectifying contact is dependent on the reverse bias voltage and frequency. Its voltage and frequency dependence is due to the particular features of a Schottky barrier, impurity level, high series resistance, interface states and interface layer between Au and ZnO. This dependence of capacitance on frequency is shown in figure 5. The higher values of capacitance at low frequency are due to the excess capacitance resulting from the interface states in equilibrium with the n-type ZnO that can follow the ac signal. At low frequency the measured capacitance is dominated by the depletion capacitance of the rectifying contact, which is bias-dependent and frequency-independent. As the frequency is increased, the total diode capacitance is affected not only by the depletion capacitance, but also the bulk resistance which is frequency-dependent and associated with electron emission from slowly responding deep impurity levels. Furthermore at high frequencies the interface states cannot contribute to capacitance because the charge at the interface states cannot follow the a.c signal, therefore the capacitance at high frequencies is low as compared to low frequencies [11].

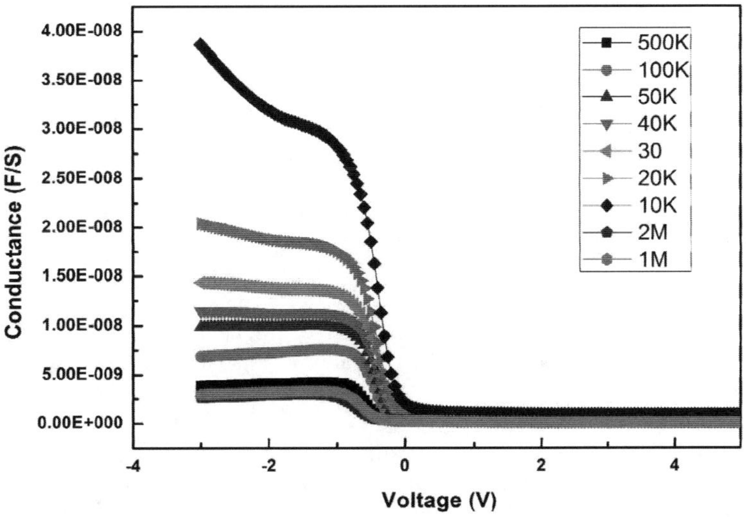

Figure 6. Conductance-Voltage (G-V) characteristics of Au / ZnO / Si Schottky diode at different frequencies.

Figure 6 shows the G - V characteristics of Au/ZnO/Si Schottky contact at different frequencies and room temperature. The conductance of device is decreased as frequency increased. This behavior of conductance is attributed to particular distribution of interface states. From this plot it is also evident that positive voltage the conductance remained almost for all frequencies. Therefore it is concluded that at negative voltages the interface states are responsible of this conductance dispersion.

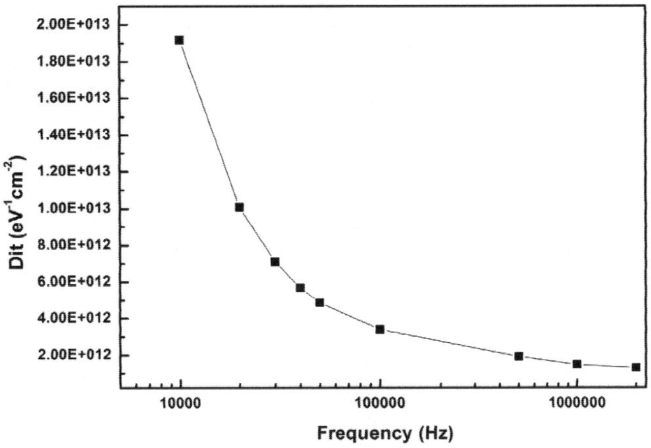

Figure 7. Frequency verses Dit plot of Au / ZnO / Si Schottky diode.

The density of interface states can be determined by Hill-Coleman method [12]. According to this method Dit can be calculated by the following formula

$$\mathrm{Dit} = 2Gc_{\max/w} / qA\ [(Gc_{\max/w}C_{ox})^2 + (1 - C_c/C_{ox})^2] \tag{6}$$

Where A is the diode area, q is electronic charge, w is angular frequency, C_{ox} is the capacitance of oxide layer in accumulation region of C-V curves, $G_{c,\,max}$ maximum conductance of G-V curve and C_c is the capacitance of diode corresponding to $G_{c,\,max}$.

Figure 7 shows the density of interface states verses frequency. The calculated value of The Dit values decrease with increasing frequency. Similar results have been published in references.

4. Conclusion

The properties of interface have been demonstrated with the help of I-V and C-V measurements of Au/ZnO/Si Schottky diode. The ideality factor and barrier height of diode was calculated by I-V measurements at room temperature with values 2.15 and 0.61 eV respectively. Frequency dependent C-V and G-V measurements performed to access the density of interface. Non ideal behavior of measured parameters suggested the presence of interface states. These interface states were calculated by famous Hill-Coleman method. The measured density of interface states is enough to strongly degrade the performance of device.

Acknowledgements

Authors are thankful to Higher Education Commission Pakistan for financial support under project No. 624/2007. The authors are also thankful to UNC-Charlotte, USA for technical support.

References

[1] Zou C W, Wang H J, Yi M.L, Li M, Liu C S, Guo L P, Fu D J and Kang T.W 2010 *Appl. Surf. Sci.* **256** 2453
[2] Asghar M, Mahmood K and Hasan M A 2012 *Key Eng. Mat.* **511-512** 132
[3] Karadeniz S, Tugluoglu N, Serin T and Serin N 2005 *Appl. Surf. Sci.* **246** 30
[4] Karadeniz S, Tugluoglu N, Serin T and Serin N 2004 *Appl. Surf. Sci.* **233** 5
[5] Cova P and Sing A 1990 *Solid-State Electon.* **33** 11
[6] Songul D, Emre G, Syedi D and Sebahattin T 2009 *Current Appl. Phys.* **9** 1181
[7] Asghar M, Mahmood K and Hasan M A 2012 *Key Eng. Mat.* **511-512** 227
[8] Card H and Rodrick EH 1971 *J. Phys.; D* **4** 1589
[9] Wenksten H V, Kaidashev E M, Lorentz M, Hochmuth H, Biehne G, Lenzer J, Gottscalch V, Pickenhain R and Grundmann M 2004 *Appl. Phys.Lett.* **84** 79
[10] Nicollian E H and GoetzbergerA 1967 *Bell Syst. Tech. J.* **46** 1055
[11] Sing A 1985 *Solid-State Electron.* **28** 233
[12] Hill W A and Coleman C C 1980 *Solid-State Electron.* **23** 987

Journal of Physics: Conference Series **439** (2013) 012032 doi:10.1088/1742-6596/439/1/012032

High efficiency interleaved bi-directional ZVS DC-DC converter

M Zafarullah Khan[1], M Mohsin Naveed[1] and D M Akbar Hussain[2]
[1] Institute of Industrial Control Systems, P.O Box, 1398, Rawalpindi, Pakistan
[2] Department of Energy Technology Aalborg University, Denmark

E-mail: mohsinnaveed30@gmail.com

Abstract. A high efficiency interleaved bi-directional ZVS DC-DC converter is presented in this paper. This converter can be operated in both buck and boost mode. CoolMOS is used as a power device to achieve low conduction losses and fast turn off. The value of inductance is selected such that the CoolMOS drain-to-source voltage always falls to zero before it turns on and ZVS is achieved. Multiphase interleaved inductors are used to achieve high power and low ripple currents. Converter is operated at 50kHz and MATLAB Simulink simulation is performed. 6kW prototype converter is implemented in buck mode and simulation results are verified.

1. Introduction

Bi-directional dc-dc converter is being used in many power applications like hybrid electric vehicle (HEV)[1], satellite[2], UPS, renewable energy systems[3] etc. In UPS, a bi-directional dc-dc converter acts as a dc link between low voltage battery source and high voltage dc bus of an inverter. Bi-directional power flows from the high voltage dc bus to low voltage battery when the battery is being charged and from low voltage battery to high voltage dc bus when the battery is supplying power. In renewable energy systems, a bi-directional dc-dc converter is used as a voltage regulator when the power is being transferred to high voltage dc bus while charging the capacitive energy source when the power is being transferred from high voltage dc bus to low voltage side. In HEV, bi-directional power flows from low voltage battery to high voltage dc bus during acceleration and from high voltage dc bus to low voltage battery during regenerative braking. A non-isolated high efficiency interleaved bi-directional zero voltage switching (ZVS) DC-DC converter is shown in Figure 1. The main advantages of this converter are high power density, high efficiency, low switching losses due to ZVS, low conduction losses due to interleaving of phase-leg currents and CoolMOS utilization, and low output voltage ripple due to coupled inductor.

Journal of Physics: Conference Series **439** (2013) 012032 doi:10.1088/1742-6596/439/1/012032

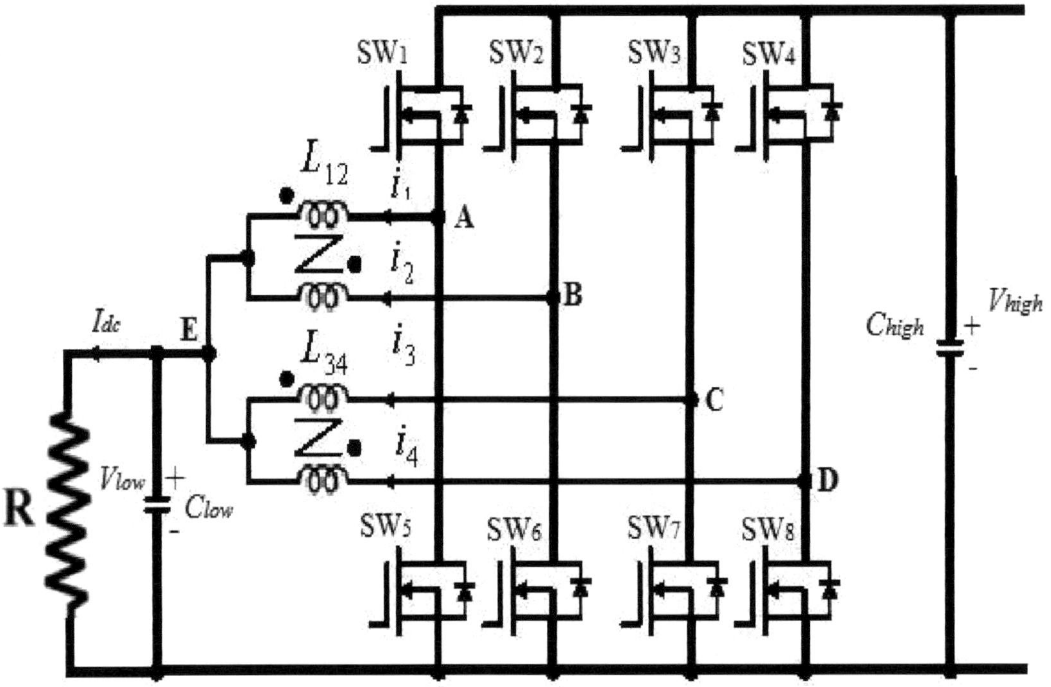

Figure 1. High efficiency interleaved bi-directional ZVS DC-DC converter.

In this paper, design, simulation and implementation of a non-isolated high efficiency interleaved bi-directional ZVS DC-DC converter is realized. Simulation of the converter operation in buck mode is presented and then results obtained are verified experimentally.

2. Design of high efficiency interleaved bi-directional ZVS DC-DC converter

2.1 Power topology

The power topology of high efficiency interleaved bi-directional ZVS DC-DC converter utilizes CoolMOS as the power device. CoolMOS has distinct benefits over other power devices like MOSFET and IGBT [4]. There is no tail current in CoolMOS as compared to IGBT hence it has fast turn off time. Whereas high voltage MOSFET has a problem of high on state resistance resulting in increased conduction losses. CoolMOS has low on state resistance at high voltage compared to MOSFET resulting in low conduction losses in forward conduction period as well as reverse conduction period of synchronous rectification hence high efficiency. The inductors of phase A and B are coupled [5, 6] together and represented by L_{12}. The inductors of phase C and D are also coupled together and represented by L_{34}. C_{low} and C_{high} serve as the smoothing capacitors. Interleaved inductors are used to minimize the ripple current flowing through C_{low}.

2.2 Complementary gating signals

Figure 2 shows the gating signals of all the power devices of high efficiency interleaved bi-directional ZVS DC-DC converter with duty cycle defined in buck mode. Phase B is 180^0 phase shifted with respect to phase A, phase C is 90^0 phase shifted with respect to phase A and phase D is 270^0 phase shifted with respect to phase A. Each phase connected to the same coupled inductor is 180^0 phase shifted. Complementary PWM is applied to the power devices of same half - bridge leg.

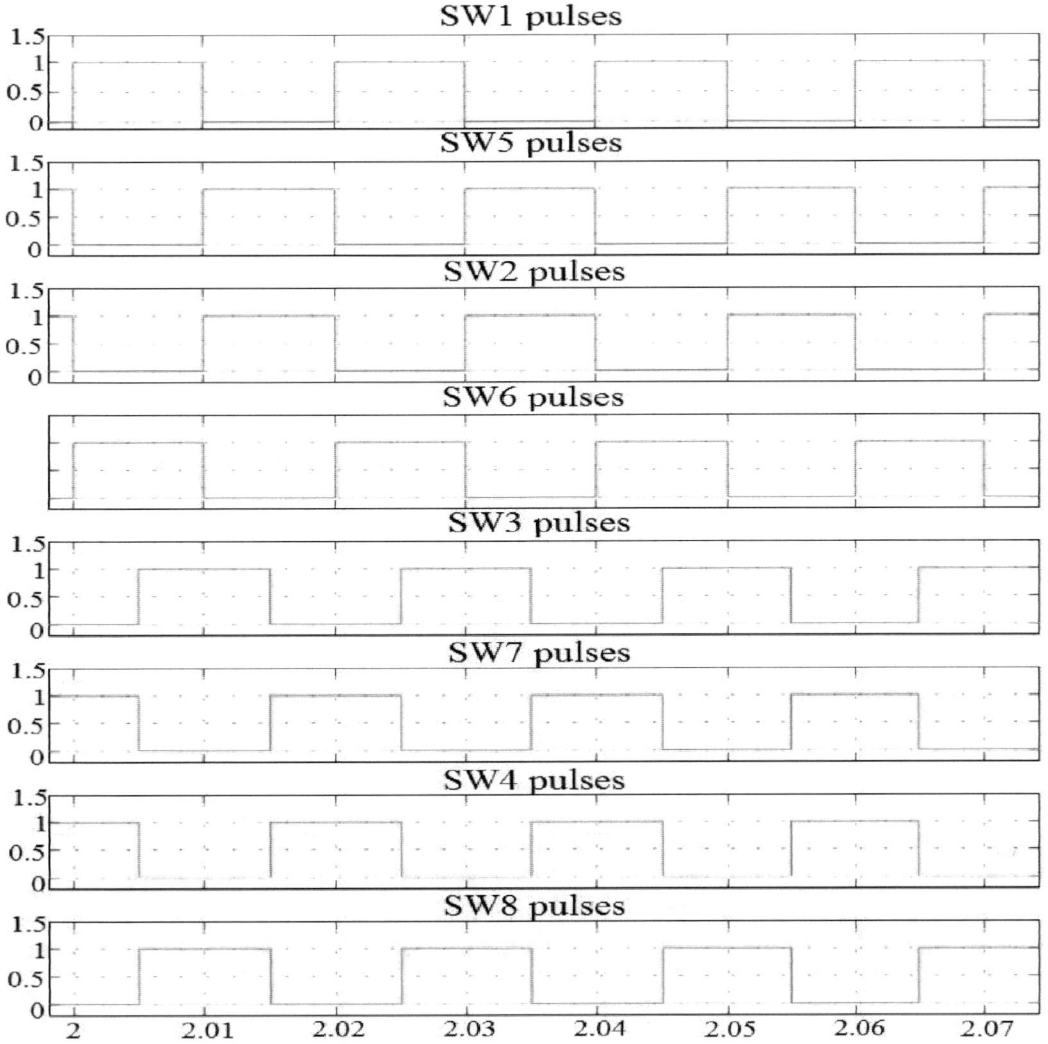

Figure 2. Gating signals of high efficiency interleaved bi-directional ZVS DC-DC converter with duty cycle defined in buck mode.

Figure 3 shows the gating signals with dead time of phase A power devices and current through the inductor connected to phase A. The inductor current varies from positive value to zero and further goes negative to achieve ZVS.

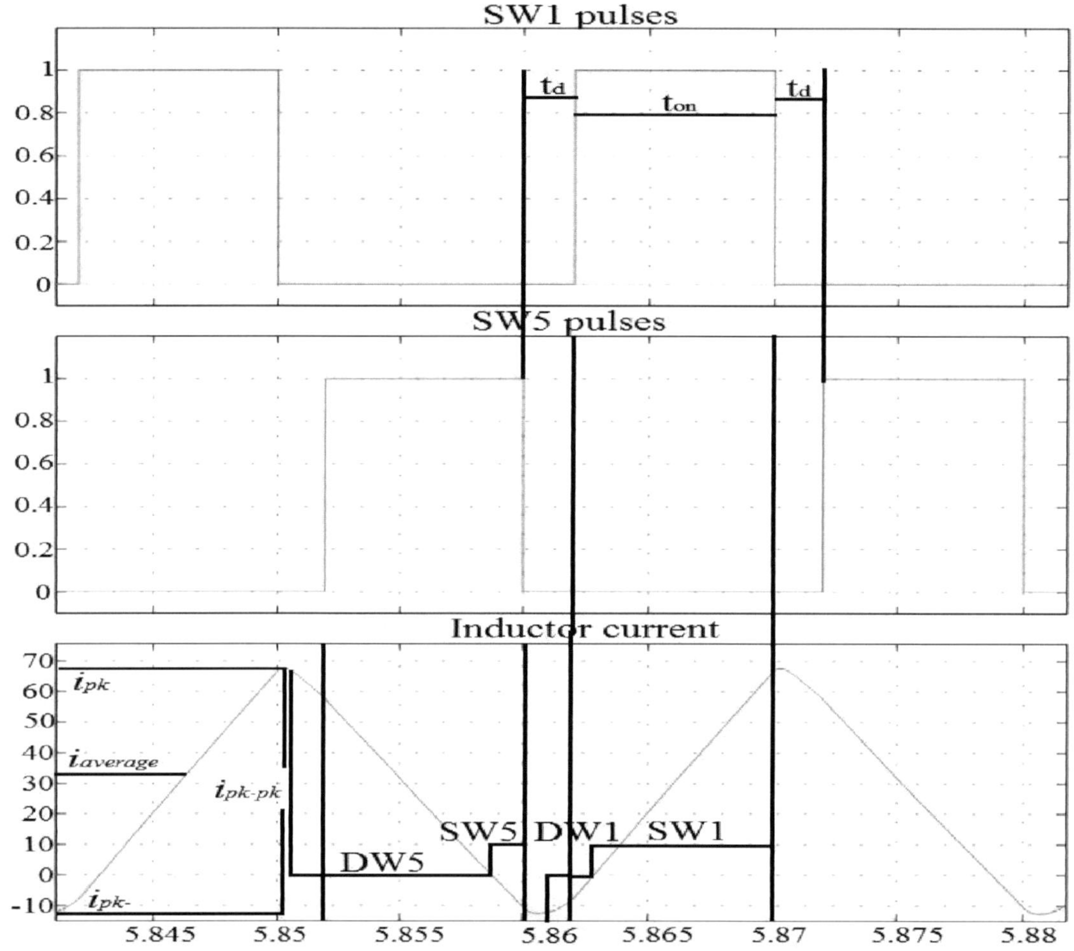

Figure 3. Gating signals and inductor current waveform.

When the power device SW1 is turned on, power flows from the high voltage side to low voltage side and inductor current starts rising. During the dead time t_d, inductor current charges the drain to source capacitance (C_{ds}) of the SW1 and discharges the C_{ds} of SW5. Then the current starts decaying through the DW5, integrated diode of SW5 and SW5 turns on in ZVS condition as the DW5 diode is freewheeling. When the current reaches zero, negative current flows through SW5 and DW5 is naturally turned off without reverse recovery loss. This negative current of SW5 charges the C_{ds} of SW5 and discharges the C_{ds} of SW1. After the complete discharge of C_{ds} of SW1, the drain to source voltage V_{ds} across the switch falls to zero and SW1 turns on in ZVS condition. ZVS operation of the devices results in zero turn on losses and high frequency converter can be easily implemented with reduced inductor size. When the power flows from low voltage side to high voltage side, the converter is operating in "boost mode". When power flows from high voltage side to low voltage side, the converter is operating in "buck mode".

2.3 Coupled inductor design to achieve ZVS

To minimize the output voltage and current ripples, the inductance need to be large enough but with large inductance converter operates in continuous conduction mode (CCM) and transient response is slow. To improve the transient response, small value of inductance should be used but this will increase the ripple content in output voltage and converter will operate in discontinuous conduction mode (DCM). It will impose stress on the output capacitor and large output capacitor will be required to filter out these ripples. In order to achieve ZVS, the converter should be operated further into synchronous conduction mode which requires the flow of negative current as shown in Figure 3. As in synchronous conduction mode, ripple current is further increased, multichannel interleaving is used to reduce this ripple current. Due to small value of inductance required in this converter, the size of the magnetic components is very small. Physical structure of two EE type magnetically coupled inductors is shown in Figure 4(a) and Figure 4(b). The coupling coefficient is close to 0 in Figure 4(a) and between 0 and 1 in Figure 4(b). Air gap is required in the coupled inductors of Figure 4(a) and Figure 4(b) to avoid saturation of the inductors.

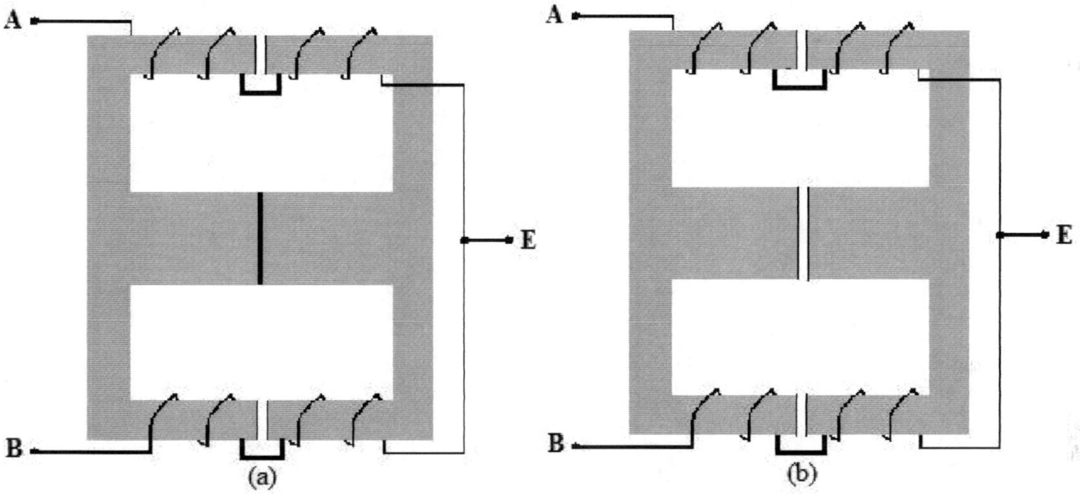

Figure 4. Physical structure of magnetically coupled inductors. (a) Non-coupling. (b) Loosely coupling.

The type of inductor shown in Figure 4(a) has certain drawbacks compared to the inductor shown in Figure 4(b). The cores are in contact at center limb and there is gap in side limbs. Gap must be filled with non-magnetic material to ensure mechanical stability. As there is no gap in the center limb, the thickness of spacers used should be precise and equal to the gap length. Moreover, the gap cannot be adjusted to achieve different inductance values. These drawbacks introduce complications in the manufacturing of inductor.

The inductor shown in Figure 4(b) has gap in each limb. The thickness of the spacers required need not to be very precise. The gap can be easily adjusted and different values of inductance can be easily made. This structure is mechanically stable and provides ease in adjustment of required inductance.

The equivalent inductance required in Figure 4(b) is given by Eq (1)

$$L_{eq} = \frac{L^2 - M^2}{L - \frac{D}{1-D}M} \qquad \text{Eq (1)}$$

Where L is the self inductance of the coupled inductor, M is the mutual inductance and D is the duty cycle defined in buck mode.

As the inductor in Figure 4(b) has equal gap in each limb its mutual inductance is given by $M = L/3$. Putting this value of M and required value of D in Eq (1), the relationship between L and L_{eq} can be obtained i.e. for $D = 0.4$, $L_{eq} = 1.14L$. The negative peak of the inductance current I_{pk-} can be obtained by the energy balance between the inductor and C_{ds} of the power device. It is given in Eq (2)

$$I_{pk-} = \sqrt{\frac{2C_{ds}{V_{high}}^2}{L_{eq}}} \qquad \text{Eq (2)}$$

Where C_{ds} is the output capacitance of the CoolMOS, V_{high} is the high side voltage and L_{eq} is the equivalent inductance.

The half of the peak-to-peak current $I_{pk\text{-}pk}$ through the inductor equals the average low voltage side current plus I_{pk-}, as shown in Eq(3)

$$\frac{1}{2}I_{pk-pk} = \frac{1}{2}\frac{V_{high} - V_{low}}{L_{eq}}DT_s = \frac{P}{V_{low}} + I_{pk-} \qquad \text{Eq(3)}$$

Here T_s is the switching period and P is low voltage side power in each phase.

This converter is normally operated at high frequency and it is important to know that at high frequency eddy current losses increases significantly which causes heating of wire. So Litz wire should be used at high frequency to fully utilize the capacity of wire.

2.4 MATLAB Simulink simulation

High efficiency interleaved bi-directional ZVS DC-DC converter has been simulated with Simulink. Self inductance of the coupled inductors is 13uH and switching frequency of the converter is 50kHz. High side voltage is 150V, low side voltage is 60V and low side current is 100A.

Figure 5 shows the drain to source voltage V_{ds} and gate to source voltage V_{gs} of SW5. It can be seen that V_{ds} drops to zero before V_{gs} goes high resulting in ZVS.

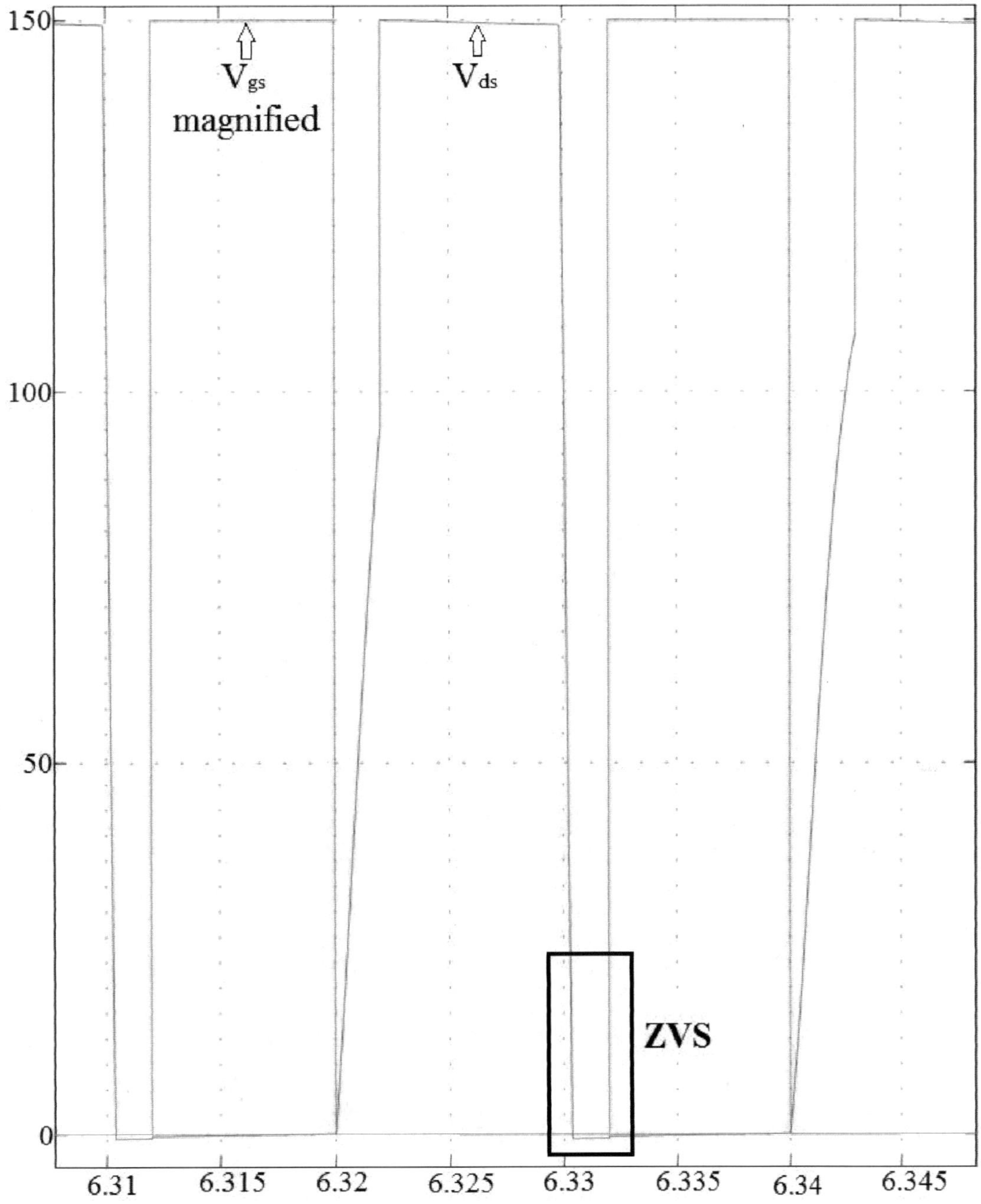

Figure 5. V_{ds} and V_{gs} of SW5 showing ZVS.

Figure 6 shows the individual four inductors currents i_1, i_2, i_3, i_4 and total peak-to-peak current $i_{total,pk-pk}$ at point E. Individual phase inductor peak current i_{pk} is 62A, negative peak current i_{pk-} is -10A, peak-to-peak current i_{pk-pk} is 72A, total peak-to-peak current $i_{total,pk-pk}$ is 7A and the average current $i_{average}$ is 96A.

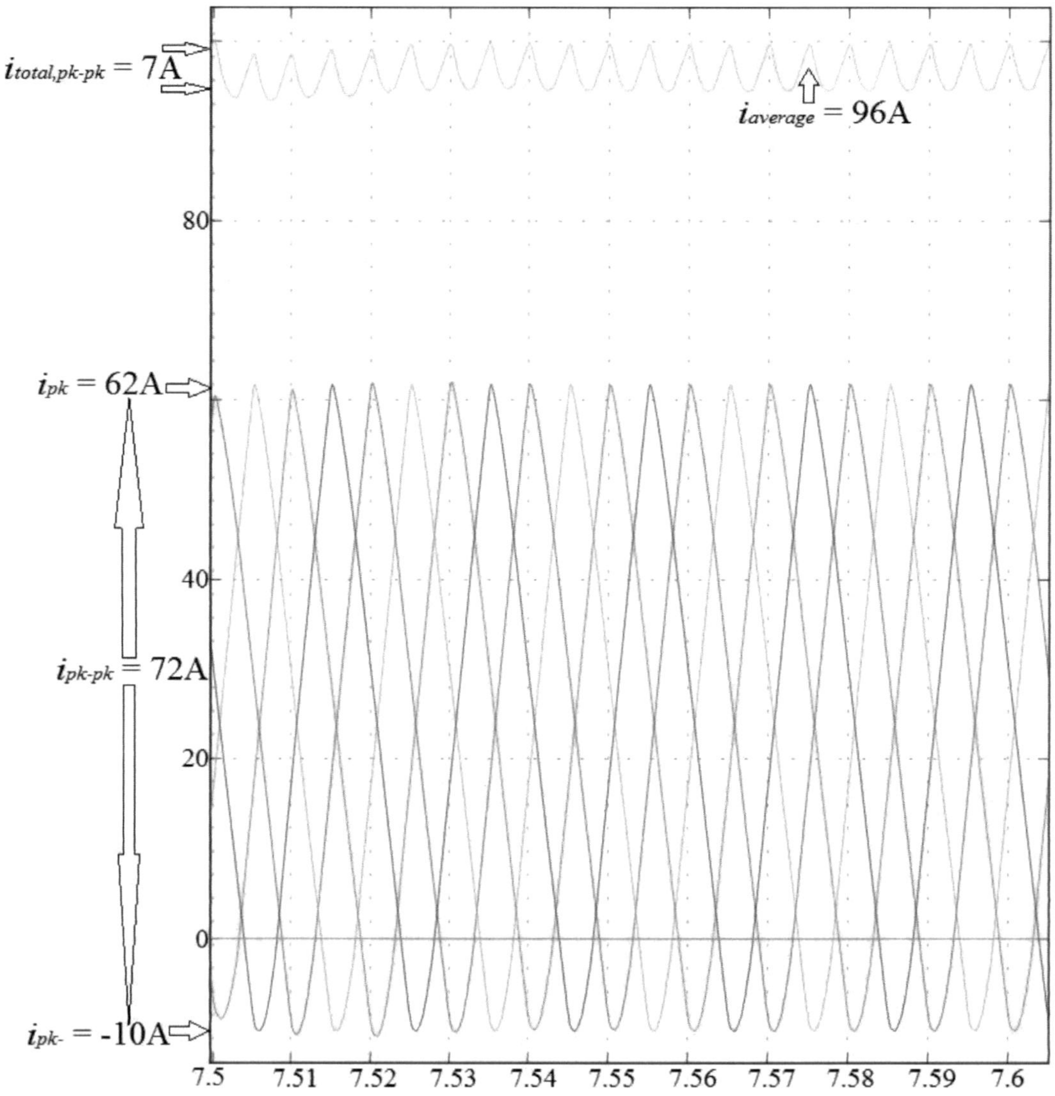

Figure 6. Individual four inductors currents and total current.

Figure 7 shows the gate-to-source voltage V_{gs}, drain-to-source voltage V_{ds} and drain-to-source current I_{ds} of SW5. V_{ds} is150V and I_{ds} is $72A_{pk-pk}$.

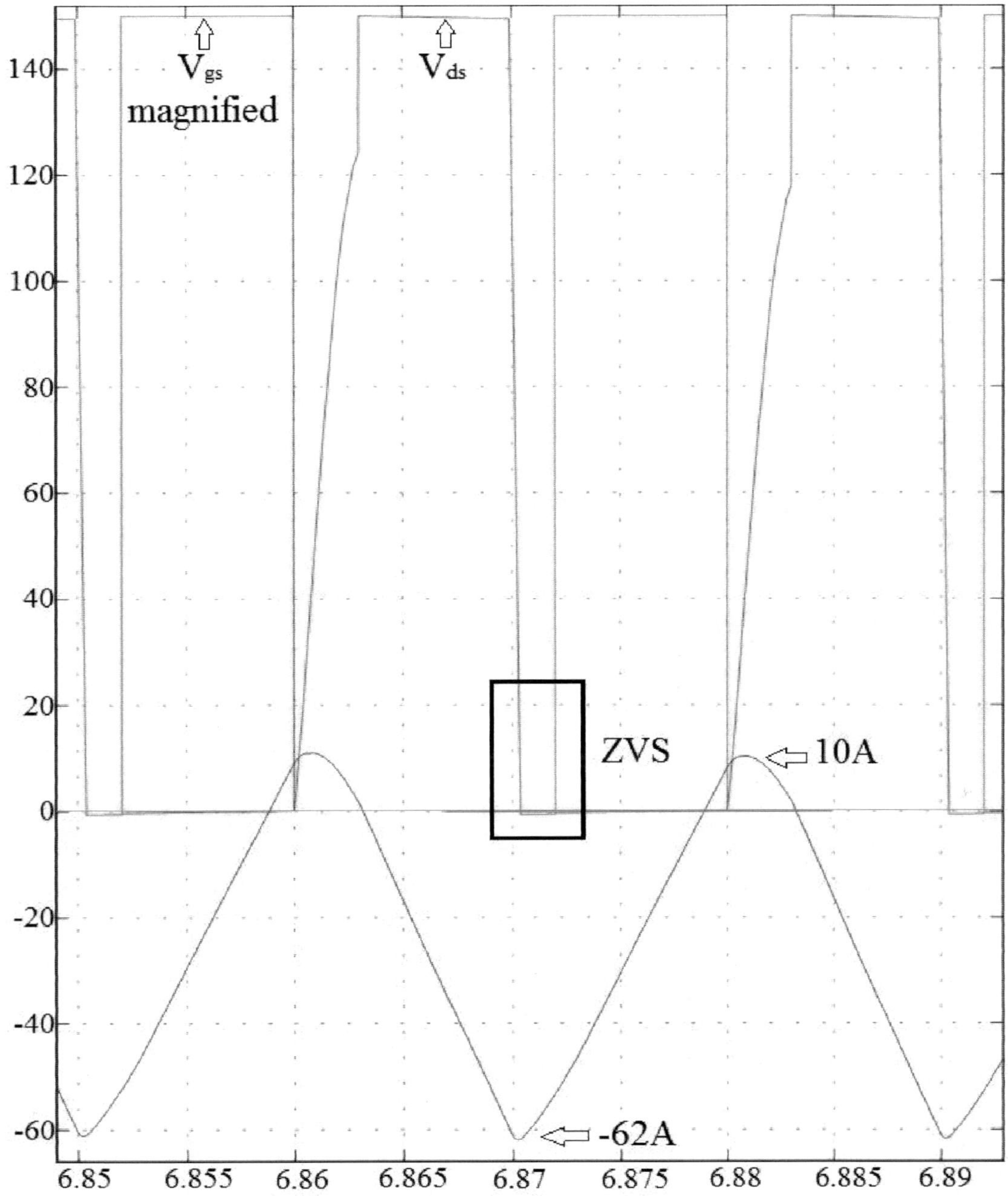

Figure 7. V_{gs}, V_{ds} and I_{ds} of SW5.

Figure 8 shows the high side voltage V_{high}, low side voltage V_{low} and output current I_{dc}. V_{high} is 150V, V_{low} is 60V and I_{dc} is 100A.

Journal of Physics: Conference Series **439** (2013) 012032 doi:10.1088/1742-6596/439/1/012032

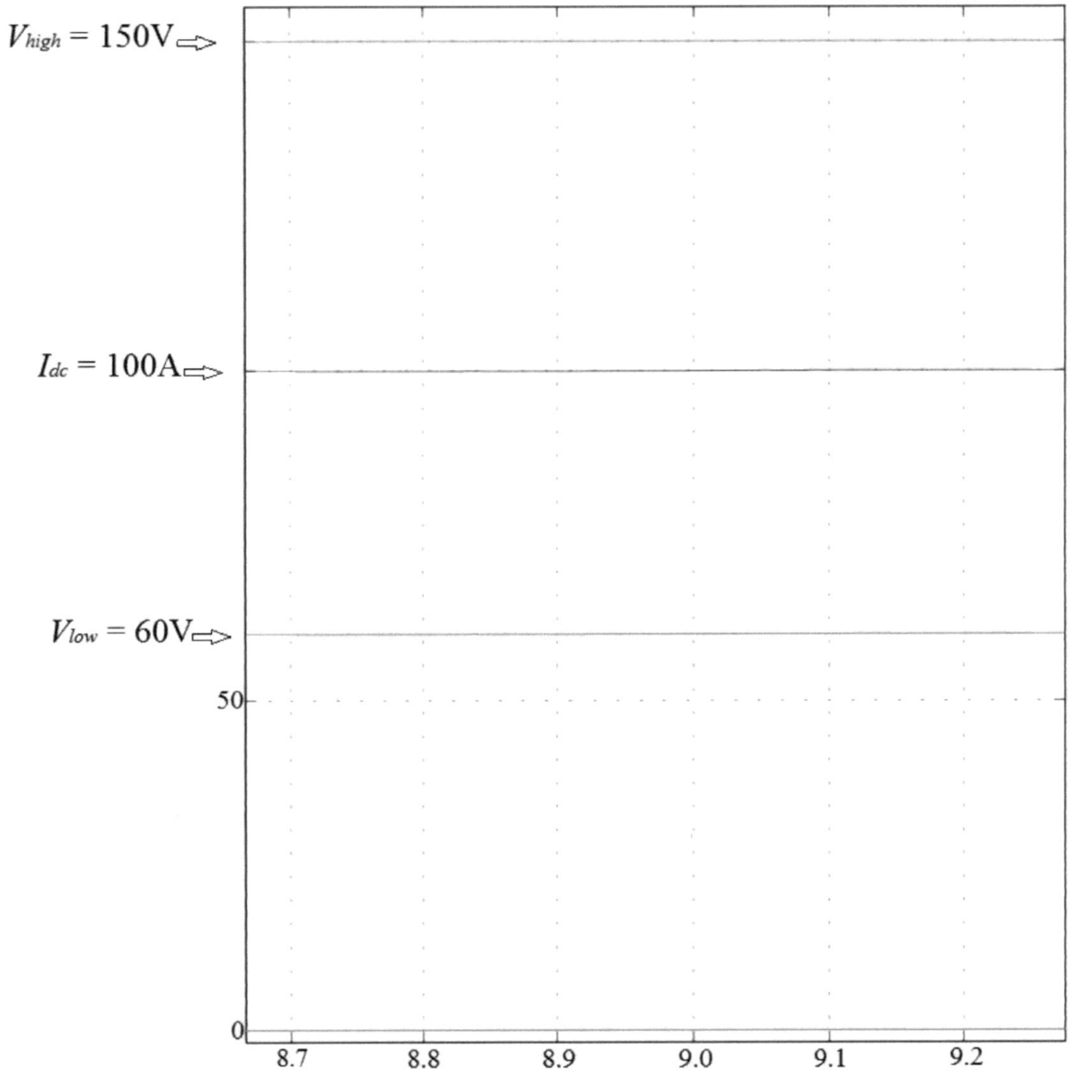

Figure 8. V_{high}, V_{low} and I_{dc}.

2.5 Experimental verification

6kW prototype of high efficiency interleaved bi-directional ZVS DC-DC converter is implemented using CoolMOS as a power device. Coupled inductors are made using ferrite core with 13uH self inductance. Switching frequency of the converter is 50kHz. High side voltage is 150V, low side voltage is 60V and low side current is 100A. Figure 9 shows the drain to source voltage V_{ds} and gate to source voltage V_{gs} of SW5. It can be seen that V_{ds} drops to zero before V_{gs} goes high resulting in ZVS.

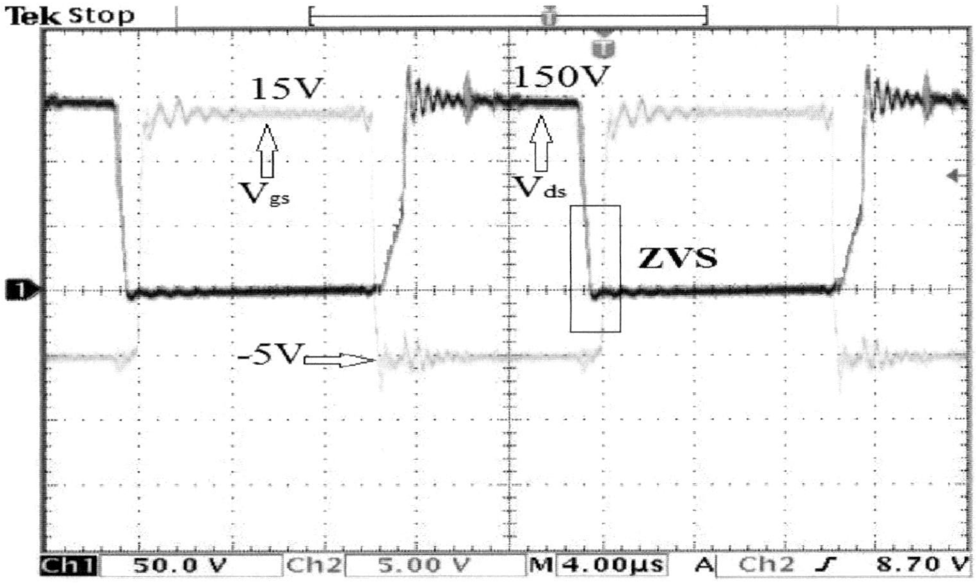

Figure 9. V_{ds} in Ch1 and V_{gs} in Ch2 showing ZVS of SW5.

Figure 10 shows the individual four inductors currents i_1, i_2, i_3, i_4 and total peak-to-peak current $i_{total,pk-pk}$ at point E. Individual phase inductor peak current i_{pk} is 62A, negative peak current i_{pk-} is -10A, peak-to-peak current i_{pk-pk} is 72A, total peak-to-peak current $i_{total,pk-pk}$ is 7A and the average current $i_{average}$ is 94A.

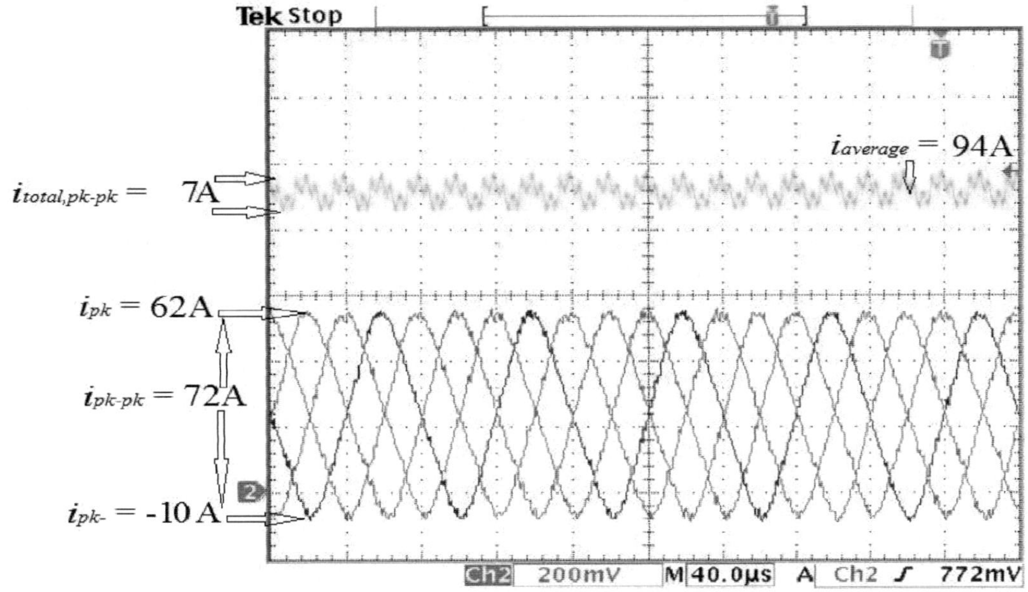

Figure 10. Individual four inductors currents and total current.

Figure 11 shows the gate-to-source voltage V_{gs}, drain-to-source voltage V_{ds} and drain-to-source current I_{ds} of SW5. V_{ds} is 150V, I_{ds} is $72A_{pk-pk}$ and V_{gs} is +15V, -5V.

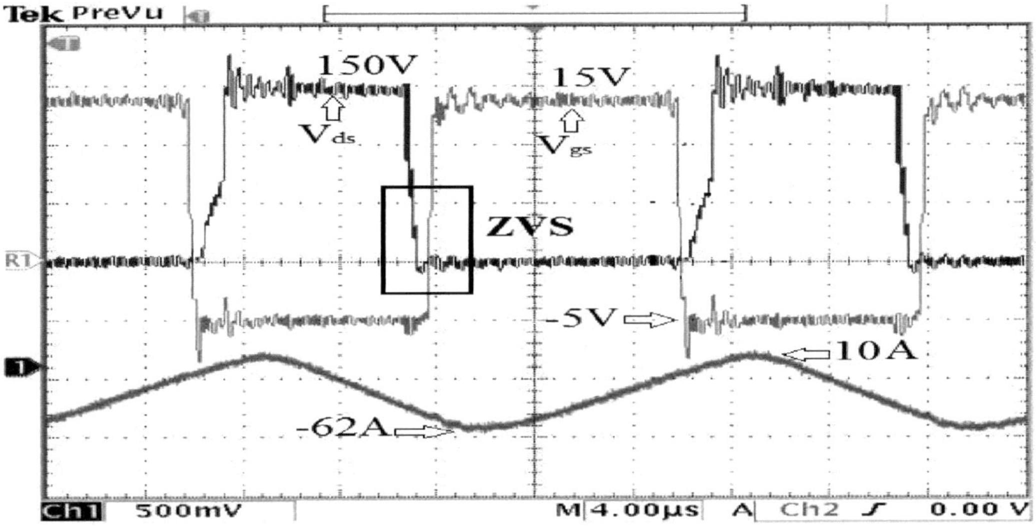

Figure 11. V_{gs}, V_{ds} and I_{ds} of SW5.

Figure 12 shows the high side voltage V_{high}, low side voltage V_{low} and output current I_{dc}. V_{high} is 150V, V_{low} is 60V and I_{dc} is 100A. Efficiency of the converter is measured to be 98%.

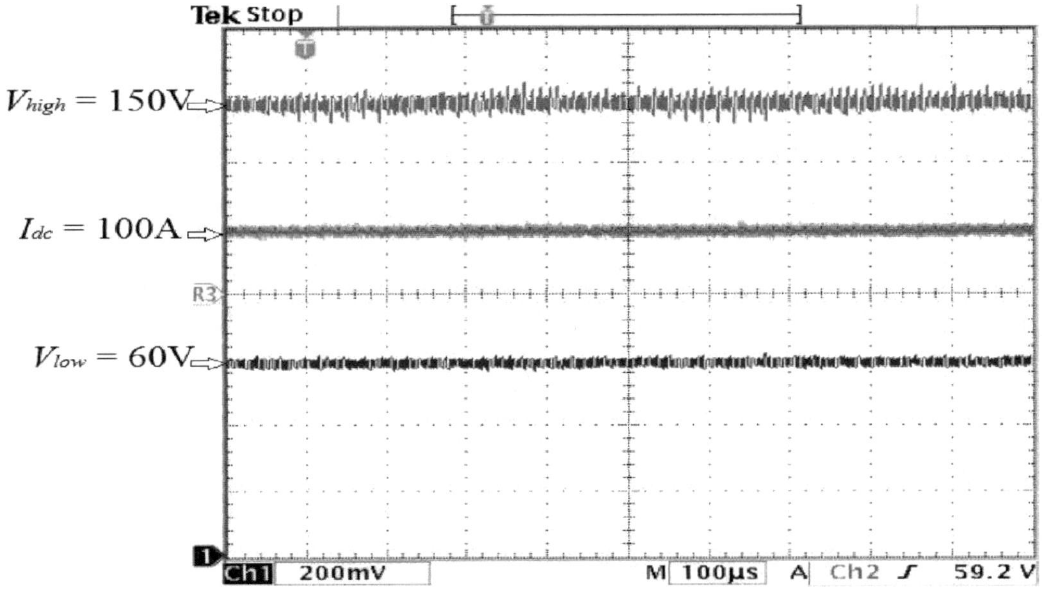

Figure 12. V_{high}, V_{low} and I_{dc}.

From all the measurements shown above it can be observed that experimental results clearly verify the simulation results.

3. Conclusions

High efficiency interleaved bi-directional ZVS DC-DC converter is simulated and realized experimentally in this paper. Following distinct features of the converter are discussed.

1. CoolMOS is used as a power device resulting in low conduction losses compared to the MOSFET and fast turn off compared to IGBT mainly due to the absence of tail current.

2. Output capacitance of the CoolMOS is used to achieve ZVS resulting in almost zero turn on loss.

3. Coupled inductors are used to cancel out low voltage side ripples in the center limb resulting in small capacitance and low ripple contents in the output voltage.

4. Four phase interleaved operation is significant for dense power applications with less stress on power devices, low conduction losses and low ripple contents in the output current.

High efficiency interleaved bi-directional ZVS DC-DC converter is designed, simulated and implemented for a 6kW system. Experimental results verified the simulated results. Its efficiency is more than 98% at full load.

References

[1] Lai J S and Nelson D J 2007 *Proc. IEEE,* vol. **95** 766
[2] Dan M Sable, Fred C Lee, and Bo H Cho 1992 *Proc. IEEE APEC* 614
[3] Khan M Z, Hussain M M, Naveed M M, and Akbar Hussain D M 2012 *Proc. IMECS*
[4] Zhang J, and Lai J S 2006 *Proc. IEEE APEC* 19
[5] Jamieson R S 1984 Negative coupled inductors for poly-phase choppers *U.S. Patent* **4** 442 401
[6] Andersen R 1993 H-driver DC-to-DC converter utilizing mutual inductance *U.S. Patent* **5** 204 809

Journal of Physics: Conference Series **439** (2013) 012033 doi:10.1088/1742-6596/439/1/012033

Nano-holes for vacuum applications

V Ierardi[1], G Firpo and U Valbusa
NanoMed labs, Physics Department (DIFI), University of Genova, Largo R. Benzi 10, 16146 Genova, Italy.

E-mail: Vincenzo Ierardi, vincenzo.ierardi@unige.it

Abstract. In order to calibrate leak-detection instruments, devices based on permeation leaks and physical leaks are used. Unfortunately, permeation leaks are very sensitive to small temperature fluctuations and can be used only with those gases for which permeating materials are available. While, physical-leaks that overcome these restrictions are susceptible to clog, in particular when the gas flow through them is in viscous regime. Furthermore, for many type of leak devices, the conductance is not predictable based on dimensional, gas species and temperature data, for this reason no estimates or only rough estimates of leak rates are possible for conditions under use where they differ from those of the calibration. Here, we present the fabrication and characterization of a new kind of physical-leak devices based on nano-holes, which overcome these problems. Nano-holes, with diameters less than or equal to 200 nm, work in molecular-flow regime up to atmospheric pressure and for this reason they do not clog. The nano-holes are manufactured by milling a silicon nitride membrane by means of Focused Ion Beam (FIB), and their shapes are characterized by both Scanning Electron Microscopy (SEM) and Atomic Force Microscopy (AFM). Because of the capability of the AFM to acquire surface three-dimensional images with very high resolution, it is a very useful tool to perform a topographic characterization of the nano-holes.

1. Introduction

Leak detection techniques are employed widely in systems in which must be guaranteed the leak-proof, such as vacuum equipment, packaging, or cryogenic storage applications. The most used devices to obtain quantitative measurements in leak detection are the Magnetic-sector or quadrupole-mass spectrometers. Generally, these spectrometers are calibrated using permeation or physical leaks. The permeation leaks are glass element immersed in a helium reservoir, which generate a gas flow that is used as reference. The leak rate obtained with these leaks is ranging from 10^{-11} to 10^{-5} Pa m^3 s^{-1}. However, the permeation leaks are very sensitive to the temperature fluctuations, have only one value of leak rate and work only for those gases for which permeating materials are available. Whereas, in the case of the physical leaks, in particular orifices, the tracer gas flows through a restriction or an aperture, generating a leak rate that is greater than or equal to 10^{-8} Pa m^3 s^{-1}. They can be used for any gas, including mixtures, and it is possible to change the flow rate tuning the inlet pressure [1]. Unfortunately, the orifices have great tendency to clog. Moreover, physical leaks work in an unknown flow regime. This means that there is need of a calibration curve in order to use it [2]. It has been a great improvement to fabricate orifices with dimension at nanometer size since these devices work in a

[1] To whom any correspondence should be addressed.

molecular-flow regime up to atmospheric pressure, and overcome all drawbacks characteristic of the physical and permeation leaks [3]. Today the nanotechnology development provides several techniques and instrumentations able to fabricate nanometer holes with precise metrology [4-6]. In this paper we show the fabrication by means of Focused Ion Beam (FIB) and the characterization by means of atomic force microscopy (AFM) and scanning electron microscopy (SEM) of nano-holes, which work as physical leaks. Moreover, we show the performance of the nano-holes in helium flow generation and a rough estimation of their conductance obtained from geometric parameters.

2. Fabrication

The nano-holes are drilled in a Silicon Nitride (Si_3N_4) membranes low stress, made by SPI [7]. The dimensions of the membranes are custom made with chip size of 5 mm × 5 mm, window size 100 μm x 100 μm, thickness 200 nm and it is able to resist at 1 atm of pressure differential, in Figure 1 it is possible to see a representation of the chip cross-section with a nano-hole, which is milled from the upper side of the membrane.

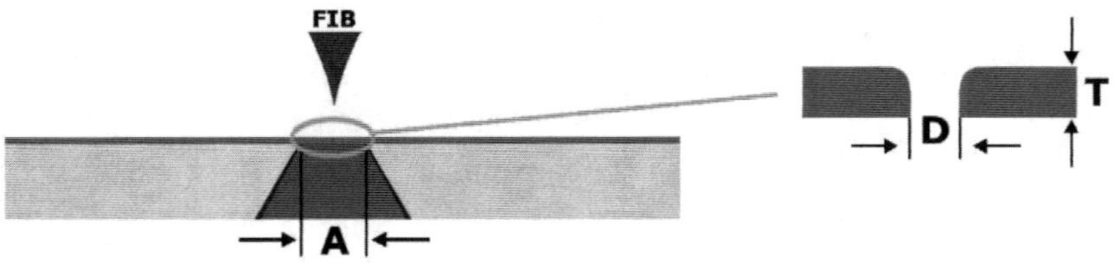

Figure 1. Schema of the cross-section of the silicon chip with the membrane of Si_3N_4. A is the window size of the membrane, T is the thickness of the membrane and D is the diameter of the nano-hole.

To drill the nano-holes we utilized a CrossBeam® workstation 1540XB model by Zeiss, which combines a SEM and a FIB. The Ga^+ ion beam has an energy range from 5 to 30 keV, current of 1-20000 pA, and a minimum spot size of 7nm. Nano-holes with a diameter around 200 nm have been drilled with a single spot of 30 keV, using a current of 100 pA and a dwell time of 2s.

The membranes, after the geometric characterization, were mounted on perforated copper disks CF16 compatible in order to test their performance in helium flow generation. The copper disks have been shaped in order to allow an effective assembling of the membrane with the disk by means of the Varian Torr-Seal glue, as depicted in Figure 2.

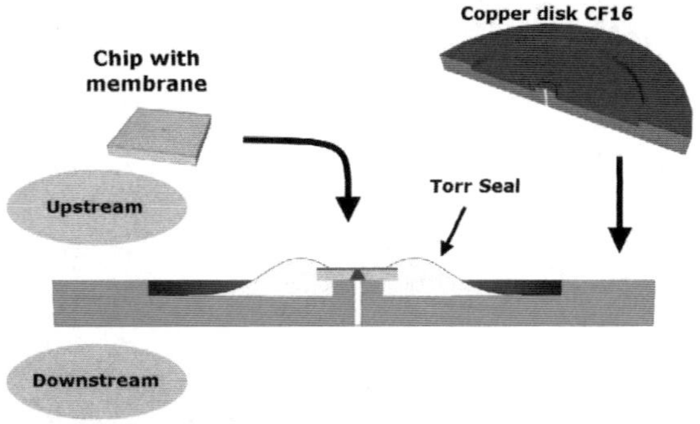

Figure 2. Nano-hole leak schema, the membranes mounted on perforated copper disk by means of Torr-Seal glue.

3. Geometric characterization

The geometric characterization of the nano-holes allows us to evaluate the conductance of the nano-holes using theoretical estimation. First of all we collected SEM images of these nano-holes just after that they were drilled. From the SEM images it is possible to measure the open area of the nano-holes with high precision.

Subsequently the nano-holes were analyzed by means of AFM, the nano-holes images were collected in tapping mode using tips Olympus OMCL AC160TS with an apical radius of 10 nm. The AFM analysis of the nano-holes has displayed some limits due to the geometry of the interaction between the AFM-tip and the nano-holes. Usually the AFM images are reconstructed based on the assumption that the AFM-tip/sample interaction happens between the apex of the tip and the sample surface, unfortunately during the AFM-imaging of the nano-holes this assumption is not always true. In others words the tip/sample interactions have different geometric configurations, which induce different effects into the image processing. In Figure 3 are displayed three different situations in function of the geometry of the interaction between the AFM-tip and the sample surface: a) the tip interacts with the flat upper portion of the sample, the interaction happens between the apex of the tip and the surface; b) the tip approaches the edge of the nano-hole the interaction start to change, but in these condition the AFM-image of the nano-hole is still reliable; c) the tip is moved past the edge of the nano-hole and the interaction happens between the lateral surface of the tip and the walls of the nano-hole, the interaction tip/sample become much more complex when at a certain point the tip fills completely the nano-hole. Taking into account of all the configurations of the interactions tip/sample, it is clear that it exists a threshold, parallel to the membrane surface, below which the AFM data are no more reliable, i.e. below this threshold the interaction does not happen more between the apex of the tip and the sample surface, but it becomes much more complex and involves the whole tip and the walls of the nano-hole. Therefore the AFM analysis gives us information on the morphology of the inlet part of the nano-holes. The AFM analysis of the inlet part of the nano-holes shows a funnel-shaped region with a Gaussian profile due to the FIB beam profile, as it is possible to see in the figure 3d.

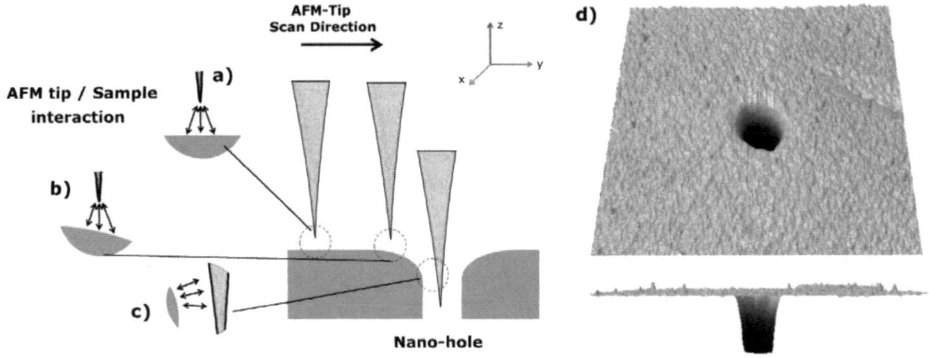

Figure 3. AFM-tip/sample interactions: a) the interactions tip/sample happen between the apex of the tip and the surface; b) the interaction begins to change compared with previously case, but the measure is still reliable; c) the interaction became very complex and the measure is no more reliable; d) an example of AFM-imaging of a nano-hole (2 μm x 2 μm image size). Below it is possible to see the funnel-shaped region of the nano-holes.

In order to obtain more information on the actual shape of the inner part of the nano-holes we have cut and analyzed the cross-sections of several nano-holes using the FIB and the SEM. To minimize the damages due to the FIB milling we have deposited platinum on the nano-holes and after the milling the nano-holes cross-sections were analyzed by means of the SEM. In Figure 4 it possible to see an example of the shape analysis of a cross-section of a nano-hole with a radius of 80 nm.

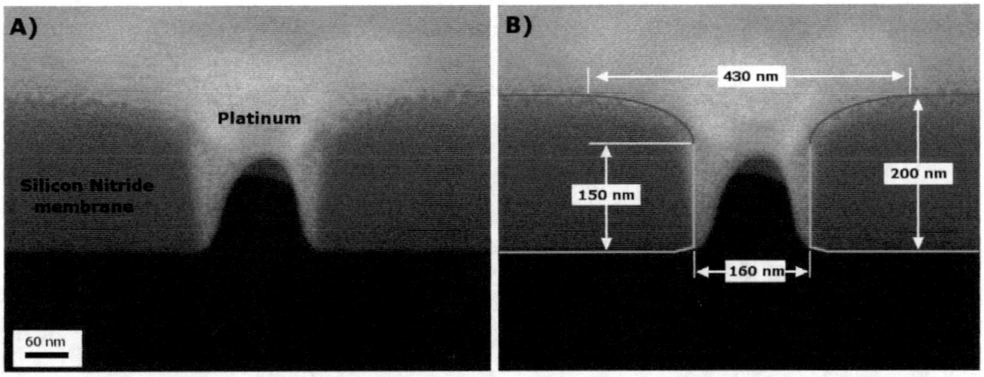

Figure 4. SEM image of the FIB cross-section of a nano-hole with a radius of 80 nm. The bevelled edge of the upper part of the nano-hole is due to the Gaussian FIB beam profile.

Since the analysis of the cross-section involves the destruction of the nano-holes, we have extrapolated the shape for the inner part of the nano-holes tested as leaks. By merging all together the information obtained from SEM, FIB and AFM analysis we were able to obtain a geometric characterization of the nano-holes, i.e. we can model the nano-hole like a tube with circular section and the upstream side funnel-shaped with a Gaussian profile.

4. Helium flow testing

The apparatus used for the calibration of the nano-holes is shown in Figure 5. It is composed of a vacuum chamber evacuated by a Navigator TV1001 Varian turbomolecular pump and equipped with a Varian ionization gauge UHV-24 and a custom-made Prisma Balzers quadrupole-mass spectrometer. In order to perform the helium flow test we have followed the same procedure described by Firpo et al. in [3].

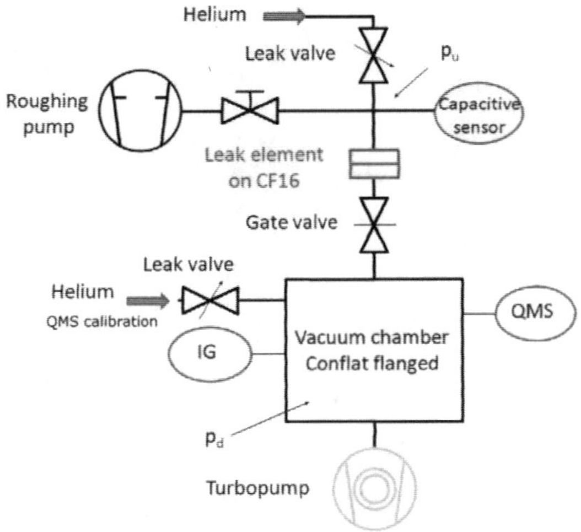

Figure 5. Experimental setup of the vacuum system for testing the nano-holes.

5. Results and discussion

In order to obtain a reliable measurement of the nano-holes by means of the SEM, we have calibrated the SEM using a calibration grating, which has been supplied by the Italian institute of metrology (INRIM). The pitch of the calibration grating used in order to calibrate the SEM is p=462.92 nm with standard uncertainty of 0.025 nm (calibration certificate n. 12-0381-02 by INRIM). After the SEM calibration the error in the SEM-images is less than 2%.

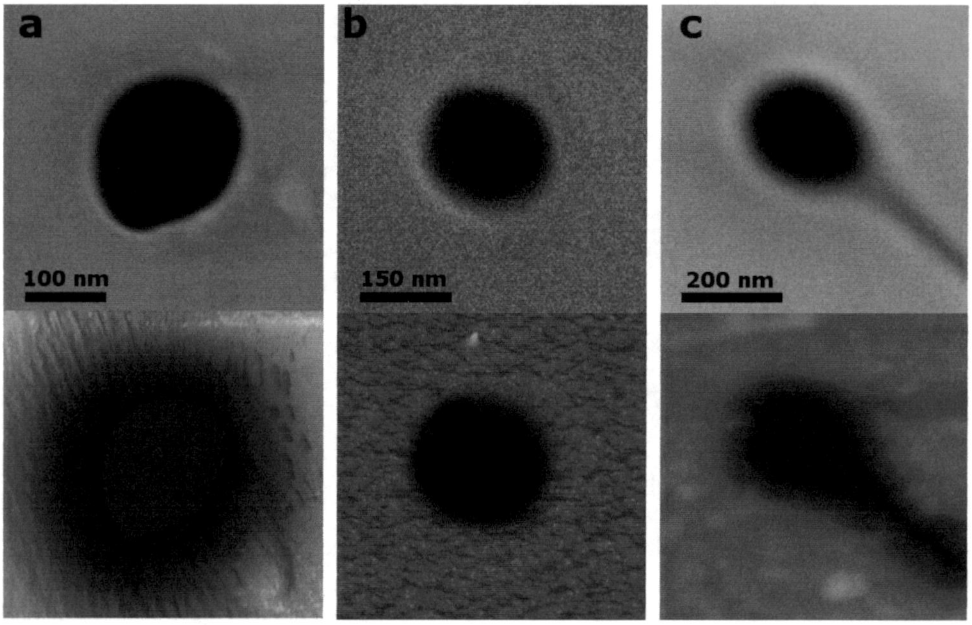

Figure 6. Inlet side images of the nano-holes investigated: above SEM-images and below AFM-images. The thickness of the membrane is the same for all nano-holes and is 200 nm, while the radii are: R_a = 93 nm, R_b = 85 nm, and R_c = 110 nm, for all values the error is around of 2%. The tail visible in the images of the nano-hole **a** is due to a temporary instability of the FIB beam.

In Figure 6 are shown the SEM and AFM-images of the nano-holes investigated. The shape and the dimensions of the open area depend on several factors, for instance the temporary instability of the ion beam or the slight fluctuation of the FIB current, because of no conductivity of the membrane, can produce the tail visible in figure 6c, ion beams slight astigmatic can give elliptic shapes rather than circular, instead slight differences in the focus of the ion beam can give different size of the open area. The radii of the nano-holes have been calculated measuring the open area of the nano-holes from the SEM-images, so we have extrapolated the radius for each nano-hole assuming that the nano-holes are perfectly circulars. The nano-holes radii are: R_a=93 nm for the nano-hole **a**; R_b=85 nm for the nano-hole **b**; and R_c=110 nm for the nano-hole **c**; for all these value the error is around 2%.

From the AFM-images is possible to obtain the profiles of the nano-holes, which are fitted with a Gaussian function. The points where finishes the funnel-shaped part and begins the straight tube are individuated by the comparison between the AFM-profile and the Gaussian function (see Figure 7A). Therefore in the case of the nano-hole **a**, we obtained that funnel-shaped part finishes about of 50 nm below the membrane surface, so the length of the straight tube part of the nano-hole is around of 150 nm. In the others two cases the inlet region has almost the same dimensions, this is reasonable since all nano-holes have been drilled using the same FIB parameters.

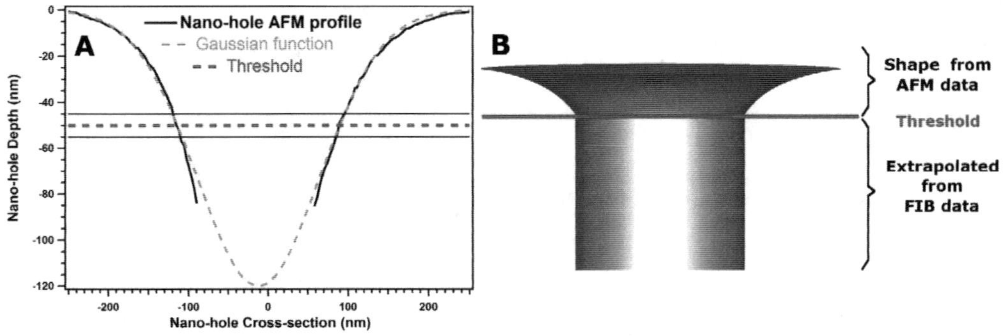

Figure 7. A) AFM profile of the nano-hole "**a**" (black line) fitted with a Gaussian function (green line), it is possible to locate the position of the threshold (red line) below which the nano-holes can be modelled like a tube with straight walls. B) nano-holes model.

The results of the helium flow measurements for the nano-holes **a**, **b**, and **c** are reported in Figure 8. The graphs show the linear dependence between the throughput and the upstream pressure up to 10^5 Pa. A fit with a power law has been added to the graphs and the exponents result to be very close to 1, confirming the linearity of the throughput in function of the upstream pressure, i.e. nano-holes with diameters around 200 nm work in molecular-flow regime.

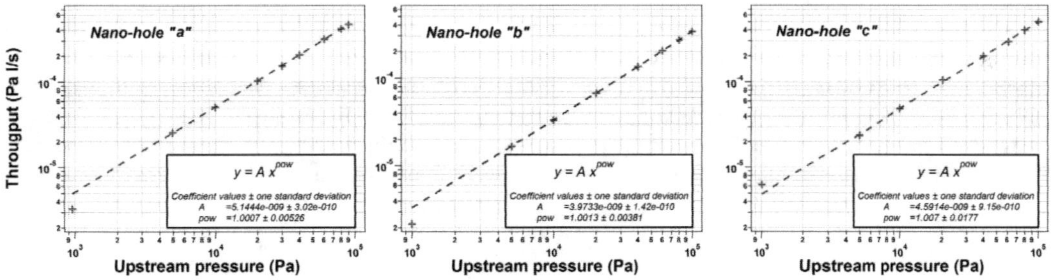

Figure 8. Measured throughput of helium as function of the upstream pressure for the nano-holes **a**, **b**, and **c**. The fit with a power law is reported into the graph.

Since the nano-holes investigated work in molecular-flow regime it possible to estimate the theoretical value of the conductance from the geometric parameters using the following equations:

$$C = \frac{\bar{c}}{4} A P \quad , \quad P = \frac{14 + 4\frac{T}{D}}{14 + 18\frac{T}{D} + 3\left(\frac{T}{D}\right)^2} \quad ,$$

where $\boldsymbol{C}$ is the conductance, $\bar{c}$ is the mean thermal speed, $\boldsymbol{A}$ is the open area, $\boldsymbol{P}$ is the transmission probability for a circular tube of length $\boldsymbol{T}$ and diameter $\boldsymbol{D}$ [8]. Therefore in order to calculate the conductances we have considered two different geometries. For the first one, we have approximated the nano-holes with straight tubes, so we have three tubes with the same length, 200 nm, but with three different radii. For the second one, we have neglected the funnel-shaped region of the nano-holes, considering only their straight tube part in the calculus of the conductances. The experimental and theoretical results are reported in table 1. As it is possible to see in the table 1, the higher discrepancy between the experimental and theoretical data is obtained for the nano-hole c, which has the most irregular shape, while in the others two cases we have a very good agreement between the data experimental and theoretical ones.

Table 1. Nano-holes conductances. The helium flow tests have been performed at the temperature of 23°C.

Experimental Cond.(l/s)		Theoretical Cond. (l/s)		Comparison (%)	
Nano-hole	$C_{Exp.}$	Tube $_{(200\ nm)}$	Tube $_{(150\ nm)}$	$(C_{Exp}-C_{T200})/C_{exp}$	$(C_{Exp}-C_{T150})/C_{exp}$
a	$5.14 \cdot 10^{-9}$	$3.89 \cdot 10^{-9}$	$4.44 \cdot 10^{-9}$	24.3	13.6
b	$4.37 \cdot 10^{-9}$	$3.38 \cdot 10^{-9}$	$3.86 \cdot 10^{-9}$	22.6	11.7
c	$4.59 \cdot 10^{-9}$	$6.39 \cdot 10^{-9}$	$7.19 \cdot 10^{-9}$	-39.2	-56.6

6. Conclusions

Thanks to the high resolution capability of the AFM which makes it possible to characterize the inlet part of the nano-holes. The AFM imaging of the nano-holes can give us a 3D reconstruction of the inlet part of the nano-holes, without destroying them. The inlet region of the nano-holes has a funnel-shaped morphology, which is generated by the FIB beam. In fact the power of the FIB beam has a Gaussian distribution. From the analysis of several nano-holes cross-sections, it is possible to see that the funnel-shaped region is connected to a duct with straight walls, this duct can be considered like a tube. Moreover, making a fit of the AFM profile of the inlet region with a Gaussian function, it is possible to find where the straight tube region of the nano-hole begins. So it is possible to approximate the nano-hole with a tube with circular section and the same open area of the nano-hole. The comparison between the experimental values and the theoretical estimations of the conductances, of the nano-holes tested, has shown that the best agreement is obtained neglecting the inlet funnel-shaped part of the nano-hole. In addition it is also evident that more regular is the actual shape of the nano-holes and more precise is the theoretical estimation of the conductances. The precise geometric characterization of the nano-holes by means of the AFM and the SEM gives the possibility to perform numerical simulation studies of these devices using their actual shape without any approximations.

Acknowledgements

This work is funded by EMRP (European Metrology Research Programme). The EMRP is jointly funded by the EMRP participating countries within EURAMET and European Union.

EMRP
European Metrology Research Programme
Programme of EURAMET

The EMRP is jointly funded by the EMRP participating countries within EURAMET and the European Union

References

[1] Thornberg S M 1991, *Journal of Vacuum Science & Technology A: Vacuum, Surfaces, and Films* **9**(1) 164

[2] Ehrlich C D and Basford J A 1995 *Journal of Vacuum Science & Technology A: Vacuum, Surfaces, and Films* **10**(1) 1

[3] Firpo G, Repetto L, Buatier de Mongeot F, Valbusa U 2009, *J. Vac. Sci. Technol. B 27(6): p. 2347*

[4] Giannuzzi L A, Stevie F A 2005 Introduction to Focused Ion Beam: Instrumentation, Theory, Thechniques, and Practice, *New York: Springer*

[5] Reyntjens S, Puers R 2011 *J. Micromech. Microeng.* **11**

[6] Volkert C A, Minor A M 2007, Focused Ion Beam Microscopy and Micromachining. *MRS Bulletin* **32** 389

[7] *http://www.2spi.com/catalog/grids/silicon-nitride-membrane-windows-xray-microscopy.php*. 2012

[8] Jousten K 2008, Handbook of Vacuum Technology, Berlin: *Wiley-VCH Verlag GmbH & Co. KGaA*

Aluminium ultrahigh vacuum system for the 3 GeV TPS synchrotron light source

G Y Hsiung[1], C C Chang[1], C L Chen[1], L H Wu[1], C M Cheng[1], C K Chan[1], Y C Yang[1], H P Hsueh[1], S N Hsu[1] and J R Chen[1,2]

[1]National Synchrotron Radiation Research Center, 101 Hsin-Ann Road, Hsinchu 30076, Taiwan
[2]Institute of Biomedical Engineering and Environmental Science, National Tsing Hua University, Hsinchu 30043, Taiwan

E-mail: hsiung@nsrrc.org.tw

Abstract. The 3-GeV Taiwan Photon Source (TPS) is a large accelerator and synchrotron light source of circumference 518.4 m. The electron storage ring of TPS requires an ultrahigh-vacuum pressure per beam current less than 2×10^{-10} Pa/mA in the beam duct to maintain a long life of the circulating beam without scattering of ions by residual gases. Aluminium alloys used for the beam ducts have a benefit of greater thermal conductivity that simplifies the structure of vacuum vessels built with the cooling components. Machining completely free of oil applied to the aluminium chambers followed by cleaning with ozonized water and welding in house provide a precise dimensional control within 0.3 mm and a clean surface with a small rate ~ 6.4×10^{-12} Pa m/s of thermal outgassing after baking at 150 °C for 24 h. The assembled ion pump with non-evaporable getter pump is capable of evacuating the chamber to a pressure < 1×10^{-9} Pa. The average pressure inside the duct is expected to be sufficiently small. The clean process to manufacture the aluminium ultrahigh vacuum system is described.

1. Introduction

Designed as a third-generation synchrotron light source with electron-beam energy 3 GeV, beam current 400 mA in multi-bunch top-up operation mode, circumference 518.4 m, beam emittance < 2 nm·rad, Taiwan Photon Source (TPS) is under construction and will be installed in 2013. The vacuum system for the electron-storage ring of TPS provides an ultrahigh vacuum of highest quality with average pressure much less than 1×10^{-7} Pa throughout the beam duct to mitigate the effects of gas scattering of the stored electron beam [1]. As the aluminium ultra-high vacuum (UHV) systems for many accelerators have been operated successfully, and as manufacturing techniques for aluminium chambers have been established during the construction of the Taiwan Light Source (TLS) operated since 1993 [2], aluminium alloys were selected for TPS for the beam ducts of the electron-storage ring. For the 3-GeV TPS, the heat load from the synchrotron radiation and the rate of outgassing from photon-stimulated desorption are much greater than for the 1.5-GeV TLS. As the impedance budget for the beam ducts is more stringent, and as the dimensional control of vacuum components is more precise, new manufacturing techniques for the aluminium beam ducts must be developed to fulfill the requirements of the UHV systems for TPS rather than TLS [3]. The design of the aluminium vacuum systems for TPS complies with the following six strategies: (1) to accept the large heat load, the absorbers are located remotely from the radiation source; (2) the beam ducts are designed with inner apertures sufficient to accommodate the dimensions of the beam stay clear; (3) the UHV pumps are located near the gas sources or the absorbers; (4) the structures of vacuum components are modified to be smooth with small impedance; (5) to maintain the clean qualities of surfaces, manufacturing was developed completely free of oil, and (6) there is no baking in situ inside the tunnel. The design of the vacuum system, the manufacturing and the test results are described in the following sections.

2. Design of the vacuum system

The vacuum system for the electron-storage ring is divided into six-fold super-period systems; each comprises four long straight sections, with lengths one 12 m and three 7 m, and four arc-cell bending sections of lengths 13 ~ 14 m. All lattice magnets are gathered in the arc-cell bending sections that yield a limited space to accommodate vacuum chambers and components. Figure 1 illustrates the assembly drawing for a typical one-sextant vacuum system with magnets. The design of the vacuum chambers must hence utilize the entire space between the magnets to accommodate the beam-dynamic apertures and to allocate photon absorbers, beam-position monitors (BPM), vacuum pumps, bellows, flanges, gate valves and supports. Besides the constraint from limited space, the manufacturing tolerance for all components must be precisely controlled and the cleanliness of the surface must be maintained through all manufacturing. The concept of a large bending chamber made of aluminium alloys, with ease of machining capability and increased thermal conductivity, was adopted to optimize the design that fulfills the specified requirements [4]. The layout drawing of a typical vacuum system of a 14-m cell appears in figure 2. To mitigate the impedance and to save space from flanges and bellows, the vacuum chambers in the cell are welded to one piece of 14-m sector with only two bellows joined; two flanges and two sector gate valves (SGV) join both ends.

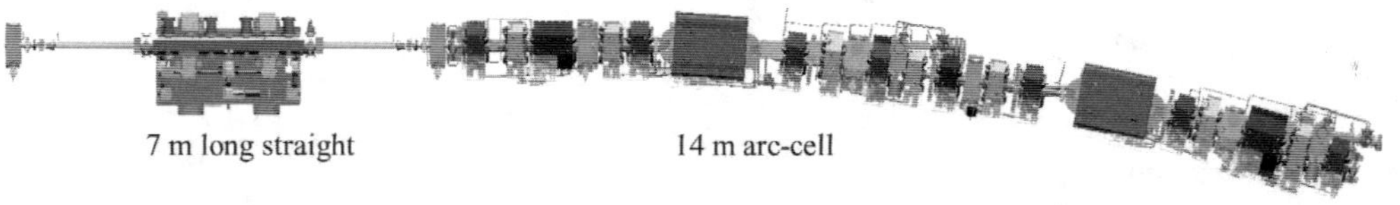

Figure 1. Assembly drawing for a typical sextant super-period vacuum system with magnets.

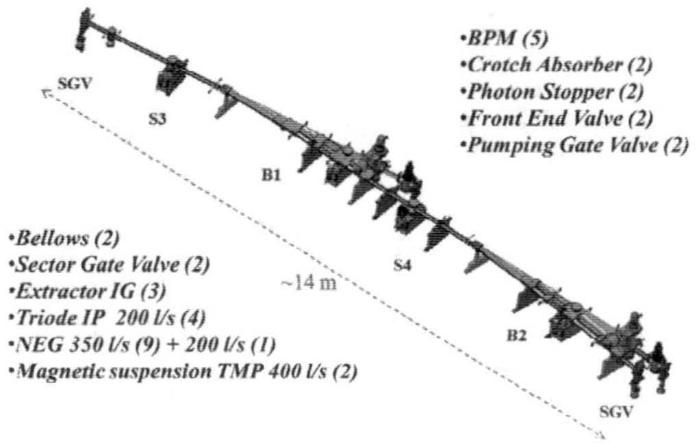

Figure 2. Layout drawing of a 14-m cell vacuum system. The beam ducts are composed of two straight chambers, S3 and S4, and two bending chambers, B1 and B2.

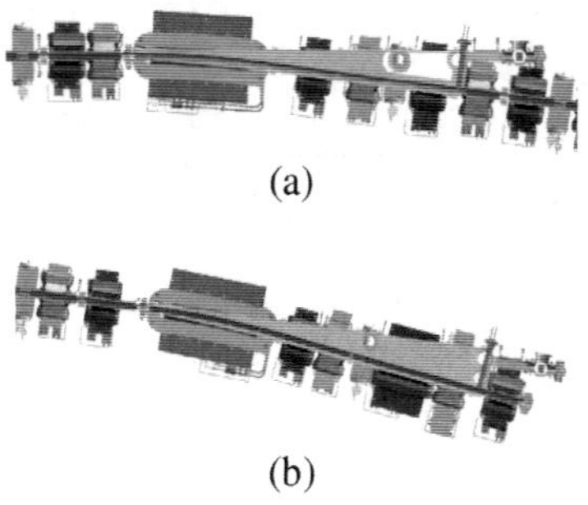

Figure 3. Section drawings inside bending chambers (a) B1 and (b) B2.

Several issues concerning the optimized design of the cell vacuum system are described as follows. The crotch absorber must be located at least 2.3 m from the source of synchrotron radiation emitted from the upstream bending magnet so that the water-cooled OFHC copper material is capable of accepting the thermal load [5]. With this design concept, the bending (B-) chamber has a triangular shape with length about 3 ~ 4 m. The crotch absorber installed downstream from the antechamber that produces molecules outgassing from photon-stimulated desorption (PSD) is confined locally to decrease the gases backfilling the beam channel. The surface of aluminium chambers is readily machined to an overall precision < 0.02 mm. Figure 3 shows section drawings inside the (a) B1- and (b) B2-bending chambers. The long channels (dark blue) represent the orbital channels for the electron beam. Two crotch absorbers and photon stoppers (orange) are located downstream of the B-chambers. The circles shown in the antechambers represent the pumping ports near the absorbers to enhance the localized-pumping efficiency. The back sides of the B-chambers are partially machined away to fit the chambers inside magnets with clearances about 1 ~ 3 mm from the magnetic poles or coils. The orbital channel made with precise machining in the aluminium B-chamber provides a finished smooth surface that minimizes the impedance.

As space between the chambers and magnets is tight, baking the cell vacuum systems in situ without conflicting with the magnets is impractical; all cell vacuum systems must therefore be baked ex situ in the laboratory to achieve an UHV without leakage before installation in the tunnel. Without baking in situ, the design of the vacuum cells can be simplified, which minimizes the quantity of bellows and flanges to mitigate the impedance, to avoid shifting of the BPM from thermal stress, to decrease the risks of leakage through the baking and to save significantly the long working hours in the tunnel for the baking. The cell vacuum systems will be transported from the laboratory to the tunnel and installed on the girders with a conveyance system [6].

The vacuum chambers in the long straight sections comprising pumping-port chambers, primary BPM ducts, bellows, taper chambers and drift chambers, will cooperate with the vacuum systems for insertion devices. All drift chambers will be made with aluminium extrusions. The long straight sections will be installed after positioning the two arc-cell vacuum systems on both ends. Baking in situ will be performed for the long straight sections after installation, which is easier than baking of the cells.

3. Manufacturing

3.1. Manufacturing of the aluminium vacuum chambers

The vacuum beam ducts for the 14-m arc-cell vacuum system are composed of two short straight (S-) chambers, S3 and S4, and two bending (B-) chambers, B1 and B2, as shown in figure 2. As the manufacturing of S-chambers differs from that of B-chambers, the four chambers must be manufactured individually and then welded to form a 14-m sector chamber. Figure 4 shows a design drawing of the four chambers to be prepared before welding. The manufacturing is described in the following subsections.

3.1.1. S-chamber. The S-chamber comprises the drift chamber, the pumping-port chamber, the BPM duct and flanges. All chambers are made of aluminium alloys for welding to one piece of the S-chamber. The straight aluminium tubes other than the BPM duct are made with extrusion. All aluminium extruded parts were cleaned with chemical solvents before welding in the clean room. The manufacturing of the S-chambers follows a procedure similar to that for TLS [7]. The aluminium extruded pipes were machined for pumping holes, cooling channel and welding edges. After the chemical cleaning, the pre-machined pumping port chamber will be inserted into the extruded pipe and welded together with a tungsten-inert-gas (TIG) method to form the drift straight chamber. The aluminium BPM duct is precisely machined with a machine free of oil similar to the manufacturing of the B-chamber to be mentioned in the next section. The drift chambers and the BPM duct will be aligned on pre-aligned supports and welded to the one piece S-chamber, S3 or S4, as shown in figure 4. Photographs of the welded S-chambers appear in figure 5(a).

3.1.2. B-chamber. The B-chambers are made on welding two halves of aluminium plate of length ~ 4 m. The aluminium plates are precisely machined with a machine under numerical control of a computer (CNC) and free of oil in a clean room (class 10000). The ambient temperature is controlled at 25±1 °C and the relative humidity at < 50 %. All machining tools must be cleaned with alcohol throughout the work. A system to spray pure alcohol (99.5 %) is developed to cool the aluminium working pieces during machining. A system for compressed air free of oil provides super-dry air to the spray system for spraying the alcohol on the working pieces with ultra-low moisture < 10 ppb. The machining of each aluminium plate is divided into three steps -- rough machining to intermediate machining to precise machining with intervening intervals 1 ~ 3 days for stress release. The overall deformation after machining is then controlled < 0.02 mm. Each aluminium plate will be stored in a clean aluminium bag filled with pure dry nitrogen gas and sealed.

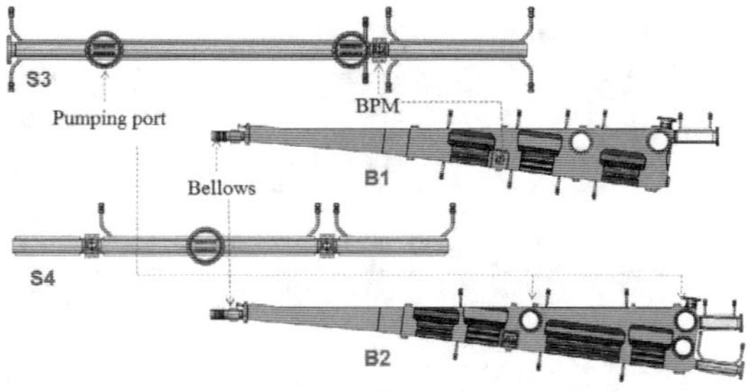

Figure 4. Design drawing of four chambers including two short straight (S-) chambers, S3 and S4, and two bending (B-) chambers, B1 and B2, to be welded to a 14-m sector of an arc-cell vacuum chamber.

Before the welding of a B-chamber, the two halves of machined plates are taken from the sealed dry bag and immersed in a clean vessel filled with the ozonized water (ozone concentration > 20 ppm) for 30 min, then dried with pure nitrogen gas spraying and promptly moved to the clean room. The machining in an environment of pure alcohol is similar to the former process for the aluminium B-chambers in TLS [2], but a recent investigation of cleaning, with ozonized water, of aluminium chambers machined free of oil provides a much smaller rate of thermal outgassing, ~ 6.4×10^{-12} Pa m/s after baking [8] and a smaller yield of photon-stimulated desorption ~ 2×10^{-5} molecules/photon [6] that is applicable for the TPS UHV system. As the cleanliness of the surfaces must be maintained and the rate of growth of oxidation on the aluminium surface must be suppressed throughout the welding process, all welding work must be done in the dry clean room (class 1000) controlled at temperature 25±1 °C and relative humidity < 50 %. The welding sequence consists of the welding of pumping port with flange, welding of half plate with pumping port, auto-welding of the B-chamber with two half plates, precise machining of the terminal ends of the side ports and welding of B-chamber with the side tube, flanges and bellows. The overall deformation of B-chambers after welding was < 0.1 mm vertical and < 0.3 mm transverse. Each step in the welding sequence must be inspected for leakage [6]. The B-chambers after welding will be promptly evacuated and sealed in vacuum. The photographs of the welded B-chambers and the welding work appear in figure 5(b) and 5(c), respectively.

3.2. Welding of the 14-m vacuum cell

Before joining the chambers, the supports for the chambers are installed on the girders and aligned to precision < 0.1 mm with a laser tracker. All major supports are made of thick aluminium plates. With a precisely machined base plane provided on the girders, the aluminium plates can be well positioned within 0.1 mm without adjustment mechanism. When those supports have been aligned and fixed on the girders, the B1 and B2 chambers are placed on the aluminium supports and fixed with the positioning guide pins near the BPM of B1 and B2 chambers, respectively. The S3 and S4 chambers are then placed on the supports. The four chambers are aligned with a laser tracker to precision < 0.1 mm, then fixed on the supports. As the surfaces of the aluminium chambers and aluminium supports near the adjacent interface are both precisely machined, the positioning of chambers on the pre-aligned supports can be done quickly without adjustment. After positioning the chambers, the TIG welding to join the four chambers to one sector is performed on site on the girders and then evacuated to test the leakage. With this efficient welding, a full cell of a 14-m vacuum chamber proved to be precisely aligned on the girders and leak-tight. The 14-m cell is then exposed to the atmosphere with pure nitrogen gas for the installation of vacuum components including ten BPM flanges, two sector-gate valves (SGV), two front-end valves (FEV), two pumping-gate valves (PGV), four ion pumps (IP), ten non-evaporable getter (NEG) pumps, three extractor ionization gauges (IG), four turbo-molecular pumps (TMP) etc., and evacuated for leak testing. Figure 6 presents photographs of the assembly works for the 14-m vacuum cell on the girders. After installation of two crotch absorbers and two photon stoppers, the vacuum system of the entire cell must be baked at 150 °C for 24 h to attain the required UHV pressure.

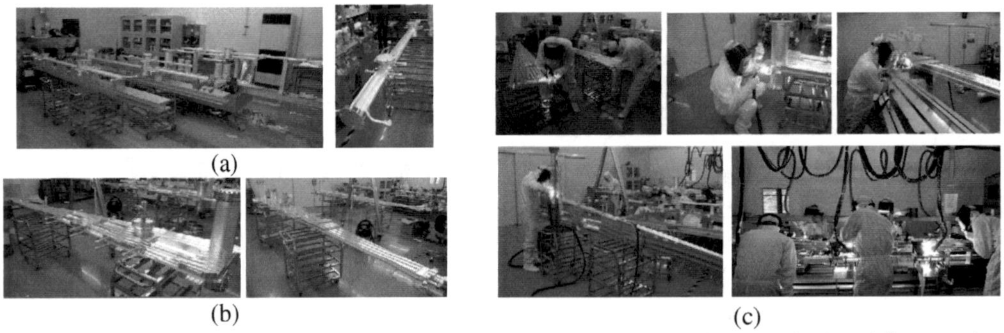

Figure 5. Photographs of (a) welded S-chambers, (b) welded B-chambers, and (c) welding works for the B-chamber.

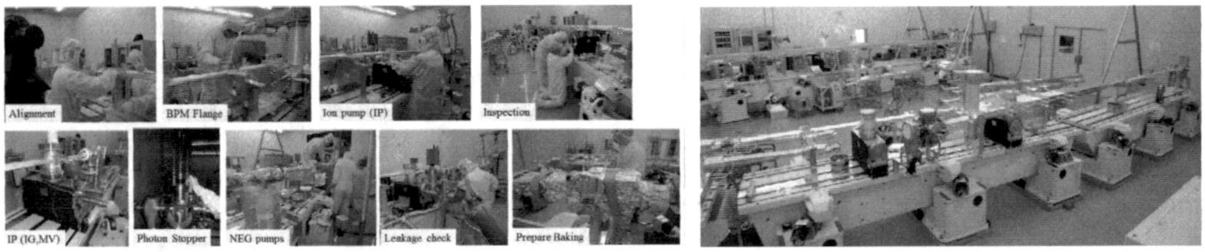

Figure 6. Assembly work for the 14-m vacuum cell on the girders.

4. Test results

4.1. Baking test for a 14-m vacuum cell

The average pressure in the beam duct is determined by the rate of thermal outgassing of the aluminium vacuum chambers and the efficient pumping performance at UHV. The baking test for a 14-m cell vacuum system is performed on reaching 150 °C and kept for 24 h for inspection of the outgassing behavior. Before cooling, the IP will be switched on to refresh the cathodes with a high-temperature ion-sputtering effect, and the gauges degassed more than 20 min; all NEG pumps will then be simultaneously activated to 450 °C for 1 h when all IP must be turned off to avoid pumping contamination from the outgassing of NEG during activation. After the NEG activations the triode IP are switched on and maintained at a fixed mode of high voltage (-7 kV) applied for more efficient discharge-cleaning of the cathodes. The entire system is then cooled uniformly. All bolts and nuts of the flanges will be tested for torque and tightened as required at temperature ~ 60 °C during cooling. Figure 7 shows a photograph of one cell baking ex situ in the laboratory, the pumping curve through the baking and the mass spectrum after baking. The ultimate pressure after baking attained 6×10^{-9} Pa, indicated in figure 7(b). The major residual gases comprising H_2, CO, H_2O and CO_2 are shown in figure 7(c).

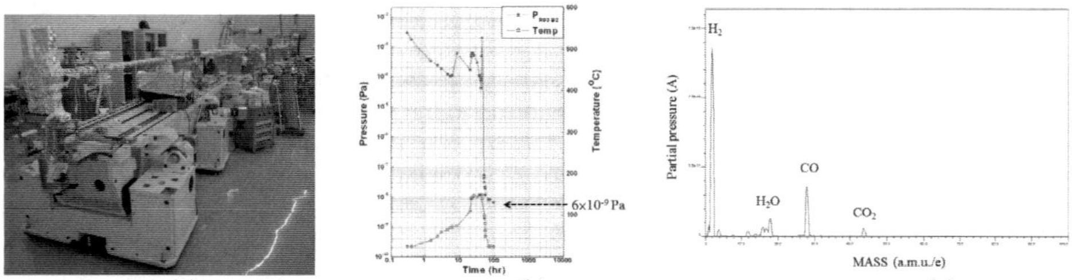

(a) (b) (c)

Figure 7. Baking ex situ of one cell in the laboratory: (a) preparation for baking, (b) pumping curve, (c) mass spectrum after baking.

4.2. Baking test for the IP and NEG pumping assembly

The combination of triode ion pump (IP) (200 L/s) and non-evaporable getter (NEG) pump (350 L/s) was designed to fit in the spaces between the magnets and located nearer the major outgassing sources from the photon absorbers to enhance the highly efficient pumping performance. An experimental system was set up, as shown in figure 8(a), for the UHV-pumping test. The ion pump was connected to the chamber at the bottom. Two side ports of the ion pump were mounted with the NEG pump and the roughing pump. To measure the pressure, an extractor ion gauge, labeled G1 in fig. 8, was installed in the chamber. Vacuum gauges G2 ~ G6 of other kinds were mounted near the chambers for comparison. The performance of pumping and the gauge measurements in the UHV were evaluated through the pumping and the baking. After evacuation of the system, baking for degassing and NEG activation, the ultimate pressure (G1) attained ~ 8×10^{-10} Pa (~ 6×10^{-12} Torr as shown in fig. 8(b)). The major residual gases comprised H_2, CO and CO_2. When the system was isolated from the turbo-molecular pump and the ion pump shut off, the pressure increased to ~ 4×10^{-9} Pa, about five times as great, with the pressure increasing from the residual outgassing of methane (CH_4). The NEG pump actively pumps all gases but noble gases such as CH_4. The pressure recovered to 8×10^{-10} Pa after switching on the IP to reveal the workable pumping at 10^{-10} Pa.

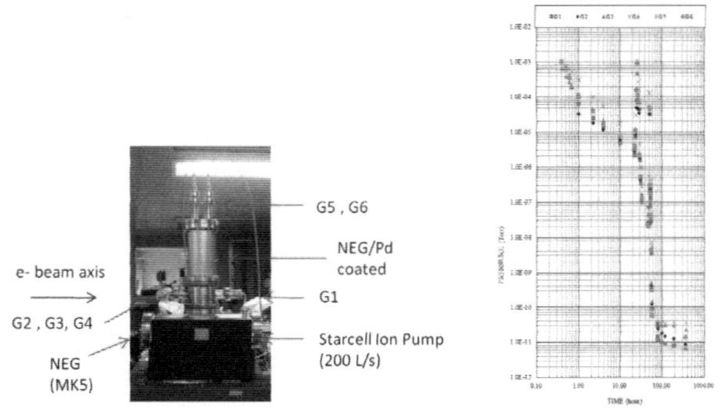

(a) (b)

Figure 8. (a) Experimental setup of vacuum chambers with ion pump, NEG pump, and gauges (G1 ~ G6); (b) Pumping curve.

5. Conclusion

The design concepts, manufacture and test results for the aluminium ultra-high vacuum system for the TPS synchrotron light source are described. CNC machining completely free of oil with a system to spray pure alcohol working for machining the aluminium B-chambers and BPM ducts in a dry clean room provides a precise control of dimension with tolerance < 0.02 mm and a clean surface. After the machining, the machined piece is promptly sealed in a clean aluminium bag filled with pure N_2 gas for storage. The aluminium parts are not removed until the welding process proceeds in the clean room. Cleaning with ozonized water applied for aluminium parts before welding provides a least rate of thermal outgassing, 6.4×10^{-12} Pa m/s. The aluminium chambers for the 14-m cell vacuum systems are welded on the pre-aligned girder supports in the clean room for stringent dimensional control and cleaning quality of the vacuum chambers. The overall tolerance for the deformation and the alignment for the 14-m vacuum chambers is controlled within 0.3 mm. All work for the welding in house and the vacuum assembly were performed in a clean room. The pumping test of an experimental system assembled with IP and NEG and a baking test for the 14-m vacuum cell have attained a pressure less than 6×10^{-9} Pa in UHV. The result proves that the design of the aluminium UHV system is acceptable for the TPS synchrotron light source.

Acknowledgements

We thank colleagues in the Precision Mechanical group for providing the pre-aligned girders for the installation work for the 14-m vacuum cells, and colleagues in the Vacuum group for their collaborative hard work on welding, assembling, carrying and baking processes.

References

[1] Hsiung G Y, Chan C K, Hsueh H P, Yang T L, Kuan C K, Chang C C, Hsu S N, Yang C Y, Chen C L and Chen J R 2007 *CP879, Synchrotron Radiation Instrumentation: Ninth International Conference* 62

[2] Chen J R, Chen G S, Wang D J, Hsiung G Y and Liu Y C 1990 *Vacuum* **41(7-9)** 2079

[3] Hsiung G Y, Chan C K, Chang C C, Chen Y B, Hsueh H P, Chen C L, Yang C Y, Shu S N, Cheng C M, Kuan C K, Sheng I C and Chen J R, 2008 *Journal of Physics: Conference Series* **100** 092014

[4] Hsiung G Y, Chan C K, Chang C H, Hseuh H P, Yang T L and Chen J R, 2006 *Proceedings of EPAC 2006* THPLS064 3433

[5] Sheng I C, Cheng Y T, Kuan C K, Hsiung G Y, Chen J R and Yang C Y, 2010 *Proceedings of IPAC'10* TUPEA079 1506

[6] Chen J R, Hsiung G Y, Chang C C, Chen C L, Chan C K, Cheng C M, Yang C Y, Wu L H and Hsueh H P 2010 *J. Vac. Sci. Technol.* A**28(4)** 942

[7] Liu Y C and Chen J R 1991 *AIP Conference proceedings* **236** 173

[8] Chan C K, Hsiung G Y, Chang C C, Chen Rouge, Yang C Y, Chen C L, Hsueh H P, Hsu S N, Liu Ivan and Chen J R 2008 *Journal of Physics: Conference Series* **100** 092025

Design and development of primary orifice plate flowmeter

M Maqsood[1], J Ali[2], A Usman[1] and M Farooq[2]
[1]National Institute of Vacuum Science & Technology, NCP Complex, Shahdara Road, P.O. Box # 3125, Islamabad – Pakistan
[2]Department of Physics, Garden Campus, Hazara University, Mansehra, Pakistan

E-mail: maqsood@ninvast.edu.pk

Abstract: Accurate and stable gas flows are very important in different applications like, the performance study of vacuum pumps, gauge calibration, leak detection and advance research in low pressure physics. Primary Orifice Plate Flowmeter (OPF) has been designed and developed indigenously. This flowmeter consists of two orifice plates. Orifice-1 (O_1) acts as a flow restriction, while orifice-2 (O_2) enables continuous pumping mode. By varying upstream pressure P_1, a change in downstream pressure P_2 is recorded, and the flow-rate is calculated from the conductance "C" of the orifice O_1 and pressure difference by the relation $Q = C(P_1 - P_2)$. Main feature of this primary OPF is that it is very simple and compact. A stable and reproducible flow-rate has been achieved in the range of 10^{-5} to 10^{-6} mbar.l/s. A variation in Q has been observed with a change in temperature of orifice and orifice diameter.

1. Introduction

Accurate measurement of the transport of gas molecules from low vacuum to high vacuum has been greatly emphasized throughout the history. Small gas flows occur at very low rate, but its impact bring significant changes in sophisticated and complex manufacturing processes and research work, like mixing of gases in different ratios, gauge calibration, fusion and particle accelerators, in electronic devices manufacturing labs and leak testing laboratories. Measurement of such micro-flow is very crucial and fluctuations in gas flow may result in errors.

In last couple of decades, rapid industrialization has taken place in Pakistan. During this period, the understanding of vacuum technology has improved remarkably. A large number of industries and research laboratories in Pakistan are using vacuum technology because of the increasing demand of innovations in industrial products. A primary flowmeter for the calibration of leak was demanded by industries and research laboratories, which motivated us to initiate a step towards the development ofa Primary Orifice Plate Flowmeter (OPF).

In the present work design parameters, performance and results of this OPF are discussed in detail. A detail of flowmeter designs is discussed comprehensively in literature [1]. OPF is a constant volume flowmeter and flow-rate for Nitrogen gas with high purity (99.99%) generated through this

[1] To whom any correspondence should be addressed.

flowmeter is presented in this paper. The compact design of OPF ensures minimum out gassing rate. Due to compact design with minimum surface to volume ratio of the OPF alongwith conflate flanges ensures minimum outgassing rate.

2. Experimental setup

A schematic diagram of the OPF is shown in figure 1. It consists of two small volumes with upstream and downstream pressure sides separated by a specially designed orifice plate "O_1". Upstream pressure side P_1 is installed with a Pirani gauge, whereas downstream side P_2 is monitored by cold cathode gauge. Both the gauges were calibrated on primary standard before installation on the OPF. Upstream side is connected with a buffer through a fine needle valve. It adjusts reference pressure in the upstream side which varies from 10^{-2} to 10^{-1} mbar. The buffer is further connected to the gas cylinder through a regulated valve.

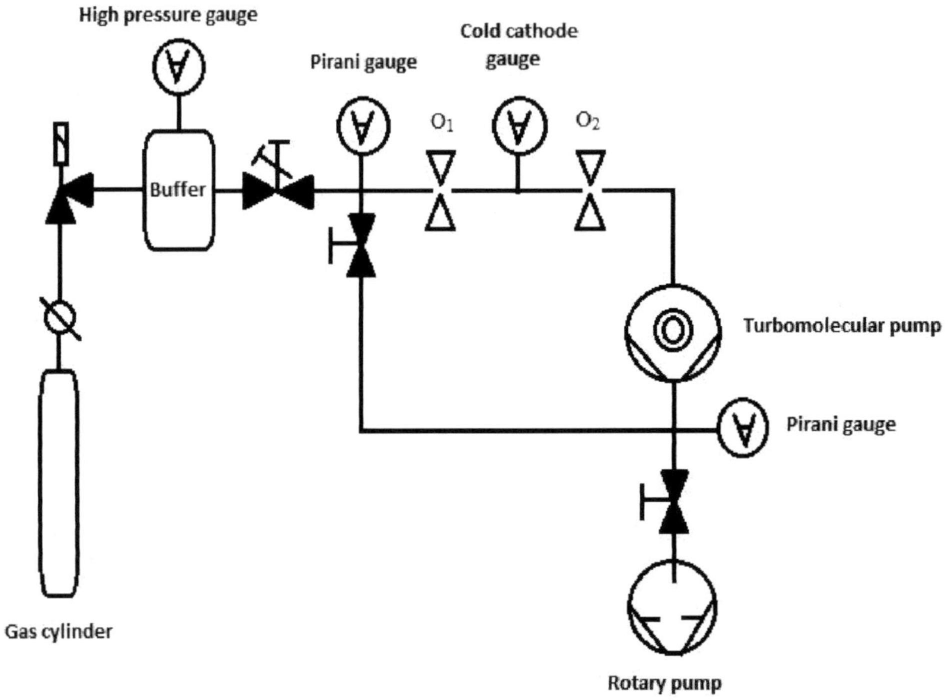

Figure 1. Schematic diagram of Orifice Plate Flowmeter.

The arrangement, shown in figure 1, helps in maintaining a desirable upstream pressure P_1. Downstream side is pumped by turbo molecular pump backed by rotary vane pump. A second orifice "O_2" of 1.7 mm diameter is installed between "P_2" side and turbo molecular pump to maintain a dynamic flow. The idea of two orifices is also reported in literature [2,3]. The flowmeter is designed such that the conductance of orifice "O_1" is kept smaller than that of the "O_2", and the conductance of "O_2" is in turn smaller to that of the pumping speed of turbo molecular pump. To minimize outgassing, orifice plates are sandwiched between conflate flanges. Two different orifice plates of orifice diameter 110μm (O_a) and 130μm (O_b) respectively have been employed at "O_1" position. These plates were specially designed for OPF. Stainless steel sheet was electron beam welded in copper seal as shown in figure 2.

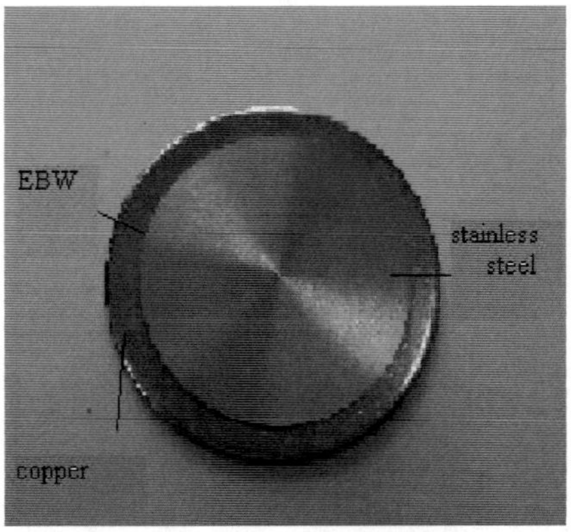

Figure 2. Orifice plate used in experiment.

Flow-rate Q of the flowmeter was calculated by the following equation (2.1) [4]:

$$Q = C(P_1 - P_2) \quad (2.1)$$

Where C is conductance of the orifice O_1 and P_1 and P_2 are upstream and downstream pressures respectively. The conductance C of orifice in the molecular regime is given by equation (2.2) [4]:

$$C = 3.64A\sqrt{\frac{T}{M}} \quad (2.2)$$

Here T is temperature, M is molecular weight of the gas and A is area of the orifice. The main objective of current work was to study the behavior of flow-rate with different orifice diameters and to study the effect of temperature on the flow-rate.

3. Results and discussions

The objective of the present work was to study the behavior of flow-rate with different orifice plates O_a & O_b. The flowmeter consists of CF flanges, CF cross, ionization gauges and copper seals with orifice plates. The whole assembly (all metal) was leak tested at 5×10^{-9} mbar.l/s and baked at high temperature (150°C) that is more than the experimental temperatures (75°C). Figure 3 shows a plot of flow-rate Q vs upstream pressure P_1 at three different temperatures 25°C, 50°C and 75°C for orifice plate O_a. The variation in upstream pressure from 2×10^{-2} to 2×10^{-1} mbar was kept same for all the three temperatures. Results show that flow-rate Q increases linearly with an increase in upstream pressure P_1 and temperature effect is also prominent as shown in figure 3. A variation in flow-rate has been observed when temperature was raised from 25°C to 75°C as shown in figure 4. Figure 3 also shows plot of flow-rate Q vs upstream pressure P_1 at three different temperatures 25°C, 50°C and 75°C for orifice plate O_b. Here, again a linear behavior of flow-rate against upstream pressure was observed as in case of orifice plate O_a.

Journal of Physics: Conference Series **439** (2013) 012035 doi:10.1088/1742-6596/439/1/012035

A comparison of the flow-rate results obtained from plate O_a & O_b is shown in figure 3. The minimum flow-rate Q_a recorded for orifice "O_a" was 3.22×10^{-6} mbar.l/s, whereas flow-rate Q_b for orifice "O_b" was 4.63×10^{-6} mbar.l/s.

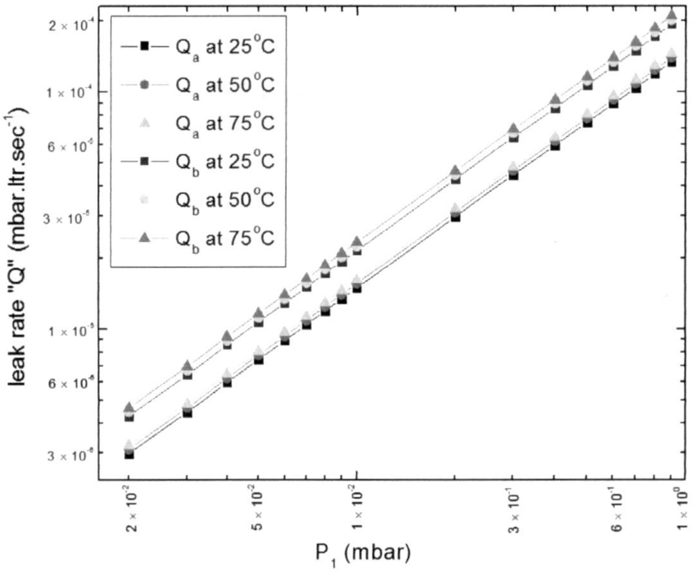

Figure 3. Flow-rate Q vs upstream pressure P_1 for "O_a" and "O_b".

Similarly, the maximum Q_a observed for orifice "O_a" was 2.97×10^{-5} mbar.l/s and Q_b for orifice "O_b" was 4.28×10^{-5} mbar.l/s.

The behavior of flow-rate with the increase in temperature has also been analyzed. It is found that the flow-rate increases linearly by 0.16% with per degree rise in temperature of orifice plate. Thus, a total of 8% rise in flow-rate is observed with an increase in temperature from 25°C to 75°C.

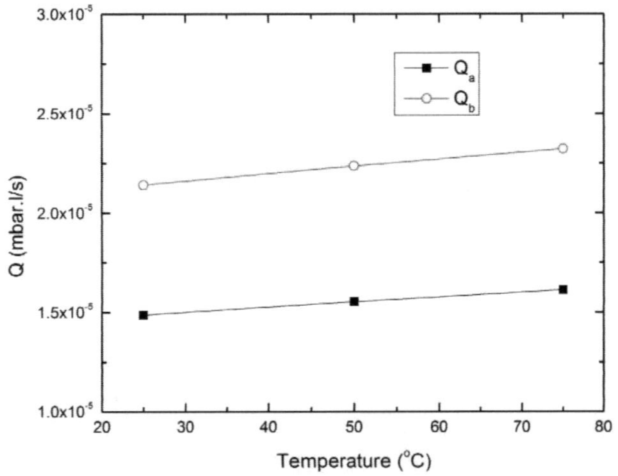

Figure 4. Flow-rate Q vs temperature for "O_a" and "O_b" at the upstream pressure of 1×10^{-1} mbar.

4. Uncertainty

Uncertainties of the different factors effecting flow-rate $Q$are evaluated by deriving standard deviation of all the parameters (pressure, temperature and hole diameter) and the expanded uncertainty using equations 4.1, 4.2 and 4.3 which shows an increase with the increase in temperature as shown in figure 5.

$$\sigma = \sqrt{\frac{1}{N-1}\Sigma(x-\bar{x})^2} \tag{4.1}$$

$$U_c = \frac{\sigma}{\sqrt{N}} \tag{4.2}$$

$$U = kU_c \tag{4.3}$$

Where,

σ is the standard deviation, N is the number of trials, x is the sample data, $\bar{x}$ is the mean value, U_c is the combined uncertainty, U is the expanded uncertainty and k is the confidence level which is 95% with $k = 2$.

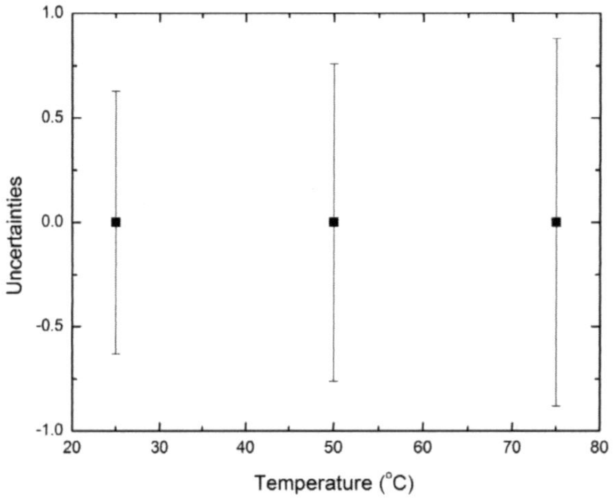

Figure 5. Uncertainties of the factors effecting Q.

5. Conclusions

In the present study it is found that, an increase of 20μm in orifice diameter results in an increase of 44% in flow-rate. Hence a minimum flow-rate can be obtained by using orifice of smaller diameter in the upstream pressure "P_1" range of 2.0×10^{-2} to 2.0×10^{-1} mbar. Also, a variation of 8% in flow-rate has been observed when temperature was raised from 25°C to 75°C. Hence 0.16% increase in flow-rate has been observed per degree rise in temperature of orifice.

References

[1] Berman A 1985 *Total pressure Measurement in Vacuum Technology* Academic Press Orlando
[2] Bergoglio M, Raiteri G, Rumiano G and Jitschin W 2006 *Vacuum* **80** 561
[3] Hufnagel H, Hartmann H K and Jitschin W 1992 *Vacuum* **43** 773
[4] Berman A, *Vacuum engineering calculations, formulas, and solved exercises* Academic Press, Inc.

Design, construction and evaluation of solarized airlift tubular photobioreactor

A Bahadur[1], M Zubair[1] and M B Khan[2*]
[1]School of Chemical and Materials Engineering (SCME), National University of Sciences and Technology, Sector H-12, Islamabad, Pakistan
[2]Centre for Energy Systems, USAID Centre for Advanced Studies
National University of Sciences and Technology, Sector H-12, Islamabad, Pakistan

[*]E-mail: principal.ces@nust.edu.pk

Abstract. An innovative photobioreactor is developed for growing algae in simulated conditions. The proposed design comprises of a continuous tubular irradiance loop and air induced liquid circulation with gas separation through air lift device. The unique features of air lift system are to ensure the shear free circulation of sensitive algal culture and induce light/dark cycles to the photosynthetic micro-organisms. The design strategy employs to model and construct a 20-liter laboratory scale unit using Boro-silicate glass tubing. The material is selected to ensure maximum photon transmission. All components of the device are designed to have flexibility to be replaced with an alternative design, providing fair chance of modification for future investigators. The principles of fluid mechanics are applied to describe geometrical attributes of the air lift system. Combination of LEDs and Florescent tube lights (Warm white) were used to illuminate the photosynthesis reaction area providing a possibility to control both illumination duration and light intensity. 200 Watt Solar PV system is designed to power up the device which included air pump (100 Watt) and illumination system (100 Watt). Algal strain *Chlorella* sp was inoculated in photobioreactor which was sparged with air and carbon dioxide. The growth was sustained in the batch mode with daily monitoring of temperature, pH and biomass concentration. The novel photobioreactor recorded a maximum experimental average yield of0.65 g/l.day (11.3 g/m^2.day) as compared to theoretical modeled yield of 0.82 g/l.day (14.26 g/m^2.day), suggesting the device can be efficiently and cost-effectively employed in the production of algal biomass for biofuels, concomitantly mitigating CO_2.

Keywords: Photobioreactors; Microalgae; *Chlorella vulgaris*; Airlift; Biomass

[*] To whom any correspondence should be addressed.

Journal of Physics: Conference Series **439** (2013) 012036 doi:10.1088/1742-6596/439/1/012036

Nomenclature

μ	Specific growth rate	day^{-1}
μ_{max}	Maximum specific growth rate	day^{-1}
X	Biomass concentration	g/L
C_{algo}	Initial algal concentration	g/L
$C_{alg(t)}$	Algae concentration at time "t"	g/L
U_{Lr}	Liquid circulation velocity	m/s
U_G	Superficial gas velocity	m/s
ϵ_r	Riser gas holdup	-
ϵ_d	Down-comer gas holdup	-
A_d&A_r	Area of riser and down-comer	m^2
h_d	Gas liquid dispersion height	m
h_L	Un-gassed liquid column height	m
K_B	Frictional loss coefficient	-
C_f	Fanning friction factor	-
Re	Reynold's Number	-
V	Culture volume	Liter
L	length of continuous irradiance loop	ft
I_{in}&I_{out}	Incident & transmitted light intensities	$W.m^2$ or Lux
I_k	Light intensity saturation constant	$W.m^2$ or Lux
b	Light path (diameter of tube)	m
a_c	spectral average absorption coefficient	m^2/kg
R_{O_2}	Vol- rate of oxygen generation by photosynthesis	$molO_2\ m^{-3}\ s^{-1}$
U_b	Bubble rise velocity	m/s
X & W	Height & width of degasser	m
U_{LH}	Velocity of liquid in degasser	m/s

1. Introduction

Micro algae biomass offers more sustainable and carbon efficient alternative for biofuels production. Aquatic microorganisms have the ability to double their mass in less than 24 hours by efficient utilization of CO_2 in presence of light for biofuels production.

The limited growth rate and photosynthetic efficiency of open raceways due to the extended radiance pathway, which results in self-shadowing and additional ecological parameters like temperature, poisonous contamination with other microbes has facilitated the design and construction of closed photobioreactors [1]. Of the many designs of closed photobioreactors tubular photobioreactors are the most promising for growing algae [2, 3]. Table-1 represents published tubular photobioreactors with their engineering analysis.

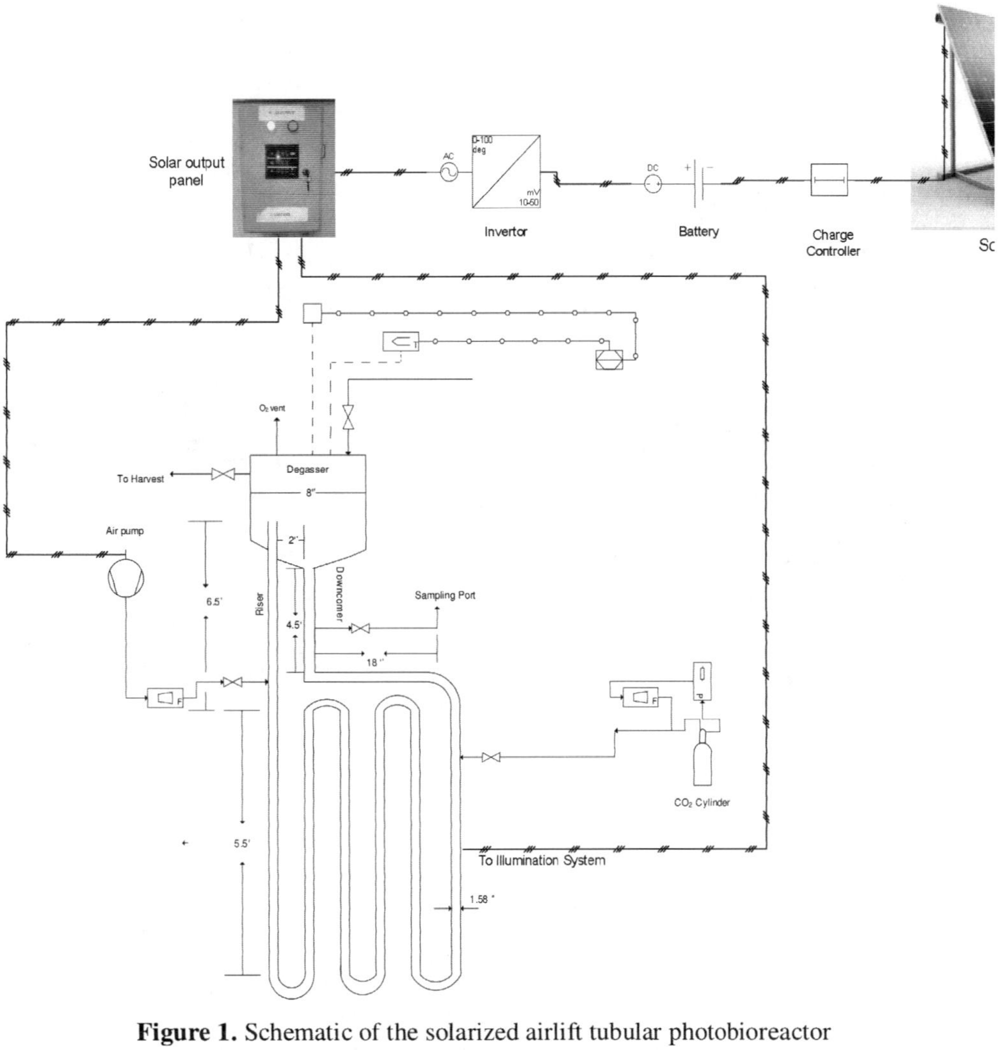

Figure 1. Schematic of the solarized airlift tubular photobioreactor

Table 1. Published tubular photobioreactors specifications

Year	Specifications	Specie Tested	Results	Investigators
1983	52 pyrex glass tubes, length: 1m, Dia: 1cm, Horizontally stacked, circulation via pump	Chlorella type green alga	Productivity: 24 $gm^{-2}d^{-1}$, light intensity: 38 W/m^2, high S/V ratio (127), oxygen accumulation not considered	[4]
1988	Area: 100 m^2, 20 polyethylene tubes, length 20 m, dia: 6cm	Porphyridiumcruentum	20-25 gm^{-2} d^{-1} Low scaling due to self cleaning via two plastic balls	[5]
1993	8 Poly carbonatemanifold tubes, Length: 20m. Dia: 3cm, Outdoor, Temp control by water spraying	Cyanobacteria	Vol- Productivity : 0.55 g $l^{-1}d^{-1}$, Reduction of head losses and lower oxygen concentrations	[6]
1995	300 L, Elevated manifold, Covered land area 12 m^2, 10 PVC tubes, Length: 25m, Dia 3 cm	Dunaliellasalina	72 gm^{-2} d^{-1}	[7]
1998	Manifold Rigid or flexible tubes, ID: 5cm, Length: 30m	A. platensis, A. siamensis, Nannochloropsis	Vol prod: 1.3 $gl^{-1}d^{-1}$, Areal Prod: 28$gm^{-2}d^{-1}$, low shear stress	[8]
2000	Airlift Serpentine 200 L, Plexiglass tubes, dia 2.5-5 cm, loop length : 98.8 m, riser length: 3-3.5 m,Dilution rate : 0.04 h^{-1}	Phaeodactylumtricornutum	Vol-productivity:1.2-1.9 $gl^{-1}d^{-1}$, Areal Productivity: 32 g $m^{-2}d^{-1}$, Degasser for effective gas exchange	[2]

Tubular photobioreactors that comprise of an airlift device is especially attractive to circulate the fluid without any moving parts [9]. Airlift device combines the function of a pump and a gas exchanger that removes the oxygen produced by photosynthesis [10]. Continuous removal of oxygen is necessary to avoid excessive buildup that inhibits photosynthesis. In addition, the photobioreactors geometry must ensure the capturing of maximum photons while minimizing the land surface occupied [11].

Effect of illumination duration on algae was studied by many researchers in literature. One study revealed that the most advantageous illumination duration for algal growth is 16 hours; however illumination for 24 hours gives a slight increase in productivity [12]. In another study growth of algal sp. *Botryococcusbraunii* was investigated at different illumination cycles, The best growth was achieved under 24 hour continuous illumination with The highest specific growth rate of 3.6 per day as compared to 0.05 under diurnal light cycles (12hr light/12 hour dark) [13]. In present study the purpose of reactor solarization is to prove the concept of exploiting solar energy to provide 24 hour continuous illumination cycle for future outdoor algae cultivation.

Here, we present a method for designing solarized airlift tubular photobioreactor. Effects of light intensity, hydrodynamics and temperature on various evaluation parameters are discussed. A photobioreactor designed using the approach outlined is proved for culture of the microalga *Chlorella Vulgaris.*

2. Photobioreactor design

A solarized airlift tubular photobioreactor is shown in figure 2. Air lift system induces circulation of fluid through the continuous tubular loop where main photosynthesis reaction occurs. The oxygen produced from the photosynthesis is separated in the air lift section when fluid returns to that section. A liquid or slurry pool at the top of the riser and down comer tubes prevents the gas bubbles from recirculating. The transparent Borosilicate glass continuous loop is designed to ensure maximum photon capturing. The diameter of tubing is selected on the basis of light beam attenuation so that dark zone is kept minimal. In addition, the light-dark fluctuations must be sufficiently robust to prevent long dark zone exposure of fluid.

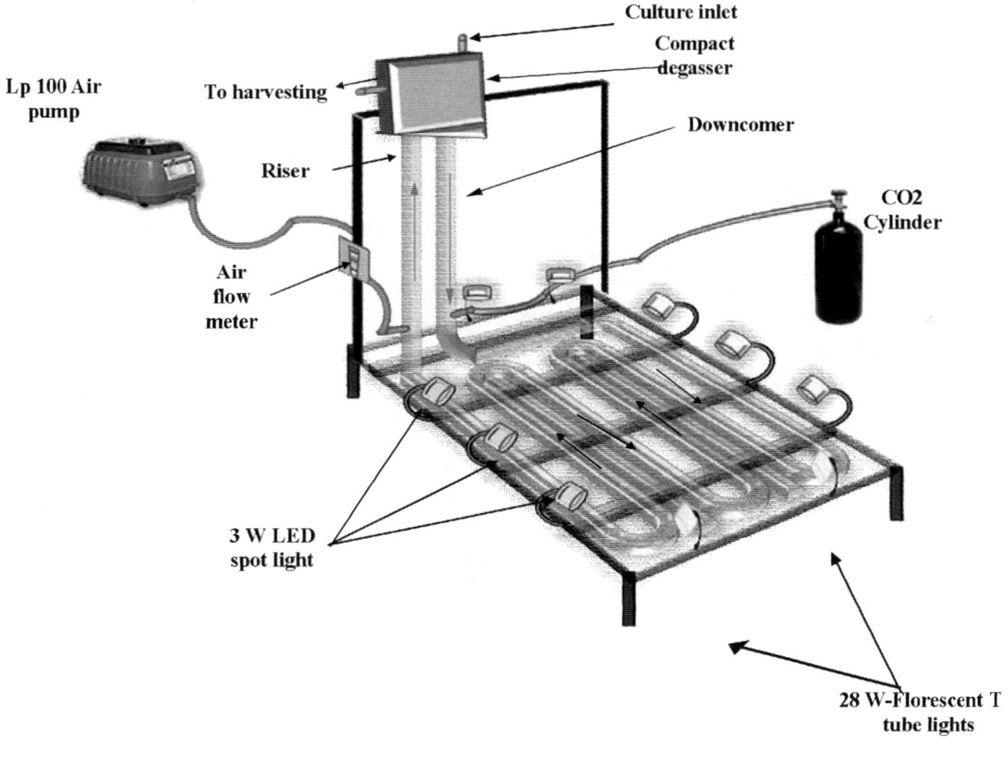

Figure 2. The photobioreactor

2.1. Continuous irradiance loop

Continuous irradiance loop is actually the photosynthetic area where illumination is provided for growth of algae. To avoid mutual shading it was proposed to construct a single continuous loop (figure 2.) so that maximum photons could be utilized in given surface area. In phototrophic cultures biomass productivity is usually control through available light intensity. In batch cultures specific growth rate of algae depends on level of irradiance, light path and algae concentration. The volumetric productivity can be easily determined as:

$$\frac{dC_{alg(t)}}{dt} = \mu C_{alg(t)} \tag{2.1}$$

Where $\frac{dC_{alg(t)}}{dt}$ is the volumetric growth rate at any given batch time t, $C_{alg\,(t)}$ is algae concentration in photobioreactor and μ is the specific growth rate of algae.

Specific growth rate is a kinetic expression for this biochemical reaction. Monod kinetic model is (Equation 2.2) used as light-limited growth kinetic model to determine specific growth rate [2].

$$\mu = \frac{\mu_{max} I_{avg}}{I_{avg} + I_K} \tag{2.2}$$

Where μ_{max} is the maximum specific growth rate, I_K is the light intensity saturation constant for specific algal specie and I_{avg} is the average light intensity on irradiance loop.

Areal productivity can be easily calculated after knowledge of volumetric productivity by employing this relation:

$$P_{alg_A} = P_{alg_V} \frac{V}{S} \tag{2.3}$$

Where V is the volume of the reactor and S is the land surface occupied by it.

Estimation of μ is important in determination of volumetric productivity. Specific growth rate requires the identification of average irradiance on tubular loop surface. Lambert-Beer's law is used to determine the average light intensity and the light gradient inside tubular photobioreactor as light attenuation along tube diameter contributes to the algae growth limitation due to light transmission and self shading phenomena [14].

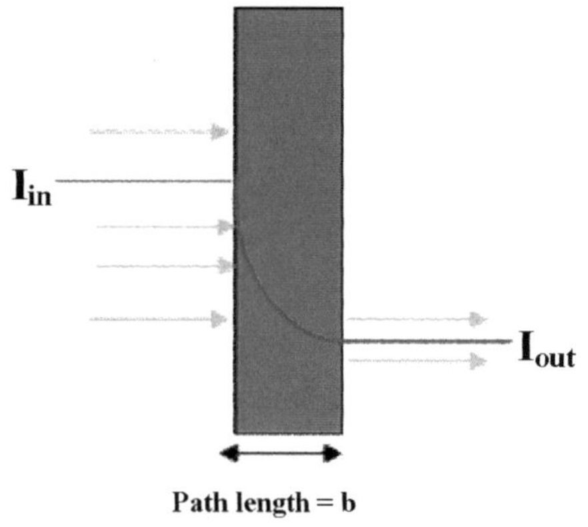

Figure 3. Light attenuation in PBR

Lambert beer's relationship along with principles of astronomy after simplification yields Equation-2.4:

$$I_{avg} = I_{in} \cdot \frac{1}{b}\left(1 - e^{-a_c \cdot C_{alg_o} \cdot b}\right) \cdot \frac{1}{C_{alg_o} \cdot a_c} \tag{2.4}$$

In above equation I_{in} & I_{out} are light intensities falling on PBR surface and coming out from PBR, a_c is the spectral average absorption coefficient and b is the path length which is the diameter of the tube in this case.

Thus from a knowledge of the characteristics parameters of the algal strain (i.e.μ_{max},a_c ,I_K) and using Equations 2.1-2.4, the biomass productivity may be determined for any combination of external irradiance and the diameter of tubes.

It is recommended that loop configuration must ensure a turbulent flow to avoid stagnation of cells in dark regions of irradiance loop [2]. Simultaneously, shear stress on algal cells will decide the upper limit on turbulence. To avoid the damage associated with shear stress, the dimensions of micro eddies generated through turbulence should be greater than the size of specific algal specie cell [15].

Certain factors restrict the maximum length of continuous irradiance loop which include liquid circulation velocity in loop, rate of photosynthesis and acceptable upper limit on DO (dissolved oxygen) concentration [10]. The maximum restricted length *L* of a continuous loop is determined by under mentioned relation:

$$L = \frac{U_L([O_2]_{out}-[O_2]_{In})}{R_{O_2}} \tag{2.5}$$

Where $U_L$is the maximum allowable fluid velocity in consideration with shear stress,$[O_2]_{In}$ is the oxygen concentration at the entrance of loop which is almost equal to saturation value when the fluid is in equilibrium with the atmosphere, $[O_2]_{out}$ is the oxygen concentration at the outlet which should be the maximum acceptable value that does not inhibit photosynthesis and R_{O_2} is volumetric rate ofoxygen generation by photosynthesis.

2.2. Airlift device with gas separator

Gas induced liquid circulation is the interesting feature of airlift system. Gas-liquid contact operations in process industries are now a days shifting to airlift system due to its simplicity and low energy consumption. The liquid slurry pool at head region of riser and down-comer (figure 4.) can be designed as effective separator for continuous removal of O_2 from culture.

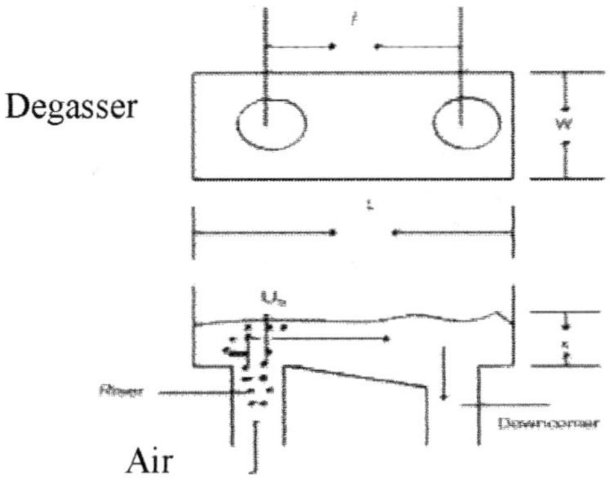

Figure 4. Airlift device

For effective separation of gas bubbles, the time taken by the bubbles to raise height '**X**' should be equal to or less than that of time taken by the fluid to traverse length '*l*'. Apply equation of continuity as same quantity of fluid passes through the regions of riser, down-comer and degasser

$$U_L A_r = U_{Ld} A_d = U_{LH} XW \tag{2.6}$$

Where A_r is the cross-sectional area of the riser tube, A_d is the mean vertical cross-sectional area of the degassing zone, U_{Ld} is the mean superficial liquid velocity in the degasser, U_{LH} is the liquid velocity in degasser tube, X is the height of liquid in degasser and W is the width of degasser. To satisfy the disengagement criterion [16], the length *'l'* of the degasser is governed by the relationship:

$$l \geq U_L A_r / U_b W \quad (2.7)$$

Where U_b is the bubble rise velocity, 0.2 m/s was used to determine length *'l'*
Superficial gas &liquid velocities and gas hold up are considered as significant hydrodynamic attributes for the design and performance of air lift and bubble column reactors. Energy balance on air lift reactor results in the equation to determine the superficial liquid velocity in the reactor [16]. "U_{Lr}" The superficial liquid velocity is found by Equation 2.8

$$U_L = \left[\frac{2gh_D(\epsilon_r - \epsilon_d)}{K_B\left\{1/(1-\epsilon_r)^2 + \left(A_r/A_d\right)^2 1/(1-\epsilon_d)^2\right\}}\right]^{0.5} \quad (2.8)$$

ϵ_r=Riser gas hold up with respect to liquid

ϵ_d=Downcomer gas hold up

A_d and A_r= Areas of the down-comer and the riser (m^2)

h_d= Gas liquid dispersion height (m)

The factor h_d is associated to the overall gas hold up inside the reactor (ϵ) and un-aerated liquid height (h_L), as explained in under mention equation [16]:

$$h_d = h_L/(1-\epsilon) \quad (2.9)$$

The following equation can be used to determine K_B which frictional loss coefficient is

$$K_B = 4C_f\frac{L_{eq}}{d} \quad (2.10)$$

$$C_f = 0.0791\,Re^{-0.25} \quad (2.11)$$

The riser gas hold up can be determine using the following governing equation [18]

$$\epsilon_r = \frac{U_G}{[0.24+1.35(U_G+U_L)^{0.93}} \quad (2.12)$$

Where

$$\epsilon = \frac{\epsilon_r A_r + \epsilon_d A_d}{A_r + A_d} \quad (2.13)$$

The equations 2.8 to 2.13 are solved through numerical iteration procedures assuming an initial value or superficial liquid velocity. The iteration process is used to model liquid circulation velocity with respect to superficial gas velocity.

3. Material and methods

3.1. Specie and nutrient Media

Samples of locally grown fresh water algal specie *Chlorella vulgaris* were taken from Pakistan Agriculture Research Council (PARC). Initially sample was inoculated in a glass column by providing specified media, light and air for its initial growth. Bold Basal Medium and Modified Bold Basal Medium were the stock solutions as mentioned in Appendix A.

3.2. Dry biomass concentration

The dry mass of algal specie *Chlorella Vulgaris* was determined manually by taking samples from PBR after specified period. A filter paper was utilized to determine the dry biomass concentration, which was previously dried in an oven at 150 °C for one hour and weighed for its initial weight. The filter papers were dried in the lab oven after filtration at 150 °C for about 2 hours. Final weight was determined by cooling the filter paper at room temperature. After filtration and before filtration difference in weights divided by the filtered sample volume provided the dry mass concentration.

3.3. Temperature and pH

TECPEL 4-Channel input DTM 319 digital temperature data logger (FIG-4.14) was used to record temperature at four different locations around PBR. Martini thermometer coupled with a type K thermocouple was also employed to determine the temperature of fluid inside the photobioreactor. The temperature was recorded by adjusting the tip of thermocouple at least two inches below the water level in the degasser. Martini pH meter Mi-160 pH meter was used to record online pH for all the photobioreactor experiments.

3.4. Light Intensity

DLM105HA digital light meter was used to measure light intensity. The units of measurement were Lux. As the reactor had light sources on its both sides, the sensor was placed on both upper and downward sides of the reactor; the readings were noted and averaged to get the light intensity.

3.5. Liquid circulation velocity

Actual gas induced liquid circulation was determined using tracer analysis. Series of experiments were undertaken at different airflow rates to determine actual velocity of the culture. A coloured piece of cotton fibre cloth whose density was almost equal to algal dry cell density inserted in photobioreactor, for a predefined distance time taken by that coloured piece was noted to get the actual velocity of the culture.

Table 2. Overview of experimental observations.

Test No	Duration	Light Intensity (Lux)	Liquid circulation velocity (m/s)	Avg. Volumetric Productivity (g/L.d)	Avg. Areal Productivity (g/m^2.day)	Nutrient Media
1	25 Nov-01 Dec	1300	0.60	0.55	9.56	Bold Basal
2	02 Dec-15 Dec	1300	0.65	0.33 [a]	5.74	Bold Basal
3	14 Jan-23 Jan	700	0.43	0.46	8	Bold Basal
4	24 Jan-30 Jan	700	0.5	0.37 [b]	6.43	Bold Basal
5	03 Feb-14 Feb	650	0.5	0.25 [c,d]	4.35	Bold Basal
6	17 Feb-27 Feb	1350	0.55	0.65	11.30	4xN Mod Bold Basal
7	01 Mar-10 Mar	1350	0.55	0.50	8.70	4xN Mod Bold Basal

a= No external carbon dioxide, b= Low mass transfer/position changed
c= very low temperature <10°C, d= Algae attached to the inner side of tube walls

4. Results and discussions

4.1. Design of Photobioreactor
Biomass productivity in a batch mode was predicted using equations 2.1-2.4 for *Chlorella vulgaris* at known maximum specific growth rate and light intensity. Batch mathematical model gave volumetric productivity of 0.82 g/l.day (Areal productivity: 14.26 g/m^2.day), this modeled yield can be compared well with the experimentally achieved maximum average yield of 0.65 g/l.day. With the increment in tube diameter, he volumetric growth rate decreases, on the other hand, the areal growth rate increases as higher culture volumes can be contained in a particular area [17]. To make a compromise between the contradictory demands of volumetric and areal productivities, available culture velocity, the energy dissipation per unit mass and local availability, a tube diameter of 0.038 m was preferred for fabrication of continuous irradiance loop.

4.2. Hydrodynamics
The mathematical model equations discussed were solved by iteration. The results revealed are for the flow rate used in all tests performed to validate PBR. The results show the expecting superficial liquid velocities in PBR with respect to superficial gas/air velocity. The model expectation for circulation time of culture through continuous tubular irradiance loop is about 16 seconds at the maximum velocity of 0.65 m/s. This period referred to as 'light side' period of complete cycle. Based on the length of the riser (2 m) and down-comer the culture from the bottom of the riser to the bottom of the down-comer takes 8 seconds, this period is 'dark side' period. The experimental observations were found in the range of ±5-10% of the modeled values. figure 5 is the graphical representation of hydrodynamics comparison.

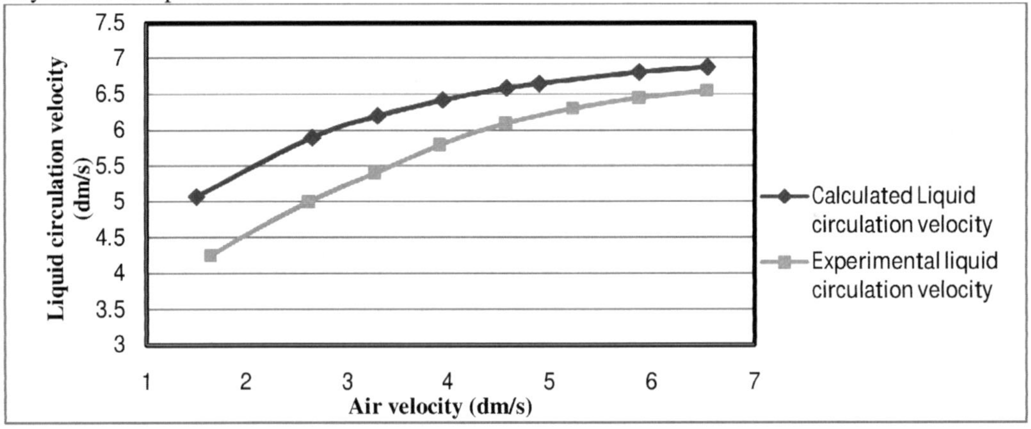

Figure 5. Calculated and experimental hydrodynamics comparison

4.3. Indoor batch culture
Performance of indoor batch culture was assessed using different irradiance, liquid circulation velocities, temperatures and Nutrient media. As analyzed from Table-2 maximum average yield was recorded at light intensity of 1350 Lux while growth rate decreased more than half of maximum average yield at 650 Lux. The experimentally obtained maximum growth rate was almost 20 % less than that of estimated using mathematical model. The major reason of this deviation is low temperature than optimal in actual conditions. Effect of liquid circulation velocity was also investigated at five different liquid circulation velocities (m/s): 0.43, 0.5, 0.55, 0.60 and 0.65. A graphical plot between liquid circulation velocities against growth rate (figure 6) reveals that

maximum growth rate was achieved at velocity of 0.55 m/s. The optimal range of liquid circulation velocity obtained from experimental evaluation is 0.47 to 0.57 m/s.

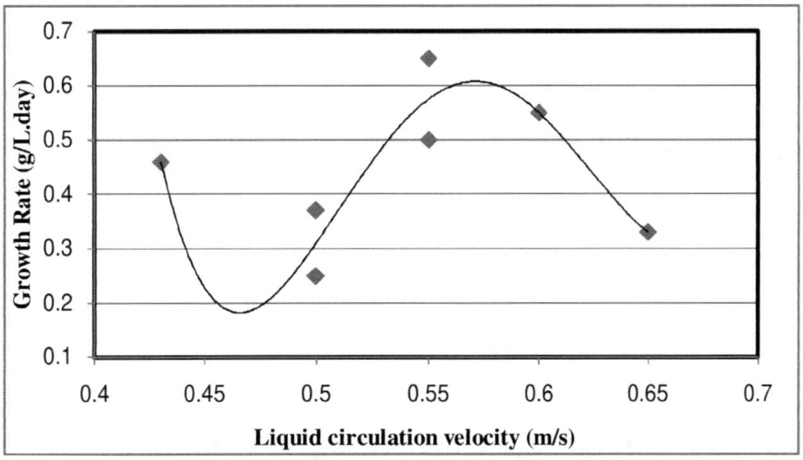

Figure 6. Effects of liquid circulation velocity on productivity of algae data were obtained in winter at different temperatures

The optimum temperature range for many algal species is 25-35 °C. In this research effects of low temperature range 10-20 °C were observed. Different studies reveal that low temperature causes the algal activity of multiplication slows down. In this study temperature below 12°C caused a considerable slow growth as depicted from the graph. A general trend in figure 7. shows that growth rate increases with increasing temperature. Maximum productivities are observed in the range of 13-20 °C.

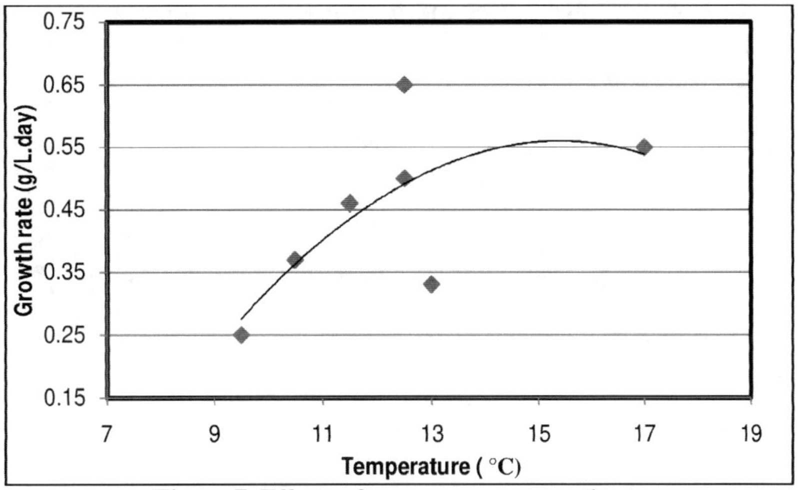

Figure 7. Effects of temperature on growth rate

Artificial illumination system was employed to study the effects of light intensity on biomass productivity. The main purpose of using artificial lights comprising warm white T-5 florescent tubes and warm white LEDs was to control both light intensity and illumination duration. It was observed by different experiments that algal growth primarily depends on the light intensity. In indoor conditions, it is believed that there was no photo-inhibition effects as controlled or limited light intensity conditions were provided. A general trend is obtained by plotting light intensity versus growth rate as described in figure 8. The investigation reveals that by increasing light intensity productivity increases.

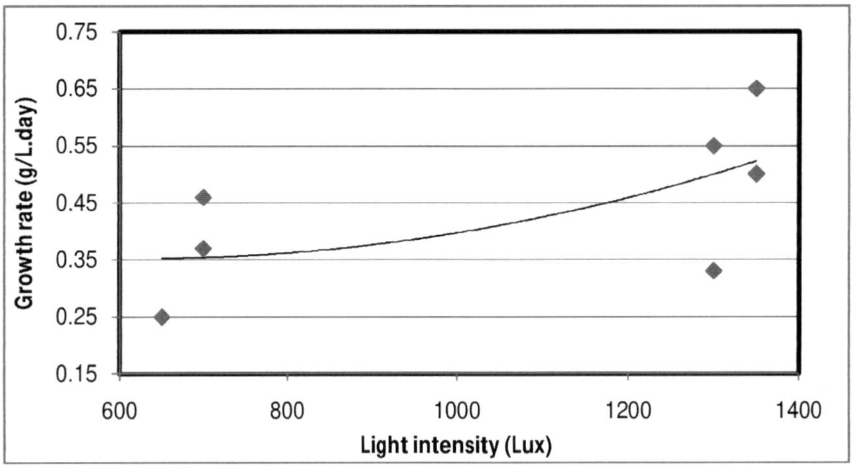

Figure 8. Light intensity versus average growth rate

Two tests 6-7 were performed using the 4xN Modified Bold Basal recipe as growth media. It can also be concluded that the use of 4xN Modified Bold Basal media instead of simple Bold Basal media resulted in the most vigorous *Chlorellasp* growth in photobioreactor. Figure 9. shows that nutrient media is vital for the growth of algae. Nitrogen, Phosphorous and Potassium are essential for the growth. Dedicated experiments are performed to validate this factor. Nitrogen-deprived culture causes the growth rate to decrease as compared to nitrogen-excess media. Nitrogen excess definitely associated with the higher protein formation and a bit lower lipid production.

A bar graph is plotted to compare the performance of PBR with different nutrient media. Test 6 & 7 were performed using 4xN Modified Bold Basal media, and all other tests 1-5 were conducted with simple Bold Basal media. It is clearly depicted from graph (figure 9.) that modified media results in the higher growth rates.

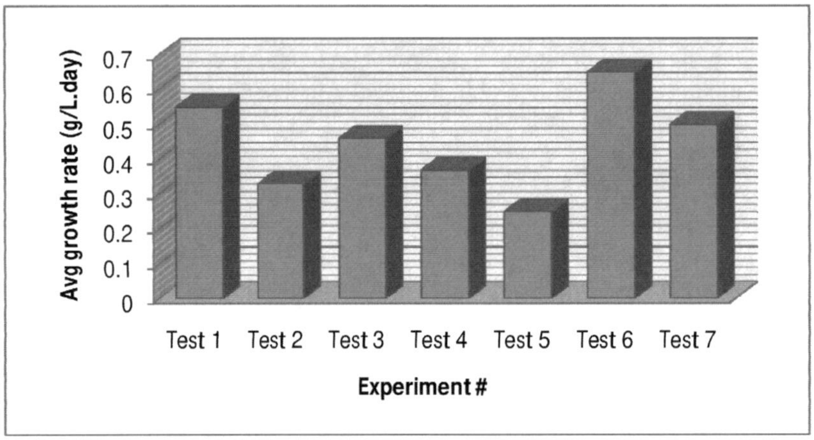

Figure 9. Effects of nutrient media on growth rate

5. Concluding comment

An engineering approach is discussed to design solarized airlift tubular photobioreactor. The novel system convenes all the design measures and restraints in terms of maintenance, mobility and flexibility for potential future amendments. The design approach effectively simulates light intensity-dependent biomass productivity, hydrodynamics that determines liquid circulation velocity in continuous loop and gas separation in airlift system. 20 L photobioreactor designed using proposed strategy was assessed in batch indoor culture of *Chlorella vulgaris*. Biomass productivity and hydrodynamics predicted from a mathematical model were in close agreement with measured experimental results. A horizontally stacked single layer continuous loop with optimal diameter of 0.038 m, length 42 ft, connected to 2 m tall airlift system. This configuration covered a land area of about 1.15 m^2. Optimal liquid circulation velocities were in the range of 0.47-0.57 m/s. The reactor practiced algae bio-fouling to the tube walls which caused the cell productivity to decline. 200 Watt Solar PV was employed to power up the reactor including 100 Watt air pump and 100 Watt illumination system.

Modeling/simulation parameters used in photobioreactor performance prediction is summarized in Table 3.

Table 3. Modeling & Simulation parameters

Factor	Unit	Value range
μ_{max}	h^{-1}	0.09
I_k	W/m^2	8
C_{algo}	g/L	0.25
I_{in}	W/m^2	15
a_c	m^2/kg	200
b	m	0.038
Illumination duration	hr	24 hr

References

[1] Watanabe Y, De la Noue J, and Hall D O 1995 Photosynthetic Performance of a Helical Tubular Photobioreactor Incorporating the Cyanobacterium Spirulinaplatensis *Biotechnology and Bioengineering* **47** 261-269

[2] Molina Grima E, Acien Fernandez F G, Garcia Camacho F and Chisti Y 1999 photobioreactors: light regime, mass transfer, and scale up *J. Biotechnol.* **70** 231–248

[3] Tredici M.R 1999 *Bioreactors, photo, Encyclopedia of Bioprocess Technology: Fermentation, Biocatalysis and Bioseparation* vol 1 (Wiley New York)s pp 395–419

[4] Pirt S J 1983 *Patent UK* 2,118,572 (to Queen Elizabeth College, University of London)

[5] Chaumont D, Thepenier C and Gudin C 1988 scaling up a tubular photoreactor for continuous culture of Porphyridiumcruentum – From laboratory to pilot plant *Elsevier Applied Science* London 199–208

[6] Richmond A, Boussiba S, Vonshak A and Kopel R 1993 A new tubular reactor for mass production of microalgae outdoors *J. Appl. Phycol.* **5** 327–32

[7] Lee Y K, Ding S Y, Low C S, Chang Y C, Forday W L and Chew P C 1995 Design and performance of an a-type tubular photobioreactor for mass cultivation of microalgae J. *Appl. Phycol.* **7** 47–51

[8] Tredici M R, ChiniZittelli G and Benemann J R 1998 *A tubular internal gas exchange photobioreactor for biological hydrogen production*, pp 391–401 (Plenum Press New York)

[9] Chisti Y, 1989 (Airlift Bioreactors *Elsevier* London)

[10] Camacho Rubio F, Acien Fernandez F G, Sanchez Perez J A, Garcia Camacho F, Molina

Grima E, 1999 Prediction of dissolved oxygen and carbon dioxide concentration profiles in tubular photobioreactors for microalgal culture *Biotechnol. Bioeng.* **62** 71–86

[11] Acien Fernandez F G, Garcia Camacho F, Sanchez Perez J A, Fernandez Sevilla J M, and Molina Grima E 1997 A model for light distribution and average solar irradiance inside outdoor tubular photobioreactors for the microalgal mass culture *Biotechnol. Bioeng.* **55** 701–714

[12] Aqua Medic, *Plankton Light Reaktor*, Operation manual, 8-14

[13] Dumrattana P and Tansakul P 2006 Effect of photoperiod on growth and hydrocarbon content of Botryococcusbraunii cultured in effluent from seafood processing plant.Songklanakarin *J. of Sci and Tech.* **28** 99-105

[14] Barbosa M J, Hoogakker J and Wijffels R H 2003 Optimisation of cultivation parameters in photobioreactors for microalgae cultivation using the A-stat technique *Biomolecular Engineering* **20(4-6)**:115-123

[15] Kawase Y and Moo-Young M 1990 Mathematical models for design of bioreactors: Applications of Kolmogoroff's theory of isotropic turbulence *Chemical Engineering Journal,* **43** B19-B41

[16] Chisti, Y and Moo-Young M 1993 Improve the performance of airlift reactors *Chem. Eng. Prog.* **89 (6)** 38–45

[17] E Molina J, Fernandez F G and Acien Y Chisti 2001 Tubular photobioreactor design for algal cultures *Journal of Biotechn.* **92** 113–131

To investigate the surface properties for increasing efficiency of solar water heater

A Hai[1], Qurat-ul-Ain
Chemical Engineering Department University of Engineering & Technology Peshawar, 25000, Pakistan

Email: haiabdul20@gmail.com

Abstract. Energy crisis is becoming a major problem in Pakistan. Renewable energy sources are used to overcome this crisis. This research is about to increase the efficiency of fin type solar water heater by modification of its surface properties. During this research work, solar water heater module is fabricated and the modification of surfaces by using bare surface, external surface coated with high conductivity paint and lead electroplating is studied and the efficiency of solar heater is observed. The temperature profile and the heat transferred is studied and it was found that paint coated surface is more efficient than bare surface, further surface electroplated with lead is more efficient than painted surface for the same ambient conditions. The average increase in the heat absorption for lead plated and paint coated surface was observed 28.57% and 10.79 % respectively.

1. Introduction

There is a major problem of energy crisis which is faced by all over the world. In order to control this energy crisis problem many researches are performed using renewable energy sources like using solar energy instead of electrical and fuel energy. Because solar energy is free of cost and environmentally friendly. In this perspective different solar water heaters are designed to use solar energy for heating of water, which is used for domestic and industrial purposes.

Solar water heater largely depends on transmittance, absorption and conduction of solar and the conductivity of the working fluid [1]. The design parameters such as plate efficiency factors have been analyzed by Hottle and Whiller, in1958 [2]. Thermal performance of collector is also dependent on the optical efficiency of glass cover, design and thermal properties of absorber plate. The maximum energy conversion of absorber plate using selective coatings to reduce radioactive losses has been analyzed by many researchers in 1979 [3-5]. Shariah et al. in 1999 studied the effect of thermal conductivity of the absorber plate on the performance of solar collectors through the transient simulation system (TRNSYS). They confirmed that the characteristic factors like fin efficiency, collector efficiency and heat removal are strongly dependent on the thermal conductivity of absorber plate '[6]'. Storage tank is a main component in solar water heater for achieving maximum efficiency, there should be minimum energy losses from the storage tank. For this purpose Colle, et al. in 2001, worked on proper insulation using various materials and various thicknesses [7]. Results showed that the heat removal efficiency of thermosyphone solar water heater is highly dependent on thermal stratification. Further experimental studies have been carried out by Chang in 2004 [8]. He evaluated

[1] To whom any correspondence should be addressed.

the heat removal efficiency during the system application phase. The overall performance rating of a thermosyphone solar heater considering the thermal performance of the system during the energy collecting phase and the system cooling loss during the cooling face has been analyzed experimentally by Chang, et al. in 2004 '[9]'.Heat transfer, friction factor coefficient of plain, microfin and twisted tape insert tube has been analyzed by A-Fahed, et al. in 1999 [10]. The collector efficiency is also accessible to increase the transfer area by adding fins [11]. Researchers have been made for special coatings on collector tube which behaves as a selective surface with low diffusivity constant for solar radiation [12].Solomon, et al. in 2006, have developed a model of a flat plate solar collection system [13].Yousef and Adnan in 2008, have investigated the effect of mass flow rate, flow channel depth and collector length on the system thermal performance and pressure drop through the collector with and without porous medium [14].

In the recent research, we observed the effect of different surface propertiesto improve heat transfer coefficient and the heat transfer rate. Heat transfer rate is evaluated for each modified surface.

2. Material and method

2.1. Experimental setup

The experiments were conducted on a lab scale fabricated solar water heater. The figure 1. shows the schematic diagram of solar water heater.

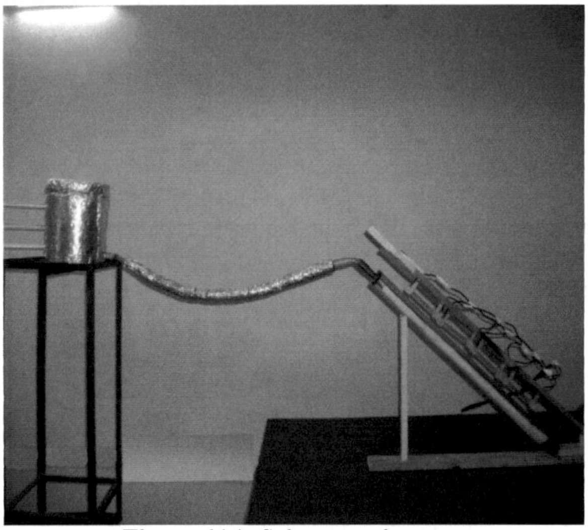

Figure 1(a). Solar water heater

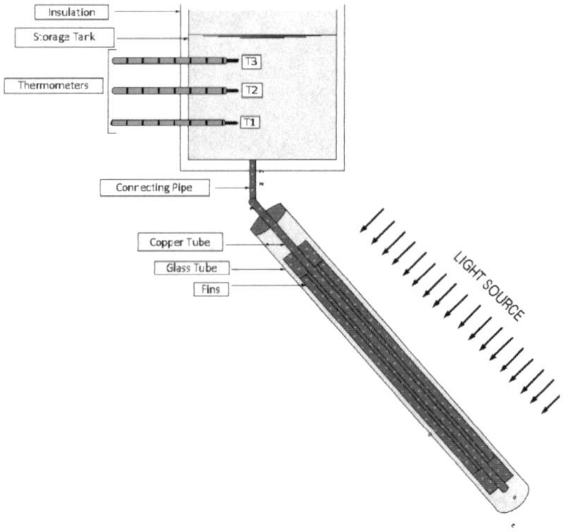

Figure 1(b). Solar water heater

2.2. Experimental procedure

Solar water heater consists of a storage tank and collector tube. Storage tank has a capacity of 5litres. The collector tube is fabricated from copper tube of having a diameter of 0.75 in and length 75cm with fins on both sides. To avoid the heat transfer from the collector tube towards the surroundings, it is fixed in a glass tube. Three thermocouples were used to find the temperature of storage tank at three different locations named T_1, T_2 and T_3.

Now for the bared copper tube, starting from a temperature of 28°C, the temperature of three thermocouples is noted at a time interval of 10minutes. Then the copper tube is covered with a black paint and again the readings are noted at the same time interval. Finally the lead electroplating is applied on the copper tube and the same procedure is repeated for the same time interval.

3. Results and discussions

Different experiments are carried out to find surface properties of solar water heater by different techniques.

3.1. Heat flux of solar water heater with bared copper tube

Figure 2. shows the effect of heat flux for simple copper tube surface. The time interval is 10minutes. We observed that with the passage of time the temperature within the storage tank increases which in turn increase the heat flux. The reason of this increase is that when the light falls on the collector surface, it is transferred to the water by convective mode of heat transfer. With the passage of time the temperature within the tank increases showing the absorption power of collectors.

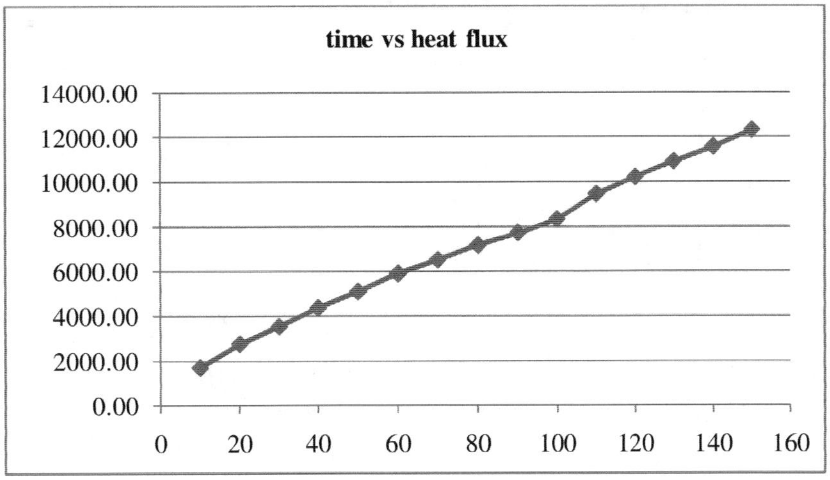

Figure 2. Time vs. heat flux for bared copper tube

3.2. Heat flux of solar water heater with copper tube coated with black paint

Figure 3. shows the effect of heat flux for collector tube (copper tube) coated with black paint. The time interval is 10minutes. We observed that with the passage of time the temperature within the storage tank increases which in turn increase the heat flux. Here the increase in heat flux is greater than the previous case where we used bared copper tube; the reason for it is that the surface coated with black paint is more absorptive for light. So the convective heat transfer rate increases due to black paint on the collector surface.

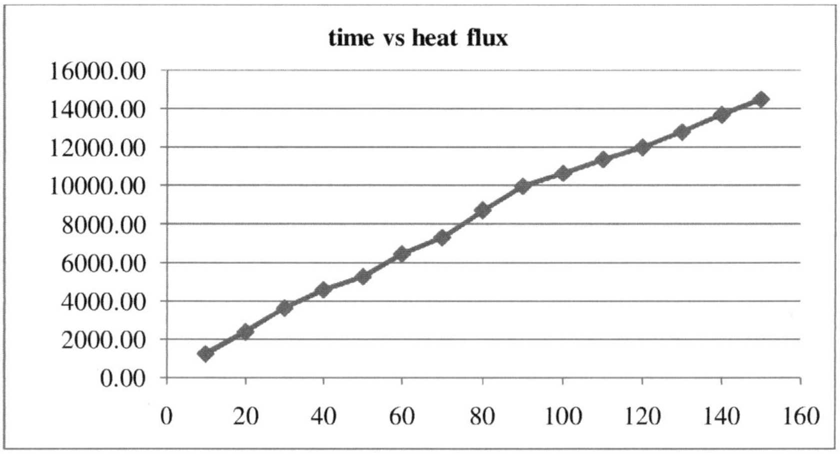

Figure 3. Time vs. heat flux for copper tube coated with black paint

3.3. Heat flux of solar water heater with copper tube electroplated with lead

Figure 4. shows the effect of electroplating on collector tube of solar water heater. In this case the collector tube is electroplated with lead. We used the lead nitrate salt and a strip of lead is used for providing lead ions at a constant rate. The current is passed and lead is coated on the copper tube (collector tube). Then the same procedure is repeated and the different readings are recorded at time interval of 10minutes. We observed that the temperature increased within the storage tank very rapidly as compared to other two cases. This shows a rapid increase of heat flux for electroplated collector tube.

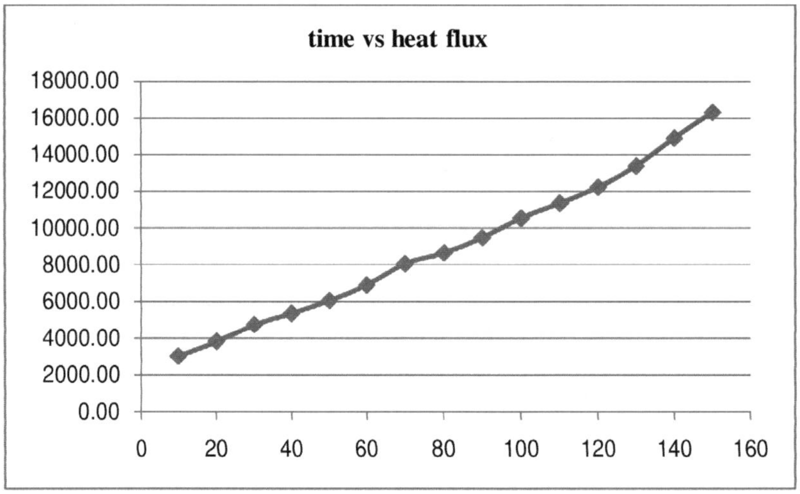

Figure4. Time vs. heat flux for copper tube electroplated with lead

3.4. Comparison of time Vs heat flux for different cases

Figure 5. shows the comparison of time vs. heat flux for different surfaces. It is proved that heat flux for copper tube coated with paint is greater than bare surface. Further the heat flux for the copper tube electroplated with lead is greater than bared copper tube as well as surface coated with paint. Table 1.shows the maximum percentage increase and average percentage increase for different cases.

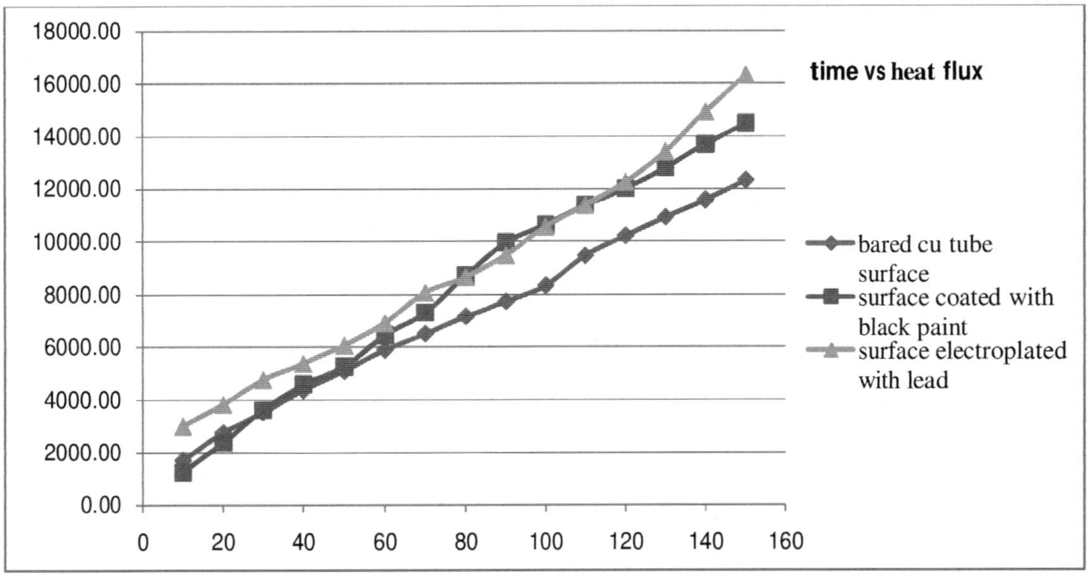

Figure 5. Comparison of time vs. heat flux for different cases

Table 1. Percentage increase for different surfaces.

Solar heater Type	Maximum percentage increase in heat flux	Average percentage increase
Surface coated with black paint	29.03 %	10.79%
Surface electroplated with lead	77.33%	28.57%

4. Conclusions

From this research, we conclude that the solar absorption in solar heater can be increase by mainly its surface modification. The electroplated surface is more efficient as compared to the surface coating with high conductive black paint. The average increase in the heat absorption for lead plated and coated surface was observed 28.57% and 10.79 % respectively.

Nomenclature

T_1	First thermometer reading
T_2	Second thermometer reading
T_3	Third thermometer reading
Q	Rate of Heat Transfer (kW)
q"	Heat Flux(kW)
A	Area (m^2)
^oC	Degree Centigrade
C_p	Specific Heat (kJ/kg.K)

References

[1] Jaisankar S, Radhakrishnan T K, Sheeba K N, Suresh S, 2009 Experimental studies on heat transfer and friction factor characteristics of thermosyphon solar waterheater system fitted with left–right twisted tapes *International Journal for Applied Engineering Research.* **83(11)** 1943–52

[2] Hottle HC, Whiller A 1958 *Proc. Conf. on the Use of Solar Energy* vol.2 Evaluation of flat-plate solar collector performance (University of Arizona) pp 74–104

[3] Hutchins M G 1979 *Spectrally selective solar absorber coatings AppliedEnergy* **5(4)** 251–326.

[4] Lampert CM 1979 Coatings for enhanced photo thermal energy collection2ndnon-selective and energy control films. *Solar Energy Materials and Solar Cells* **1** 319

[5] Cost worth J L, Raghunathan K, 1980 *technical progress report* Development of selective surfaces USA Department of Energy Contract no. DE-AC04-78CS-34293,

[6] Sharia A M, Rousan A, RousanKh K, Ahmad AA 1999 Effect of thermal conductivity of absorber plate on the performance of a solar water heater *Applied Thermal Engineering* **19(7)** 733–41

[7] Colle S, Abreu S L, Glitz K, Colle F 2001 Optimization of the auxiliary heating and water storage insulation of a low cost domestic hot water heating system with an electric shower In *Proceedings of ISES Solar World Congress*

[8] Chang J M 2004 Characteristic heat removal efficiency for thermosyphon solar water heaters During the system application phase *Journal of Solar Energy Engineering* **126(6)** 950–6

[9] Chang J M, Leu J S, Shen M C, Huang B J 2004 A proposed modified efficiency for thermosyphon solar heating systems *Solar Energy* **76(6)** 693–701

[10] Al-Fahed S, Chamra L M, Chakroun W 1999 Pressure drop and heat transfer comparison for both microfin tube and twisted-tape inserts in laminar flow *Experimental Thermal Fluid Sciences***18(4)** 323–33

[11] NaphonĤ 2005 on the performance and entropy generation of the double-pass solar air heater with longitudinal fins *Renew Energy* **30** 1345–57.

[12] Hachemi A 1999 Technical note comparative study on the thermal performances ofsolar air heater collectors with selective and nonselective absorber-plate *Renew Energy* **17** 103–12

[13] Solomon R, Marumo R, Garebamono A 2006 Modelling and simulation of a flat platesolar collector using neural networks *6th International conference on modelling, simulation, and optimization* (Acta Press Anahei) p. 25–30

[14] Yousef B A A, Adam N M 2008 Performance analysis for flat plate collector with and without porous media **19** 32–42

Comparative analysis of mapping burned areas from landsat TM images

A Mazher
Institute of Remote Sensing and Geographic Information System, Peking University, Yifu No.2 Building (New Geology Building), Beijing, China

E-mail: abeermazhar@yahoo.com

Abstract. Remote sensing is a major source of mapping the burned area caused by forest fire. The focus in this application is to map a single class of interest, i.e. burned area. In this study, three different data combinations were classified using different classifiers and quantitatively compared. The adopted classifiers are Support Vector Data Description (SVDD), a one-class classifier, Binary classifier Support Vector Machines (SVMs) and traditional Maximum Likelihood classifier (ML). At first, the Principal Component Analysis (PCA) was applied to extract the best possible features form the original multispectral image (OMI) and calculated spectral indices (SI). Then the resulting subset of features was applied to the classifiers. The comparative study has undertaken to find firstly, the best possible set of features (data combination) and secondly, an effective classifier to map the burned areas. The best possible set of features was attained by data combination- II (i.e., OMI information). Furthermore, the results of the SVM showed the high classification accuracies than ML. Experimental results demonstrate that even though the SVDD for mapping the burned areas doesn't showed the higher classification accuracy than SVM, but it shows the suitability for the cases with few or poorly represented labelled samples available. The parameters should be further optimized through the use of intelligent training for improving the accuracy of SVDD.

1. Introduction

Land cover change is regarded as one of the most important variables of global change affecting ecological systems [1]. Wild fires are a key ecological disturbance factor of natural ecosystems, affecting the distribution of the land use and land cover and threatening environmental systems and infrastructure worldwide. Thus, information on the spatial distribution of the burned areas is of vital importance and required by both environmental scientist and policy makers. Satellite remote sensing is an important source for mapping burned areas. Burned patches are very complex to detect automatically due to their spatial and spectral diversity caused by the severity of the fire and other factors. Landsat thematic Mapper (TM) and Enhanced Thematic Mapper plus (ETM+) images have been widely used in mapping the burned areas at local and regional scales [2], [3]. The existing methods for mapping burned areas in these studies showed the relatively high accuracy for the specific sites, but didn't present any generalized algorithm applicable to other sites with different input conditions. The global algorithms have also been developed for coarse resolution images by considering the wide range of burning conditions but the high accuracy from these algorithms are not sure in general. An alternative approach to enhance the discrimination of burned areas is to use several spectral indices. Some studies used the most sensitive input bands to discriminate between burned and unburned pixels [4]-[11]. The two problems, so far has been observed from these studies firstly, the methods used to discriminate the burned areas are case dependent and secondly, only spectral indices (SI) were used to discriminate burned areas, but not the original multispectral image (OMI) information. No study did the comparison of SI and OMI information to map the burned areas. Therefore, the first objective of this study is to compare the OMI, SI and the combination of these two to analyze how OMI and SI perform in mapping the burned areas individually and in combination. Secondly, to see whether the additional information of SI to the OMI improves the accuracy of a classifier. Three classifiers have been used for the comparative analysis.

The conventional supervised classifiers (e.g. Maximum Likelihood) typically requires the analyst to train the classifier on each and every class that is in study area to satisfy the assumption of an exhaustively defined set of classes, even most are of no interest for the analysis. The conventional classifiers aim to optimize the classification accuracy over all the land cover classes rather than focus on that specific class of interest. Thus, when a specific class is of interest, it is preferable to adopt an alternative to the standard multiclass classifiers. One approach is to adopt binary classifiers to separate the class of interest from all the other classes [12], [13]. Further refinement to this approach is to use one-class classification. A series of one-class classifiers have been introduced to map those remotely sensed images, where only a specific class is of interest to improve the classification accuracy [14]. One-class classifiers require no statistical assumption regarding the distribution of the data and it only focuses on the class of interest for selecting the training samples. Hence by comparing the capacity of Conventional Maximum Likelihood (ML), Binary Support Vector Machine (SVM) and One-class Support Vector Data Descriptor (SVDD) for optimizing training in the classification, the efficient monitoring of burned areas can be achieved.

The rest of the paper is organized in four sections. The classifiers are described in Section III under the methodology after introducing the study site in Section II. The results obtained and discussion revealed in Section IV. Finally Section V summarizes the observations and the concluding remarks to end this paper.

2. Study Site

The Landsat 5 Thematic Mapper (TM) images of the test site before and after the fire acquired on July 05, 2009 and October 25, 2009 respectively were obtained from U.S. Geological Survey (USGS). Figure 1 shows the dataset covered some part of Los Angeles, California. The six non-thermal spectral bands with a spatial resolution of 30m were acquired for both the images. These images are located on the satellite path and row 041 036. The spectral band 6 was discarded while doing the analysis.

Figure 1. Location of the study site. (a) Before Fire and (b) After Fire.

3.Methodology

Many authors have recognized that burned patches do not present a specific spatial pattern [15] and therefore the discrimination of these areas should be based on either the spectral or temporal contrast or on both. The detection of the burned areas is not an easy task as the wide variety of spectral characteristics contributes within the burned territory. The methodology adopted in this study is to extract the features appropriate for mapping burned areas and to find an effective classifier in terms of accuracy. The methodology adopted in this study mainly consists of three steps: (a) extraction of the best possible features (b) comparison of classification methods and (c) assessment of accuracy.

3.1. Feature extarction

In this study, both original multispectral images and derived spectral indices were used. Three data combinations were compared: (1) derived spectral indices (Data-I); (2) original multispectral data only (Data-II); (3) combination of both spectral indices and multispectral data (Data-III).

The spectral data from bitemporal Landsat 5 TM images contain 12 non-thermal multispectral bands. The indices used in this study were Normalized Difference Vegetation Index (NDVI), Burned Area Index (BAI), Normalized Burn Ratio (NBR), Burned Area Index Improved ($BAIM_s$), MIRBI, GEMI, NBR_L and $BAIM_L$.The indices NBR_L and $BAIM_L$ use the long SWIR variation instead of Short SWIR variation of the NBR and BAIM. The number of spectral indices from bitemporal images is 16 (8 for each). Thus,the third data combination contains 28 bands (12 from bitemporal multispectral imagesand 16 from bitemporal spectral indices). To reduce the dimensionality of each data combination and choose the best subset of the original features, Principal component analysis (PC) was adopted in this study.

3.2. Selection of trainig pixels

The training pixels acquired for the classification of the Landsat TM data set follows a manual selection procedure, whereby randomly selected samples from each land cover class were usedas training data set. As the class of interest (burned class) is relatively smaller in size than the non-burned class, 9,807 pixels out of total 31,411 pixels were selected from burned class, while 21,604 from the other land cover classes were used as samples of non-burned class. The training data set was later used to train the ML and SVM classifiers. However, the SVDD used only the samples for burned area (target class) (i.e. 9,807 pixels). Using this training data set, a variety of the parameter combinations (γ,α) were tested to attain the optimal parameter combination setfor the SVDD classification. The aim was to find the set of parameters that provide the effective hyper-sphere (represent the burned areas) in feature space.

3.3. Image classification

Three classification methods are adopted to classify the burned areas from the Landsat TM imagery. Maximum Likelihood (ML) is the parametric classifier based on the assumptions of the normality of data distribution for each class and exhaustively selected set of classes. For mapping the land cover areas, number of studies used the ML classifier as a benchmark to compare its classification accuracy with the other newly developed classifiers. It is considered as a standard approach to thematic mapping from the remotely sensed imagery. In real life applications hardly the nature of the distribution is known. It is preferable to use the non-parametric classifiers that are free from assumptions. For the sake of comparison of the classification accuracies and to validate the suitability of the non-parametric classifier for mapping the burned areas in this study, ML classification is conducted.

Support Vector machines (SVM) for the classification of multispectral remote sensing images gains a sky-scraping attention now a day [16], [17]. Vapnik introduced the concept of the SVM for the first time [18]. SVM is a non-parametric binary classifier that locates the optimal hyper plane between the two classes to separate them in a new high-dimensional feature space by taking into account only the training samples that lie on the edge of the class distributions known as support vectors. Support vectors are the samples that lie closest to the decision boundary; this means that only the extreme cases are used by the SVM to separate the classes. This is the primary benefit of using SVM over the other classifiers.

In addition SVM results in high classification accuracies and holds very good generalization abilities. In general, the SVMs are much more effective than other conventional non-parametric classifiers in terms of classification accuracy, computational time and stability to parameter setting. Moreover, it shows the less sensitivity to the problem occur by the curse of dimensionality. The details regarding the SVM execution can be found in [17], [19]. This study only deals with the application of the SVM.

The one-class classification method used in this study is Support Vector Data Descriptor (SVDD). The term one-class classification is coined from [14]. One-class classification proved to be valuable in variety of research areas. It does not require the parametric assumption to be fulfilled that is the chosen set of classes is exhaustively defined. In this way only the training samples from the class of interest are chosen to train the classifier resulting in reducing the amount of training data as well as high accuracy of a classifier will be anticipated. The SVDD is a one-class classifier based on the principles of the SVMs [20]. Analogous to SVM, it too only uses the training set (support vectors) from the class of interest but instead of optimizing a hyper plane between two classes it calculates an optimal hyper sphere with the smallest radius that makes a tighter boundary around the target class and separates it from novel data (outlier data) or any other possible class. Therefore, it models (or trains) only the boundary of a target class instead of modeling a complete density distribution. In this study the SVDD is used to train the classifier with the different model parameter combinations to reach the optimal accuracy level. All the three classifiers were used to classify an image by using original multispectral information, spectral indices and the combination of these two.

3.4. Accuracy assesment

The confusion matrix was used to assess the accuracy measures for all the three classification procedures by using the ground truth pixels as the reference. The Overall accuracy (OA), Kappa coefficient (K), Producer's accuracy (PA) and the User's accuracy (UA) were used. The overall accuracy is the percentage of all validation pixels correctly classified, whereas the users and producers accuracy provide information about the commission and omission errors associated with the individual classes, respectively. Unlike the overall accuracy, Kappa takes into account the possibility of agreements occurring by chance in a random classification [21].The accuracies of the both the burned and the outlier class (non-burned area) were evaluated but the particular attention was focused on the class of interest (burned area).

4. Results and discussion

The output from three classifications is the binary map illustrating the burned areas for all three situations as shown in figure 2. Only the SVDD classification maps with the optimum parameters $\gamma = 10$ and $\alpha = 5\%$ is shown here.

By visually comparing the results shown in figure2 and original images figure 1, it is easily found that the ML significantly overestimated the burned area, i.e. many non-burned areas were wrongly identified as burned areas. The SVDD also slightly overestimated the burned areas for the Data-I but underestimate for Data-II and Data-III. However, the binary SVM obtained relatively accurate classification results, using different data combinations (figure 2).

To compare the accuracy measurements the confusion matrix calculated for ML, SVDD and SVM for all data combinations are shown in Table 1, Table 2 and Table 3, respectively. The best classification result in this study was obtained by the binary classifier SVM for theData combination-IIas shown in the middle of figure 2, which resulted in the Kappa coefficient of 92.58%. Moreover, it showed the better tradeoff between the producer's and user's accuracy in comparison to the other two classifiers for the burned class.

The following discussion concerns the details and the comparison of the classification results used for mapping the burned areas in this study.

Journal of Physics: Conference Series **439** (2013) 012038 doi:10.1088/1742-6596/439/1/012038

Data Combination- IData Combination- II Data Combination- III

SI OMISI &OMI

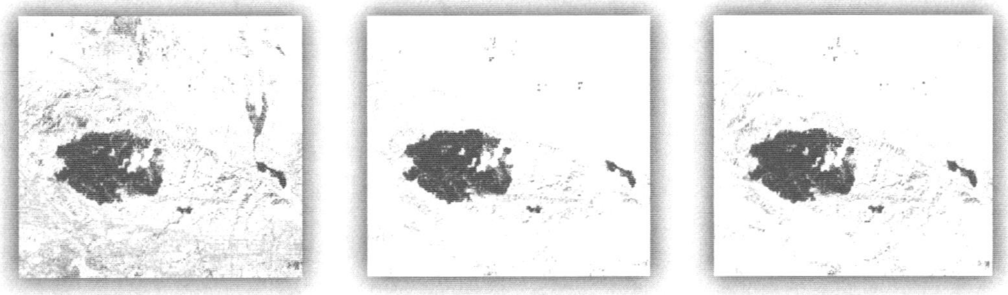

(a)Conventional ML

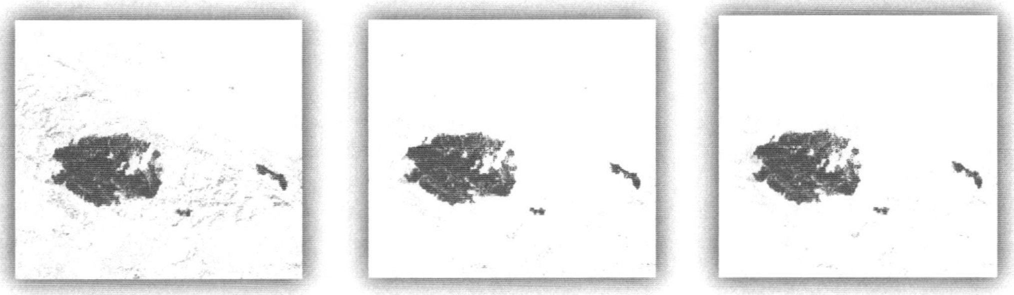

(b) One-class SVDD

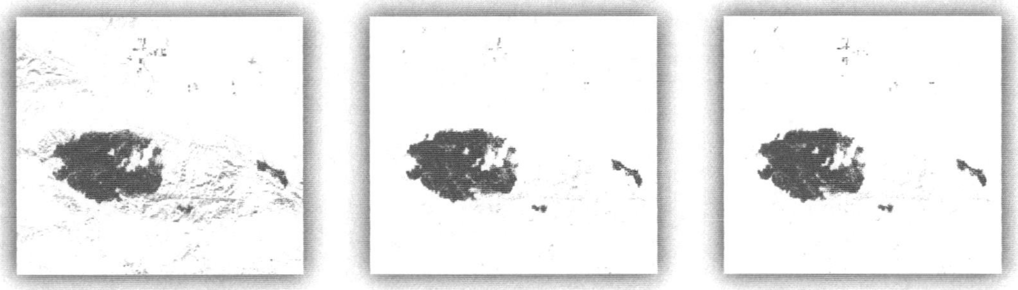

(c) Binary SVM

Figure 2. Classification results. Data combination-I shows the classification results by using spectral indices (SI), Data combination-II shows the classification results obtained by the original multispectral image (OMI) and Data combination-III shows the results obtained by combining the spectral indices and the original multispectral image (combined SI and OMI) for all the three classifiers.

Table 1. Confusion Matrix for Conventional ML.

(%)	Data-I		Data-II		Data-III	
	Spectral Indices		Multispectral image		Combined	
	Burned	*Unburned*	*Burned*	*Unburned*	*Burned*	*Unburned*
PA	88.63	91.93	92.23	98.77	93.58	97.79
UA	46.79	99.02	85.74	99.37	77.23	99.48
OA	91.68		98.29		97.47	
K	57.08		87.94		83.26	

The accuracy of the ML classifier was calculated for all three data combinations as it provides the benchmark for the assessment of the SVM based classifications. The classification result of ML by using original multispectral information only, shows the best Kappa (87.94%) among all the three data combinations. Moreover, the user's accuracy (i.e. 85.74%) is also higher than the other two data combinations, which shows that the commission error is least for Data-II (14.26%) as compared to the Data-I (53.21%, i.e. 1-46.79%) and Data-III (22.77%, i.e., 1-77.23%). It can be further observed from table I that the users accuracy for Data-I that is ML classification by using spectral indices only showed the commission error of 53.21%, which means that the burned area is largely over estimated. Hence, the use of spectral indices for classifying the burned areas is not recommended in this study. Even though the commission error is least for Data-II as compared to others but still 14.26% is the high percentage of classifying non-burned pixels as burned pixels. It can also be evident from (figure 2(a)).

From this, we conclude that the ML classification approach is improper to map the burned areas in this study, since the class of interest is substantially over estimated by all the three data combinations. In order to monitor any loss in the environmental and the ecosystem, an accurate classification of the burned area is of crucial interest.

Table 2. Confusion Matrix for One-class SVDD.

(%)	Data-I		Data-II		Data-III	
	Spectral Indices		Multispectral image		Combined	
	Burned	*Unburned*	*Burned*	*Unburned*	*Burned*	*Unburned*
PA	88.42	97.55	85.46	99.585	84.13	99.58
UA	74.32	99.06	94.27	98.84	94.18	98.74
OA	96.87		98.54		98.44	
K	79.07		88.87		88.04	

A real potential for the saving in resources was demonstrated through the use of a one-class SVM i.e. SVDD, as it trains the classifier by using the samples only from the class of interest (burned area). In this case study, confusion matrix for the SVDD of optimal parameter combination is shown in table 2. The best result in terms of accuracy is attained by Data-II (using original multispectral image only) for one-class SVDD. The Kappa coefficient for this is 88.87% showing the high agreement with the burned areas. Also the high user's accuracy (94.27%) indicates that the error of commission is quite low as compared to the other data combinations. Moreover, Data-II by SVDDclassification showed the good tradeoff between users and producers accuracy.

Table 3. Confusion Matrix for Binary SVM.

(%)	**Data-I**		**Data-II**		**Data-III**	
	Spectral Indices		Multispectral image		Combined	
	Burned	*Unburned*	*Burned*	*Unburned*	*Burned*	*Unburned*
PA	91.00	96.83	92.64	99.49	92.79	99.47
UA	69.69	99.26	93.62	99.41	93.39	99.42
OA	96.40		98.98		98.98	
K	77.00		92.58		92.54	

The binary SVM obtained the Kappa value of 92.58% for Data-II, higher than the ML and SVDD in general. The user's and the producer's accuracies for the burned class,are also high, i.e. 93.62% and 92.64%, (table 3). By using only the original multispectral image (Data-II) in SVM classification, the commission and omission errors both are reduced by 6.38% and 7.36%,respectively.

In this case study the user's accuracy is given weight-age since it indicates the capacity of the classifier for identifying the occurrences of the outlier class on the ground. In this regard, SVM showed much higher accuracy than the ML and SVDD method. Moreover, SVM showed the best classification results for Data-II.

It has been observed that the accuracy measurements for Data II and Data III in table 2 are not significantly different. The use of original multispectral image and the combineddataset provides more or less the same classification accuracy measurementsfor mapping burned areas. Moreover, the same pattern can be observed in table 3.

5. Conclusion

In this case study the Data-II i.e. the original multispectral image (OMI) showed the best features to discriminate the burned and outlier (un-burned) classes. The spectral indices calculated in this study do not have significant impact on the burned areas classification accuracy.Other significant spectral indices and vegetation indices should be taken into account to extract the more suitable features appropriate for mapping the burned areas.Moreover, the Binary classifier SVM showed the potential for the accurate mapping of burned areas. In general SVM and SVDD both provide better accuracy than the parametric ML classifier. However, the SVDD performs not well than the SVM in this case study, although SVDD has an advantage of using the training samples only from the class of interest. To get a reliable classifier by only having few labeled pixels is very difficult. In particular for one-class SVM i.e. SVDD, this task is even more harder because the free parameters of the model need to be finely adjusted, but no clear criterion can be adopted. Further work is required to find an algorithm for the optimization of parameters. Moreover, the different remotely sensed images can be used to further rely on the findings of this study.

References

[1] Otukei J Rand Blachke T 2010 *Int. J. Appl. Earth Obs. Geoinf.* **12** S27-S31

[2] Chuvieco E, Martín M P and Palacios A 2002 *Int. J. Remote Sens.* **23** 5103 – 5110

[3] Mitri G H and Gitas I Z 2004 *Int. J. Remote Sens.* **25** 2863 –2870

[4] Barbosa P M, Gregoire J M and Pereira J M C 1999 *Remote Sens. Environ.* **69** 253 –263

[5] Chuvieco E, Engle field P, Trishchenko A P andLuo Y 2008 *Remote Sens. Environ.* **112** 2381 – 2396

[6] Fernández A, Illera P and Casanova J L 1997 *Remote Sens. Environ.* **60**1 53 – 162

[7] Kasischke E S and French N H F 1995 *Remote Sens. Environ.* **51** 263 – 275

[8] Martín M P 1998 *Cartografía e inventario de incendios forestales en la Península Ibérica a partir de imágenes NOAA- AVHRR*Ph.D. dissertation (Universidad de Alcalá, Alcalá de Henares)

[9] Martín M P, Gómez I and Chuvieco E 2005 *Proc. of the 5th International Workshop on Remote Sensing and GISapplications to Forest Fire Management* : Fire Effects Assessment(Paris: Universidad deZaragoza)

[10] Pereira J M C 1999 *IEEE Trans. Geosci.Remote Sens.* **37** 217 – 226

[11] Smith A M S, Drake N A, Wooster M J, Hudak A T, Holden Z A and Gibbons C J 2007 *Int. J. Remote Sens.* **28** 2753 – 2775

[12] Boyd D S, Sanchez-Hernandez C and Foody G M2006 *Int. J. Remote Sens.* **27** 2631–2644

[13] Foody G M,Mathur A, Sanchez-Hernandez C and Boyd D S2006 *Remote Sens. Environ.* **104** 1–14

[14] Moya M, Koch M and Hostetler L 1993 *International Neural Network Society* 797-801

[15] Smith A M S, Wooster M J, Powell A K and Usher D 2002 *Int. J. Remote Sens.* **23**1 733-1739

[16] Hermes L, Frieauff D, Puzicha J and Buhmann J M 1999 *Proc. Geoscience and Remote sensing Symposium* (Hamburg, Germany)

[17] Huang C, Davis L S and Townshend J R G 2002 *Int. J. Remote Sens.* **23** 725-749

[18] Vapnik V N 1999 *IEEE Transactions on.* **10** 988-999

[19] Sanchez-Hernandez C, Boyd D S, Foody G M 2007 *IEEE Trans. Geosci. Remote Sens* **45** 1061–1073

[20] Tax D M J 2001 *One class classification*Ph.D. dissertation (Delft Univ. Technol Netherlands)

[21] Brown D G, Lusch D P and Duda K A 1998 *Geomorphology* **21** 233-250

Journal of Physics: Conference Series **439** (2013) 012039 doi:10.1088/1742-6596/439/1/012039

An optimal design of coreless direct-drive axial flux permanent magnet generator for wind turbine

D Ahmed[1,3] and A Ahmad[2]
[1]Department of Electrical Engineering (DOEE), University of Engineering and Technology, Taxila, Pakistan
[2]University of Engineering and Technology, Taxila, Pakistan

E-mail: daniyal.ahmed@uettaxila.edu.pk

Abstract. Different types of generators are currently being used in wind power technology. The commonly used are induction generator (IG), doubly-fed induction generator (DFIG), electrically excited synchronous generator (EESG) and permanent magnet synchronous generator (PMSG). However, the use of PMSG is rapidly increasing because of advantages such as higher power density, better controllability and higher reliability. This paper presents an innovative design of a low-speed modular, direct-drive axial flux permanent magnet (AFPM) generator with coreless stator and rotor for a wind turbine power generation system that is developed using mathematical and analytical methods. This innovative design is implemented in MATLAB / Simulink environment using dynamic modelling techniques. The main focus of this research is to improve efficiency of the wind power generation system by investigating electromagnetic and structural features of AFPM generator during its operation in wind turbine. The design is validated by comparing its performance with standard models of existing wind power generators. The comparison results demonstrate that the proposed model for the wind power generator exhibits number of advantages such as improved efficiency with variable speed operation, higher energy yield, lighter weight and better wind power utilization.

1. Introduction

Renewable energy sources are those which are naturally replenished, for example, geothermal, solar, wind, and tidal. Wind power, being a readily and naturally available source, is growing significantly worldwide. Presently, the worldwide installed wind power capacity is around 200 GW. It is anticipated that by 2019, it would raise above 1000 GW indicating that wind power is really an important power source [1]. The key units in the conversion of wind energy to electrical energy are the wind power generators. The overall performance of the wind power generation system depends on the conversion efficiency of the power generators. These generators are run by wind turbines which are often made taller to utilize advantages of wind at higher altitudes such as uniformity, steadiness and high speed. For better stability and improved performance, these turbines require more complex and flexible towers along with lighter weight and smaller sized nacelles [2].

Significant advancements have been made in the field of wind power generators since the inception of utilization of wind for production of electrical energy. The commonly used generators are IG, DFIG, EESG and PMSG. IGs are simple, rugged and cost effective fixed speed generators that require no separate excitation source [3]. DFIGs with a partially rated rotor side converter are the mainstream technology in the market for wind turbines [4]. The DFIG operates at partial variable speed, which is

[3] To whom any correspondence should be addressed.

about ±30 % above and below the rated speed. EESG and PMSG both are synchronous generators that can operate at full variable speed. These generators do not require reactive magnetizing current in comparison to IG and DFIG. EESG and PMSG are connected to the grid through a full scale power electronic converter and therefore are decoupled from the grid. The generators can operate at a wide variable frequency range for optimal operation and provide the option to implement gearless operation [5]. EESG requires DC excitation current for its operation which is provided from the grid through a rectifier. On the other hand, PMSGs use permanent magnets in place of field windings and hence are the most reliable wind power generators as they are independent from the grid and can operate even at the lowest wind speed of 3 m/s [6]. PMSG is further divided into Radial Flux Permanent Magnet (RFPM) and Axial Flux Permanent Magnet (AFPM) generators. The AFPM generator has higher power density and can operate more efficiently at low speeds as compared to its radial counterpart. Moreover, for same power ratings, AFPM has smaller volume and lower mass as compared to RFPM [7]. All the features of AFPM generator listed above make it the generator of choice for wind power generation system.

In the literature, many authors have worked on the full rated power converter system being used for PMSG-based wind conversion systems, for example, Xibo Yuan *et al.* [8] has worked on H-bridge converters and neutral point clamped converters [9] for PMSG wind power generation system. These are medium voltage, high power converters which have been cascaded in series resulting in the elimination of the grid side step-up transformer. These series cascaded converters have high reliability and fault-tolerant ability. At another place, a robust and reliable grid power interface system for wind turbines using a PMSG has been considered where an integration of a generator-side three-switch buck-type rectifier and a grid-side Z-source inverter is employed as a bridge between the generator and the grid [5]. At another place [10], a matrix converter has been used along with PMSG system which eliminates the bulky storage capacitors resulting in a direct ac-ac power conversion system. Some authors in [11], [12] and [13] have replaced the conventional transformer with a high frequency, all-solid-state power electronic transformer resulting in better power quality output and increased power density. Other authors have also worked on modeling and designing of PMSG, for example, Ming Yin *et al.* [14] has presented the complete model of a variable speed wind turbine system with radial flux PMSG for supplying power to the grid through HVDC link. The modeled generator is a high speed generator that requires a gear box for its integration with the wind turbine system. In [15] and [16], authors have also presented their models of PMSG for variable speed wind turbine generation systems. In [17], the author has designed an AFPM machine with flux switching capability resulting in the improvement of power density and reduction in cogging torque with better performance. In [18], the author has designed and optimized a nine-phase AFPM generator with concentrated winding for direct-drive wind turbine. The goal for optimization was to minimize the net mass of the generator along with maximizing its power factor. In [19], AFPM generator with concentrated windings has been used in wind energy conversion system. In [20], work has been done to improve the power quality of distributed winding AFPM generator for wind turbine systems. In [21], a high-speed AFPM generator has been optimally designed. The structure consists of single coreless stator in between twin rotors having steel core. However, the design and modeling of low-speed modular direct-drive AFPM generator with coreless stator and rotor for a wind turbine power generation system has not been reported. This paper, therefore, proposes such a system. Figure 1 shows the model of the proposed design of wind turbine generating system which has been prepared using the software Sketch Up [22]. The model is validated by making a nose to nose comparison with existing wind power generators (IG, DFIG & EESG). The comparison is made on the basis of wind power utilization and total energy yield over a fixed time span under similar wind conditions through simulations using the software Simulink/ MATLAB [23].

Journal of Physics: Conference Series **439** (2013) 012039 doi:10.1088/1742-6596/439/1/012039

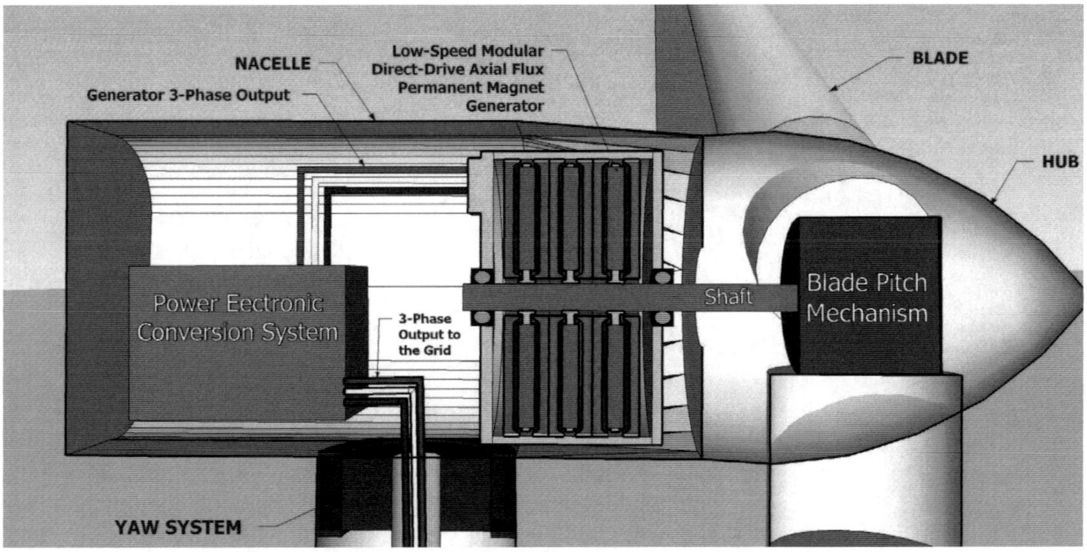

Figure 1. Wind Turbine Generating System Model using the Proposed AFPM Generator

2. Proposed generator design

Different topologies of AFPM generators exist in literature. Some designs have single stator and rotor while some have multiple stators and rotors. Depending on their configuration, AFPM generator can be classified as, single-sided, double-sided and multi-stage (modular) generator. These topologies are further classified on the basis of the type of stator and rotor being used, for example, coreless or cored generator, slotted or slotless generator etc.

The proposed generator consists of a multi-disc structure having three modules. Each module consists of two identical stator discs (armature) and single rotor disc (field). The stators are connected in series and the three modules are connected in parallel to increase the net output power. Both the rotor and the stator are coreless that is, they have non-ferromagnetic cores. A 2D model of this proposed design is shown in Figure 2. The coreless structure minimizes the attractive forces between stator and rotor and also reduces the cogging effect in torque, that is, torque pulsations are minimized at zero stator current. Such a complete coreless design results in reduction of overall mass of the generator and increase in the efficiency. With this design, direct drive low speed power generation with maximum wind power utilization is possible. Moreover, with a coreless structure the losses are also minimized.

In each module, the twin stator discs are fixed to the frame and are the stationary part of the machine. The stator coil is a three phase double layered toroidal type winding that is embedded in a resin or a plastic of high mechanical integrity which together make up a single stator disc. Similarly, in each module, the rotor disc, consisting of NdFeB rare-earth permanent magnets and its supporting non-magnetic material (aluminum or plastic) structure, is the rotating part which is directly connected to the shaft. Since this design does not include any ferromagnetic material in its core hence it is free from cogging effect and there are no core losses as well.

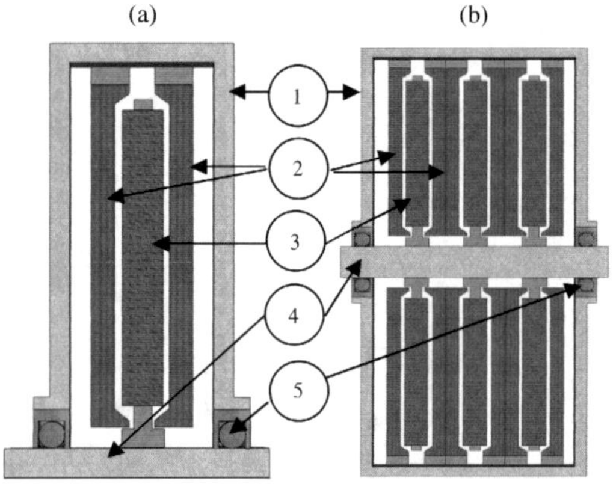

Figure 2. Proposed Coreless AFPM Generator (3 modules assembly): (a) single module (segment), (b) three module assembly; 1 – frame, 2 – twin coreless stator windings, 3 – permanent magnets (coreless rotor), 4 – shaft, 5 – bearing

The proposed design offers a high power or high torque density which has been achieved by arranging the permanent magnets in Halbach array [7]. In this arrangement, the angle between magnetic polarization vectors of two adjacent magnet pieces is 90°. By using this array, the magnetic field strength of the rotor disc is enhanced on the front side and diminished to zero (approximately) on the backside allowing the magnets to be attached to the non-ferromagnetic structure. The pattern of the magnetic flux lines for this arrangement, using the software FEMM (Finite Element Method Magnetics) [24], is depicted in Figure 3.

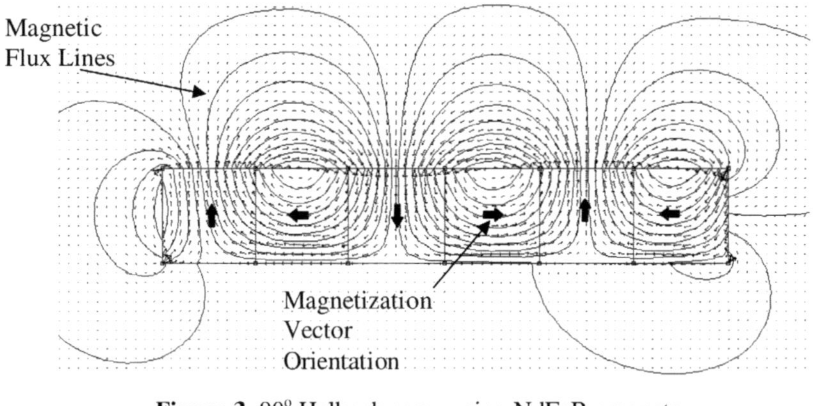

Figure 3. 90° Halbach array using NdFeB magnets

The magnetic field produced as a result of this pattern is a sinusoidal field. The peak value of magnetic flux density $B_{PM,peak}$ at front side of the rotor disc is:

$$B_{PM,peak} = 0.9 \times B_{remanent} \times \left(1 - e^{-62.83 * h_{PM}}\right) \quad (1)$$

In equation 1. $B_{remanent}$ is the remanent flux density and h_{PM} is the axial height of the permanent magnets. This value increases exponentially with increase in the height (depth) of a permanent magnet piece. In the proposed design, an optimized value has been evaluated by varying the height of the magnet keeping in consideration the efficiency, the desired power density and the cost of permanent magnets. The induced EMF E_{field} in a single stator is given as

$$E_{field} = 0.2441 \times p_{pp,mod1} \times \phi_{field,PM} \times N_{1s} \times \omega_{e,gen} \quad (2)$$

where $p_{pp,modl}$ is the total number of pole pairs in a single module, $\phi_{field,PM}$ is the magnetic flux produced by the permanent magnets, N_{1s} is the number of turns per phase per stator disc for a single module and $\omega_{e,gen}$ is the electrical angular speed of the generator. If I_{stator} is the phase stator winding current of single module, then the electromagnetic torque T_{em} developed is given by

$$T_{em} = 0.371 \times n_{ph} \times p_{pp,\mathrm{mod1}} \times \phi_{field,PM} \times N_{1s} \times I_{stator} \quad (3)$$

In equation 3. n_{ph} is the phase number of the generator. The electromagnetic apparent power for a single module $S_{em,mod1}$ induced in the two stator windings is thus given as

$$S_{em,\mathrm{mod1}} = n_{ph} \times (2 \times E_{field}) \times I_{stator} \quad (4)$$

The outer diameter $D_{out,gen}$ and the inner diameter $D_{in,gen}$ of the generator are calculated using the following sizing equations.

$$D_{out,gen} = \left(\frac{21.87 \times 10^3 \times P_{out,gen_{\mathrm{mod1}_{max}}}}{\omega_{gen,rpm_{max}} \times B_{PM,peak} \times A_{M,peak}} \right)^{\frac{1}{3}} \quad (5)$$

$$D_{in,gen} = D_{out,gen} \times 0.577 \quad (6)$$

where $A_{M,peak}$ is the peak value of the line current density of a single module. Since the relative recoil permeability of the NdFeB permanent magnets is unity [7], the d-axis and q-axis stator inductances (L_d and L_q) in the synchronous reference frame are equal and are calculated as:

$$\mathrm{L}_{stator} = \mathrm{L}_d = \mathrm{L}_q = n_{ph} * \mu_0 * \frac{1}{\pi} * \left(\frac{1.0492 * N_{1s}}{p_{pp,\mathrm{mod1}}} \right)^2 * \left(\frac{R_{out,gen}^2 - R_{in,gen}^2}{g_{air}} \right) \quad (7)$$

where μ_0 is the permeability of free space ($0.04\pi \times 10^{-6}$ H/m), g_{air} is the air gap (0.001 m) and $R_{in,gen}$ & $R_{out,gen}$ are the inner and outer radii of the generator. The design parameters of the proposed generator are shown in table 1.

3. Wind turbine and proposed generator dynamic model

The complete wind power generation system, as shown in figure 1, comprises the proposed low-speed modular AFPM generator coupled directly (without gearbox) to the wind turbine through a mechanical shaft. The output electrical power of the generator is fed to the grid through a power electronic conversion system. This section presents the mathematical models of the wind turbine, the drive train and the proposed AFPM generator. Since the scope of current research is designing and modeling of AFPM generator only, therefore, the models of power electronic conversion system and its control have not been discussed here. In this section, the mathematical models of the mechanical systems have been obtained using the Newton's law of rotational motion and those of electrical systems have been obtained using Kirchoff's Current and Voltage laws [25].

3.1. Wind Turbine

Before modeling the wind turbine, a suitable site needs to be selected since its meteorological data (wind speeds, air density, other land features etc) is required for modeling. Gharo, Pakistan has been selected as suitable site for the proposed system. The wind speeds selected for validation purposes are 12 m/s, 9 m/s, 6 m/s and 3 m/s. The variable wind speed input V_{wind} to the system has been modeled as follows using the step function:

$$\begin{aligned} V_{wind} = 12*u_0(t) - 3*u_{50}(t) - 3*u_{60}(t) - 3*u_{60}(t) - 3*u_{70}(t) + 3*u_{80}(t) \\ + 3*u_{90}(t) + 3*u_{100}(t) - 12*u_{150}(t) \end{aligned} \quad (8)$$

The mechanical power extracted from the upstream wind V_{wind} by the turbine, that is, the turbine output power $P_{out,turbine}$ is:

$$P_{out,turbine} = \frac{1}{2}\frac{\pi}{4} D_{blade}^2 \rho_{gharo} C_p V_{wind}^3 \quad (9)$$

where D_{blade} is the blade diameter of the turbine, ρ_{gharo} is the air density at Gharo site and C_p is the power or performance coefficient of the turbine and its theoretical maximum value that can be achieved is 0.59 [26]. The air density at Gharo has been calculated using:

$$\rho_{gharo} = \rho_0 - 1.194 \times 10^{-4} E_{gharo} \tag{10}$$

where, ρ_o is the air density at sea level (1.225 kg.m^{-3}) and E_{gharo} is the Gharo site elevation from sea level (1750 m). The turbine output torque $T_{out,turbine}$ can be calculated from turbine output power using the following relation:

$$T_{out,turbine} = \frac{P_{out,turbine}}{\omega_{m,turbine}} \tag{11}$$

Table 1. Design parameters for one module of the proposed 90kW AFPM generator

Parameter	Symbol	Value
Nominal output power	$P_{out,gen_{mod1\,max}}$	30 kW
Generator nominal angular speed in rpm	$\omega_{gen,rpm_{max}}$	25 rpm
Generator nominal angular speed in rad/s	$\omega_{m,gen_{max}}$	2.618 rad/s
Line voltage (Y connection)	$V_{ll-2s,mod1_{nom}}$	840 V
Number of phases	n_{ph}	3
Stator resistance	R_{stator}	0.425 Ω
Stator inductance	L_{stator}	0.0013 H
Number of pole pairs	$p_{pp,mod1}$	32
Pole pitch	p_{pitch}	0.05 m
Axial height of permanent magnet	h_{PM}	0.0138 m
Axial width of permanent magnet	w_{PM}	0.0191 m
Axial length of permanent magnet	l_{PM}	0.0191 m
Magnetic flux density peak value in the air gap	$B_{PM,peak}$	0.653 T
Outer diameter of generator	$D_{out,gen}$	1.596 m
Inner diameter of generator	$D_{in,gen}$	0.922 m
Average diameter of generator	$D_{ave,gen}$	1.259 m
Magnetic flux excited by permanent magnets	$\Phi_{field,PM}$	0.0086 Wb
EMF Constant	K_{emf}	3.235 V.s
Moment of inertia	J_{gen}	5.99 Kg.m^2

Where $\omega_{m,turbine}$ is the mechanical rotational speed of the turbine. The actual value of C_p depends on the tip speed ratio ζ and the pitch angle α_o. The equation for C_p, based on the variable speed turbine characteristics of [27], is

$$C_p = 0.73 * \left(\frac{151}{\zeta_i} - 0.58\ \alpha_0 - 0.002\ \alpha_0^{2.14} - 13.2 \right) * e^{\frac{-18.4}{\zeta_i}} \tag{12}$$

where, $\zeta_i = \dfrac{1.0}{\left(\dfrac{1}{\zeta + 0.02\alpha_0} - \dfrac{0.03}{\alpha_0^3 + 1}\right)}$ and $\zeta = \dfrac{\omega_{m,turbine} * R_{blade}}{V_{wind}}$

The Simulink model of the wind turbine is shown in figure 4. The parameters of the Turbine are shown in Table 2.

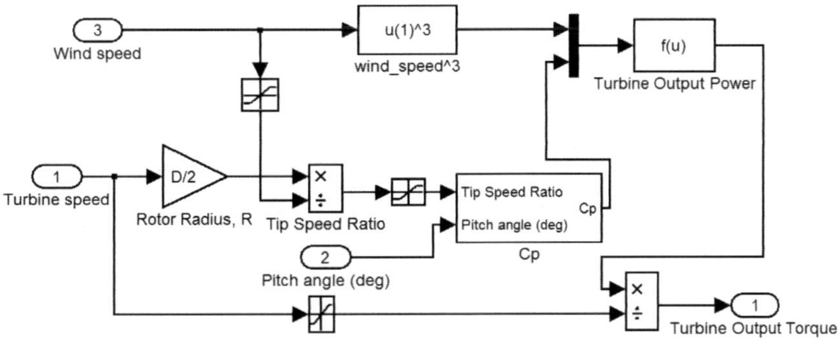

Figure 4. Wind turbine model

Table 2. Parameters for wind turbine

Parameter	Symbol	Value
Nominal Turbine Output Power	$P_{out,turbine_{max}}$	90 kW
Nominal Wind Speed	$V_{wind_{max}}$	12 m/s
Turbine Rotor Diameter	D_{blade}	15 m
Air Density at Site (Gharo)	ρ_{gharo}	1.01605 Kg/m3
Site(Gharo) Elevation	E_{gharo}	1750 m

3.2. Drive train

The wind turbine generation system uses a single mass drive train dynamic model. The inertia of the system (sum of inertias of blades, hub, shaft and generator) is J_{sys} and its angular velocity is $\omega_{m,gen}$. The losses due to friction in the proposed model are minimal since the gear box has been removed. The dynamic behavior of the drive train can be explained by the following differential equation:

$$J_{sys}\frac{d\omega_{m,gen}}{dt} = T_{e,gen} - T_{out,turbine} - F_{sys}\omega_{m,gen} \tag{13}$$

The Simulink model of the drive train is shown in figure 5 where as its parameters are shown in table 3. In this research, the combined damping effect of the turbine and the shaft has not been considered.

Table 3. Parameters for drive train.

Parameter	Symbol	Value
Lumped Moment of inertia of complete system	J_{sys}	74.22 Kg.m^2
Lumped Friction factor of complete system	F_{sys}	0. 001189 N.m.s

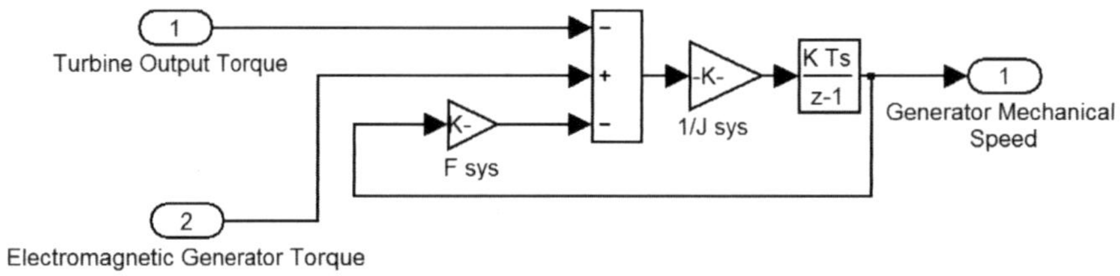

Figure 5. Drive train model

3.3. Proposed AFPM generator

The proposed direct-drive, modular AFPM generator has been designed in section 2 and its parameters, for a single module, are shown in table 1. Each module of the generator can be electrically described using the equivalent circuit shown in figure 6. The three phase voltage equation of the generator is as follows:

$$v_{abc} = \boldsymbol{e}_{abc} - R_{abc} i_{abc} - L_{abc} \frac{di_{abc}}{dt} \tag{14}$$

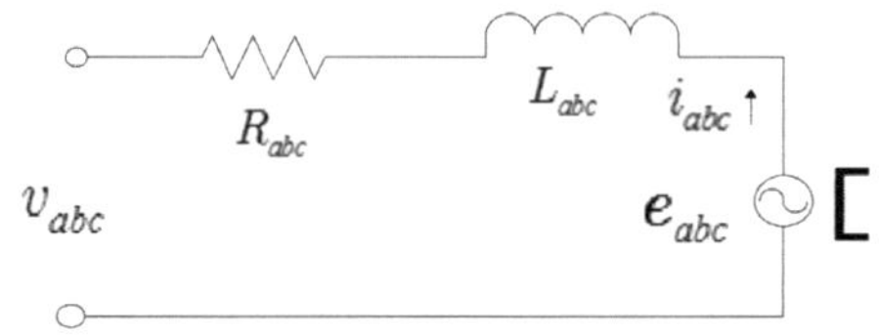

Figure 6. Equivalent Circuit of AFPM Generator

In equation 14. e_{abc} is the induced EMF and υ_{abc} is the terminal voltage in the twin stators of a single generator module. In order to linearize the time varying inductances L_{abc}, the three phase dynamic equation at (14) can be converted into *dq0* or synchronous reference frame using the Park's transformation [28], as follows:

$$f_{dq0} = k_s f_{abc} \tag{15}$$

$$where,\ k_s = \frac{2}{3}\begin{bmatrix} Cos\theta_{e,gen} & Cos\left(\theta_{e,gen} - \frac{2\pi}{3}\right) & Cos\left(\theta_{e,gen} + \frac{2\pi}{3}\right) \\ Sin\theta_{e,gen} & Sin\left(\theta_{e,gen} - \frac{2\pi}{3}\right) & Sin\left(\theta_{e,gen} + \frac{2\pi}{3}\right) \\ \frac{1}{2} & \frac{1}{2} & \frac{1}{2} \end{bmatrix}\ and\ \theta_{e,gen} = \int \omega_{e,gen} * dt$$

In above equation, $\theta_{e,gen}$ is the electrical angular displacement of the generator. After the transformation, the generator dynamic equations are:

$$\begin{aligned} v_q &= \boldsymbol{e}_q - R_{stator} i_q - \omega_{e,gen} L_d i_d - L_q \frac{di_q}{dt} \\ v_d &= \boldsymbol{e}_d - R_{stator} i_d - \omega_{e,gen} L_q i_q - L_d \frac{di_d}{dt} \\ v_0 &= \boldsymbol{e}_0 - R_{stator} i_0 - L_0 \frac{di_0}{dt} \end{aligned} \tag{16}$$

The zero sequence quantities can be neglected since the proposed generator is electrically balanced [28]. Moreover, the d-axis reference frame is always aligned along the permanent magnet flux

Journal of Physics: Conference Series **439** (2013) 012039 doi:10.1088/1742-6596/439/1/012039

position, making the EMF in d-axis of the stator to be zero. Therefore, the dynamic equations of the generator model become:

$$v_q = K_{emf}\,\omega_{e,gen} - R_s i_q - \omega_{e,gen} L_d i_d - L_q \frac{di_q}{dt}$$
$$v_d = -R_s i_d - \omega_{e,gen} L_q i_q - L_d \frac{di_d}{dt} \tag{17}$$

Since in (7), $L_d = L_q = L_{stator}$, the above equations in reduced form are:

$$\frac{di_q}{dt} = \frac{1}{L_{stator}} K_{emf}\,\omega_{e,gen} - \frac{R_s}{L_{stator}} i_q - \omega_{e,gen} i_d - \frac{1}{L_{stator}} v_q$$
$$\frac{di_d}{dt} = -\frac{R_s}{L_{stator}} i_d - \omega_{e,gen} i_q - \frac{1}{L_{stator}} v_d \tag{18}$$

Furthermore, the electromagnetic torque $T_{e,gen}$ of the proposed generator is:

$$T_{e,gen} = 4.85 * p_{pp,\mathrm{mod1}} * i_q \tag{19}$$

The Simulink model for single module of the proposed generator is shown in figure 7.

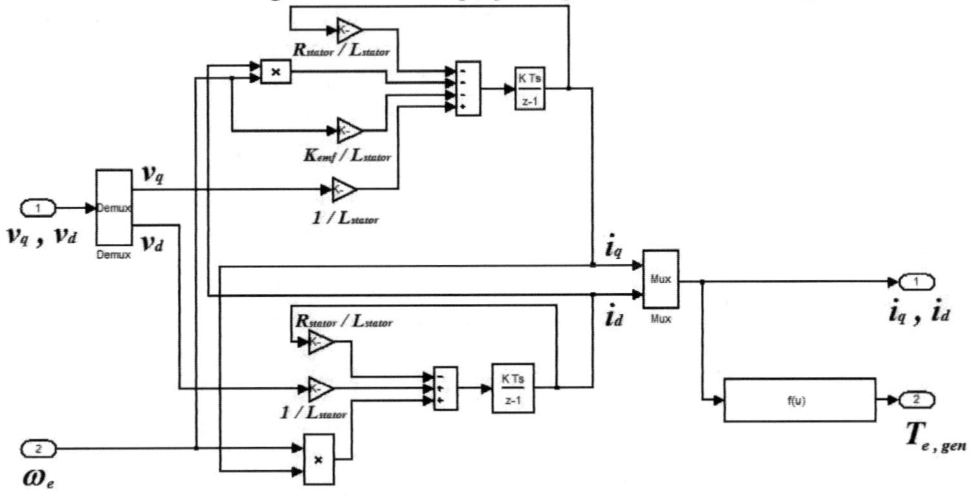

Figure 7. Electrical model of single module

4. Results and discussion

This section presents the comparative results of simulated active & reactive power outputs, energy yield outputs and rotor angular velocities of various wind power generators under variable wind speeds during their operation in wind turbines. The wind power generators under comparison comprise of the induction generator (IG), the doubly-fed induction generator (DFIG), the electrically excited synchronous generator (EESG) and the proposed direct-drive modular axial flux permanent magnet (AFPM) generator. The wind turbine power generation system model for the proposed AFPM generator has been constructed using MATLAB/Simulink. The standard MATLAB/Simulink models for the existing wind power generators, available at [23], have been used for the comparison studies.

In order to make a nose to nose comparison on the basis of wind power utilization and total energy yield, all the wind generators being used have same power ratings of 90 kW and are simulated using the same wind speed model. The variable wind speed model used as input for all wind turbine generators is shown in figure 8. Simulation results of the comparative studies carried out are depicted in following sub-sections. All the results, under each sub-section, are plotted on same magnitude scales. Moreover, their upper and lower limits have been mentioned explicitly.

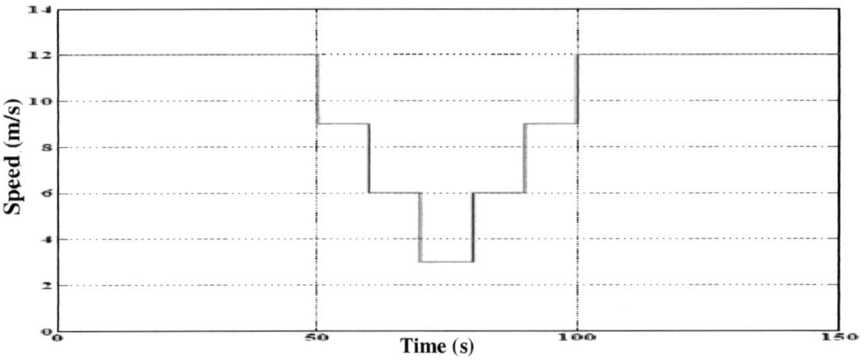

Figure 8. Wind Speed Input

4.1. Mechanical rotational speeds (rad/s)

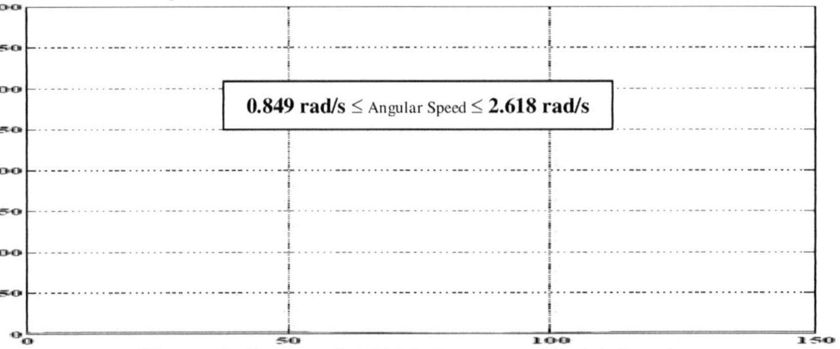

Figure 9. Proposed AFPM Generator Model Result

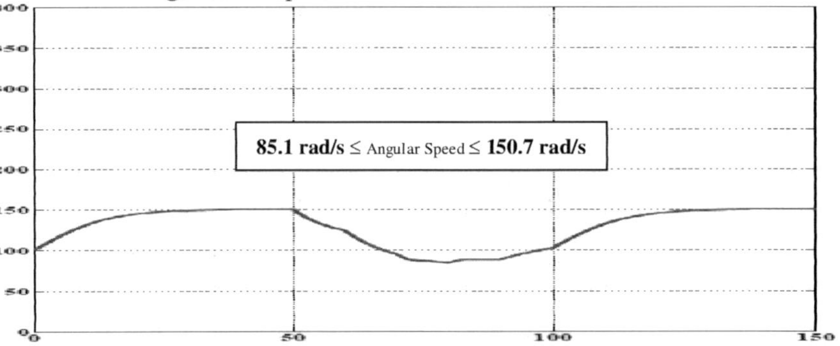

Figure 10. Standard DFIG Model Result

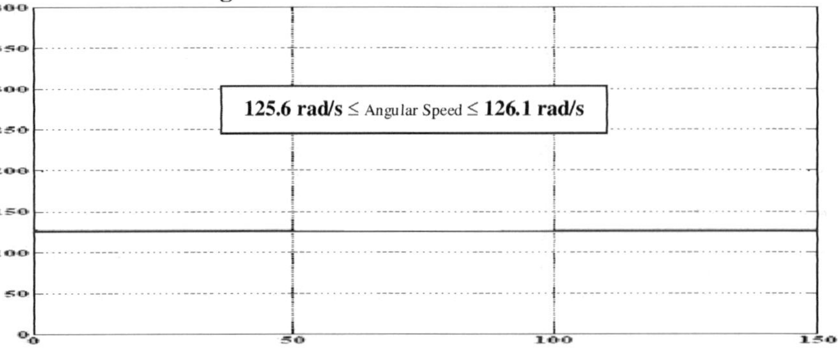

Figure 11. Standard IG Model Result

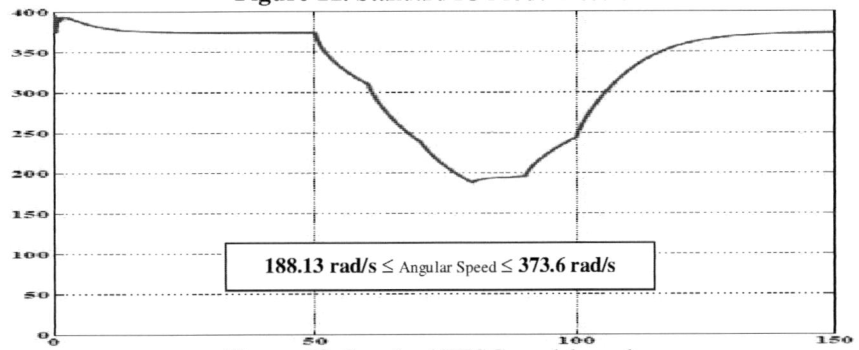

Figure 12. Standard EESG model result

4.2. Active power delivered (Watts)

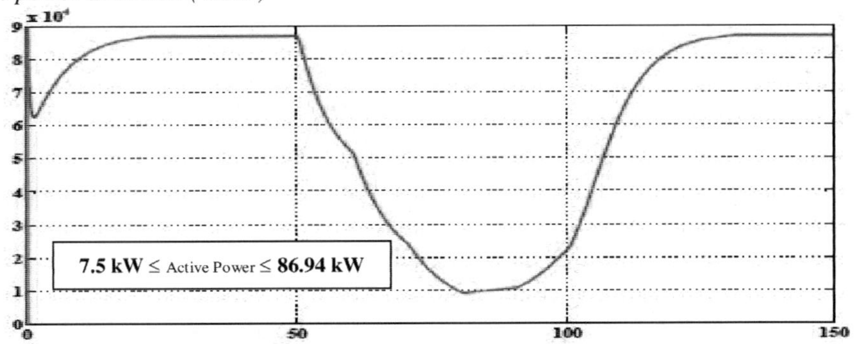

Figure 13. Proposed AFPM generator model result

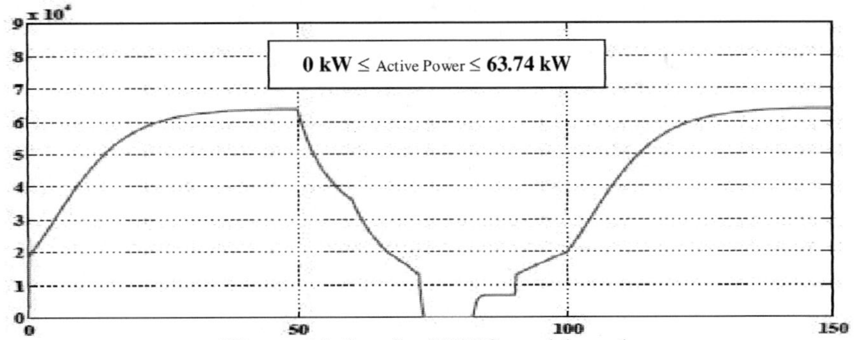

Figure 14. Standard DFIG model result

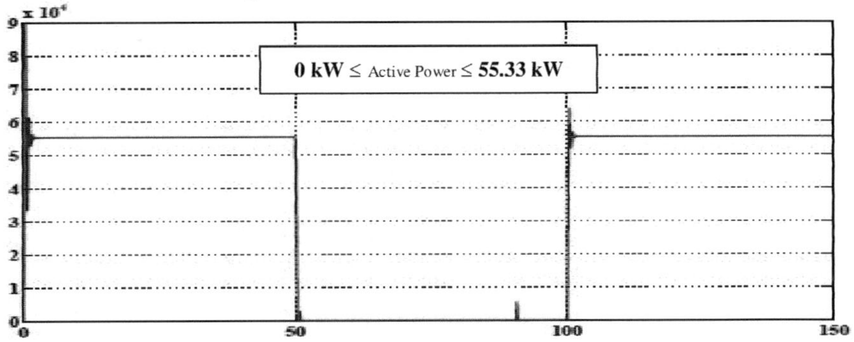

Figure 15. Standard IG model result

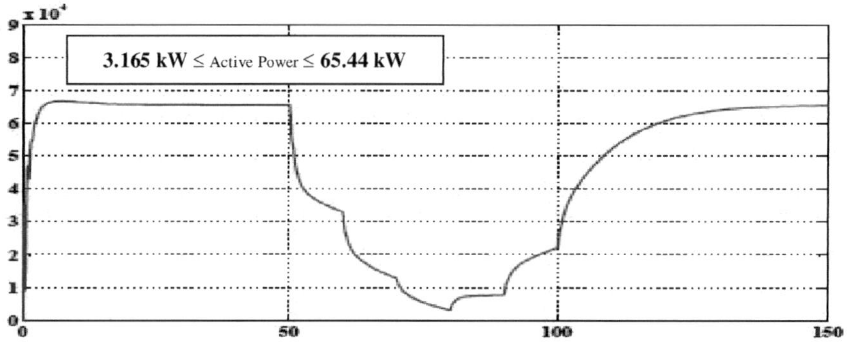

Figure 16. Standard EESG model result

4.3. Reactive power delivered/absorbed (VARs)

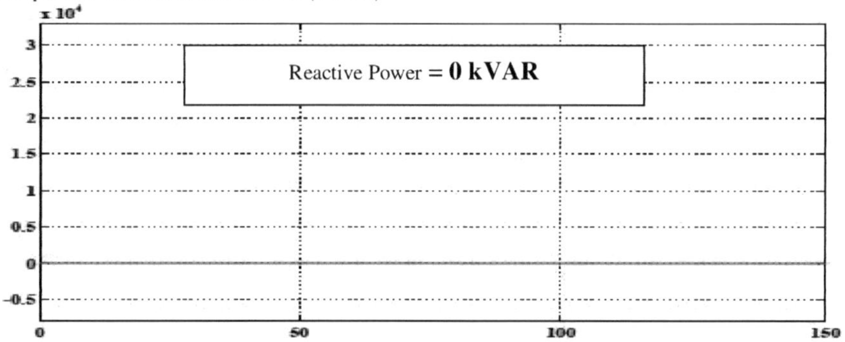

Figure 17. Proposed AFPM generator model result

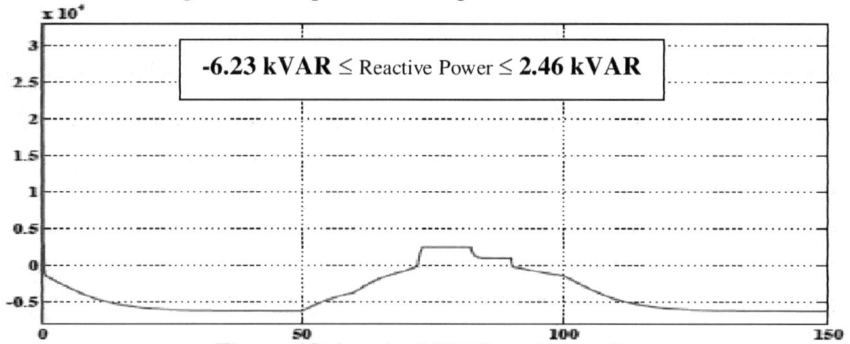

Figure 18. Standard DFIG model result

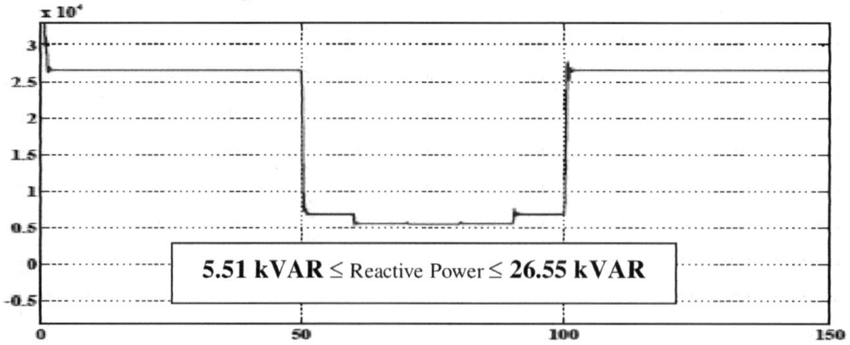

Figure 19. Standard IG model result

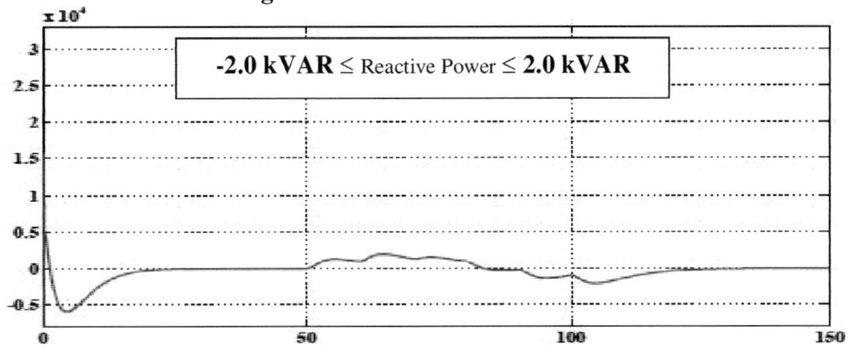

Figure 20. Standard EESG model result

4.4. Energy yield (Watt-hours)

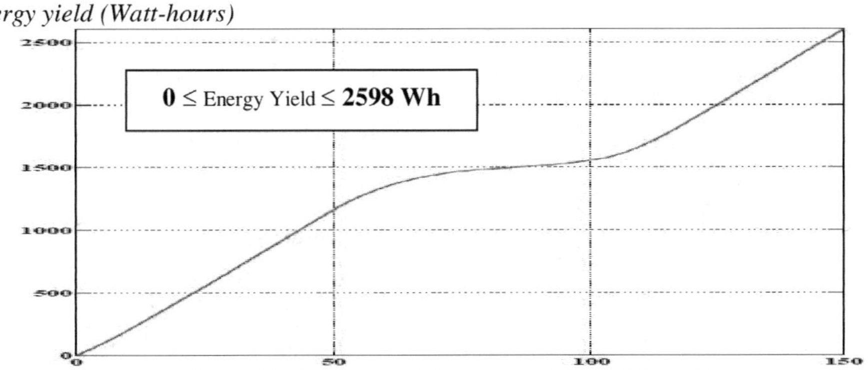

Figure 21. Proposed AFPM generator model result

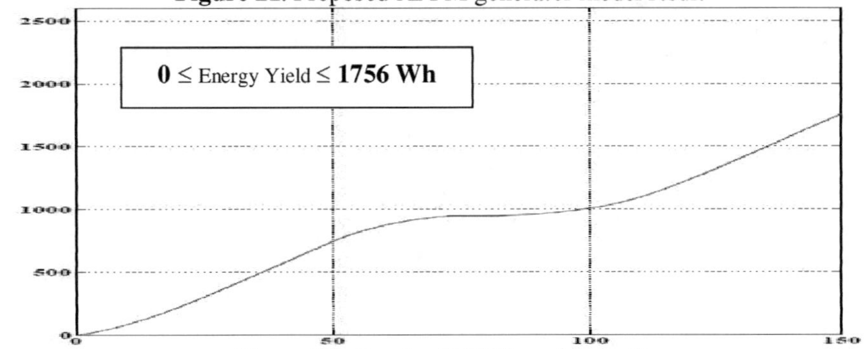

Figure 22. Standard DFIG model result

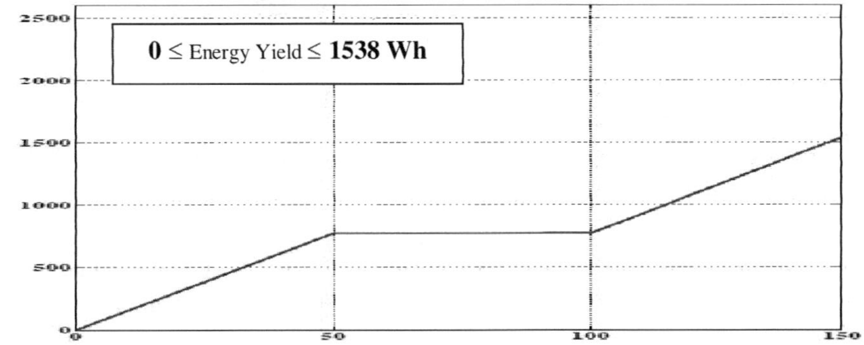

Figure 23. Standard IG model result

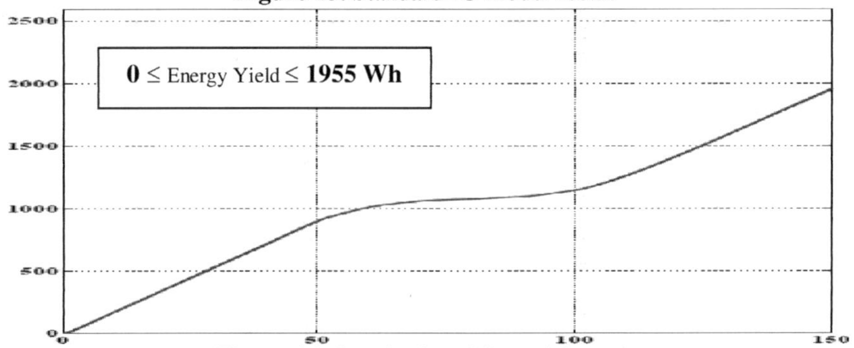

Figure 24. Standard EESG model result

4.5. *Performance comparison*

The innovative design of a low-speed modular, direct-drive AFPM generator with coreless stator and rotor for a wind turbine power generation system has been developed and implemented using dynamic modeling techniques in MATLAB/Simulink. This proposed design has been compared with standard existing wind turbine generator models. The efficiency comparison with variable speed operation and the total energy yield comparison for depicting better wind power utilization have been shown in table 4 and table 5 given below.

Table 4. Efficiency Results.

Wind Speed (m/s)	Percentage improvement in efficiency of the Proposed AFPM Generator compared to:		
	DFIG	IG	EESG
12	26.69%	36.36%	24.73%
9	31.06%	IG stalled	34.62%
6	46.68%	IG stalled	42.62%
3	DFIG stalled	IG stalled	66.33%

Table 5. Energy Yield Results.

Percentage improvement in Energy Yield of the Proposed Generator compared to:		
DFIG	IG	EESG
32.41%	50.80%	24.75%

The comparison results demonstrate that the proposed model for the wind power generator exhibits improved efficiency with variable speed operation, higher energy yield and better wind power utilization. Moreover, the proposed generator's maximum efficiency at the rated operating conditions is 96.6% as compared to the value of 95.7% of the model of AFPM generator proposed in [21]. The proposed design has also reduced the weight of the nacelle since it uses non-ferromagnetic cores that are lighter in weight compared to the ferromagnetic material and provides a gearless turbine system.

5. Conclusion

In this paper, an innovative design of a low-speed, direct-drive AFPM generator is proposed. The proposed model has a modular structure with coreless stator and rotor. It was proved through simulation results that the proposed model has improved efficiency for variable speed operation, compared to the standard models of existing wind power generators. Moreover, it was also proved that the proposed model exhibits number of other advantages such as higher energy yield, lighter weight and better wind power utilization.

Acknowledgment
The authors wish to thank Mr. Hammad Shaukat, Mr. Atif Anwer and Mr. Rizwan Masood for their valuable contribution towards preparation of this manuscript.

References

[1] Blaabjerg F, Iov F, Terekes T, Teodorescu R and Ma K 2011 Power Electronics - Key Technology for Renewable Energy Systems *2nd Conf. on Power Electronics, Drive Systems and Technologies* (Tehran, Iran: 16-17 February 2011) pp 445-66
[2] Wang Y A and Xiao D M July 2011 Prototype design for a high-voltage high-frequency rectifier transformer for high power use *IET Power Electronics* **4** 615-23
[3] Li H and Chen Z 2008 Overview of different wind generator systems and their comparisons *IET Renewable Power Generation* **2** 123-38
[4] Blaabjerg F and Chen Z 2006 *Power Electronics for Modern Wind Turbines* (USA: Morgan & Claypool)
[5] Zhang S, Tseng K J, Vilathgamuwa D M, Nguyen T D and Wang X Yu January 2011 Design of a robust grid interface system for PMSG-based wind turbine generators *IEEE Transactions on Industrial Electronics* **58** 316-28
[6] Cheng K W E, Lin J K, Bao Y J and Xue X D 2009 Review of the Wind Energy Generating System *8th Int. Conf. on Advances in Power System Control, Operation and Management* (Hong Kong, China: 8-11 November 2009) pp 1-7
[7] Jacek F. Gieras, Rong-Jie Wang and Maarten J. Kamper 2008 *Axial Flux Permanent Magnet Brushless Machines* vol 2 (Springer)
[8] Yuan X, Li Y, Chai J and Ma M 2009 A Modular Direct-Drive Permanent Magnet Wind Generator System Eliminating the Grid-Side Transformer *13th European Conf. on Power Electronics and Applications (*Barcelona, Spain: 8-10 September 2009) pp 1- 7
[9] Yuan X, Li Y and Chai J 2010 A Transformer-less Modular Permanent Magnet Wind Generator System with Minimum Generator Coils *25th Annual IEEE Applied Power Electronics Conf. and Exposition* (Palm Springs, California, USA: 21-25 February 2010) pp 2104-10
[10] Gupta R K, Castelino G F, Mohapatra K K and Mohan N 2009 A Novel Integrated Three-Phase, Switched Multi-Winding Power Electronic Transformer Converter for Wind Power Generation System *35th Annual Conf. of IEEE* (Porto, Portugal: 3-5 November 2009) pp 4481-6
[11] Basu K, Somani A, Mohapatra K K and Mohan N 2010 A Three-Phase AC/AC Power Electronic Transformer-Based PWM AC Drive with Lossless Commutation of Leakage Energy *Int. Symp. on Power Electronics Electrical Drives Automation and Motion* (Pisa, Italy: 14-16 June 2010) pp 1194-200
[12] Basu K and Mohan N 2010 A Power Electronic Transformer for Three Phase PWM AC/AC Drive with Loss Less Commutation and Common-mode Voltage Suppression in 36th Annual Conf. on IEEE Industrial Electronics Society (Glendale, Arizona, USA: 7-10 November 2010) pp 315-20
[13] Banaei M R and Salary E 2011 Power Quality Improvement Based on Novel Power Electronic Transformer *2nd Power Electronics, Drive Systems and Technologies Conf.* (Tehran, Iran: 16-17 February 2011) pp 286-91
[14] Yin M, Li G, Zhou M and Zhao C 2007 Modeling of the Wind Turbine with a Permanent Magnet Synchronous Generator for Integration *IEEE Conf. on Power Engineering Society General Meeting* (Tampa, Florida, USA: 24-28 June 2007) pp 1-6
[15] Venkatraj S and Mohan G September 2011 Modeling of wind farms with variable speed direct

Journal of Physics: Conference Series **439** (2013) 012039 doi:10.1088/1742-6596/439/1/012039

driven permanent magnet synchronous generator wind turbines *International Journal of Research and Reviews in Electrical and Computer Engineering* **1** 982-90

[16] Rolan A, Luna A, Vazquez G, Aguilar D and Azevedo G 2009 Modeling of a Variable Speed Wind Turbine with a Permanent Magnet Synchronous Generator *IEEE Int. Symp. on Industrial Electronics* (Seoul, Korea: 5-8 July 2009) pp 734-9

[17] Lin M, Hao L, Li X, Zhao X and Zhu Z Q October 2011 A novel axial field flux-switching permanent magnet wind power generator *IEEE Transactions on Magnetics* **47** 4457-60

[18] Brisset S, Vizireanu D and Brochet P May/June 2008 Design and optimization of a nine-phase axial-flux PM synchronous generator with concentrated winding for direct-drive wind turbine *IEEE Transactions on Industry Applications* **44** 707-15

[19] Gerlando A D, Foglia G, Iacchetti M F and Perini R September 2011 Axial flux PM machines with concentrated armature windings: design analysis and test validation of wind energy generators *IEEE Transactions on Industrial Electronics* **58** 3795-805

[20] You Y M, Hwang K and Kwon B 2010 Optimal Design of Distributed Winding Axial Flux Permanent Magnet Synchronous Generator for Wind Turbine Systems *14th Biennial IEEE Conf. on Electromagnetic Field Computation* (Chicago, Illinois, USA: 9-12 May 2010) p 1

[21] Wang R J, Kamper M J, Westhuizen V D and Gieras J F January 2005 Optimal design of coreless stator axial flux permanent-magnet generator *IEEE Transactions on Magnetics* **41** 55-64

[22] Trimble SketchUp Software for 2D/3D Modeling: http://www.sketchup.com/

[23] MATLAB and Simulink Software for Technical Computing: http://www.mathworks.com/

[24] Finite Element Method Magnetics Software for Two-Dimensional Planar Electromagnetic Modeling: http://www.femm.info/wiki/HomePage/

[25] Franklin G F, Powell J D and Naeini A E 2002 *Feedback Control of Dynamic Systems* (New Jersey, USA: Prentice Hall)

[26] Patel M R 1999 *Wind and Solar Power Systems* (USA: CRC Press)

[27] Muyeen S M, Tamura J and Murata T 2009 *Stability Augmentation of a Grid-connected Wind Farm* (Springer)

[28] Krause P C, Wasynczuk O and Sudhoff S D 2002 *Analysis of Electric Machinery and Drive Systems* IEEE Press Series on Power Engineering (USA: John Wiley & Sons)

Proposed applications with implementation techniques of the upcoming renewable energy resource, The Tesla Turbine

M Usman Saeed Khan[1, 2, 3], M Irfan Maqsood[2], Ehsan Ali[2], Shah Jamal[2] and M Javed[2]
[2]Univesity College of Engineering and Technology,
The Islamia University of Bahawalpur, Pakistan

[3]E-mail: usmansunshine@gmail.com

Abstract. Recent research has shown that tesla turbine can be one of the future efficient sources of renewable energy. Modern techniques used for designing of tesla turbine have given optimum results regarding efficiency and applications. . In this paper we have suggested fully coordinated applications of tesla turbine in different fields particularly in power generation at both low level and high level generation. In Energy deficient countries the tesla turbine has wide range of applications and it can play an important role in energy management system. Our proposed applications includes,

- the use of tesla turbine as renewable energy resource
- using tesla turbine in distributed generation system
- use of tesla turbine at home for power generation
- use of tesla turbine in irrigation channels
- using tesla turbine in hybrid electric vehicles

All applications are explained with the help of flow charts and block diagrams and their implementation techniques are also explained in details. The results of physical experiments and simulations are also included for some applications.

1. Introduction

In this age of technology and revolutions the world is progressing very fast in each and every field. With the fast development in industrial and transportation sectors the demand of energy is increasing exponentially. The population of world is increasing rapidly and the life style of the people is also getting modernized. All these facts are leading the world to a serious energy crisis. To meet the technological and social revolutions power sector has also to grow with the demands of the consumers. Oil, gas, coal and other conventional fuels are the major sources of energy in these days and are being used extensively for the purpose of power generation. The rapid consumption of these energy resources is creating an alarming situation for the future of energy requirements and challenging both the developed and progressive countries to find out the energy resources for their future needs. World primary energy consumption has grown 2.5% in the year 2011 as compared with the year 2010 [1]. No doubt the reserves of petroleum and gasoline are increasing as new reserves are being discovered in different countries but still the rate of increase in the reserves of these energy resources is far less than the rate of increase in the consumption of these resources in industrial, transport and power generation

[1] To whom any correspondence should be addressed.

sectors. Another drawback of conventional fuel is that the prices of the petroleum are increasing to a very high level. Figure 1 shows the estimated increases in price and the consumption of energy resources in the future [2]. In this graph the units used for oil prices are dollars per barrel and for energy consumption British thermal units are used. The figure clearly shows how the energy consumption and fuel prices are going to increase in the future. Demand of electricity is also increasing rapidly and it has increased from 439 million tonnes of oil equivalent to 1441 million tonnes of oil equivalent during the period of 1973 to 2009 [3]. This has been shown in figure 2 as a share of different electricity consumption sectors. Emission of Carbon dioxide and greenhouse effect is yet another big issue of the conventional fuels.

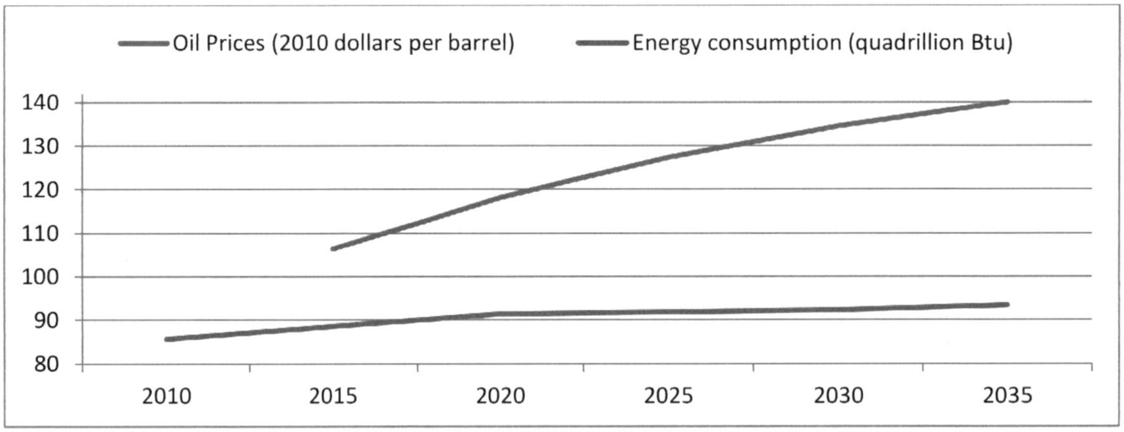

Figure 1. Estimated world energy consumption and Oil prices in future, 2010-2035

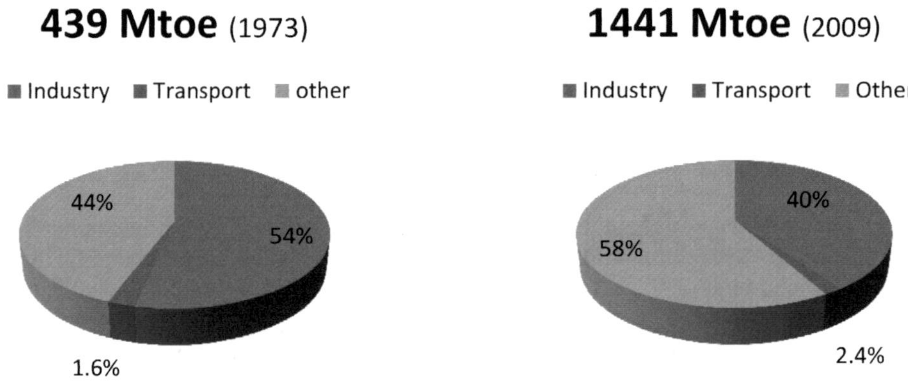

Figure 2. World electricity consumption in 1973 and 2009

The only solution to these crises is to convert the fuels from conventional to the renewable. Only the renewable energy resources can compensate for the growing demands of the energy consumption. Tesla turbine is an upcoming renewable energy resource as it is catching the attention of researchers and market holders. Although it has been invented long ago but due to the technical and political issues it was out of market for a long time. People were unaware of the advantages and applications of this turbine. The main thing of tesla turbine is that it is a bladeless turbine. That is why it is also called boundary layer turbine. It has many advantages and its use in many applications can make it much beneficent and thus it can reduce the demand of conventional fuels in the future. Some important applications of tesla turbine have been proposed and explained in details in the following paragraphs.

2. Proposed applications of the Tesla Turbine

Previously the applications of the Tesla turbine were in a very limited range. The use of this turbine was limited to the applications which required high speed but low torque. Many researchers has tried to elaborate the wide spread advantages of this turbine but still the tesla turbine was not able to catch the attention of patents for its commercialization because of its narrow range of applications. This was due to the fact that there were some certain defects in its design. But now after the advancements in the design of this remunerative turbine it can become a promising source for the future needs in many aspects of the world particularly it can serve to compensate for a part of the future energy crises of the world. The new design of tesla withdraws the defects of the turbine and makes it more beneficial by improving the torque of the turbine to a great extent and thus making it able to be used in a vast range of applications. Here we have proposed some useful applications of the tesla turbine which are explained in the following paragraphs with some suggestions for the practical implementation of these applications.

2.1. Renewable energy resource

Like any other rotating machine Tesla turbine can also be used to rotate the rotor of a generator and thus it can be used for the generation of electric power. Usually Tesla turbine uses the water or pressurized air as a fuel thus its running cost is lowest than any other engine or turbine and its fuel is abundantly available everywhere in the world. The use of tesla turbine in generation of electricity is proposed as it can be implemented on home level as well as on higher level. At home level it can be installed in the pipe lines of the daily use water tabs and on higher level they can be installed in irrigation channels, streams of waters and tube wells in agriculture areas and in the down-stream of water in hilly areas. A general scheme for implementation of tesla turbine as a renewable energy resource is explained here which is shown in figure 3. There are many choices of using different fuels, depending upon the type and place of application the most suitable green fuel can be used which can be any fluid like water or air. This green fuel in form of fluid is definitely required to be filtered. Then to increase its speed it is passed through a pressurizer. After that a very fast and thin stream of fuel is injected to the discs of Tesla turbine which in return rotates the rotor of generator and thus we can get the electric power without any pollution and side effects.

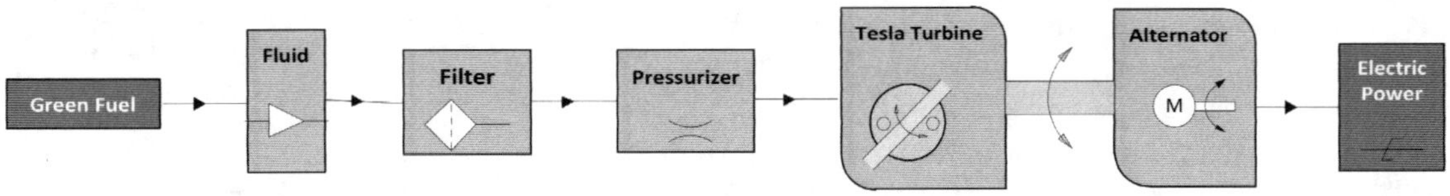

Figure 3. Implementing Tesla turbine as a renewable energy resource

In this way the Tesla turbine can be used in the wind turbine applications and particularly Tesla turbine will be very beneficial in the geothermal power plants where hot steam from the earth is used as a fuel. So the viscous steam will be a good fuel to the tesla turbine and it will give much better efficiency.

2.2. Distributed generation system

Another important application of Tesla Turbine is to use it in the distributed Generation system. It is known that the adhesive properties of the fluid are increased with the increasing temperature so the torque of the Tesla turbine can be increased by using the warm fuel [4]. In distributed generation the heat obtained from any generating source can be used to heat up the fuel of the Tesla turbine and thus make it more efficient. Particularly in case if fuel cell generation is being used than the heat which is a side product of fuel cell can be used to increase the temperature of water or the air (fuel for tesla turbine). Similarly in the conventional power plants a lot of heat is wasted as a side product. This heat can be utilized to heat up the fluid and provide this fluid to the tesla turbine. The implementation of this process is explained with the help of block diagram in figure 4. In this figure the waste heat of a thermal power plant and the heat obtained from the operation of fuel cell, is used to rise the temperature and thus cohesive capability of the fuel for the tesla turbine.

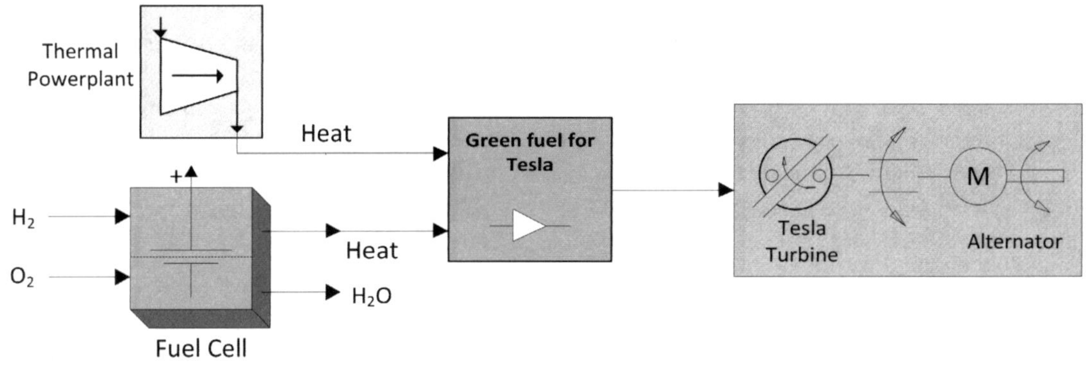

Figure 4. Implementing Tesla turbine as a renewable energy resource

Distributed generation system is a new but popular technology in the modern electric power system particularly the DGS technology is being implemented in the smart grids. In this way tesla turbine can become an important component of the smart grids which are a modern technology based power system.

2.3. Power generation at home level

In homes, tesla turbine can be rotated by the water which is circulating by pressure from water reservoir tanks to our common use water tabs. To increase the pressure of water the pressurizer can be used. Thus the turbine will rotate at high speed and if its shaft is connected with the small generating unit via gears, a sufficient amount of power can be generated to charge the batteries. Then using inverter circuit we can get our desired AC voltage.

Thus the running cost of Tesla Micro Generation System will be zero as free of cost fuel is used to run the turbine and even the water is not wasted and is used without any loss. This application can be very useful in remote areas where electric utilities cannot reach. This application can also be implemented in communities, residential colonies and other institutes like universities etc. where a common large scale reservoir tank is used to supply the whole colony or the institute. Thus in such areas tesla turbine can be used at a relatively large scale and a large amount of electric power can be generated at a very low cost. The typical implementation scheme for this application has been shown in figure 5. More than one turbine can be used easily without sacrificing the performance of the any turbine.

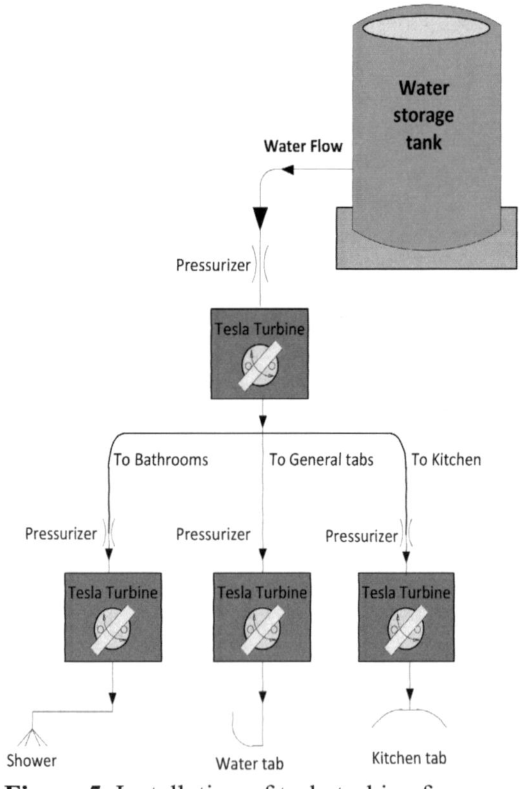

Figure 5. Installation of tesla turbine for power generation at home level

2.4. Use of Tesla turbine in irrigation channels

Another proposed application of tesla turbine is especially for rural areas and agriculture lands where a large number of tube-wells and canals exist. The tesla turbines of relatively large size can be rotated by the extremely heavy pressure of the water of tube-wells and canals to generate relatively large amount of electricity which can be provided to homes. Also this turbine can rotate a motor of many horsepower which can run another tube-well. Hence a huge amount of energy can be used without any serious cost. The implementation technique for this application is very simple. The good thing in this application is that this turbine does not require the high head of water. Implementation scheme of application of tesla turbine in canals and other irrigation channel has been shown in figure 6. It has been shown that multiple turbines can be installed in a single canal at different places both in series and parallel. Some arrangement must be made to make the path or water a bit narrow or slope like so that the speed and pressure of the water can be increased.

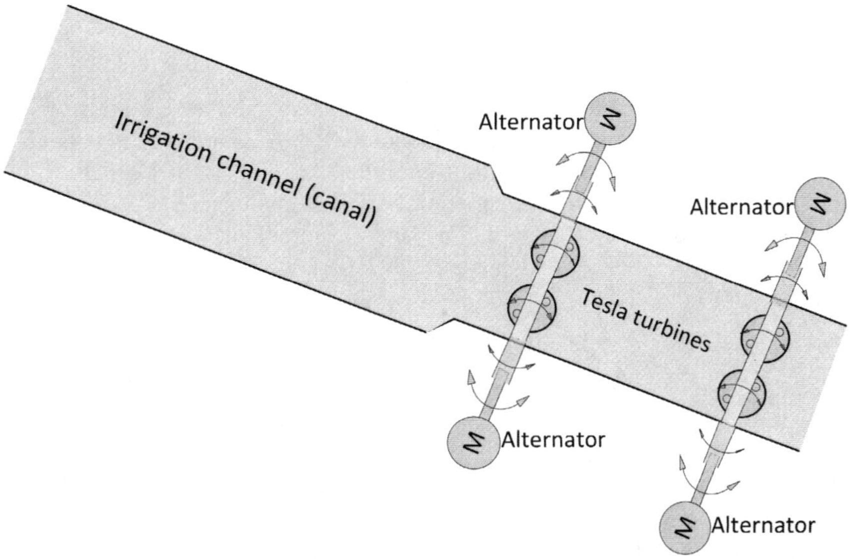

Figure 6. application of multiple tesla turbines in irrigation channels

2.5. Using Tesla turbine in hybrid electric vehicles

Due to the high emission of carbon dioxide gas from transportation sectors the use of conventional fuels is becoming unlikely and electric or hybrid electric vehicles are making their ways to the market. As the electric vehicles do not use the conventional fuels so they are environment friendly and also the electric vehicles produces very low noise as compared to the vehicles having internal combustion engine. Hybrid electric vehicles are those vehicles which have a combination of both the internal combustion engine and the electric propulsion engine. Further there can be two types of hybrid electric vehicles. One in which the internal combustion engine is used to move the vehicle directly or in other case it is only used to work as a generator and recharge the storage batteries of the electric vehicle. Tesla turbine can be an efficient green fuel for the recharging of the batteries. The implementation technique for the using of tesla turbine in a hybrid electric vehicle is explained with the help of figure 7. Initially the vehicle will be started with the help of internal combustion engine (ICE) or if the batteries are already charged than it can also be started electrically. Once the vehicle is started the air faced by vehicle will be collected by an air receiver. After the filtration and compression of this air it will be sent to the tesla turbine via some special pressure nozzles at a certain angle and pressure. Thus the viscous and cohesive effect of air with the plates of turbine will rotate the turbine and so the alternator will produce AC electric power which is used to charge the batteries after the rectification process.

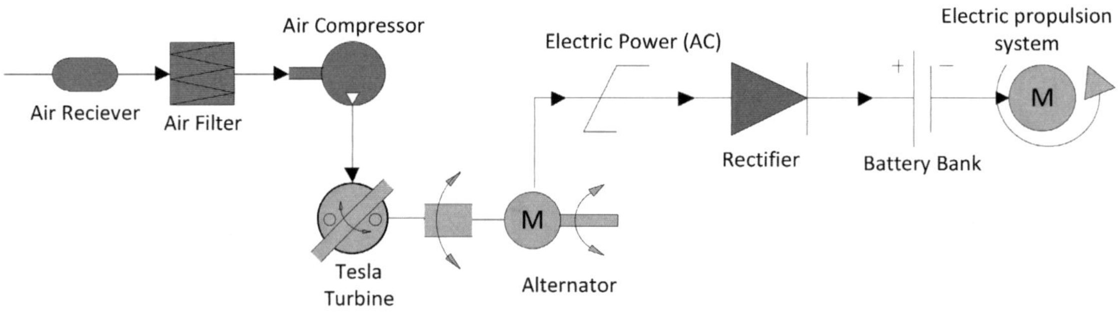

Figure 7. Electrical system of a hybrid electric vehicle using Tesla turbine.

3. Conclusion

The outcome of the complete research is that' due to the expected future energy crisis, it is not supposed to lose any type of energy anywhere. Use of renewable energy resources is need of the hour. Generation of power form solar and wind has the limitation that they are not reliable for whole day generation so adding the tesla turbine in the generation scheme will make the power system much more reliable and that too without any fuel cost. The only requirement is to use the turbine in many specific applications so that the optimum advantage can be taken from this bladeless and improved turbine.

4. References

[1] *BP Statistical Review of World Energy,* June 2012. p 02
[2] *Annual Energy Outlook 2012 with Projections to 2035*, June 2012, DOE/EIA-0383(2012) p 104
[3] *2011 Key World Energy Statistics*, International Energy Agency, 9, rue de la Federation 75739 Paris Cedex 15, p 35
[4] Matej Podergajs, *Seminar: The Tesla Turbine,* University of Ljubljana Faculty of Mathematics and Physics, March 2011
[5] Piotr Lampart Lukasz Jędrzejewski, *Investigations Of Aerodynamics Of Tesla Bladeless Microturbine*, Journal Of Theoretical And Applied Mechanics 49, 2, pp. 477-499, Warsaw 2011
[6] Mark G. Breiter Kar. Pohlhausen, *Laminar Flow between Two Parallel Rotating Disks,* Applied Mathematics Research Branc, Aeronautical Research Laboratory Office of Aerospace Research United States Air Force
[7] Bryan P. Ho-Yan *Tesla turbine for pico hydro applications* Guelph Engineering Journal, (4), 1 - 8. ISSN: 1916-1107, year 2011.

Journal of Physics: Conference Series **439** (2013) 012041 doi:10.1088/1742-6596/439/1/012041

Numerical climate modeling and verification of selected areas for heat waves of Pakistan using ensemble prediction system

S Amna[*1], N Samreen[1], B Khalid[1], A Shamim[1]
[1] Department of Environmental Sciences, International Islamic University, Islamabad, Pakistan

[*]*E-mail: shahzadi.bses169@iiu.edu.pk*

Abstract. Depending upon the topography, there is an extreme variation in the temperature of Pakistan. Heat waves are the Weather-related events, having significant impact on the humans, including all socioeconomic activities and health issues as well which changes according to the climatic conditions of the area. The forecasting climate is of prime importance for being aware of future climatic changes, in order to mitigate them. The study used the Ensemble Prediction System (EPS) for the purpose of modeling seasonal weather hind-cast of three selected areas i.e., Islamabad, Jhelum and Muzaffarabad. This research was purposely carried out in order to suggest the most suitable climate model for Pakistan. Real time and simulated data of five General Circulation Models i.e., ECMWF, ERA-40, MPI, Meteo France and UKMO for selected areas was acquired from Pakistan Meteorological Department. Data incorporated constituted the statistical temperature records of 32 years for the months of June, July and August .This study was based on EPS to calculate probabilistic forecasts produced by single ensembles. Verification was done out to assess the quality of the forecast t by using standard probabilistic measures of Brier Score, Brier Skill Score, Cross Validation and Relative Operating Characteristic curve. The results showed ECMWF the most suitable model for Islamabad and Jhelum; and Meteo France for Muzaffarabad. Other models have significant results by omitting particular initial conditions.

1. Introduction

The terms temperature, atmospheric composition wind speed and direction, pressure and density determine the initial condition of atmosphere at a particular instant. The averages of these parameters are taken into account to determine the climate of an area. Intensity of the radiations emitted from the sun and features of the earth plays a fundamental determining role [1].The changes in topography and local features occur within the atmosphere or over the earth surface in a systematic way and orderly pattern, hence creating an identifiable pattern of climatic differences from place to place [2].

Climate, over along interval of time, determines the landscape, and the plants and animals of a region. Over geological time, climate has helped to shape mountains, build up the soil, determine the nature of the rivers, and build flood plains and deltas [3].

Hence, climate is decisive to the world. Significant change in global climate has been witnessed around the world. Though, it is not a new experience in the Earth's history. During the long geological period, extreme weather events have always been the part of earth's climate. Climate varies from decade to decade, reliable averages of the frequency and magnitudes of extreme events require weather observations over longer periods. The studies of this long period observation of climate have shown the rapid increase in the temperature of the earth. However, the present change in climatic conditions has pronounced effect. The rationale behind this cause is the amplified concentration of greenhouse gases from the value of about 280 ppm to 379 ppm in 2005 resulting from the anthropogenic activities. This has resulted in the boost in the global temperature by 0.76° C during the last century [4].

Consequently there is a dire need to understand the basics of climate; the changes occurring in it. Developing countries are the most vulnerable to present climate change impacts due to the unavailability of resources to adapt: socially, technologically and financially, as at this time developing countries are facing lack of food and water [5].

Changes in some types of extreme events have already been observed in various parts of the world because of the global climate change, for example, increases in the frequency and intensity of heat waves and heavy precipitation events [6].

Climate change has threatened the world as whole. Amongst the most susceptible region of the world, South Asian Region is of utmost significance. Countries in this region have geographical and ecological assortment. Pakistan is the second largest country of the South Asia, where the variations in geography greatly shapes the climate of the country. Lying at extreme north-west corner of the Indo-Pak subcontinent between latitudes 24°N to 37°N and longitudes 60°E to 75°E, occupies sub-tropics and partially temperate region of the world where the Climatic conditions vary from hot, dry summers to mild winters [7].

Pakistan being a developing country is more prone to the impacts posed by the changing climate because of its agrarian based economy. As a consequence of climate change, there is a rise in mean temperature of 0.6-1.0°C throughout the country as well as 0.5 to 0.7% increase in solar radiation over southern half of the country. In addition to this water logging, desertification of land, growing frequency of pest attacks and weather related disasters has been observed through out the country [7]. With the little increase in temperature there is a significant increase on the extreme events. In recent years, the number of hot days has been increased resulting in the frequent occurrence of the heat waves throughout the country. Heat waves are referred as the above normal hot weather for days to week [6-7].

As a result of excessive heating, heat islands are formed in the areas where the non-reflectivity, water resistance surfaces and vegetation cover is low. Building materials, such as stone, concrete, and asphalt incline heat at the surface. In urban areas the effect of heat island enhances the impact of heat waves [8-9] and in addition lack of vegetation reduces heat lost due to evapotranspiration [10]

Urban areas have high population densities so the energy demand is also high resulting in the high demand of air conditioning during summers [8] that results in the increased local and regional air pollution. The emissions, from power generation, in turn increases absorption of radiation [9] and create inversion layers. Hence, slow dispersion of pollutants produced in urban areas occurs [11].

Focusing the climate of Pakistan, there is a great variation in the summer and winter temperatures [12]. The prolonged high temperature and high humidity is considered as weather hazard. The high heat index can cause a health risk to the people engaged in outdoor activity even for small time period [13].

In the present situation in which world has turned into global village and the frequency of extreme event causing lot damage to the life and infrastructure, the forecasting of various constituents and factors of climate and weather has become an important affair. In particular, it has now become a significant tool in order to prepare the man for the incoming disruptions in the climatic system of the earth that may be in form of increased rainfall or extreme temperatures.

Hence, regarding the importance of climate forecasting in the present scenario, this study stresses on the establishment of the early warning system in Pakistan to determine the frequency of extreme event i.e., heat waves.

Numerous climate models are operating around the world that computes the climatic conditions and gives prediction about the future climate. These can either be global circulation model (GCM) or a regional Circulation model (RCM). The most reliable global climate models used, now a day, are European Centre for Medium Range Weather Forecast (ECMWF), Max Plank's Institute (MPI), United Kingdom Meteorological Office (UKMO), Meteo France, and European Centre for Medium Range Weather Forecast Re-Analysis-40 (ERA-40). GCMs simulate large-scale patterns of seasonal variation. Pakistan, lack its own Global circulation model for the climate forecasting, therefore, one of the basic aims to purse the current research is to suggest the best suitable General Circulation Model for accurate seasonal weather prediction for the selected areas.

In 2011, the predictability of the 2010 floods in Pakistan was studied using the three data sets of TRMM, GPCP data and CMORPH analyzed the data for 51 members of ECMWF EPS. The study showed the ability of the ECMWF EPS to predict the extreme event in the country in order to establish an early warning system. Therefore such a system can help to have hydrological forecasts in Pakistan to anticipate for the losses from devastating floods [14].

The Numerical Weather Predication of the climate models depends upon the sensitivity of initial condition and the uncertainties associated with the forecast model being used [15]. To find out the level of uncertainties probability is calculated from these forecasts. Based on the historical data observations, statistical techniques are mostly used to determine Probabilistic forecast. It provides the estimation of climatic risks incorporating the uncertainties into it [16].

The inaccurate initial condition and insufficiency of numerical models turn into a source of forecast errors. These errors classified as; initial condition errors and model errors, provide the forecast with some uncertainty, limits the quality of forecast. As a response to these limitations Ensemble Forecast system is used to predict the probability distribution of forecast. Ensemble forecasting predicts the probability distribution of the state of atmosphere for future, in a quantitative manner [15]. In the current study, the validity of the climate model and Ensemble forecast has been checked in order to verify if the selected models can give skillful forecast for the selected areas.

To quantify the relationship between the observation and the probabilistic forecasts verification is carried out in order to access the quality of the forecast. A forecast is considered to be deficient until the success of it is not determined [17]. To verify the probabilistic forecasts, Use of Relative Operating Characteristic Skill Score, Brier score and Brier Skill Score is done. The quality assessment helps to find and spot out the strengths and weakness of a forecast. It also helps in determining the best suitable verification technique by comparison of different methods [18]. Brier score measures the probability of the forecast whether it will occur or not. It is sometimes hard to interpret therefore converted into brier skill score [19].Brier skill score, being a standard measure, is used to access the skill of a forecast. It is applied to both single model and multi model ensemble. The Relative Operating Characteristic (ROC) is a graph of sensitivity (y-axis) vs. 1– specificity, where the area under the curve shows the forecast's accuracy; a verification method to check the skill of the forecast by comparing the hit rate and the false-alarm rate [20]. The series of threshold for the probabilistic forecast is plotted between the false alarm rate and the hit rate [21].

2. Materials and methods

Three stations were selected to access the variability of the temperature over Pakistan in order to determine the occurrence of heat waves and to check out the validity of the selected climate models in these areas to give accurate forecast.

2.1. Islamabad

The capital city of Pakistan lies 33° 42' N, 73° 10 E'. The climate of Islamabad varies from an average daily low of 1° C in January to an average daily high of 40° C in June. Half of the annual rainfall occurs in July and August, averaging about 255 millimeters in each of those two months. Temperature reaches to -4o C in winters and to 50 0 C in summers [22] However the climatic change that's has been occurring throughout the globe has affected the climate of Islamabad, too. The occurrence of several heat waves has been reported in the capital city. A major heat wave was reported on June 21, 1994, when the temperature reached to 46.5°C. The recent episode of the event was recorded in May-June 2012, when the temperature touched the 50° C mark on thermometer.

2.2. Muzaffarabad

Muzaffarabad, the capital of the Pakistan Held Jammu & Kashmir, is situated at the confluence of the Jhelum & Neelum Rivers, lies at 32° 44' N latitude and 74° 54' E Longitude. The mean maximum and minimum temperatures during the month of July are about 35°c and 23°c; and in January 16°c and 3°c respectively. The average annual precipitation of the district is 1511mm [23]. Heat waves has been occurring in the past in Muzzaffarbad, however their frequency has been increased in last five years. Recently in July 2012, heat waves touched the city when temperature reached 43 °C.

2.3. Jhelum

Jhelum lies on the right bank of right bank of river Jhelum ranging from at 32°56′ of North latitude to 73°44′ in East longitude. During the summers, Jhelum experience extremely hot and humid weather and warm and generally dry in winter. The temperature reaches to 45.7 °C (114.3 °F) from April to June, whereas in winter the minimum temperature recorded is 1.8 °C (35.2 °F) [23]. A major Heat wave occurred during June 2005 in Jhelum, resulting in the rise in water level in river Jhelum. Recently, in 2010, extreme heat waves have been reported when the temperature of the city has reached to 48°C.

3. Method

Research was carried out in a systematic way. The major steps involved were data acquisition, data analysis, data validation evaluation and assessment. By using Microsoft Excel 2010 and SPSS v20.0, the preparation of data for analysis was done. 32 years temperature data, for three areas was collected to validate, evaluate and assess for the seasonal hindcast. For the analysis, Microsoft Excel 2010 spread sheets comprising real time data of PMD and simulated data of the climate models were created.

The first step for the analysis was to calculate ensemble forecast, using nine initial condition of each of the five climate models used in the study that includes; ECMWF, MPI, Meteo France, ERA-40 and UKMO. To determine the percentiles upper and lower tercile (0.667) threshold values were used. The percentile values were then used to find out the climatological probability of real time data of PMD, simulated data of ERA-40 and probability of respective climate model. Using the probability obtained for each of the nine initial conditions, probabilistic forecast was computed by using all the nine initial condition and by using them alternatively. These were then compared with the probabilistic forecast of PMD and ERA-40. Brier score was calculated for reliability, resolution and uncertainty and brier skill score to check the skill of the probabilistic forecast.

$$BS= 1/N\sum (A_i - B_i)^2$$

Where A_i is the probability that was forecast,
And B_i the actual outcome of the event at instance,

The brier skill score being the mean squared error in probability space was calculated as;

$$BSS= 1- BS/ BS_{\text{reference}}$$

BSS was calculated for all the Initial condition and then for alternatively used initial condition with ERA-40 and PMD. It was taken as, if Brier skill score is positive it indicates a skill forecast, however if a BSS is negative it shows the poor forecast. To assess the accuracy of the probabilistic forecast, Cross validation was done. For which graphs were plotted. SPSS v20.0 was used to verify the probabilistic forecast by using Relative Operating Characteristic (ROC). If the ROC curves have hit rate of 1.0 and no false alarms, probabilistic forecast could be perfectly predictable by the forecast system. If it falls in the false alarm region, the forecast would be poor.

4. Results and discussions

By the use of ensemble prediction system, the seasonal forecast of heat waves for selected areas for the months of May, June and July were computed with one month lead time April. The heat waves were presented by "upper tercile". The results obtained provided with the following conclusions.

For Islamabad the average of real time observation of PMD for 32 years i-e., 1969-2000 for heat waves came out to be **29.48**, while the average of ERA-40 is **16.88**. By using these values climatological probability obtained from PMD was **0.34.**

Table 1. Results evaluated from climate models for Islamabad.

Model	Ensemble forecast	Probabilistic Forecast			Climatological probability		Reliability	Resolution
		All conditions	Excluding Zero	Excluding Nine	PMD	ERA-40		
ECMWF	18.10	0.33	0.34	0.33	Skill	Skill	High	Low
Meteo France	15.09	0.33	0.32	0.34	Skill	Poor	High	Low
MPI	19.96	0.33	0.34	0.27	Skill	Skill	High	Low
UKMO	13.23	0.30	0.32	0.33	Poor	Poor	High	Low

For Jhelum the average of real time observation of PMD for 32 years i-e., 1969-2000 for heat waves came out to be **31.4**, while the average of ERA-40 was **32.5**. Climatological probability obtained, by using these values, from PMD was **0.34.**

Table 2. Results evaluated from climate models for Jhelum.

Model	Ensemble forecast	Probabilistic Forecast			Climatological probability		Reliability	Resolution
		All conditions	Excluding Zero	Excluding Nine	PMD	ERA-40		
ECMWF	30.8	0.33	0.34	0.34	Skill	Skill	High	Low
Meteo France	31.51	0.33	0.32	0.34	Skill	Poor	High	Low
MPI	30.26	0.33	0.33	0.32	Poor	Poor	High	Low
UKMO	30.78	0.33	0.34	0.32	Poor	Poor	High	Low

The average of real time observation of PMD for Muzaffarabad from 1969 to 2000 i.e., 32 years for heat waves obtained was **28.10**, while the average of ERA-40 was **8.06**. By using these values climatological probability obtained from PMD was **0.34.**

Table 3. Results evaluated from climate models for Muzaffarabad.

Model	Ensemble forecast	Probabilistic Forecast			Climatological probability		Reliability	Resolution
		All condition	Excluding Zero	Excluding Nine	PMD	ERA-40		
ECMWF	10.64	0.33	0.33	0.33	Poor	Poor	High	Low
Meteo France	8.72	0.33	0.32	0.34	Skillful	Skillful	High	Low
MPI	14.93	0.33	0.35	0.27	Skillful	Poor	High	Low
UKMO	7.08	0.33	0.33	0.34	Skillful	Skillful	High	Low

ERA-40 project for forty five year (45) global time series has been completed in 2003. This analysis is used widely to find out the climatic changes around the globe [24]. These provides with the useful recognition of original observational errors. As the forecast made by the global climate models, especially ECMWF, is almost accurate therefore the observational errors are small [25]. More or less, similar results were obtained while comparing the ensemble forecast made by the models used in the study. The outcomes show Ensemble forecast being close to the ERA-40., is a good forecast but poor as compared to PMD, concluding ERA-40 to be an accurate analysis to compare ensemble forecast observational average.

To study severe weathers event in United Kingdom, a system through the output of the ECMWF to give a probabilistic forecast for applying the calibration to the EPS model was established. The output was modified for the probabilistic early warnings and was verified over 2 winter seasons. The probabilistic skill scores were applied. The results gave the probabilistic forecast to be an ideal application for ensemble prediction system. They suggested the EPS work in a better way for the medium range, and may need other perturbation strategies for successful short-range ensemble prediction [26].

In particular for this study, Probabilistic forecast was calculated through EPS by using the nine initial conditions and then by using them alternatively for the forecast of extreme event. Climate models are not subtle to the initial conditions. However, change in the initial value problem and boundary value problem are the basic reasons of change in accuracy of the probabilistic forecast for the area under study [27]. With all the initial conditions and alternatively using them the probabilistic forecast were calculated. And the results formulated from table 1, table 2 and table 3, states that by eliminating those initial conditions of each of the climate models that are not suitable for the areas under study; an almost accurate forecast probabilistic forecast can be obtained.

Probabilistic forecasts during forecasting include uncertainty. To verify the reliability, resolution and uncertainty of probabilistic forecast, Brier Score was used. Brier skill score results that the probabilistic forecast of ECMWF, MeteoFrance and MPI is the skill forecast for Islamabad while using the alternative initial condition; all the forecasts were skill forecasts when Brier Skill Score was calculated against PMD except for UKMO that has poor skill for PMD. To validate the reliability, resolution and uncertainty of probabilistic forecast against ERA-40, Brier score was again applied and it results that the probabilistic forecast of ECMWF and MPI to be skill full while stating the forecast of MeteoFrance and UKMO to be poor for Islamabad.

For Jhelum, probabilistic forecast of ECMWF and MeteoFrance is the skill forecast against PMD while using the alternative initial condition to calculate the probabilistic forecast; except for MPI and UKMO. When Brier score was applied against ERA-40 for Jhelum, it results that the probabilistic forecast of ECMWF to be skill, however the forecast of MeteoFrance, MPI and UKMO to be poor for Islamabad and for Muzaffarabad, probabilistic forecast of MeteoFrance, MPI and UKMO is skill forecast, while using the alternative initial condition to calculate the probabilistic forecast against PMD; except for ECMWF that has poor skill. Brier score against ERA-40, results that the probabilistic forecast of MeteoFrance and UKMO to be skill while stating the forecast of ECMWF and MPI to be poor for Muzaffarabad.

Reliability is one of the most important aspects to verify the quality of the forecast. The research carried out worldwide on the global climate models, focusing ECMWF, on the extreme events has shown the higher reliability [28]. In the current study, probabilistic forecast of all the four models is reliable by using all nine initial conditions and by using them alternatively also the Brier skill score shows that the Probabilistic forecast has the low resolution when calculated against the PMD and ERA-40 each of the four models.

However there are few factors that cause slight changes in the results given by these models. Therefore, to measure the accuracy and the performance of predictive model, cross validation was applied. Analyzing the cross validation graphs from the four models the results obtained verified the cross validation approach to give forecast almost similar to the forecaster's prediction in addition to take objectivity in account.

Journal of Physics: Conference Series **439** (2013) 012041 doi:10.1088/1742-6596/439/1/012041

Table 4. Cross validation results.

Model	Result	Extreme Event Years
	Islamabad	
ECMWF	Accurate forecast	1981, 1993, 1994.
Meteo France	Almost Accurate forecast	1970, 1982, 1990, 1993, 1994, 1998.
MPI	Accurate forecast	1970, 1980, 1990, 1993, 1998.
UKMO	Accurate forecast	1981, 1984, 1990, 1994, 2000.
	Jhelum	
ECMWF	Accurate forecast	1981, 1993, 1994.
Meteo France	Almost Accurate forecast	1970, 1982, 1990, 1993, 1994, 1998.
MPI	Accurate forecast	1970, 1980, 1990, 1993, 1998.
UKMO	Accurate forecast	1981, 1984, 1990, 1994, 2000.
	Muzaffarabad	
ECMWF	Accurate forecast	1980, 1984, 1993, 1995 2000
Meteo France	Almost Accurate forecast	1970, 1973, 1988, 1990, 1995
MPI	Accurate forecast	1969, 1970, 1980, 1994, 2000
UKMO	Accurate forecast	1969, 1981, 1982, 1986

Cross validation results shows all the models have almost accurate forecast for selected areas. It has been studied that the high model accuracy is the lack of predictive ability of the model. However, if proper cross-validation is applied a fine forecast can be obtained [29]. While the current study showed that the scattered graph of cross validation of MeteoFrance is more scattered as compared to the graphs of other three models. The value of the years causing more scattering represents the occurrence of extreme events in these years. If these years of extreme event be eliminated the forecast can be made more accurate. Hence the accuracy of the cross validation is in accordance with the studies carried out previously.

To measure the skill of the forecast the area under the curve of relative operating characteristic was considered. The ROC curves were plotted to check the skill of the probabilistic forecasts obtained from the models against the climatological probability of PMD. The ROC was plotted for the probabilistic forecasts obtained from the climate models against the climatological probability of PMD and ERA-40 for each of the five models using all the nine initial conditions and by eliminating the initial condition 'zero' and 'nine'

Table 5. Relative operating characteristic output for Islamabad.

Model	Eliminated Initial Condition	Area Under The Curve	Status
		0.810	Skill forecast
ECMWF vs. PMD	0	0.825	Skill forecast
	9	0.818	Skill forecast
ECMWF vs. ERA-40		0.848	Skill forecast
		0.758	Skill forecast
Meteo France Vs. PMD	0	0.751	Skill forecast
	9	0.721	Skill forecast
Meteo France Vs. ERA-40		0.693	Skill forecast
		0.745	Skill forecast
MPI vs. PMD	0	0.742	Skill forecast
	9	0.723	Skill forecast
MPI vs. ERA-40		0.721	Skill forecast
		0.682	Skill forecast
UKMO vs. PMD	0	0.665	Skill forecast
	9	0.688	Skill forecast
UKMO vs. ERA-40		0.682	Skill forecast

Table 6. Relative operating characteristic output for Jhelum.

Model	Eliminated initial conditions	Area Under the Curve	Status
		0.777	Skill forecast
ECMWF Vs. PMD	0	0.781	Skill forecast
	9	0.747	Skill forecast
ECMWF vs. ERA-40		0.723	Skill forecast
		0.712	Skill forecast
Meteo France Vs. PMD	0	0.695	Skill forecast
	9	0.684	Skill forecast
Meteo France Vs. ERA-40		0.641	Skill forecast
MPI vs. PMD		0.576	Poor forecast
	0	0.571	Poor forecast
	9	0.587	Poor forecast
MPI vs. ERA-40		0.387	Poor forecast
		0.613	Skill forecast
UKMO vs. PMD	0	0.604	Skill forecast
	9	0.628	Skill forecast
UKMO vs. ERA-40		0.636	Skill forecast

Table 7. Relative operating characteristic output for Muzaffarabad.

Model	Eliminated initial conditions	Area Under the Curve	Status
		0.617	Skill forecast
ECMWF Vs. PMD	0	0.558	Poor forecast
	9	0.587	Poor forecast
ECMWF vs. ERA-40		0.706	Skill forecast
		0.600	Skill forecast
Meteo France Vs. PMD	0	0.628	Skill forecast
	9	0.578	Poor forecast
Meteo France Vs. ERA-40		0.628	Skill forecast
		0.452	Poor forecast
MPI vs. PMD	0	0.450	Poor forecast
	9	0.407	Poor forecast
MPI vs. ERA-40		0.738	Skill forecast
		0.645	Skill forecast
UKMO vs. PMD	0	0.665	Skill forecast
	9	0.654	Skill forecast
UKMO vs. ERA-40		0.545	Poor forecast

Skill scores from linear error in probability space and relative operating characteristics has been compared to find out the skill of the climate model to predict the future state. For this the Australian seasonal rainfall forecast for summers and winters hindcast for 27 years has been done. Different skill score showed different skills. However the skill of the ROC was found to be the most skill full [30].

Therefore, by evaluating ROC for each of the five models for our study, the skill forecast has been confirmed, as Area under the cure (AUC) for most of the curve is found to be above 0.5 except for MPI for Jhelum and Muzaffarbad. It has been concluded that the forecast of the ECMWF is most skill full for each of the selected areas.

5. Conclusion

5.1 Islamabad

The most accurate probabilistic forecast was obtained by eliminating initial condition 'zero' when alternatively using the initial conditions to calculate the Probabilistic Forecast by using single model ECMWF and MPI. While for MeteoFrance most accurate probabilistic forecast was obtained by eliminating initial condition 'ninth' and for UKMO by using all the initial conditions.

The Probabilistic forecast calculated from ECMWF, Meteo France, MPI and UKMO and verified against Brier Score shows that it has low resolution and high reliability for Islamabad. Cross validation shows that the forecast is accurate. Whilst, ROC verifies the probabilistic forecast of each of the four models as a skill forecast for Islamabad.

5.2. Jhelum

The most accurate probabilistic forecast was obtained by eliminating initial condition 'zero' and 'ninth', when alternatively using the initial conditions to calculate the Probabilistic Forecast by using single model ECMWF. For MeteoFrance initial condition 'ninth' and by eliminating initial condition 'zero' for the determination of probabilistic forecast of the models MPI and UKMO.

The probabilistic forecast calculated from each of the four models and verified against Brier Score shows that it has low resolution and high reliability for Jhelum. The forecast is found to be accurate through cross validation and skill when verified against ROC for ECMWF, MeteoFrance and UKMO while for Jhelum, MPI it has poor skill.

5.3 Muzaffarabad

The probabilistic forecast was obtained by using all the initial conditions and alternatively using the initial conditions to calculate the Probabilistic Forecast by using single model ECMWF come out to be same. For MeteoFrance and UKMO, however the most accurate forecast was obtained by eliminating initial condition 'ninth' when alternatively using the initial conditions to calculate the probabilistic forecast and by using all the initial conditions for MPI.

The probabilistic forecast calculated and verified against Brier Score shows that it has low resolution and high reliability for Muzaffarabad for each of the five climate models. Cross validation shows the forecast to be accurate and ROC verified it to be skill forecast for Muzaffarabad for all four model except for MPI, which has a poor skill for Muzaffarbad.

6. Recommendations

The rapidly changing climate is resulting in the occurrence of extreme events that causes human life loss as well as infrastructure damage. Therefore climate and weather forecast system based on numerical modeling system incorporating ensemble prediction system should be developed in Pakistan in order to minimize the loss. On the basis of forecasts made by climate models, adaptations should be made by the concerned authorities for the extreme events that are likely to be more overstated in the future. Therefore, for climate forecasting, the ECMWF model is recommended to be used for the seasonal prediction of heat waves for Islamabad by eliminating the initial condition 'zero' for Jhelum by omitting initial condition 'zero' and 'ninth' and for Muzaffarabad with all the nine initial conditions. It is also suggested to use Meteo France by omitting initial condition 'ninth' for heat waves for all the three selected areas. MPI is proposed to be used by eliminating initial condition 'zero' for Islamabad and Jhelum while with all the initial conditions for Muzzaffarbad. For Islamabad UKMO is suggested to be used for seasonal predication by using all the initial conditions; for Jhelum, by omitting initial condition 'zero' and for Muzaffarbad by eliminating initial condition 'ninth'.

The climate models recommended includes almost all the significant physical developments that regulate climate and climate change, provides assurance in using these modeling system to predict the extreme events in the districts studied. As the probabilistic forecast provides the basis for the development of early warning system, the forecasters must use the seasonal forecast of the models used for the three districts being studied for developing an early warning system by giving the accurate prediction of extreme climatic events and natural hazards to protect the most vulnerable population of the country.

References

[1] Thorpe A J 2004 *MSMM: Extratropical Cyclones -Dynamical Concepts*

[2] Rohli R V and Vega A J 2008 *Climatology* (MA: Jones and Bartlett) p 466

[3] Pittock B A 2009 *Climate Change : The Science, Impacts and Solutions* (Collingwood: CSIRO PUBLISHING) p 23

[4] Parry M L, Canziani, O F, Palutikof J P, van der Linden P J and Hanson C E 2007 Technical Summary. *Climate Change 2007: Impacts, Adaptation and Vulnerability. Contribution of Working Group II to the Fourth Assessment Report of the Intergovernmental Panel on Climate Change* (Cambridge: Cambridge University Press) p 976

[5] Ngigi S N 2009 *Climate Change Adaptation Strategies: Water Resources Management Options for Smallholder Farming Systems in Sub-Saharan Africa* (New York: The MDG Centre for East and Southern Africa of the Earth Institute at Columbia University) p 189

[6] Meehl G A *et al.* 2007 *Climate Change 2007: The Physical Science Basis. Contribution of Working Group I to the Fourth Assessment Report of the Intergovernmental Panel on Climate Change* ed S Solomon *et al.*(Cambridge: Cambridge University Press, UK and New York, NY, USA) chapter 10

[7] Farooqi A B, Khan A H and Mir H 2005 *Pakistan J. Meteorol.* **2** 11-22
[8] Landsberg E H 1981 *The Urban Climate* (Maryland: Academic Press)
[9] Oke T R 1982 *J. Royal Meteorol. Soc.* **108** 1-24.
[10] Lougeay R, Brazel A and Hubble M 1996 *Geocarto Int.* **11** 79-89.
[11] Sahashi K, Hieda T and Yamashita E 2004 *J. Atmospheric Environ.* **30** 531-535.
[12] Chaudhry Q Z and Rasul G 2004 *Science Vision* **9** 59-66
[13] Zahid M and Rasul G 2012 *Pakistan J.Meterol.* **12** 85-94
[14] Webster P J, Toma V E and Kim H M 2011 *Geophys. Res. Lett.* **38** 1-5
[15] Leutbecher M and Palmer T N 2008. *J. Comp. Phys.* **227** 3515-3539
[16] Jackon D D and Kagan Y Y 2000 *J. GeoPhy.Int.* **143** 438-453
[17] Jolliffe I T and Stephenson D B 2003 *Forecast Verification: A Practitioner's Guide in Atmospheric Science (*Chichester: Wiley and Sons)
[18] Murphy A H and Winkler D S 1986 Good *J. Appl. Meteor.* **7** 751-758.
[19] Wilks D S 1995 *Statistical Methods in the Atmospheric Sciences.* (New York: Academic Press) p 465
[20] Mason I 1982 *Aust. Met. Mag.* **30** 291-303.
[21] Mason S J and Garaham N E 2002 *Q. J. R. Meteorol. Soc.* **128** 2145-2166
[22] Hameed A 2007 Severe Storms On Dated 23rd July 2001 Islamabad Pakistan *4th European Conf. on Severe Storms* (Trieste , ITALY, 10 - 14 September 2007)
[23] Ghaffar A and Javid M 2011 *J .Animal & Plant Sciences.* **21** 107-110
[24] Santer B D *et al.* 2004 *J. Geophys. Res.* **109** 119
[25] Hollingsworth A, Shaw D B, Lonnberg P, Illari L and Simmons A J 1986 *Mon. Wea. Rea.* **114** 861-879
[26] Legg T P and Mylin K R 2004 *J. Wea. Forecasting.* **19** 891-906
[27] Neelin D J 2011 *Climate Change and Climate Modeling* (New York: Cambridge University Press) p 304
[28] Wood A 2006 *Medium Range Weather Prediction-The European Approach* (New York: Springer) p 286
[29] Cheung F K T and Skitmore M 2006 *Building and Environment,* **41** 1973-1990
[30] Hudson D, Alves O, Hendon H H and Marshall A G 2011 *Q. J. R. Meteorol. Soc.* **137** 673–689

Journal of Physics: Conference Series **439** (2013) 012042 doi:10.1088/1742-6596/439/1/012042

Hyperspectral image classification using Support Vector Machine

T A Moughal
Laboratory of Complex Systems and Intelligent Control, School of Mathematical Sciences, Beijing Normal University, Beijing, China

E-mail: tauqir_mughal@yahoo.com

Abstract. Classification of land cover hyperspectral images is a very challenging task due to the unfavourable ratio between the number of spectral bands and the number of training samples. The focus in many applications is to investigate an effective classifier in terms of accuracy. The conventional multiclass classifiers have the ability to map the class of interest but the considerable efforts and large training sets are required to fully describe the classes spectrally. Support Vector Machine (SVM) is suggested in this paper to deal with the multiclass problem of hyperspectral imagery. The attraction to this method is that it locates the optimal hyper plane between the class of interest and the rest of the classes to separate them in a new high-dimensional feature space by taking into account only the training samples that lie on the edge of the class distributions known as support vectors and the use of the kernel functions made the classifier more flexible by making it robust against the outliers. A comparative study has undertaken to find an effective classifier by comparing Support Vector Machine (SVM) to the other two well known classifiers i.e. Maximum likelihood (ML) and Spectral Angle Mapper (SAM). At first, the Minimum Noise Fraction (MNF) was applied to extract the best possible features form the hyperspectral imagery and then the resulting subset of the features was applied to the classifiers. Experimental results illustrate that the integration of MNF and SVM technique significantly reduced the classification complexity and improves the classification accuracy.

1. Introduction

Land cover is an elementary variable that impacts on and links many parts of the human and physical environments [1]. Thus, information on the spatial distribution of the land cover classes is of vital importance for the investigation of environmental processes. Satellite remote sensing techniques are widely used for the environmental monitoring. Hyperspectral imagery is a valuable source from which one can extract detailed information about earth surface phenomena and objects. In fact, the sensors are characterized by a very high spectral resolution that usually results in hundreds of narrow spectral channels. Remote sensing images acquired by hyperspectral sensors, such as the widely used Airborne Visible Infrared Imaging Spectrometer (AVIRIS) and Hyperspectral Digital Imagery Collection Experiment (HYDICE) sensors, have shown their usefulness in numerous earth observation applications [2],[3]. HYDICE sensor data is used in this study for land cover classification. Hyperspectral Digital Imagery Collection Experiment (HYDICE) sensor is an experimental earth imaging instrument with very high spectral resolution. This sensor provides the sample spectrum in 210 spectral channels, nominally 10 nanometers wide covering the spectral range from 400 to 2500 nanometers.

The analysis of the hyperspectral data is not a trivial task. In particular, many factors made the analysis quite complex, such as the large spatial variability of the hyperspectral signature of each land cover class, atmospheric effects; and the curse of dimensionality [1]. Therefore, the main objective of this study is to search for an effective classifier by extracting the best possible features appropriate for the land cover classification. Three classifiers have been used for the comparison purpose.

Maximum Likelihood (ML) is a very popular parametric classifier, being widely used in pattern recognition and image classification [4]. It usually acquires higher classification accuracy compared to other traditional classification approaches. It assumes that each band is normally distributed and the chosen training samples are comprised of exhaustively defined set of classes. For hyperspectral data with tens of hundreds of spectral bands, the efficient training pixels (exhaustively defined) to locate for the discrimination of land cover classes is not an easy task. Whereas, the classification accuracy of ML classifier is based on the accurate selection of the training samples. Thus, for the hyperspectral imagery with poorly represented labelled training samples, it is preferable to adopt an alternative to the standard multiclass classifier. One approach is to adopt a Spectral Angle Mapper (SAM) technique, which compares each pixel in the image with every endmember for each class instead of comparing it to the training pixels [5]. The constraint to this classifier is, while selecting the endmembers; it does not take into account the sub-pixel values and becomes more problematic for the heterogeneous earth's surface. Further improvement in the classification accuracy can be attained by the development of Support Vector Machines (SVMs), which makes a breakthrough of classification for hyperspectral data. SVM do not require an estimation of the statistical distribution of classes to carry out the classification task, whereas it only defines the classification model by exploiting the concept of margin maximization by taking into account only few training pixels. SVM is an effective method of statistical learning theory, compared with the traditional classification methods; it is suitable for small samples learning, besides, it has better generalization ability and high efficiency for learning [6]. By comparing the three classifiers i.e. Maximum Likelihood (ML), Spectral Angle Mapper (SAM) and Support Vector Machine (SVM), the efficient monitoring of land cover classification can be achieved.

The rest of this paper is organized in four sections. Section II describes the data description and area under study. Section III describes different classifiers that can be used for the analysis purpose. Section IV deals with the experimental phase of the work and summarizes the observations. Finally, Section V presents the concluding remarks and future directions to complete this paper.

2. Data description and study area

The hyperspectral data that will be used for the analysis in this study is taken by the airborne Hyperspectral Digital Imagery Collection Experiment (HYDICE) sensor available in the student CD-ROM [7]. It was collected for the Mall in Washington, DC, with 210 bands covering 0.4–2.4 μm spectral region. This image has very high spatial resolution of about 2.8 m. The HYDICE image size is 307 × 1280 pixels. The dataset contains 1280 scan lines with 307 pixels in each scan line. There are seven information classes in the Washington, DC data includes road, grass, shadow, trails, trees, roof and water. Training and test samples are available for this scene.

Table 1. HYDICE sensor data information classes and samples.

Classes	Training Samples	Testing Samples
Trails	183	42
Shadow	223	48
Trees	136	192
Grass	970	300
Water	1,172	83
Roof	1,263	147
Roads	651	197

Figure 1. HYDICE sensor data.

3. Methodology

The methodology adopted in this study consists of two steps. First, to extract the best possible features appropriate for the classification of land cover classes and second, to find an effective classifier in terms of accuracy. Three classification methods have been proposed for comparative analysis in this study i.e. Maximum Likelihood, Spectral Angle Mapper and Support Vector Machine.

3.1. Data pre-processing

To get rid of the curse of dimensionality by selecting the appropriate features for the classification of HYDICE sensor data, Minimum Noise Fraction (MNF) serves the solution in this study. At first, the HYDICE sensor data contains 191 hyperspectral bands. MNF was adopted to reduce the dimensionality and computational requirements for the further processing of the HYDICE sensor data for the land cover classification.

The training samples were already available for the HYDICE sensor data as shown in table 1. There were seven classes of interest i.e. Trails, Shadow, Trees, Grass, Water, Roof and Roads that needed to be classified. The training samples were used to train ML and SVM classifiers to categorize the seven land cover classes from HYDICE sensor data for the comparative analysis.

3.2. Classification Techniques

Three classification techniques are adopted in this study to classify the land cover classes from the HYDICE sensor data.

3.2.1. Maximum Likelihood (ML). Maximum Likelihood (ML) is the parametric classifier based on the assumptions of normally distributed data for each class and exhaustively selected set of classes. For classifying the land cover classes, number of studies used the ML classifier as a benchmark to compare its classification accuracy with the other newly developed classifiers [8]. It is considered as a standard approach to thematic mapping from the remotely sensed imagery. In real life applications hardly the nature of the distribution is known. It is preferable to use the non-parametric classifiers that are free from assumptions. For the sake of comparison of the classification accuracies and to validate the suitability of the non-parametric classifier for classifying the land cover classes in this study, ML classification is conducted.

*3.2.2. Spectral Angle Mapper (SAM).*Spectral Angle Mapper (SAM) is a supervised classification algorithm, which utilizes spectral angular information for classification of hyperspectral image data [9]. It permits the rapid classification by calculating the spectral similarity between the image spectrums to reference reflectance spectra [10]-[12]. The refrerance spectra can either be attained from the field measurements or taken directly from the image.The reference spectra for this study is taken from image directly. SAM measures the spectral similarity by calculating the angle between the image and referance spectra, treating them as vectors in the n-dimensional feature space. The smaller angles between the two spectrums indicate high similarity and viceversa.The solar illumination factors does not affect this classifier. Moreover, It is very powerful classifier as it contains the influence of the shading effects to highlight the target reflectance characteristics. The major drawback faced by this classifier is that it assumes the endmembers chosen to classify an image by representing the pure spectra of a reference material, whereas the earth's surface is heterogenous in many ways and consists of mixed pixels. This classifier is adopted in this study because in gerneral, the spectral mixture problem decreases with higher spatial resolution images like HYDICE.

3.2.3. Support Vector Machine (SVM). The Support Vector Machine (SVM) is a classification method based on the statistical information of remote sensing images [13]. Recently, particular attention has been devoted to support vector machines (SVMs) for the classification of hyperspectral remote sensing images [14]-[15]. SVM is a non-parametric binary classifier that locates the optimal hyper plane between the two classes to separate them in a new high-dimensional feature space by taking into account only the training samples that lie on the edge of the class distributions known as support vectors. Moreover, it does not require the assumption of normality and is insensitive to the curse of dimensionality. SVMs have often been found to provide higher classification accuracies than other widely used pattern recognition techniques, such as the maximum likelihood. Furthermore, SVMs appear to be especially advantageous in the presence of heterogeneous classes for which only few training samples are available. The SVMs were originally developed to solve binary classification problems. The implementation of SVMs in multiclass classification problem is possible by formulating SVMs directly as a multiclass optimization problem, but as the number of the classes needed to be classified increases, the number of parameters to be estimated increases, it in return affects the SVMs classification performance in terms of accuracy. All the three classifiers were used to classify an image to investigate for an effective classifier.

3.3. Accuracy Assesment
The confusion matrix was used to assess the accuracy measures for all the three classification procedures by using the available ground truth pixels. Also the overall accuracy (OA), Kappa coefficient (K), producer's accuracy (PA) and the user's accuracy (UA) was keenly observed. The overall accuracy is the percentage of all validation pixels correctly classified, whereas the user's and producer's accuracy provide information about the commission and omission errors associated with the individual classes, respectively. Unlike the overall accuracy, Kappa takes into account the possibility of agreements occurring by chance in a random classification [16].

The spectrum after removing the noise and extracting the suitable features acquired from the Minimum Noise fraction (MNF) were classified by the three above mentioned methods and their classification accuracies can be compared by analyzing the confusion matrices.

4. Results and Discussion
The output from three classification techniques has shown in figure 2 illustrating the seven land cover classes. By visual inspection of the classification results shown in figure 2 compared to the original image shown in figure 1, it can be seen that the ML classification significantly overestimated the Roof class and underestimated the Shadow class. Whereas, the SAM classification highly over estimated all the classes except the Water and Grass class. This is due to the reference spectra that had taken directly from the image. As the image was heterogeneous and reference spectra did not take into account the sub pixel information, therefore most of the classes are misclassified. However, the SVM classifier obtained relatively accurate classification results for all the land cover classes.

The results in terms of classification accuracy obtained by the three classifiers are summarized in table 2, table 3 and table 4. The SVM exhibited the best Overall Accuracy (OA) of 78.39%, i.e., the percentage of correctly classified pixels among all the test pixels considered, with a gain of 9.81% and 9.91% over the classical ML and SAM classifiers, respectively. Also the Kappa coefficient (K) is 0.7335 for the SVM, which is higher as compared to the ML (0.6074) and SAM (0.6163) classifiers. Moreover, for the comparative analysis of three classifiers, figure 3 and figure 4 shows a bar chart for the Producer's accuracy (PA) and User's accuracy (UA) in percentage of ML, SAM and SVM.

The following discussion concerns the quantitative and comparative analysis of three different classification techniques for hyperspectral image, including parametric and non-parametric approaches.

The accuracy of ML classifier was calculated as it provides the benchmark for the assessment of SVM classifier. The most critical class for the analysis in table 2 is the shadow class. The classification result of ML for shadow class shows the omission error of 93.75% higher as compared to the SAM (41.67%) and SVM (37.50%); hence ML is omitting the shadow pixels from the shadow class and under estimates this class. This can be better visualized by figure 3. It can be further observed from table 2 that the commission error for the roof class is also higher i.e. 58.62% in comparison to the SAM (42.51%) and SVM (45.03), which means that roof class is largely over estimated. Hence the use of ML classifier for classifying the land cover classes is not recommended in this study.

In order to monitor the environmental changes appropriately, search for an effective classifier for the classification of land cover is of crucial interest. Thus, for the comparative analysis, the confusion matrix for the SAM classifier is calculated. Table 3 shows that most of the classes are over estimated by using SAM classifier except the Grass and Water class, which shows the user's accuracy of 86.79% and 97.18% respectively.

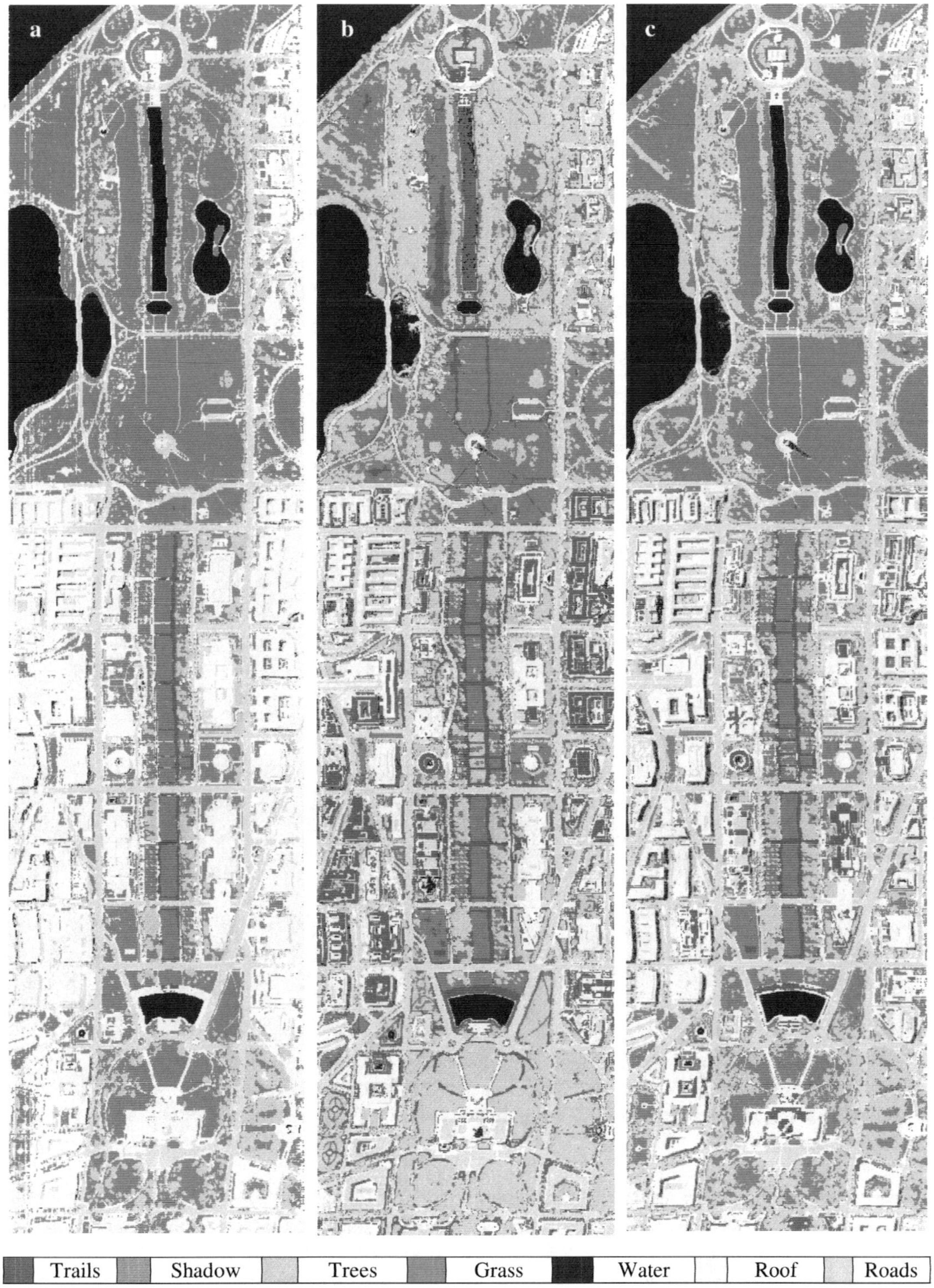

Figure 2. Classification results. (a) ML classification (b) SAM classification and (c) SVM classification.

Table 2. Confusion Matrix for Conventional ML.

	ML	Actual								
		Trails	**Shadow**	**Trees**	**Grass**	**Water**	**Roof**	**Roads**	**Total**	**PA (%)**
Predicted	**Trails**	19	0	0	0	0	0	0	19	45.24
	Shadow	0	3	0	0	0	0	0	3	6.25
	Trees	0	1	94	5	0	0	5	105	48.96
	Grass	1	0	75	266	0	1	1	344	88.67
	Water	0	1	0	0	83	0	0	84	100
	Roof	22	22	23	29	0	144	108	348	97.96
	Roads	0	21	0	0	0	2	83	106	42.13
	Total	42	48	192	300	83	147	197	1009	**K = 0.6074**
	UA (%)	100	100	89.52	77.33	98.81	41.38	78.30		**OA=68.58%**

Table 3. Confusion Matrix for SAM Classification.

	SAM	Actual								
		Trails	**Shadow**	**Trees**	**Grass**	**Water**	**Roof**	**Roads**	**Total**	**PA (%)**
Predicted	**Trails**	39	0	0	23	0	25	7	94	92.86
	Shadow	0	28	0	0	11	1	3	43	58.33
	Trees	0	1	154	90	0	0	1	246	80.21
	Grass	0	0	22	184	0	3	3	212	61.33
	Water	0	2	0	0	69	0	0	71	83.13
	Roof	3	0	3	3	0	96	62	167	65.31
	Roads	0	17	13	0	3	22	121	176	61.42
	Total	42	48	192	300	83	147	197	1009	**K = 0.6163**
	UA (%)	41.49	65.12	62.60	86.79	97.18	57.49	68.75		**OA=68.48%**

Table 4. Confusion Matrix for SVM Classification.

	SVM	Actual								
		Trails	**Shadow**	**Trees**	**Grass**	**Water**	**Roof**	**Roads**	**Total**	**PA (%)**
Predicted	**Trails**	33	0	0	0	0	24	0	57	78.57
	Shadow	0	30	1	0	4	1	5	41	62.50
	Trees	0	1	158	25	0	0	1	185	82.29
	Grass	0	0	21	254	0	5	2	282	84.67
	Water	0	5	0	0	79	0	0	84	95.18
	Roof	9	0	0	20	0	105	57	191	71.43
	Roads	0	12	12	1	0	12	132	169	67.01
	Total	42	48	192	300	83	147	197	1009	**K = 0.7335**
	UA (%)	57.89	73.17	85.41	90.07	94.05	54.97	78.11		**OA=78.39%**

The confusion matrix acquired for the SVM shows the best overall accuracy (78.39%) as well as the Kappa coefficient (0.7335) for the classification of land cover classes as shown in table 4. Moreover, the SVM shows the good tradeoff between the user's and producer's accuracies for all the land cover classes.

The results reported in table 4 confirm the superiority of the multiclass SVM in terms of both overall accuracy and Kappa coefficient. The good classification performance demonstrated by successful machine learning technique i.e. the support vector machine (SVM), using spectral signatures as input features, have improved by the incorporation of intelligent feature extraction technique (MNF), which reduces the dimensionality of the data to the right subspace without losing the original information that allows for the separation of the information classes.

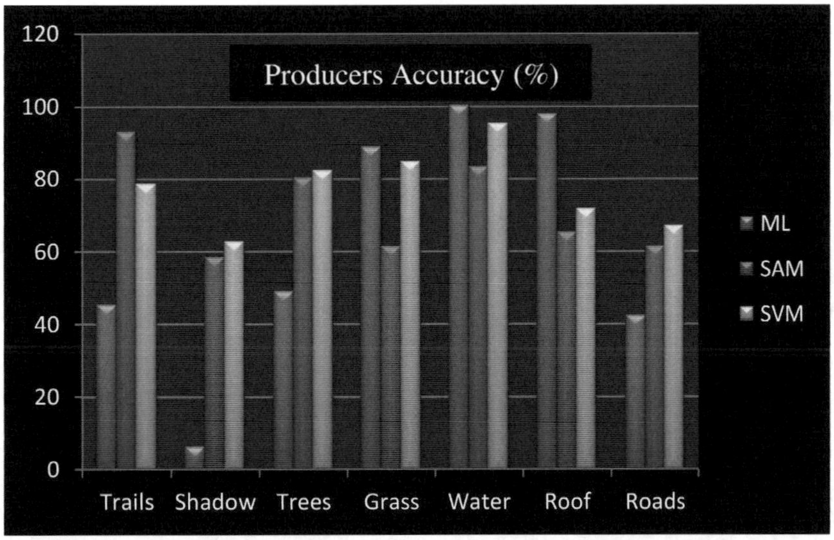

Figure 3. Producer's Accuracy (%) for Maximum likelihood, Spectral Angle Mapper and Support Vector Machine classifiers.

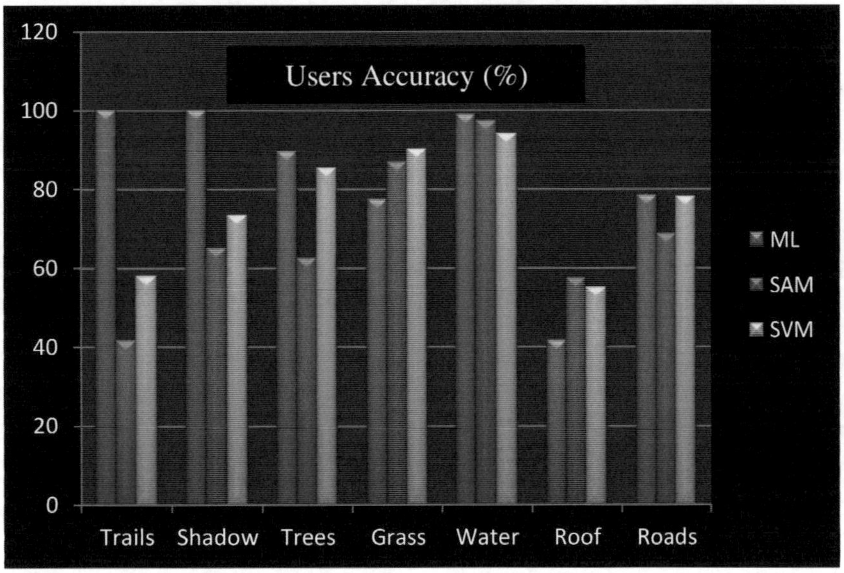

Figure 4. User's Accuracy for Maximum likelihood, Spectral Angle Mapper and Support Vector Machine classifiers.

Figure 3 and figure 4 shows the extreme classification behavior for some classes classified by the ML classifier. For example, the Shadow class shows the low producer's accuracy in comparison to a very high user's accuracy. Such a behavior of a classifier is not desirable as there is a need to trade off between the omission and the commission error. Moreover, SAM also shows the high producer's accuracy with a very low user's accuracy for the Trail class. Overall the SVM provides the better classification accuracy results for all the land cover classes.

4. Conclusion and Future Direction

In this study, the problem of the classification of hyperspectral remote sensing data using Maximum Likelihood (ML), Spectral Angle Mapper (SAM) and Support Vector Machine (SVM) have been addressed. In order to assess the effectiveness of the land cover classification methodologies, the main objective considered was to search for an effective classifier by extracting the best possible features using MNF, appropriate for the land cover classification. The results obtained from the HYDICE sensor dataset showed that SVM is much more effective than other conventional classifiers (i.e., the ML and the SAM classifier) in terms of classification accuracy, computational time, and stability to parameter settings.

Another important aspect to be pointed out is the intrinsic good generalization capability of SVM, which stems from the selection of the hyperplane that maximizes the geometrical margin between classes. In a hyperspectral context, the maximum margin solution allows to fully exploit the discrimination capability of the relatively few training samples available.

The major drawback to this solution is that the large spatial variability of the hyperspectral signature of each information class given the limited information present in the training set adversely affects the classification accuracy. However, it is worth noting that to solve the problem of the spatial variability of the hyperspectral signature of classes effectively, good generalization properties of the classifiers should be coupled with other techniques.

Further research will be done from the view of improving the classification accuracy and reducing the calculation time.

References

[1] King R B 2002 *Land cover mapping principles: a return to interpretation fundamentals*. Int. J. Remote Sens. **23** 3525-3546
[2] Zhang J, Zhang Y and Zhou T 2001 *Proc. International conference on* **1** 882-885
[3] Benediktsson J A, Palmason J A and Sveinsson J R 2005 IEEE *Trans. Geosci. Remote Sens.* **43** 480-491
[4] Jia X P and Richards J A 1994 IEEE *Trans. Geosci. Remote Sens.* **32** 274-281
[5] Girouard G, Bannari A, Harti A E and Desrochers A 2004 Istanbul Turkey XXth ISPRS 599 – 604
[6] Melgani F and Bruzzone L 2004 IEEE *Trans. Geosci. Remote Sens.* **42** 1778-1790
[7] Landgrebe D A 2003 *Signal Theory Methods in Multispectral Remote Sensing* Hoboken, NJ: John Wiley & Sons
[8] Huang C, Davis L S and Townshend J R G 2002 *Int. J. Remote Sens.* **23** 725-749
[9] Kruse F A, Lefkoff A B, Boardman J W, Heidebrecht K B, Shapiro A T, Barloon P J and Goetz A F H 1993 *Remote Sens. Environ.* **44** 145-163
[10] Crosta A P, Sabine C and Taranik J V 1998 *Remote Sens. Environ.* **65** 309-319
[11] Schwarz J and Staenz K 2001 *Can. J. Remote Sens.* **27** 216-224
[12] Hunter E L and Power C H 2002 *Int. J. Remote Sens.* **23** 2989-3008
[13] Cortes C and Vapnik V 1995 *Support-Vector Networks, Machine Learning* Lorenza Saitta Netherlands Kluwer Academic Publisher, Boston **20** 273-297
[14] Vapnik V N 1998 *Statistical Learning Theory*. New York: Wiley
[15] Burges C J C 1998 *A tutorial on support vector machines for pattern recognition, Data Mining Knowedge Discovery* **2** 121–167
[16] Brown D G, Lusch D P and Duda K A 1998 *Geomorphology* **21** 233 – 250

Journal of Physics: Conference Series **439** (2013) 012043 doi:10.1088/1742-6596/439/1/012043

Modern improved and effective design of boundary layer turbine for robust control and efficient production of green energy

M Usman Saeed Khan[1]**, Ehsan Ali, M Irfan Maqsood and Haq Nawaz**
Univesity College of Engineering and Technology, The Islamia University of Bahawalpur, Pakistan

E-mail: usmansunshine@gmail.com

Abstract. Boundary layer turbine, although invented long before but has failed to commercialize and beat the bladed turbines due to certain lacks. In this paper we have proposed and suggested some techniques to improve the design of the boundary layer turbine such that to take the optimum advantage of its efficiency, compensating all of its draw backs. A test model of proposed boundary layer turbine has been physically made and tested under different conditions. The experiments using different flow rates, different discharge rates and changing many other factors has been made and the results of these experiments are included in this paper which verifies the improvements in the output power obtained from boundary layer turbine system. This modified turbine can thus become an efficient way of producing green energy like any other renewable energy resource.

1. Introduction

The trend of using renewable energy resources is increasing day by day due to the fact that conventional fuels are decaying with the time and the price of these fuels are becoming so high that they are not being liked for economical operation of power generation plants. Renewable energy resources are also providing an advantage of being environment friendly and the emission of carbon is very low in case of using renewable energy resources in the industries, transportation and power generation plants. Many types of renewable energy resources are being implemented in the different parts of the world. The research area and the implementation techniques regarding these sources of energy is becoming most popular topics of this age. Particularly the implementation of these sources in the field of power generation is considered to be the need of hour for the green future of this world. Solar and wind energy techniques are most common sources which are rapidly being implemented to compensate the fast growing energy demands. Although both of these sources are environment friendly but still they are not considered to be fully reliable if they are installed alone without any other source of energy. The reason is that these resources can never work all the day. Photovoltic cells will generate the energy only when the sun light is available and the wind turbines will work only when a strong wind is blowing in the environment. That is why it is must to find out the renewable energy resources which is able to deliver its output all the day. Hydro turbines are also providing good results to compensate the energy requirements but these turbines are being used only for bulk level generation. As the present Francis and Kaplan water turbines require the specific head to work that is

[1] To whom any correspondence should be addressed.

why it is impossible to use these turbines at low level i.e. home level generation. Our proposed design for the new tesla turbine is of great importance in this regard, it can be implemented not only for higher level generation but also it is equally efficient for low level power generation. Previously the tesla turbine was used only for very low level applications like low height pumping and in small motors etc. but the new optimized design of tesla turbine will make it more efficient to be used in different applications of power generation. Tesla turbine uses water or air as its fuel so it is one of the greenest source for the generation of power. Global warming is increasing very rapidly and the world will face an enormous danger if the rate of increase in the emission of CO_2 was not controlled. Power generation is a one of the major source to destroy the environment with the other sources like transportation and industries. If we look at the figures of carbon dioxide emission then we comes to know that it has been increased twice from 1973 to 2009. These figures are given below in figure 1.In 1973 the emission of CO_2 in all the regions of the world was a total of 15624 million tonnes which increased to 28999 million tonnes in the year of 2009 [5]. Tesla turbine will thus help in reducing the carbon dioxide emission. The new designed proposed for the tesla turbine with some experimental results on a prototype model has been explained in the ongoing paragraphs.

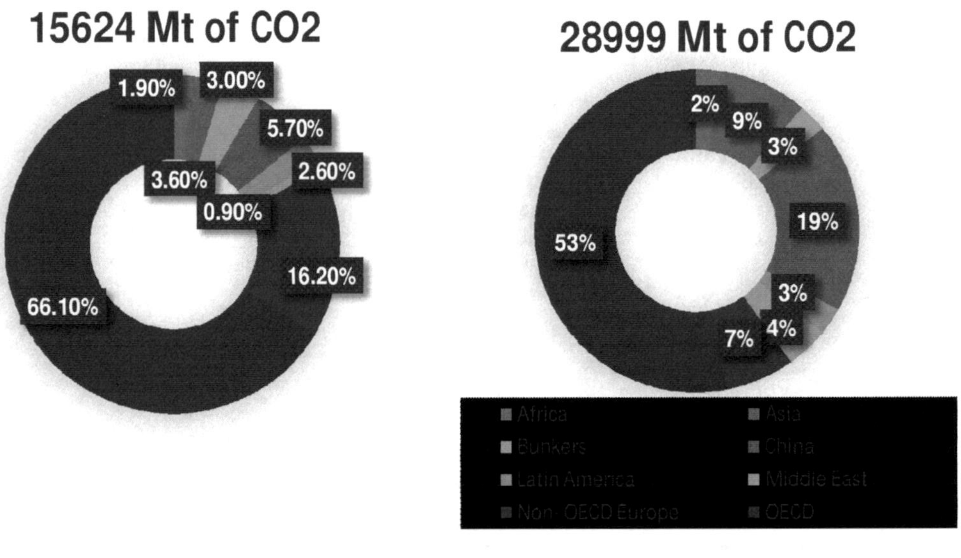

Figure 1. Comparison of the CO_2 emission in 1973 and 2009 in different regions of the world

2. Proposed design and working principle

Nikola Tesla, a well-known inventor was a great achiever in the fields of electrical generation, transmission and utilization. But his invention was not limited to these specific fields. Tesla invented a new type of turbo machinery having a different type of working principle and design. He designed a multi-disc rotor for his turbine which was spaced along a common shaft. These flat discs were very simple in construction and the experiment on Tesla's turbine gave an extra ordinary result regarding the speed of the turbine. A huge speed was obtained but when the torque was measured it was very much discouraging. At the same time the bladed turbine came out in the markets and made their place instead of the tesla turbine. Thus tesla turbine remained far behind and it was deprived from the attention of the stake holders although only a few changes were required to remove the weakness of the turbine.

2.1. Principle of operation of the boundary layer turbine

As it is clear from the name that the basic working principle of operation of the turbine is the boundary layer effect between the fluid and the walls of the rotor discs. Unlike other conventional turbines this turbine does not have any kind of blades. The fluid when enters between the spaces of the discs then adhesive forces tends the water to slow down its speed as the water molecules tends to stick with walls of the discs at that time. The viscosity of the fluid also helps to decrease the speed of water further more. But the water coming after word is in speed so it exerts a force which compel the discs to move in the direction of the water. That was the basic working principle of the conventional turbine. The experiments revealed that the conventional turbine was not efficient enough to rotate a heavy generator due to low torque. The new proposed design of tesla will combine the characteristics of both the bladed turbine and the tesla turbine. Its working principle is the combination of the principles of both these type of turbines i.e. it will work on the boundary layer effect as well as the pushing effect of the water.

2.2. Construction and designing

The basic design of the new turbine is similar to the conventional tesla turbine with some additional components in the rotor of the turbine and the special nozzle design. A number of particular size discs depending upon the generation requirement are stabilized on the shaft at a specific distance apart from each other. Some holes are made near to the centre of the each disc which acts as spill ways from where the fluid gets out of the turbine. The new addition in the design is that a special blade type strip is mounted on the top of the each disc. The purpose of which is to enhance the torque of the turbine.

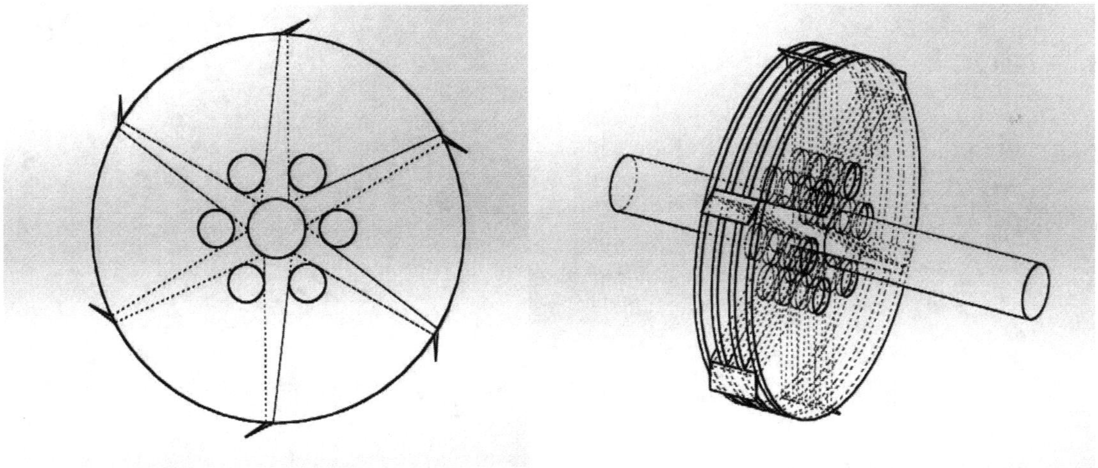

Figure 2. Two and three dimensional models of the proposed tesla turbine.

Three dimensional model of the new proposed design of the tesla turbine has been made with the help of a simulation and designing software Solid Works. The diagram of this model has been shown in figure 2 and figure 3. Another suggested improvement in the design of the turbine is to use the specific design nozzles. A separate nozzle must be applied for the each spacing between the discs instead of using a common nozzle for all the discs. This will increase the fluid pressure and amount of the working fluid will also increase. Thus the efficiency of the turbine will increase to a great extent. The round holes near the shaft are the spill ways to exit the water from the turbine. The strips mounted on discs are to increase the push of the water and thus to increase the torque of the turbine.

Journal of Physics: Conference Series **439** (2013) 012043 doi:10.1088/1742-6596/439/1/012043

Figure 3. Three dimensional models of the proposed tesla turbine.

3. Experimental analysis

A prototype model of the new tesla turbine was made according to the design which has been discussed in details in the previous paragraphs. The rotor of this newly designed boundary layer turbine is shown in figure 4. It can be seen that there are strong strips mounted on each disc. These strips will not only increase the torque but also they will collect the maximum fluid and will send it to the spaces between the discs. After the experiments in different condition the new turbine was proved to be very efficient than its conventional design. The experiments was carried out using specific pressure of water and the results recorded has been shown in the table 1.

Figure 4. Rotor of the physically made prototype new design tesla turbine.

Table 1. Average results determined from the multiple experiments

Sr. No	**Parameter**	**Value**
1	Average Power	100 watt
2	Average Voltage	18 volt
3	Average RPM	140
4	Average Current	5 A

Another important characteristic of this turbine is that its speed control is very robust and easy. The relation between speed and voltage of turbine is linear so any voltage level can be acquired easily. Figure 5 shows the relation between the speed and voltage of the turbine.

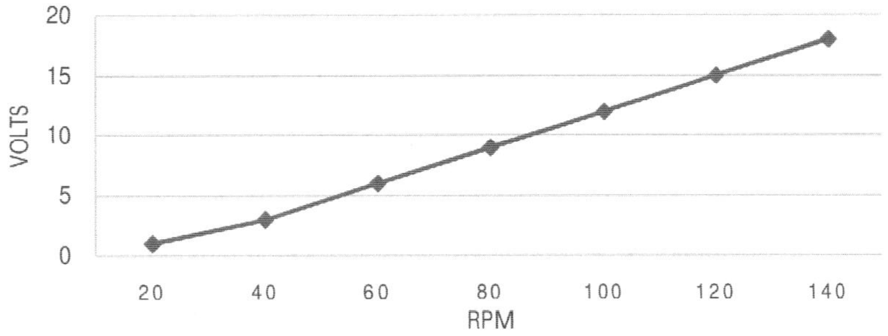

Figure 5. Relation between the speed and voltage of turbine.

4. Implementation techniques for the proposed boundary layer turbine

After the improvement in the design of the tesla turbine, it is necessary to use the proper implementation techniques for the application of this turbine so that the optimum advantages can be taken from the performance of this turbine. Using proper techniques, the maximum efficiency can be obtained which will be probably near to the theoretical values of the tesla's efficiency. The angle of the nozzle providing the fluid to the rotor of tesla turbine is very important. Nozzle must be easily adjustable so that it can adjusted any time to any angle according to the conditions and requirements. The space between the discs of the rotor are also an important factor and it must be properly adjusted depending upon the size and the area of the turbine. Another important factor is the exit way of the fluid. The torque of the turbine is very much dependent on the exit way of the turbine.

Here is an example of the implementation technique which has also been used in the experiments on the prototype model of the tesla turbine.

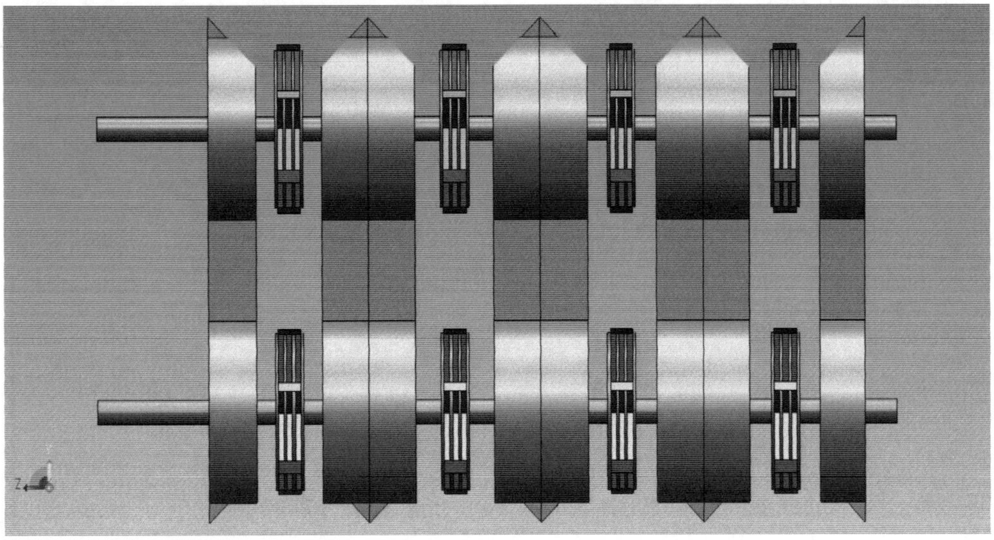

Figure 6. Implementation technique for the application of tesla turbine in canal and rivers (top view)

The tesla turbine can be used in the way of the water flow like canals and river but to increase the speed and torque some special measures are to be taken. A proper place should be made where turbine can be adjusted such that the maximum water flows though the rotor discs of the turbine and that too on the most suitable angle with an appropriate speed. Some arrangement are necessary to increase the speed of the water. For this the way of the water can be narrowed near the turbine and also the path of the water can be made in a slope shape to increase the speed and thus the energy of the water. A group of tesla's rotor discs can be connected in series on a common shaft for the generation of electricity at higher level. In a single canal this turbine can be used as many times as the requirement of power generation. For this purpose the system of tesla turbine must be arranged at different places on certain distance so that the performance of one turbine will not affect its subsequent turbine.

5. Conclusion

Rapidly increasing demands of the power generation are asking for fast development in the area of renewable energies. The generation of home level energy and use of distributed generation system is the key necessity of the present age. So the idea to return back the out of market turbine again to the practical implementation will be a great milestone in the way of achieving the green energy. For the revival to tesla turbine it was necessary to withdraw its limitation. The new design of tesla turbine which uses both the pressure pull and the boundary layer effect, has been proved very economical and efficient design and it can be used in the future to get rid of the polluted power generation.

6. References

[1] Rice W and Emiritus P 1991Tesla turbo machinery,*Proc. IV international Nikola Tesla Symposium* (Belgrade: Org. of Serbia)
[2] Luis San AndresL 2006 Introduction to Pump Rotor dynamics *In Design and Analysis of High Speed Pumps* (France: RTO) Paper 9 pp 1-26
[3] Yan B P Ho 2011Tesla Turbine for Pico Hydro Applications *Guelph Engineering Journal* **4** 8
[4] Cairns W M J 2003*The Tesla Disc Turbine* vol 2 (England: Camden Miniature Steam Services)
[5] 2011 *Key World Energy Statistic2011*, (Paris: international energy agency)
[6] Lampartand P Jędrzejewski Z 2011Investigations of Aerodynamics of Tesla Bladeless Microturbines *Journal of theoretical and applied mechanics* **49** pp 477-499
[7] Hayes J A 1994 *Tesla's Engine, A Dimension for Power* (Wisconcin: TEBA)
[8] Breiter M G and Pohlhausen K 1962 *Laminar Flow between Two Parallel Rotating Disks* (US Air force: Applied Mathematics Research Branch)

Journal of Physics: Conference Series **439** (2013) 012044 doi:10.1088/1742-6596/439/1/012044

Correlation of meteorological parameters and remotely sensed normalized difference vegetation index (NDVI) with cotton leaf curl virus (CLCV) in Multan

A Ahmed, A Akhtar, B Khalid and A Shamim
Department of Environmental Sciences, International Islamic University, Islamabad Pakistan

E-mail: indus_10@hotmail.com, kh_bushra@yahoo.com

Abstract. Climate change and weather has a profound effect on the spread of Cotton Leaf Curl Virus (CLCV) which is transmitted by whitefly. Climate change is altering temperature and precipitation patterns, resulting in the shift of some insect/pest from small population to large population thus effecting crops yield. To find out the relationship between the weather conditions, outburst of CLCV and changes in Normalized Difference Vegetation Index (NDVI) values due to the outburst of CLCV, a study was carried out for tehsil Multan. Data was acquired for the months of June, July, August and September for the year 2010. Regression analysis between CLCV and meteorological conditions as well as between CLCV and NDVI was performed. Meteorological parameters included temperature, humidity, precipitation, cloud cover, wind direction, pan evaporation and sunshine hours. NDVI values were calculated from SPOT satellite imagery (1km) using ArcMap10 and WinDisp v5.1. Correlation coefficients obtained in most of the cases were acceptable however the significance F and P-value were higher than their critical value at 95% level of significance. Therefore significant correlation was found only between CLCV and temperature and between CLCV and PAN evaporation during the month of July.

1. Introduction

Multan district is situated in the southern part of the Punjab province. The district of Multan consists of four tehsils which include Multan Cantonment, Multan Sadar, Shujabad and Jalalpur Pirwala. Multan district has an extreme climate. In summers the temperature reaches to 49°C while in winter it drops to 1°C. The average rainfall in Multan district is 127 mm. Multan tehsil has a very fertile soil. Agriculture of Multan consists of crops, fruits, vegetables etc. Wheat, cotton and sugarcane are the main crops grown here while important fruits grown here include mangoes and citrus [1]. Mangoes and cotton crop grown here play an important role in foreign exchange earnings of Pakistan.

Cotton crop is one of the main crops of Pakistan which is grown in the Kharif season and plays very significant role in the development of Pakistan's economy [2]. Cotton crop contributes to about 2% of GDP and it is considered to be the life line of Pakistan's economy [3]. Cotton crop provide oil, fibre, fuel and animal feed etc. It supplies raw material to 1200 ginning units, 180 spinning units, about 470 textile mills and 50 vegetable oil mills [4].

Although this cash crop is very important however yields are still very low and stagnant in Pakistan for the last several years as compared to other cotton growing countries. Factors which are responsible for the stagnant cotton production includes excessive precipitation at sowing time, high temperature and its variations at the flowering stage [5]. In Pakistan, annual mean surface temperature has been rising consistently since the start of 20th century [6]. This is altering optimum weather conditions required for crop growth for instance many crops are under heat stress like wheat, sugarcane and cotton [7]. Alteration in weather conditions affects the insects/pests populations, thus affecting crop yields. One of the most significant reasons of low production of cotton in Pakistan is due to the presence of number of insects/pests and diseases at different periods of its growth and development which results in vast economic losses [8].Amongst the sucking insect pests, whitefly, jassid and thrips are considered very dangerous for crops in Pakistan [9].

Whitefly is the main cause of Cotton Leaf Curl Virus (CLCV). CLCV was reported for the first time in Nigeria in 1912 [10]. After that it was reported in 1926 in Tanzania [11] and in 1934 in Sudan [12]. In Pakistan CLCV was observed for the first time near Multan on cotton plants in 1967 [13]. The emergence of the disease at seedling phase badly affect flowering, development of boll, growth and maturation, yield of cotton seed and quality of fibre [14]; [15]. It reduces length, strength and elongation of fiberupto 3.44%, 10% and 10% respectively [16].

As insects and pests damage crops theses damages can be detected through remote sensing. Satellite imagery is useful to detect alterations in plant development. Therefore to detect changes in the study area, SPOT satellite imagery was used (1km resolution). SPOT satellite images were used to calculate NDVI (Normalized Difference Vegetation Index) values of cotton crops. NDVI is a widely used measurement method to determine the density of vegetation.

Therefore this study was carried out with the following scope:

- To find out the relationship of weather conditions and development of CLCV in Multan sadar and Multan cantonment cotton field areas (Multan tehsil).
- To calculate NDVI values of cotton crop in Multan tehsil in order to find out changes in NDVI values with CLCV development.

2. Materials and Methods

2.1. Data acquisition

1. Relevant literature was acquired from International Islamic University Islamabad (IIUI) digital library.
2. Four months meteorological data i.e. June, July, August and September (2010) was acquired from Pakistan Meteorological Department (PMD). The data included daily temperature, humidity, precipitation, cloud cover, wind direction, PAN evaporation and sunshine hours for the selected four months.
3. Per month three images (Twelve images) of SPOT 1km resolution satellite imagery were acquired from PMD for the year 2010.
4. CLCV data for the selected months i.e. June, July, August and September (2010) was acquired from Pakistan Cotton Research Institute (PCRI), Multan.

2.2. Methods

Relevant literature was reviewed. Hundred points were taken from Multan cantonment and Multan sadar cotton field areas (Multan tehsil) using software Google earth. Meteorological and CLCV data for Multan tehsil was acquired from PMD and PCRI respectively.

SPOT satellite imagery was used to calculate NDVI values of cotton crops with the help of software ArcMap 10 and WINDISP 5.1. For this the selected hundred points were plotted on a map, converted to shape file format using software ArcMap 10. Then shape file was converted to .bna format. After that NDVI values were extracted using .bna file and SPOT satellite images with the help of WINDISP 5.1 software.

After that regression analysis was performed and for this separate spread sheets were used. Every spread sheet contained spots of CLCV and one parameter for a month. Each month was divided into ten days interval and average was calculated for each interval, except for precipitation (sum was taken for precipitation) and then regression analysis was performed using Microsoft excel 2010. Thus correlation was calculated between spots of CLCV and meteorological parameters and as well as between CLCV and NDVI.

Figure 1. Map of Pakistan highlighting the study area.
Source: Pakistan Meteorological Department

3. Results and discussion

It is predicted that about 20-40% losses of cotton crop take place annually as a result of different insects and pests of cotton [17]. Among different diseases CLCV is a dangerous disease of cotton crop which is transmitted by whitefly. It affects the quality and yields of cotton crop. Climate change is altering weather conditions and changing pest/insects population thus affecting crop growth. Therefore finding out this relationship was very important. Thus regression analysis was performed to find relationship between CLCV and weather conditions and as well as between CLCV and NDVI. The results obtained in this research study shows that CLCV is being affected by weather. Results acquired are discussed in subsequent section.

Table 1. Regression analysis of CLCV with meteorological parameters and NDVI.

		June	July	August	September
CLCV and mean air Temperature	R Square	0.93	0.99	0.49	0.20
	Significance F	0.15	0.03	0.50	0.70
	P Value	0.15	0.03	0.50	0.70
CLCV and PAN Evaporation	R Square	0.82	0.99	0.01	0.95
	Significance F	0.27	0	0.97	0.13
	P Value	0.27	0.03	0.97	0.13
CLCV And Cloud Cover	R Square	0.54	0.91	0.14	0.27
	Significance F	0.47	0.18	0.75	0.64
	P Value	0.47	0.18	0.75	0.64

CLCV And Precipitation	R Square	0.94	0.82	0.02	0.89
	Significance F	0.15	0.27	0.90	0.21
	P Value	0.15	0.27	0.90	0.21
CLCV And Relative humidity	R Square	0.05	0.89	0.08	0.52
	Significance F	0.84	0.20	0.81	0.48
	P Value	0.84	0.20	0.81	0.48
CLCV And Wind	R Square	0.76	0.45	0.48	0.88
	Significance F	0.31	0.53	0.51	0.22
	P Value	0.31	0.53	0.51	0.22
CLCV And Sunshine Hours	R Square	0.66	0.61	0.001	0.89
	Significance F	0.39	0.42	0.97	0.20
	P Value	0.39	0.42	0.97	0.20
CLCV And NDVI	R Square	0.40	0.21	0.03	0.44
	Significance F	0.56	0.69	0.87	0.53
	P Value	0.56	0.69	0.87	0.53

In this research relationship between temperature and CLCV was found significant in Multan station during the month of July. In Africa, CLCV was affected by temperature [18]. According to [19] CLCV increased in the range of maximum and minimum temperatures of 33-45 ^{0}C and 25-30 ^{0}C respectively. However, during June, August and September (2010) insignificant relationship was obtained between CLCV and air temperature. According to [20], there exists non-significant correlation between weekly maximum air temperature and whitefly (cause of CLCV).

The relationship between CLCV and PAN evaporation was found significant in Multan during July. However, insignificant relationship between CLCV and PAN evaporation was found in June, August and September in Multan due to high significance F and P value as mentioned in table 1.

A study was carried out in five districts of Punjab and found that the correlation between rainfall and whitefly population was non-significant [20]. Similarly poor correlation was found between weekly rainfall and CLCV development [19]. Likewise the relationship obtained in this research between CLCV and cloud cover and between CLCV and precipitation during June, July, August and September (2010) in Multan was insignificant because significance F and P value were greater than critical value 0.05.

A poor correlation of humidity with CLCV development was found [19] and non-significant correlation was obtained between relative humidity and white fly population [20]. We also found similar trends in our research. No significant relationship was established between CLCV and relative humidity in Multan during all selected four months.

Wind speed is a significant aspect which governed the activity of whitefly population [21]. However a non-significant correlation between wind and whitefly population was found according to a study carried out in five districts of Punjab [20]. Similarly in this research, the relationship between CLCV and wind was found insignificant in Multan during June, July, August and September because significance F and P value were greater than critical value 0.05.

During 2000, significant positive correlation was found between sunshine hours and CLCV [8]. However [20] observed that there was non- significant correlation between whitefly and sunshine. Similarly results obtained between CLCV and sunshine hours in this research were insignificant due to high significance F and P value, both were greater than critical value 0.05 during June, July, August and September (2010) in Multan station.

Normalized Difference Vegetation Index (NDVI) has been widely used for the estimation of biomass of plant [22], growth, development, spatial density distribution [23]. In this research the relationship between CLCV and NDVI in selected area was found to be insignificant during June, July, August and September (2010).

4. Conclusion

The objective of this study was to determine the relationship of meteorological parameters and NDVI with CLCV. For this regression analysis was performed to find the relationship of meteorological parameters and NDVI with CLCV. Value of correlation coefficient obtained was acceptable in most cases however values of significance F and P were mostly greater than their critical value i.e. 0.05. These variations in the results might have occurred due to lesser amount of data. Since the data was available only for four months i.e. June, July, August and September (2010) and only three observations within a month. Therefore significant relationship was only obtained between:

- CLCV and air temperature during July 2010.
- CLCV and PAN evaporation during July 2010.

If long term CLCV data and high resolution imagery is available then results would be much better than the present one. Therefore more research should be done with complete data set.

Appendix

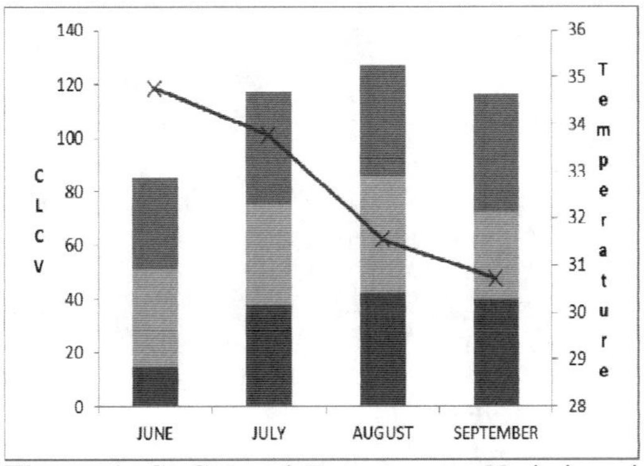

Figure 1. CLCV and Temperature Variations in June, July, August and September, 2010.

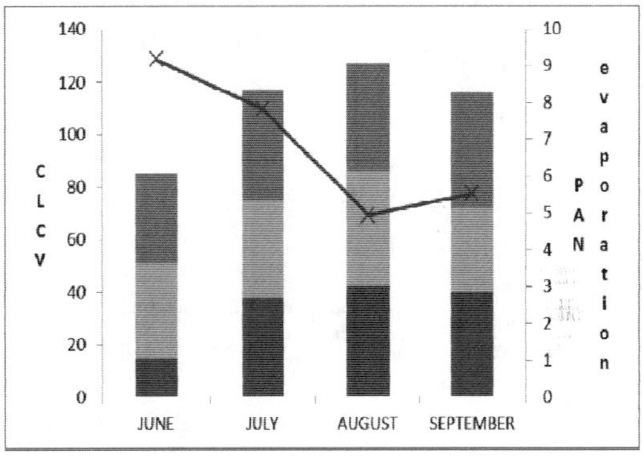

Figure 2.CLCV and PAN Evaporation Variations in June, July, August and September, 2010.

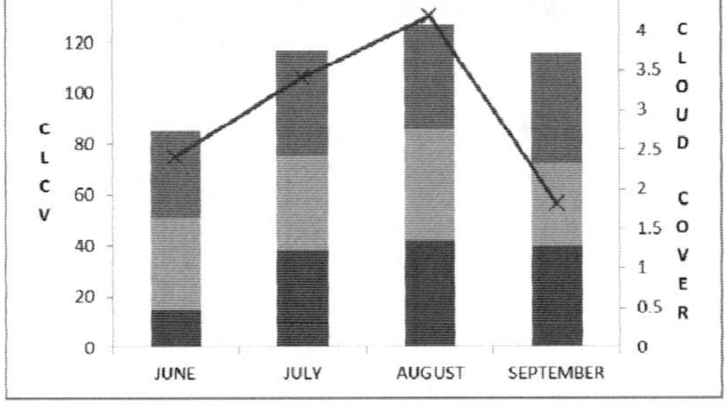

Figure 3. CLCV and Cloud Cover Variations in June, July, August and September, 2010.

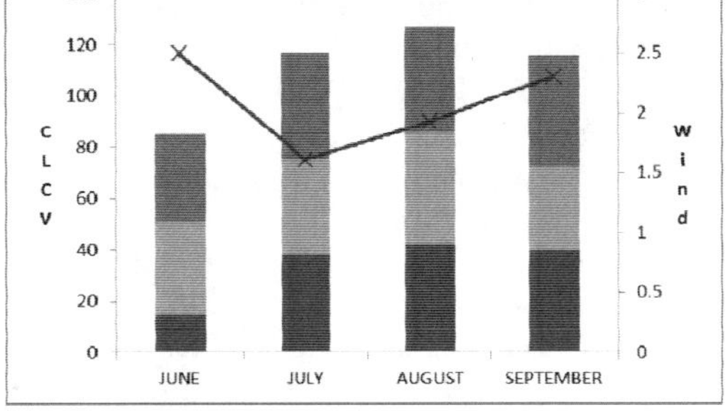

Figure 4. CLCV and Wind Variations in June, July, August and September, 2010.

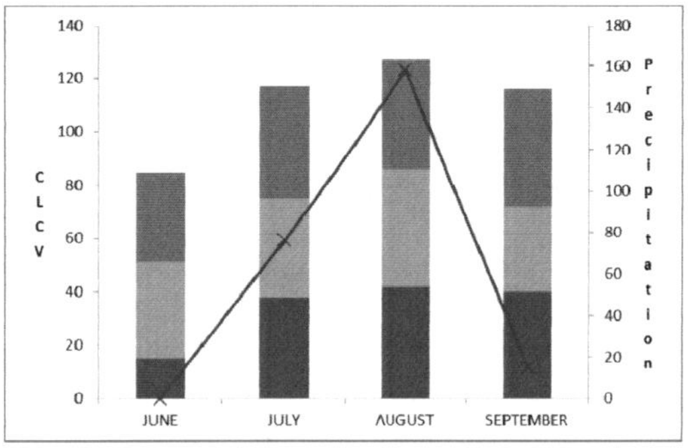

Figure 5. CLCV and Precipitation Variations in June, July, August and September, 2010.

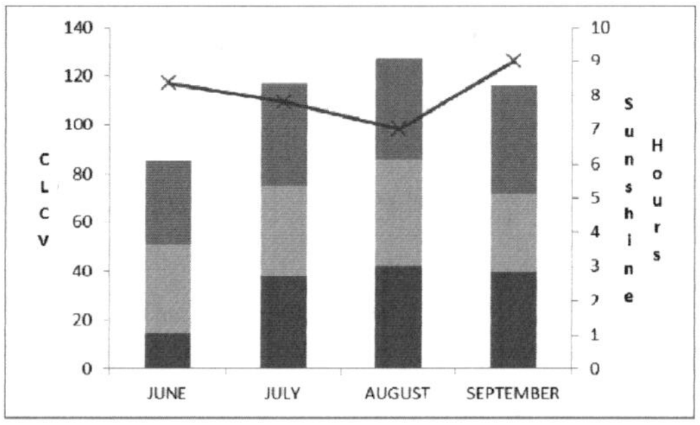

Figure 6. CLCV and Sunshine Hours Variations in June, July, August and September, 2010.

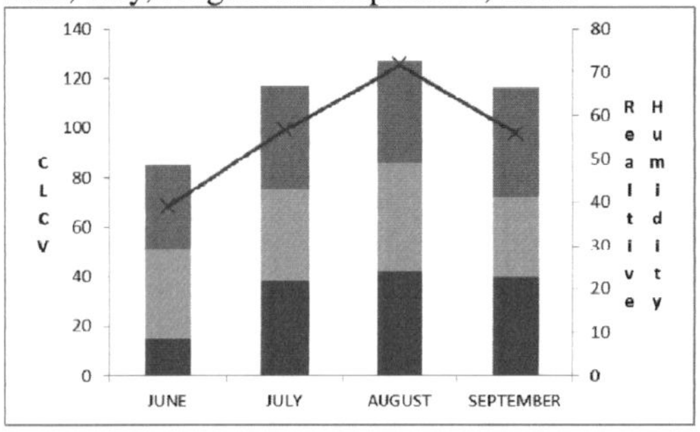

Figure7.CLCV and Relative Humidity Variations in June, July, August and September, 2010.

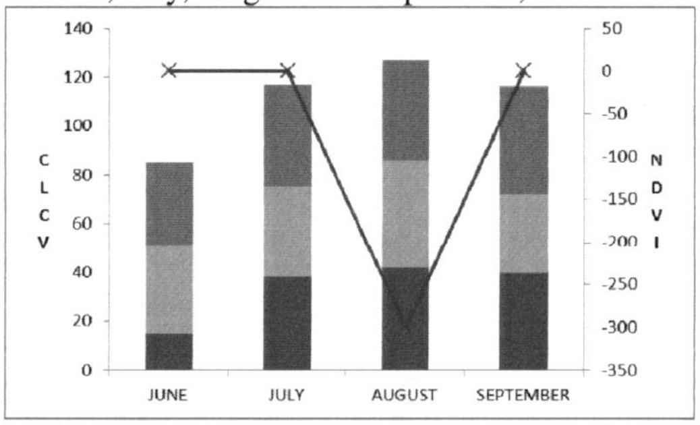

Figure 8. CLCV and NDVI Variations in June, July, August and September, 2010.

References

[1] Government of Punjab 2009 *Pre investment Study on Multan District*

[2] Naheed G and Rasul G 2010*Pakistan Journal of Meteorology* **6** 75-84

[3] Chang M A and Khan C W S 2007 *Sustainable Cotton Production in Pakistan* WWF Pakistan

[4] Ahmad M and Battese G E 1997 A Profit Analysis of the Incidence of the Cotton Leaf Curl Virus in Punjab Pakistan *The Pakistan Development Review* **36** 155-69

[5] Ahmad M, Khan N, Mohammad F, Khan S A, Munir I, Bibi Z and Shaheen S 2011 *Pak. J. Bot.* **43** 1713-18

[6] Farooqi A B, Khan A H and Mir H March 2005 *Pakistan Journal of Meteorology* **2** 11-21

[7] Ministry of Environment 2003 *Pakistan's Initial National Communication on Climate Change* Government of Islamic Republic of Pakistan

[8] Parveen R, Fani I, Islam N, Haider S, Chohan S and Rehman A 2010 *Eur. J. Sci. Res.* **42** 614-21

[9] Mohyuddin A I, Jilani G, Khan A G, Hamza A, Ahmad I and Mahmood Z 1997 *Pakistan J. Zool.* **29** 293-98

[10] Ferquharson C O 1912 *A report of the Mycologist* Agri. Dept. Nigeria

[11] Jones G H and Mason T G 1926 *Ann Bot.***160** 722

[12] Bailey, M. A. 1934. *Sudan Empire Cotton Gr. Rev.* **11** 280

[13] Hussain T and Ali M 1975 A review of cotton diseases in Pakistan *Pak Cottons* **19** 71-86

[14] Rehman H U, Khan W S, Khan M U D and Shah M K N 2000 *Field Crop Res.* **69** 251-57

[15] Monga D, Chakrabarty P K and Kranthi R 2011 *Cotton leaf Curl Disease in India-recent status and management strategies* Presented in 5th meeting of Asian Cotton Research and Development Network (Lahore, Feb 23-25)
[16] Ahmed Z 1999 Prospects and bottlenecks of cotton crop in Pakistan *The Pak Cotton Grower* **3** 6-7
[17] Aslam M M, Razaq, N A, Saeed and Ahmad F 2004 *Int.J. Agric. and Biol.* **6** 39-41
[18] Farooq A, Farooq J, Mahmood A, Shakeel A, Rehman A, Batool A, Riaz M, Shahid M T H and Mehboob S 2011 *Aust. J. Crop Sci.* **5** 1823-31
[19] Khan M A, Mirza J H and Ahmed S 1998 *Pak. J. Phytopathol.* **10** 5-8
[20] Akhtar K P, Hussain M and Khan A I 2002b *Pak. J. Phythopathol.* **15** 1-4
[21] Colvin J L, Fishpool D C, Fargette D, Sherington J and Fariquent C *Entomol. Res.* **88**369-78
[22] Prince S D 1991 *Int. J. Remote Sens.* **12** 1301-12
[23] Purevdorj T S, Tateishi R and Ishiyam T *Int. J. Remote Sens.* **19** 3519-35

Journal of Physics: Conference Series **439** (2013) 012045 doi:10.1088/1742-6596/439/1/012045

Energy conservation in developing countries using green building idea

Akram Rashid[1,*] and Ijaz Mansoor Qureshi[1]
[1]Department of Electrical Engineering, Air University, Islamabad, Pakistan

E-mail: akram@mail.au.edu.pk

Abstract. Green buildings uses processes that are environmentally responsible and resource-efficient throughout a building's life-cycle. In these buildings Certain energy conservative and environment friendly steps are considered and implemented from design, construction, operation, maintenance and renovation. In present era no doubt new technologies are constantly constructed and used in creating greener structures, energy efficient buildings. The common objective is to reduce the overall impact of the built environment on human health using available energy efficiently. To increase the efficiency of the System or the building, Onsite generation of renewable energy through solar power, wind power, hydro power, or biomasscan significantly reduce the environmental impact of the building. Power generation is generally the most expensive feature to add to a building. Any how power generation using renewable sources that is Solar system may further enhance energy conservation ideas. Power Factor improvement can also be another source of efficient tool for efficient use of Electrical Energy in green buildings. In developing countries a significant amount of Electrical Energy can be conserved and System efficiency as a whole can be increased by Power Factor correction. The reverse flow of power can be locally engaged instead of creating extra stress and opposition to the existing grid lines.

1. Introduction

Energy crises is appearing to be a bottleneck in the Economic development of countries across the globe. There is enormous increase in the Global demand of Energy because of Industrialization and population growth. Energy crises are more severe because of the deficit in Demand and Supply. In order to cope with the present energy crises in especially in developing countries low energy buildings can be a best solution. To minimize the primary energy consumption, the many passive and active day lighting techniques are the first solution to apply. Day-lighting with light emitting diode-LED lamps, Solar Powered circulation systems can also be employed in houses, Xenon and Halogen lamps can be used in large buildings and industries. In spite of all these the most immediate and more efficient solution to meet the present energy crises in the developing countries is to improve Power Factor. Since the Power Factor is purely frequency dependent hence frequency and transient analysis are the main topics of this research. In this research the green building idea is strongly related with the industries/buildings/houses offering less opposition rather facilitating the incoming power for its efficient usage.

Journal of Physics: Conference Series **439** (2013) 012045 doi:10.1088/1742-6596/439/1/012045

2. Problem statement

Energy is the most important tool in the development of any country. It is just like a blood circulation in the life line of a country. In developing countries mostly the energy crises are because of old energy distribution system. The industrial/household energy receiving systems are unable to receive the required energy efficiently. Since the loads are not purely resistive rather to say are more inductive and hence they become the source of reverse power flow. This reverse Power flow gives a significant amount of resistance to the incoming power and create excess amount of heat and power loss within the power transmission lines and system itself. Their systems work inefficiently and this way the inefficient systems are casting long shadows in the Economic development of Developing countries. This system loss as a whole can be well tackled by analyzing power factor considering its frequency dependence nature.

3. System model

Power is a precious resource. Power transfer to the load is maximum when the load impedance is equal to the complex conjugate of the impedance of the source. The loads of reactive elements give energy back and forth between the load and the source. Because these elements store and release energy, this ability of the reactive elements causes a much pressure/opposition on the energy distribution system. The instantaneous system power is

$$p(t) = \frac{1}{2}VmI_m Cos\phi + \frac{1}{2}V_m I_m Cos(2\omega t + \phi) \tag{1}$$

When the system is purely resistive the above equation reduces to

$$p(t) = \frac{1}{2}V_m I_m (1 + Cos 2\omega t) \tag{2}$$

This confirms that energy is being transferred constantly from source to load. Power is maximized when Cos ϕ =1 and the power above is apparent one.

$$S = \frac{1}{2}V_m I_m = V_{rms} I_{rms} \tag{3}$$

When $\phi = \pm 90^0$ and $Cos \pm 90^0 = 0$ the circuit is purely reactive and purely reactive load dissipate no power. In this case the energy transferred to the load during positive alternation is returned to the source during negative alternation. We sum up with the equation

$$p(t) = \frac{1}{2}V_m I_m Cos(2\omega t \pm 90^0) \tag{4}$$

It is very much clear that in Load is partly resistive and partly reactive. So the Power factor $Cos\phi$ lies between 0 and 1. This shows that Power factor is frequency dependent and hence the power is frequency dependent. Power factor is leading when the load is Capacitive and lagging when the load is Inductive. The power delivered to the load is

$$P_L = \frac{1}{2}R_L \frac{|V_{oc}|^2}{(R_{eq} + R_L)^2 + (X_{eq} + X_L)^2} \tag{5}$$

Journal of Physics: Conference Series **439** (2013) 012045 doi:10.1088/1742-6596/439/1/012045

On average reactances absorb zero power; the reactive term can be eliminated easily. This shows

$$X_L = -X_{eq} \tag{6}$$

This would lead to reduce the power delivered to the load. In this case the above equation clearly shows that for maximum power transfer we are left with the only condition

$$R_L = R_{eq} \tag{7}$$

From equations 6 and 7 we conclude that maximum power to the load to be transferred require the following condition to be valid

$$Z_L = Z^*_{eq} \tag{8}$$

So there will be no reactive power from load to the distribution network and there will be a sufficient power conservation.

For capacity requirements and system efficiency Complex Power plays a vital role. The instantaneous power as per equation (1) can be written as

$$p(t) = V_{rms} I_{rms}[Cos\phi + Cos2\omega tCos\phi - Sin2\omega tSin\phi] \tag{9}$$

ϕ is Phase difference between voltage and current. We can introduce the quantities in equation (9) as following.

$$P = V_{rms} I_{rms} Cos\phi \tag{10}$$

$$Q = V_{rms} I_{rms} Sin\phi \tag{11}$$

From equations (10) and (11) we can get the instantaneous power as

$$p(t) = p_R(t) - p_X(t) \tag{12}$$

and

$$Q = X(\omega) I^2_{rms} \tag{13}$$

When we will get Q as a positive quantity the load to which the power is going to be transferred will be Inductive and if negative the load will be Capacitive. As a power Engineer this can be recognized that inductive load consume while capacitive load produce reactive power. From equations 10 and 11 the Q will be as

$$Q = P\tan\phi = p\tan(\pm Cos^{-1} pf) \tag{14}$$

Positive is used when load is inductive with pf lagging and negative when load is capacitive with pf leading. For efficient power transmission the pf is closer to unity. The specifications are given in terms of load Impedance. So in a given industrial/household load $Z = R + JX$. In this case we are interested to find out X_P. Which means that when it is connected in parallel with Z, the power factor of the composite load will be raised to a new power factor. This way we obtain the required reactance by the following equations as

$$X_P = \frac{R^2 + X^2}{R\tan(\pm Cos^{-1} pf_{new}) - X} \tag{15}$$

Positive sign is if pf_{new} is lagging and –sign if leading. If in above equation multiply by I^2_{rms} , and placing we will get the equation given below

$$X_P = \frac{V^2 rms / p}{\tan(\pm Cos^{-1} pf_{new}) - \tan(\pm Cos^{-1} pf_{old})} \tag{16}$$

Here P is the average power and pf_{old}, pf_{new} are the power factors before and after corrections. Let us consider Example

Suppose $\phi = 45^o$ and then $pf = Cos 45^0 = 1/\sqrt{2}$ =0.707. Using this value in the equation below

$$I_{rms} = \frac{P}{V_{rms} \times pf} \tag{17}$$

We obtain I_{rms} which shows that current capacity of the system have to be increased by 41.4% to accommodate the reactive power exchanges which are present with the low power factor. Moreover if we consider the resistance of the power line as

$$R_{line(pf=0.707)} = \sqrt{2} I_{rms(pf=1)} = 1.414 I_{rms(pf=1)} \tag{18}$$

$$R_{line} I_{rms}{}^2{}_{(pf=0.707)} = 2 \times R_{line} I^2{}_{rms(pf=1)} \tag{19}$$

This shows that line power losses are double because of the reactive exchanges. The possible Capacitor locations for power factor improvements are shown in figure 1 below.

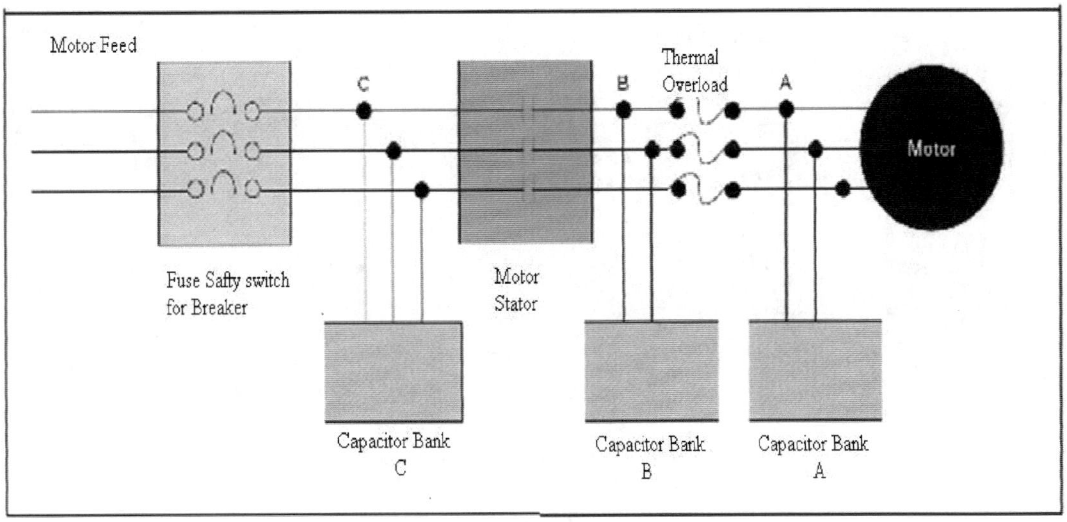

Figure 1. Possible capacitor locations for power factor improvements.

4. Results and conclusions

As shown in figure 1 a Hampden type 3-phase motor, 220V, 1.4 Ampere, 1725 RPM and 1/3 HP motor is used. For power factor correction, the Inductor or Capacitor Banks are used. Three basic power factor correction banks in series and parallel with the grid lines respectively. Inductor bank is connected in series with the line terminals and hence it get small current ripples with its characteristic easier to implement. This series Inductor forces the current to have same shape as the input voltage is.

This gives the less input impedance to the incoming power because of the improvement in the Power factor.

To understand how the input voltage influence the Power Factor we have to analyze the Harmonics. The Harmonics of the input voltage create 90° lead in the corresponding Harmonic at the Inductor current as shown in figure 2 below.

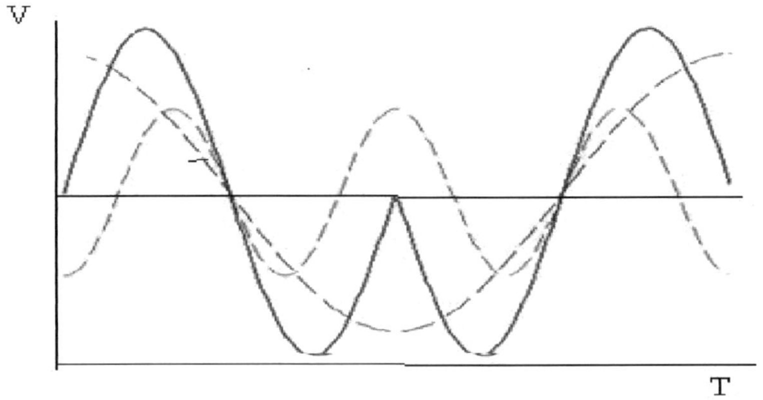

Figure 2. Harmonics of input voltages.

When the capacitors are added to proper location to an Industry the resonance current is increased more than twice as shown in the figure 3 below

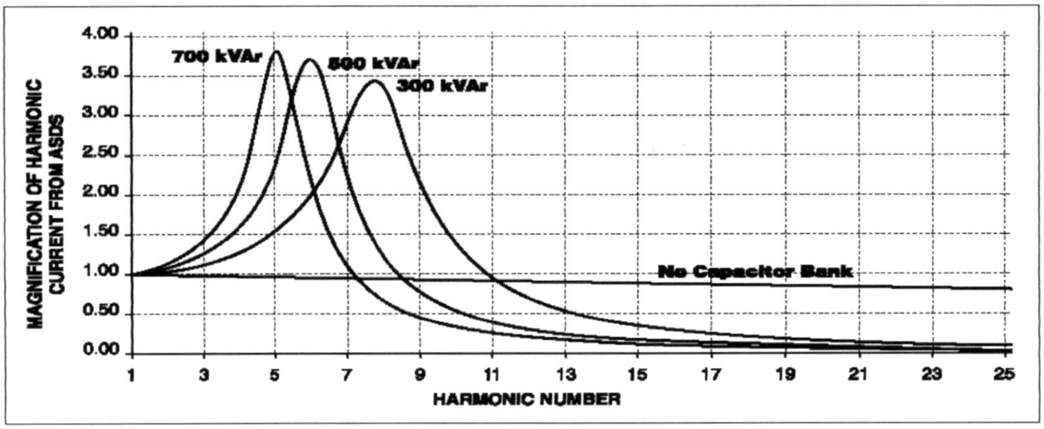

Figure 3. Magnification of harmonic current when standard capacitors are added.

Table 1. Harmonic components in input voltage.

n / Vrms	2	4	6	8	10	12
100	60	12	8	4	3	2
250	150	50	30	20	10	6

Table above shows that input voltage consists of large number of harmonics. The Higher the voltage the more the line current is influenced and hence the power factor.

Figure 4 below shows efficiency of the motor is increased when the power factor is improved. In the beginning the efficiency increased sharply.

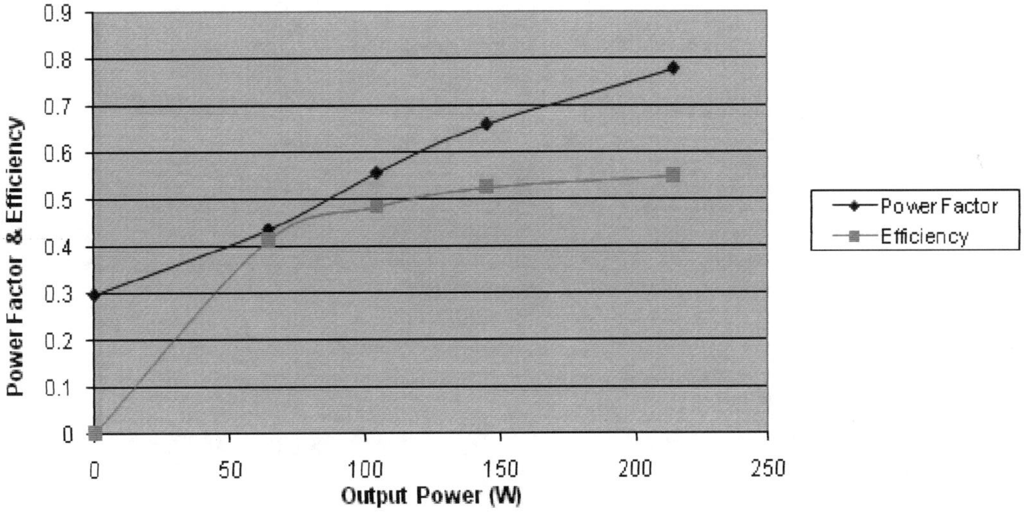

Figure 4. Efficiency increased with the improvement in the power Factor.

For power factor improvement of our system Relays are used Relay coils are energized from SATEC meter when the power factor drops below a certain point

- Line current is also monitored to ensure no switching occurs when no load is present
- Capacitors are added to the circuit in parallel with the motor
- Placed in parallel to maintain the same line voltages into the motor

Power factor can be varied over any desired range to avoid electric utility company charges Project system maintains a power factor of above 0.98. The complete sequence is shown in the flow chart below

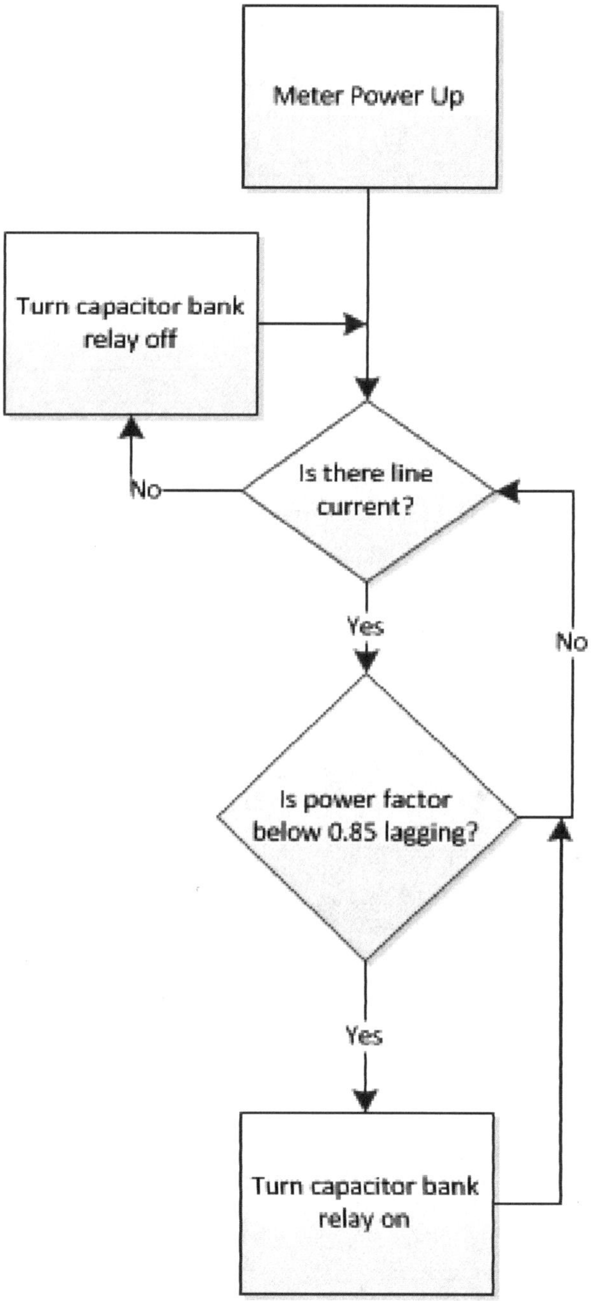

Figure 5. Sequence of improving power factor.

Figure 6 below shows the relation between source power factor and the motor power factor and in turn the efficiency of the motor.

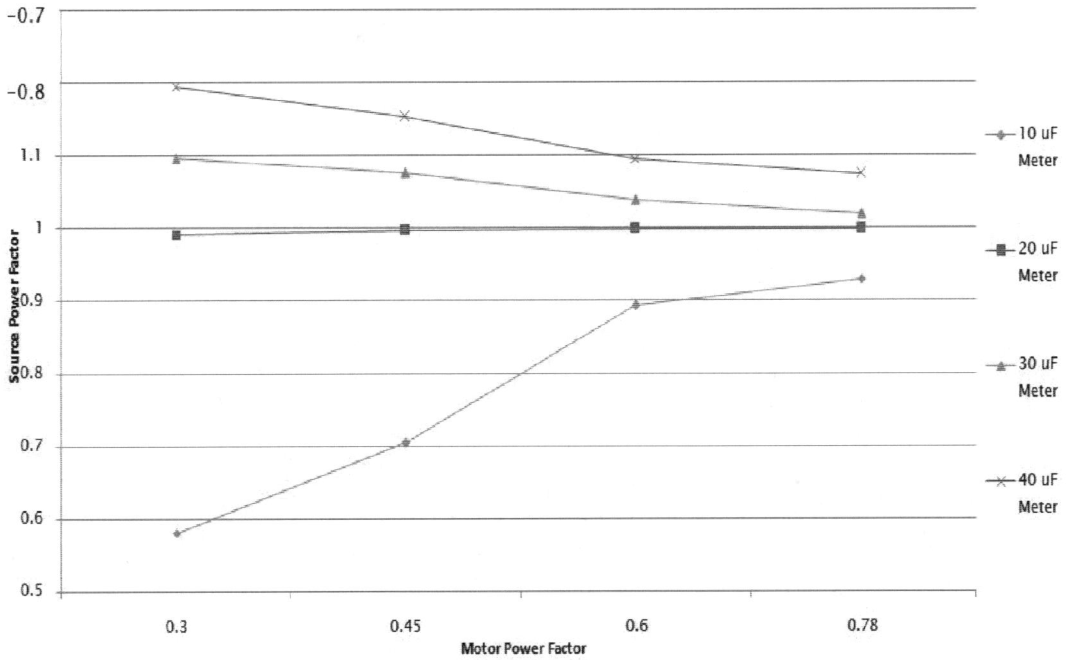

Figure 6. Motor power factor increasing constantly when the source power factor is unity.

Figure 7 below shows that maximum power will be transferred when power factor will be equal to unity and hence the reverse current flow will be equal to zero as indicated by red line.

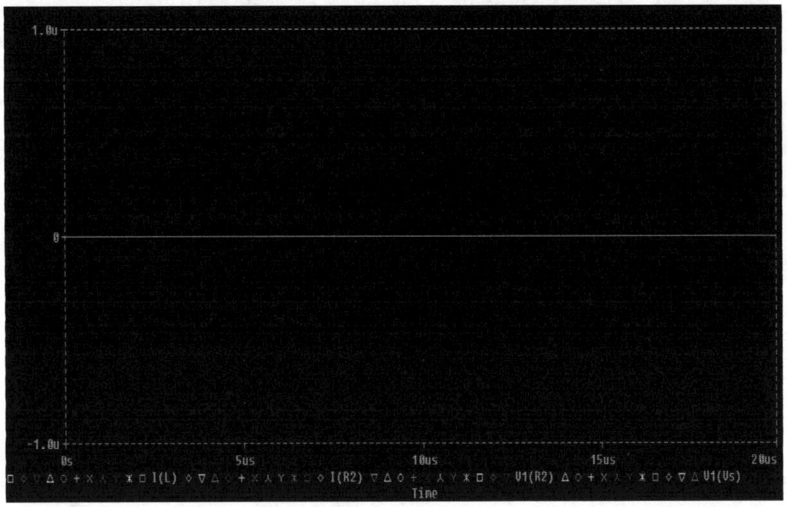

Figure 7. Unity power factor restrict zero reverse current flow to power transmission lines.

Journal of Physics: Conference Series **439** (2013) 012045 doi:10.1088/1742-6596/439/1/012045

References

[1] Sun Jing 2010 *IEEE International Conference on Electrical and Control Engineering* (Print ISBN: 978-1-4244-6880-5) p 4070
[2] Fang Lin Luo, Hong Ye 2007 *IEEE International Conference on Industrial Electronics Society*, Print ISBN: 1-4244-0783-4 1578
[3] Tong Ziyu Ding Wowo 2011 *IEEE International conference on Electric Technology and civil Engineering*, 2856
[4] Zhao Y and Zeng J 2004 *7th International Symposium on Parallel Architectures, Algorithms and Networks (ISPAN'04)*
[5] Yang S, Wang M and Jiao L 2004, *IEEE Congress on Evolutionary Computation, CEC'04*, Vol. 1, 320
[6] Ciriaco F, Abr˜ao T and Jeszensky P 2006 *Journal of Universal Computer Science*, Vol.12, No.4
[7] Akram Rashid, 2010, *World Academy of Science Engineering* Paris France issue 67 p. 1223
[8] Akram Rashid, Zahooruddin, Ijaz Mansoor Qureshi 2011, *IEEE International Conference on Machine learning and Cybernetics* Gulin, Guangxi, China, Vol. 4, p 1756
[9] Jinrong Qian "Advance Single stage Power Factor Correction Techniques" Virginia Polytechnic and state University, Department of Electrical Engineering Sep. 1995 Blacksburg Virginia University, Department of Electrical Engineering Sep. 1995 Blacksburg Virginia
[10] Akram Rashid, Zahooruddin, Ijaz Mansoor Qureshi 2012 *International Conference on Pakistan's Energy options Technology, Availability Affordability and role of regional grids*, Islamabad-Pakistan
[11] Malik R 2003 *IEEE Applied Power Electronics Conference Proceedings*, Feb., Vol. 2, p 202
[12] Lee F C 2000 *Power Engineering Society Summer Meeting,* Vol. 2, p 1229
[13] Osifchin N 2000 *IEEE International Telecommunications Energy Conference* p 1
[14] BCHP 2003 General public benefits," http://www.bchp.org/public-benefit.html
[15] Frantz G 2000 *Micro, IEEE,* issue: 6, Nov./Dec., Vol.20, p 52
[16] Lee F C, Lecture notes for ECE5254 at Virginia Tech, Spring 2003 http://www.cpes.vt.edu/public/courses/EE5254/contents.html.
[17] Zhou C and Jovanovic M M 1992, *International High Frequency Power Conversion (HFPC) Conf.* 209

Journal of Physics: Conference Series **439** (2013) 012046 doi:10.1088/1742-6596/439/1/012046

Wireless electricity (Power) transmission using solar based power satellite technology

M Maqsood[1] **and M Nauman Nasir**[2]
[1] National Institute of Vacuum Science & Technology, NCP Complex, Shahdara Road, PO Box # 3125, Islamabad, Pakistan
[2] Department of Electrical Engineering, University of Engineering & Technology, Taxila, Pakistan

E-mail: maqsood@ninvast.edu.pk

Abstract. In the near future due to extensive use of energy, limited supply of resources and the pollution in environment from present resources e.g. (wood, coal, fossil fuel) etc, alternative sources of energy and new ways to generate energy which are efficient, cost effective and produce minimum losses are of great concern. Wireless electricity (Power) transmission (WET) has become a focal point as research point of view and nowadays lies at top 10 future hot burning technologies that are under research these days. In this paper, we present the concept of transmitting power wirelessly to reduce transmission and distribution losses. The wired distribution losses are 70 - 75% efficient. We cannot imagine the world without electric power which is efficient, cost effective and produce minimum losses is of great concern. .This paper tells us the benefits of using WET technology specially by using Solar based Power satellites (SBPS) and also focuses that how we make electric system cost effective, optimized and well organized. Moreover, attempts are made to highlight future issues so as to index some emerging solutions.

1. Introduction

The present electricity generation system is not very efficient in terms of energy transfer. About 20 to 30% energy is lost during the distribution of the electricity. Therefore the scientists are working on the projects to improve the ultimate electricity supply. Scientists are looking for alternate and efficient technologies to provide 100% electricity transfer. The change and development in the various fields have brought more client satisfaction and output. Therefore the wireless transmission of electricity is also on move [1].

In 1864, James C. Maxwell predicted the existence of radio waves by means of mathematical model. In 1884, John H. Poynting realized that the Poynting vector would play an important role in quantifying the electromagnetic energy. The prediction and evidence of the radio wave in the end of 19th century was start of the wireless power transmission. During the same period of Marchese G. Marconi and Reginald Fessenden who are pioneers of communication via radio waves. Nikola Tesla is known as the father of wireless transmission. The most famous wireless technology known as Wardenclyffe tower also known as the Tesla tower [1 - 3] is the first was designed merely for wireless transmission of electricity. He made electric coil which was a 3 feet diameter ball at its top. He fed

300 kW power to tesla coil resonated at 150 kHz. The RF potential at the top sphere reached 100 MV. Unfortunately he failed due to diffusion in all directions [1, 3].

2. Wireless Electricity Transmission (WET) technology

Wireless power transmission [4, 5] is a process that takes place in any type of system in which electrical current is conveyed from a power source to an electrical load. What makes this process unique is that there is no usage of any type of wiring to connect the system to a source of power.

Wireless electricity (Power) transmission basically is the transmission of electricity with the help of microwaves and there is no need to use cables, towers and grid stations [4, 5]. There are three methods or approaches which can be developed. These are as given below:

2.1. Short range (Induction)

This ranges few centimetres e.g. transformer in which transfer takes place due to mutual induction [4, 5].

2.2. Moderate range (Adaptive Inductive Coupling)

Witricity [2] (Adaptive Inductive Coupling) wireless power transfer technology can be used to charge the electronic objects automatically. The ability of our technology to transfer power safely, efficiently, and over distance can improve products. This principle of wireless electricity works on the principle of using coupled resonant objects for the transference of electricity to objects without the use of any wire [4, 5].

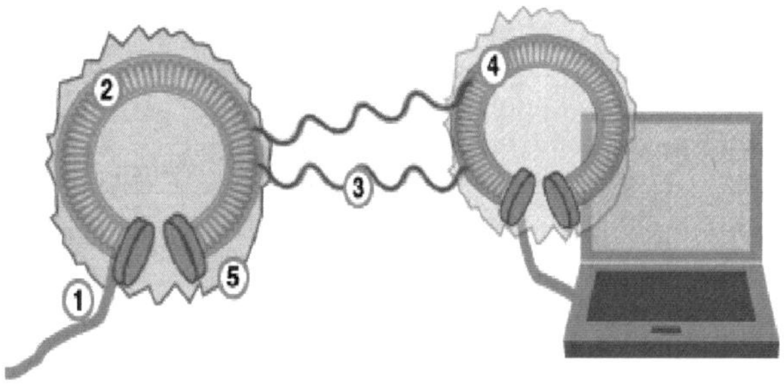

1. Power from mains to antenna, which is made of copper
2. Antenna resonates at a frequency of about 10MHz, producing electromagnetic waves
3. Tails' of energy from antenna 'tunnel' up to 2m (6.5ft)
4. Electricity picked up by laptop's antenna, which must also be resonating at 10MHz. Energy used to re-charge device
5. Energy not transferred to laptop re-absorbed by source antenna. People/other objects not affected as not resonating at 10MHz

Figure 1. Working of Wireless Electricity.

2.3. Long range

Plans for wireless power involve moving electricity over a span of miles. Long distance wireless power is the technology of sending power to earth [4, 5]. There are many new techniques but we use only two here.

2.3.1. By using Solar Power Satellite (SPS). This task is often completed by using solar power satellite (SPS) [6, 7], placed in high earth orbit. This satellite converts the sunlight into energy; this energy is composed of microwaves. These microwave signals are transmitted to an antenna on ground/Main grid station (MGS). From MGS these waves are transferred to BGS (Base grid station) so called rectenna which convert microwaves into DC electricity. There will be energy receiver box or energy router in each home. The information of the electricity or power required for each home will be available with the grid station. At the grid station the electricity will be converted into energy packets likewise internet data packets and the header of that energy packet will include the address of the energy receiver that is mounted on the wall of the house of consumer. The energy packet will then reach the energy receiver and will be stored in that energy receiver after that the consumer can use that stored energy any time he wants. We used the same concept as we do in telecom sector and through this act we can buy electricity according to our need.

2.3.2. By using WET Technology (Without SPS). Another method is very simple. We can produce electricity at MGS by (hydropower, Thermal ,Wind, Solar) and convert this electricity into microwaves by using inverse rectennas or Microwave generator and transfer it to base grid station(BGS) so called RECTENNA through transmitting Antennas and from here electricity is transmitted to home wirelessly likewise as explained in section 2.3.1.

2.4. Component detail of WET Technology
The Primary components of Wireless Power Transmission are Microwave Generator, Transmitting antenna and Receiving antenna (Rectenna).

2.4.1. Microwave Generator. The microwave transmitting devices are classified as Microwave Vacuum Tubes (magnetron, klystron) and Microwave Power Module (MPM)) and Semiconductor Microwave transmitters and amplifiers (GaAs MESFET, SiC MESFET, HFET, and InGaAs). Cooker type Magnetron is widely used for experimentation of WPT.

2.4.2. Transmitting antenna. The slotted wave guide antenna, micro strip patch antenna, and parabolic dish antenna are the most popular type of transmitting antenna. The slotted waveguide antenna is ideal for power transmission because of its high aperture efficiency (> 95%) and high power handling capability.

2.4.3. Rectennas. A rectenna is a rectifying antenna; a special type of antenna that is used to directly convert microwave energy into DC electricity. This concept is discussed in SPS detail.

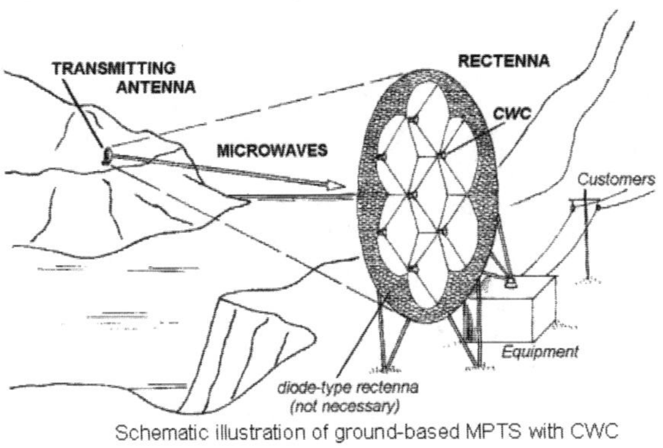

Figure 2. Detail of MPTS.

2.5. Component details of WET using SPS system

The primary components of WET using SPS involve microwaves, solar power satellite, rectenna, dengyo, etc.

2.5.1. Microwaves. Microwaves [3] are electromagnetic waves with wavelengths ranging from as long as one meter to as short as one millimeter, or equivalently, with frequencies between 300 MHz (0.3 GHz) and 300 GHz. For Wireless power transfer we use high power microwaves namely 1-10GHz radio-waves [3, 4, 8].

2.5.2. Microwave power transmission (MPT). Microwave power transmission (MPT) [7, 9] is the use of microwaves to transmit power through outer space or the atmosphere without the need for wires. It is a sub-type of the more general wireless energy transfer methods. Microwaves are widely used for point-to-point communications because their small wavelength allows conveniently-sized antennas to direct them in narrow beams, which can be pointed directly at the receiving antenna [4, 5]. This allows nearby microwave equipment to use the same frequencies without interfering with each other, as lower frequency radio waves do. Microwave Power transfer (MPT) [2.45 GHz or 5.8GHZ] of ISM band is used. (ISM= Industry, Science and Medical).

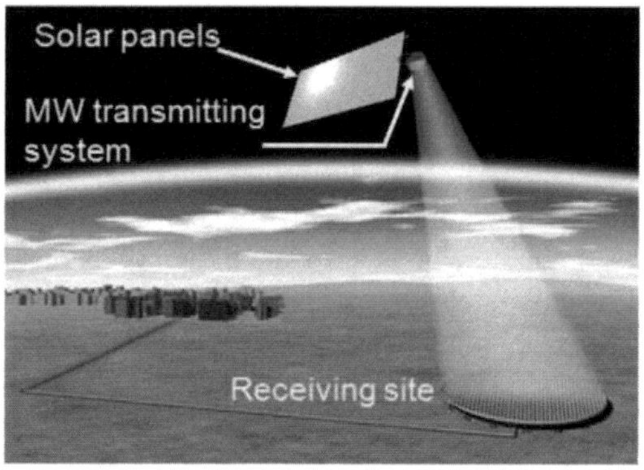

Figure 3. Microwave Power Transfer.

2.5.3. Solar Power Satellites. Space-based solar power (SBSP) [6, 7] is the concept of collecting solar power in space for use on Earth. It has been in research since the early 1970s. SBSP also introduces several new hurdles, primarily the problem of transmitting energy from orbit to Earth's surface for use. SBSP designs generally include the use of some manner of wireless power transmission. It mainly consists of three segments Solar energy collector (To convert Solar energy into DC current), DC to microwave converter and Large antenna array to beam down (Microwave) power to ground [6, 7].

- Solar energy collector (Photovoltaic cell, solar thermal turbine).
- DC to microwave converter (Microwave tube system and /semiconductor system).
- Antenna

SPS will be expected to operate in 2025-2030.

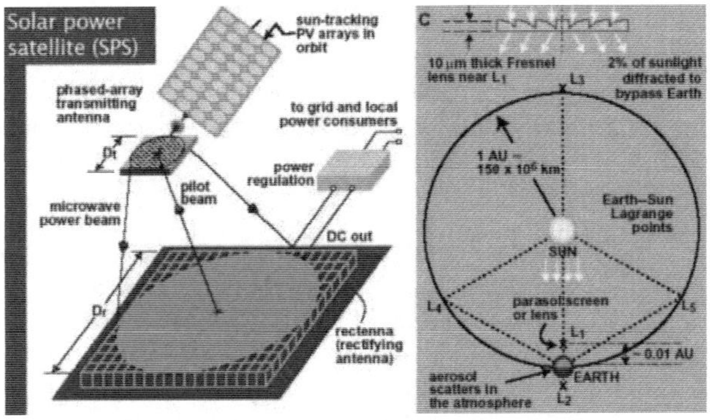

Figure 4. SPS detailed review.

2.5.4. Rectennas. A rectenna [7, 9] is a rectifying antenna, a special type of antenna that is used to directly convert microwave energy into DC electricity. Its elements are usually arranged in a multi element phased array with a mesh pattern reflector element to make it directional. Rectennas are being developed as the receiving antennas in proposed microwave power transmission schemes, which transmit electric power to distant locations using microwaves. Rectennas are used in RFID tags; the energy to power the computer chip in the tag is received from the querying radio signal by a small rectenna. One possible future application is a receiving antenna for solar power satellites.

A simple rectenna element consists of a dipole antenna with a Schottky diode placed across the dipole elements. The diode rectifies the AC current induced in the antenna by the microwaves, to produce DC power. Schottky diodes are used because they have the lowest voltage drop and highest speed and therefore waste the least amount of power due to conduction and switching. Large rectennas consist of an array of many such dipole elements.

Rectennas are highly efficient at converting microwave energy to electricity. In laboratory environments, efficiencies of over 85% have been observed. Some experimentation has been done with inverse rectennas, converting electricity into microwave energy, but efficiencies are much lower - only in the area of 1%. Rectenna conversion efficiencies exceeding 95% have been realized. The efficiency of first Rectenna in 1963 was 50% at output 4W DC and 40% at output 7W DC respectively.

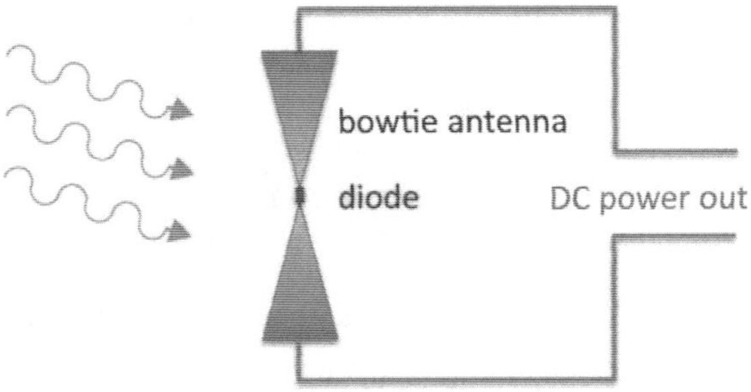

Figure 5. Rectenna.

2.5.5. Dengyo

Dengyo newly developed a microwave-band rectenna and a rectifier circuit and achieved a power conversion efficiency of 90% or higher in the 2GHz band. Dengyo has its strengths in antenna and filter technologies. And the company utilized its know-how in the development of the new technology though it did not disclose details. By using a low-loss filter, it reduced the reradiating of spurious waves from the rectifier circuit to -50 dBc or less.

With the high conversion efficiency, it becomes possible to convert electric waves that have not so far been used by devices using high frequency waves into electricity. Specifically, when the new technology is applied to a microwave heating device, energy that has been discharged as heat can be recycled. And the power consumption of such a device can be reduced by up to 40%.

Because the new technology uses electric waves, electricity can be fed to a distant place. Also, by combining a wireless sensor and the rectenna, it becomes possible to collect data by activating the sensor only when it is necessary. As a result, it eliminates the needs for power wires and primary batteries for wireless sensors.

Nihon Dengyo Kosaku Co Ltd (Dengyo) announced May 16, 2011, that it has developed a technology to convert electric wave into electricity with high efficiency. Specifically, Dengyo achieved a conversion efficiency of 90% or higher by using a rectenna technology that converts the energy of microwave into direct-current (DC) electricity, it said. The company expects that the new technology will be used to reduce the power consumption of microwave heating devices, harvest energy, etc. This time, Dengyo newly developed a microwave-band rectenna and a rectifier circuit and achieved a power conversion efficiency of 90% or higher in the 2GHz band.

Figure 6. Dengyo.

3. Latest researches

Scientists [10] lit a 60-watt light bulb from a power source 7 feet (2 meters) away with their new technique, with no physical connection between the source and the appliance. The researchers have dubbed their concept "WiTricity," as in "wireless electricity."

3.1. First commercial wireless electricity experiment in Japan

Currently Japanese scientists are set to test the largest wireless electricity transmission ever attempted in a Tesla like spectacle that is sure to capture a great amount of attention and spark strong interest and support for a technology that could change the World. The event is to take place at the Tokyo Tower, the largest man made structure in Japan, at 1100 feet tall.

The night-time experiment is meant to illuminate the top spire of the mammoth steel structure to demonstrate the use of the first wireless electricity transfer system in the World. The test is designed to transfer about 1200 watts of power at a range of 100 feet and will be a first of its kind use of system Japanese scientists are developing to transmit power at distances they hope could reach 300 feet using a science that is based on magnetically coupled resonance.

4. Merits, demerits and applications

4.1. Merits
The major benefits are as follows:

- Remove physical infrastructure "Grids and Towers" [1].
- Cost effective (Remove cost of towers and cables).
- During rain and after natural disaster it is often hard to manage cables and towers so it removes this.
- Losses during transmission and distribution can be removed.
- Microwaves (electricity) are more environments friendly. It does not involve emission of carbon gases.
- Electricity bills using conventional supply can be cut to very low.
- Zero fuel cost [1].

4.2. Demerits

- Biological effects associated with the wireless transmission of electricity due to the high frequency microwave signals is the first demerit of this technology [1, 7].
- This project is a onetime expense but it involves a lot of initial expenditure. It is expected that the price of receive electricity through this technology would decline overtime [7].
- This technology is limited to the use of few technologies like solar satellites and Tesla grid.
- The transmission of electric current through this mode is susceptible to security risks like cyber war fare [1].

4.3. Applications
There are many applications [7]. Example include; fuel free airplanes, fuel free electric vehicles, moving robots and fuel free rockets, battery charging, car charging, remote control, game controller, headsets, sensors, computers, laptop charging, television and many more.

5. Conclusions
This paper provides an overview about the most under discussion technology nowadays. In this paper we, not only tried to overcome the current issues but also discussed the pros and cons of WET technology.

Through this paper we have tried to inculcate the concept of Wireless electricity Power transmission in the mind of readers that by utilizing the power of this technology fully, we can get an efficient, cost effective, wireless(free from wires),losses free environment. Fossil fuel electric power plants generate greenhouse gases which are responsible for global warming. Wireless electricity transmission can overcome these problems well. In the nutshell, this technology will change the concept of Electricity

Journal of Physics: Conference Series **439** (2013) 012046 doi:10.1088/1742-6596/439/1/012046

in near future and make people able to acquire the most efficient, effective, pollution free and healthy way of getting electricity.

References

[1] Nikola Tesla 1905 *Electrical World and Engineer The Transmission of Electrical Energy Without Wires as a Means for Furthering Peace,* p 21

[2] Gernsback, Hugo 1919 *Nikola Tesla and His Achievements, Electrical Experimenter* p 615

[3] Nikola Tesla *Nikola Tesla on His Work with Alternating Currents and Their Application to Wireless Telegraphy, Telephony and Transmission of Power* p 126

[4] Dave Baarman and Joshua Schwannecke 2009 *Understanding Wireless Power* p 1

[5] Landis, Geoffrey A 2006 *IEEE 4th World Conference on Photovoltaic Energy Conversion* 2

[6] Glaser P E 1968 *Power from the Sun: It's Future,* 162 p 957

[7] Landis, Geoffrey A 2006 *IEEE 4th World Conference on Photovoltaic Energy Conversion* 2 1939

[8] Nikola Tesla 1904 *The transmission of electric energy without wires* p 22

[9] William C Brown 1969 *Microwave to DC Converter*

[10] MIT News 2007 *Goodbye wires*

Remediation of explosive-polluted soil in slurry phase by aerobic biostimulation

Baoping Xin, Mengyue Shen, Hina Aslam, Feng Wu
Department of Environment and Energy, School of Chemical Engineering and Environment, Beijing Institute of Technology, Beijing 100081, P. R. China

E-mail: xinbaoping@bit.edu.cn

Abstract. There is a great volume of polluted soil by 2,4,6-trinitrotoluene (TNT) manufacturing wastewater containing dozen of nitrocompounds in China. In this study, biostimulation was used for remediating the explosive-polluted soil in aerobic bioslurry by monitoring the removal of total organic carbon (TOC). The results showed that the pulp density had almost no effect on TOC removal; whereas the acetone addition evidently improved remediation efficiency of the polluted soil by intrinsic microorganism, and the TOC removal increased from 25% to 38.4% when dose of acetone increased from 0% to 4% (v/v). The maximum TOC removal of 49.1% was achieved through further adjusting pH at 9.0 and temperature at 30 °C. The second order reaction fits well removal dynamics of TOC under the optimum conditions. With the average conditions, liquid phase TOC decreased from 3404 to 3144 mg/L and solid phase TOC dropped from 1022 to 104 mg/L, leading to toxicity decline by 35%; the optimum condition witnessed 48.9% of TOC removal from 4500 to 2300 mg/L in liquid phase, causing toxicity drop by 62%.

1. Introduction

Nitroaromatic explosives have been extensively used for both military and civilian purposes for a much long time [1, 2]. The explosives can accumulate at the sites and surrounding soils during production, packing, storage, application, ordnance demilitarization, demolition procedures, and destruction of outdated and faulty ammunition, resulting in serious soil pollution by the target explosive compounds [1, 2]. The high toxicity, mutagenicity, and recalcitrance of these energetic compounds have enabled the remediation of explosive-polluted soil to be more urgent [1, 2]. Among such compounds, 2,4,6-trinitrotoluene (TNT) is a predominant contaminant [3, 4]. TNT contamination of soil occurs because of the manufacture, loading, assembly, packing and military related activities [5, 6]. Due to the potential adverse effect of the compound on human health and ecosystem, remediation of the TNT-polluted soil has been draw growing concerns in recent years [2,3,6], especially the bioremediation processes based on aerobic or anaerobic TNT transformation were developed by monitoring the residue, metabolism and fate of TNT in soils or soil slurries[7, 8, 9, 10, 11, 12].

In addition of the polluted soils by direct settlement and mixture of TNT as a target pollutant, quite a few explosive-polluted soils originated from TNT manufacturing wastewater in China. TNT production generated a massive volume of wastewater, known as red water, pink water and yellow water [13, 14, 15]. TNT manufacturing wastewater is characteristic of high COD, high chromaticity and high toxicity, containing dozens of organic pollutants including raw materials, intermediates, by products, final products and their isomers, sometimes even more than hundred of ones; however, the content of TNT in the TNT manufacturing wastewater is always rather low [16, 17, 18]. In the early 10-20 years of TNT production in the last century, many of TNT manufacturing wastewater was discharged into man-made soil ponds and pits without any treatment by TNT manufacturing factories of China for self purification. After evaporation and infiltration of water, these massive of organic pollutants accumulated in the bottom of pits and ponds and permeated into the surrounding soils, leaving vast volume of explosive-polluted soils in China. The explosive-polluted soils by TNT manufacturing wastewater greatly differ from the TNT-polluted soils by direct settlement and mixture of TNT. The former covers dozens of organic pollutants rather than TNT alone. In fact, the concentration of TNT in the explosive-polluted soil is very low. Under the circumstances, obviously it is not incompetent to assess remediation efficiency of the polluted soil through monitoring residual concentration of TNT; in contrast, the total organic carbon (TOC) is more suitable to reflect the removal efficiency of nitro-compounds pollutants as an integrated index. Moreover, the resulting treated soil should receive the biological toxicity test to evaluate the ultimate remediation effect. However, there are few reports about remediation of the polluted soils by TNT manufacturing wastewater based on TOC analysis and corresponding toxicity test.

Bioremediation is a low-cost, efficient, environmentally friendly way for remediation of polluted soil [2]. The most widely used bioremediation procedure is biostimulation through addition of carbon source, electron donors, inorganic nutrients such as N and P to stimulate degrading activity of intrinsic microorganisms which has adapted to the pollutants [6]. Solid phase bioremediation methods such as composting and land farming have been used to treat TNT-contaminated soils. However, the prolonged incubation time due to the limited mass transfer in solid phase has forced researchers to resort to slurry phase bioremediation where a mixture of contaminated soil, water and co-substrates is treated, indicating higher removal efficiencies as compared to solid phase [19, 20, 21]. In the present works, remediation of the explosive-polluted soils by TNT manufacturing wastewater in slurry phase by aerobic biostimulation was performed for the first time. For evaluating the potential of slurry biostimulation in remediation of explosive-polluted soils by TNT manufacturing wastewater, four aspects studies were conducted, 1) removal efficiencies of the organic pollutants from soil by slurry biostimulation under different conditions based on TOC analysis; 2) removal dynamics investigation of the organic pollutants from soil; 3) removal mechanisms exploration of the organic pollutants from soil; 4) toxicity decrease evaluation of the polluted soils using luminescent bacteria toxicity test.

2. Materials and methods

2.1. Soil samples

The explosive-polluted soil was collected from the discarded settling ponds for storing the TNT manufacturing wastewater from a TNT manufacturing factory, Gansu province, northwest China. The non-polluted soil was also collected from the same areas as background control. The both of soils were air-dried and sieved through a 0.25-mm sieve and stored in closed containers at room temperature for the property exploration and bioremediation experiments.

2.2. Remediation of polluted soil in slurry phase by aerobic biostimulation

2.2.1 Addition of acetone to improve bioavailability of explosives

The slurry was prepared by mixing 50 g of the polluted soil with 100 mL of deionized water in 250 mL flasks, i.e. solid-to-liquid ratio being 1:2 (w/v). The pH value of the slurry was adjusted to 7.0 with 0.5 mol/L NaOH or H_2SO_4; subsequently, different volume of acetone was added into the above slurry to final doses of 0%, 2%, 4% and 8% (v/v), respectively. The slurries with varied amount of acetone were incubated at a shaker (25 °C, 120 rpm) as aerobic biostimulation. During remediation of polluted soil by biostimulation, 2 mL of slurry was sampled periodically, and then extracted with enough fresh acetone to release all the organic matters into the aqueous solution, followed by volatilizing the acetone as organic extraction agent from the aqueous solution at a high temperature of 56.2 °C for 20 hours. The resulting solution was analyzed to determine the total organic carbon (TOC) to valuate removal efficiency of total contents of organic explosive pollutants in the soil under different acetone doses.

2.2.2 TOC removal efficiency of slurry under different initial pH

The slurry containing 50 g of soil, 100 mL of deionized water and 4% of acetone was prepared. The slurries were then adjusted with 0.5 mol/L NaOH or H_2SO_4 to pH 5.0, 7.0, 9.0 and 10.5, respectively. The slurries with varied initial pH were incubated at a shaker (25 °C, 120 rpm) to start the aerobic biostimulation remediation process and the TOC of slurries were monitored after extraction by enough acetone and following volatilization of acetone as described above.

2.2.3 TOC removal efficiency of slurry under different temperatures

The slurry containing 4% of acetone was prepared and adjusted to pH 9.0, followed by incubation at 120 rpm at different temperatures (20 °C, 25 °C, 30 °C and 35 °C). During incubation with varied temperatures, the TOC of slurries recorded to reflect the effect of temperatures on remediation efficiency of the explosive-polluted soil by aerobic biostimulation.

2.2.4 TOC removal efficiency of slurry under different pulp densities

The slurries with different pulp densities were prepared by mixing 100 mL of deionized water with 25, 50 and 100 g of polluted soil, respectively. The slurries were then supplemented with acetone to final concentration of 4% (v/v) and adjusted to pH 9.0. The slurries with different pulp densities were incubated at shaker (30 °C, 120 rpm) for aerobic degradation of explosives. The TOC of slurry were measured based on the above procedures to explore remediation efficiency of the polluted soil by aerobic biostimulation under different pulp densities.

All the experiments were carried out in triplicate. The sterile soil by autoclave (121.3 °C, 30 min) received the same procedures as the non-treated soil to serve as controls for assessing the possible abiotic remediation. The non-explosive soil also received the same as the explosive-polluted soil to serve as another controls for assessing the removal of non-explosive TOC in soil by biostimulation.

2.3. Mechanisms responsible for remediation improvement under optimum conditions

The remediation performance of slurry phase containing 50 g of polluted soil and 100 mL of deionized water under optimum conditions (4% of acetone, pH 9.0, 30 °C) by aerobic biostimulation was compared with that under average conditions (no addition of acetone, pH 7.0). During remediation, the pH value and ORP value of both slurry bioreactors were detected, the concentrations of NO_2^- and NO_3^- of the liquid phases of both slurry bioreactors were monitored, the TOC of both solid phase and liquid phase of both slurry bioreactors were measured, the total number of bacteria of both slurry bioreactors was counted.

2.4. Toxicity testing of remediated slurry under optimum and average conditions

2 ml of slurry was sampled periodically from both bioreactors, and then mixed with 4 ml of acetone for at least 4 hours at 120 rpm a shaker for extraction, followed by centrifugation at 10000 rpm for 10 min to remove the solid soil. Subsequently, the supernatant was heated at 56.2 $^{\circ}$C for at least 20 hours for completely volatilize acetone, the resulting solution was analyzed for toxicity test using the Freshwater Luminescent Bacterium Vibrio-qinghaiensis sp. based on the method described by [22]. The dilution factor for 50% of luminescent inhibition as EC50 value was obtained for reflecting the toxicity of liquid sample.

2.5. Analysis methods

The TOC of the soil was measured using the methods described by [23], the TN of the soil was determined according to [24], the pH of the soil was monitored based on description of [25]. For reflect the pollution characteristic of soil by TNT manufacturing wastewater, the soil was extracted with acetone solution (acetone/water, 1/1 in volume) at 10 g soil/50 mL, and the acetone-extracted fraction was analyzed by TOC meter (1020A, OI, USA) after the acetone was volatilized from the liquid solution to obtain the TOC originating from the explosive wastewater; the acetone-extracted fraction was analyzed by TON meter (IL500, HACH, USA) to assess the soluble TON from the explosive nitro-compounds; the dose of TNT of the acetone-extracted fraction was also measured using HPLC methods described by [26] to examine the concentration of the polluted soil.

The pH value of slurry phase was measured directly using a precise pH meter, the ORP value was determined by portable ORP meter. The concentrations of NO_2^- and NO_3^- of liquid phases of slurry bioreactor were analyzed using an ion chromatograph (ICS-1500, Dionex, USA) after removal of the solid matters from slurry samples by centrifugation. The TOC value of slurry phase known as the total TOC was analyzed with TOC meter (1020A, OI, USA) after all the organic matters of the slurry samples were extracted into liquid solution by enough acetone and subsequently the acetone was volatilized from the liquid solution. The TOC of liquid phase of slurry bioreactor was achieved with the same procedure as the total TOC except for the extraction step by acetone. The TOC of solid phase of slurry bioreactor was obtained by deducting TOC of liquid phase from the total TOC. The number of bacteria was obtained by colony count method grown in LB media.

3. Results and discussion

3.1. Properties of the explosive-polluted soil by TNT manufacturing wastewater

The TOC and TON of the non-polluted soil were relatively low, suggesting that the soil from northwest China was infertile (Table 1). In contrast, the TOC and TON of the polluted soil were much higher (Table 1), demonstrating that the soil was seriously polluted by nitro-organic compounds. Because the nitro-explosives including TNT were easily extracted from the soil using acetone, the acetone-extracted fraction could reflect the pollution by nitro-compounds from TNT manufacturing wastewater more accurately. The very high TOC and TON of the acetone-extracted fraction of the polluted soil further demonstrated that the soil suffered from the strong explosive pollution (Table 1).

Journal of Physics: Conference Series **439** (2013) 012047 doi:10.1088/1742-6596/439/1/012047

Although the polluted soil might contain dozens of nitro-compounds from the TNT manufacturing wastewater, TNT was not detected in the acetone-extracted fraction using HPLC method (Table 1), it was because that TNT in the TNT manufacturing wastewater is generally low and TNT transforms into other compounds in the soil. It was concluded that the polluted soil by TNT manufacturing wastewater was different from the TNT-polluted soil. As a result, TOC rather than residual TNT was used to assess the remediation efficiency of the polluted soil in the present studies. There were quite a few of studies about remediation of TNT-polluted soil based on monitoring residual TNT [2, 3, 6]; however, there was no report about remediation of polluted soil by TNT manufacturing wastewater based on TOC analysis.

3.2. Improved effect of addition of acetone on TOC removal from slurry

Bioremediation of explosive-polluted soil has been draw increasing attentions in recent dozens of years [2]. Bioremediation includes biostimulation and bioaugmentation, the former was accomplished by stimulating the growth and activity of the intrinsic microorganisms to enhance biodegradation of organic contaminants through addition of exogenous oxygen and inorganic nutrients; the latter was carried out by injection of the competent biodegrading microorganisms to further accelerate the biodegradation of pollutants, besides the exogenous oxygen and inorganic nutrients [6]. Because the high cost of growing certain biodegrading microorganisms and the weak competitiveness of exogenetic microorganisms in fighting the intrinsic ones, biostimulation possesses a greater application future in treating polluted soil compared with bioaugmentation [27, 28]. Therefore, in this present works the biostimulation was utilized to remediation the explosive-polluted soil in aerobic bioslurry due to the high efficiency of aerobic biodegradation and the greater mass transfer of bioslurry reactor. Several crucial parameters affecting biostimulation such as acetone addition, pH adjustion, temperature variation, pulp density change were optimized for remediation of the polluted soil based on TOC analysis.

Table 1. The properties of the polluted soil by TNT manufacturing wastewater and of the non-polluted soil as background.

Items	TOC of soil (mg/g)	TON of soil (mg/g)	pH of soil	TOC of extracted fraction (mg/g)	TON of extracted fraction (mg/g)	TNT of extracted fraction (mg/g)
Non-polluted soil	14.42	2.29	8.4	1.37	0.21	0
Explosive-polluted soil	31.74	8.33	8.5	14.92(13.55)*	5.87 (5.66)*	0

* The data in the brackets are the actual TOC and TON of extracted fraction from the explosive pollutants.

The TOC variation of soil in the slurry as a function of time under different doses of acetone was illustrated in Figure 1. Addition of acetone evidently improved the remediation of polluted soil by intrinsic microorganism. The maximum removal efficiencies of TOC increased from 25% to 38.4% when dose of acetone increased from 0% to 4% (v/v); however, higher 8% of acetone did not achieve further increase in removal efficiency. In the bioremediation of organic compounds-polluted soil, surfactants are usually utilized to release pollutants from soil for improving the bioavailability. In explosive-polluted soil, the explosives are subject to be strongly adsorbed by the soil particles, resulting in low bioavailability and poor bioremediation efficiency. In the current study, the cheap acetone as extraction agent was used to dissolve the explosives from soil particle for degradation by the intrinsic microorganism. The results showed that 4% of acetone is sufficient for improving bioavailability of the polluted soil, achieving an increase of 13.4% in removal efficiency of TOC from 25% to 38.4%. Moreover, it was speculated that 4% of acetone did not harm the activity of the intrinsic microorganism, which was proved in the following studies about bioremediation mechanisms.

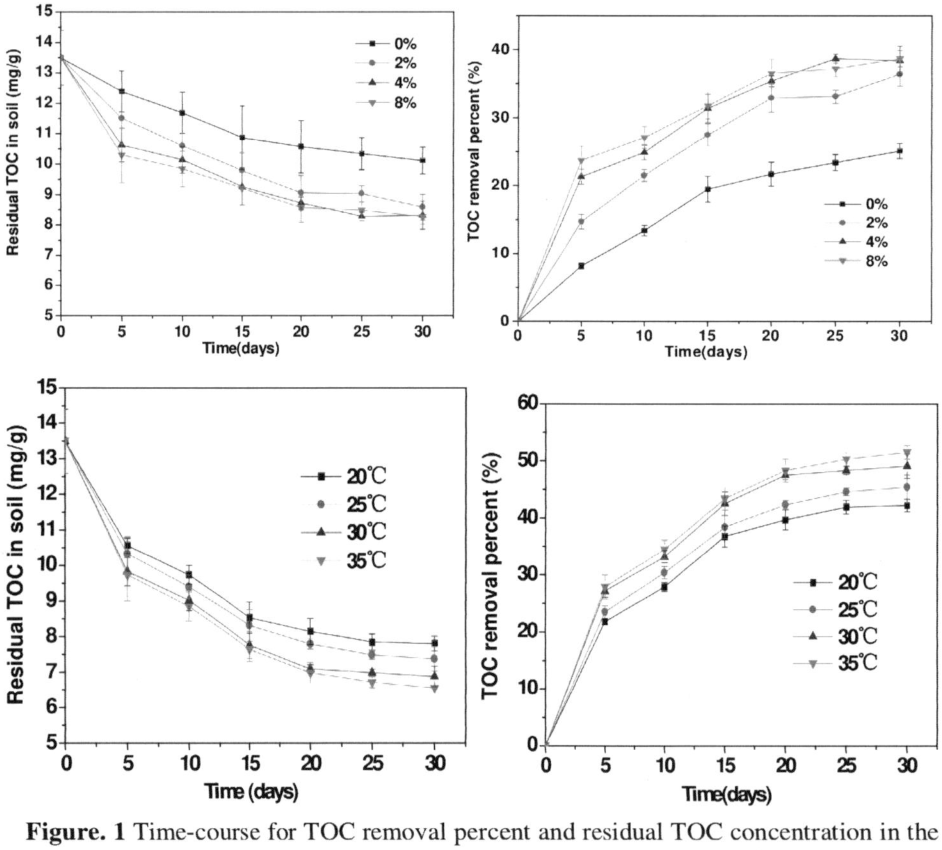

Figure. 1 Time-course for TOC removal percent and residual TOC concentration in the soil of bioslurry in the presence of different volume of acetone (v/v)

3.3. TOC removal efficiency of slurry under different initial pH

The initial pH value has a moderately influence on TOC removal, and the initial pH 9.0 achieved the maximum TOC removal of 45.4% after 30 days of incubation, witnessing an increase of 7.0% in removal efficiency of TOC from 38.4% at pH 7.0 (Figure 2). The results suggested that the weak alkaline of pH 9.0 was fit for the growth of intrinsic bacteria which played an important role in TOC removal; whereas lower pH of acidic condition such as 5.0 or higher pH of alkaline condition such as

10.5 adversely affected activity and growth of intrinsic bacteria, resulting in lower removal efficiency of TOC.

3.4. TOC removal efficiency of slurry under different temperatures

The temperature exhibits a relatively small effect on TOC removal. Although increase in temperature promoted TOC removal, the removal efficiency of TOC rose from 45.4% to 49.1 when temperature went up from 25 °C to 30 °C, only 3.7% increase. The highest temperature of 35 °C achieved the maximum TOC removal of 51.5% , but the 30 °C was recommended as the optimum temperature for bioremediation of the polluted soil due to the lower energy requirement for maintaining lower temperature.

3.5. TOC removal efficiency of slurry under different pulp densities

The pulp density has no effect on TOC removal. The four pulp densities ranging from 25% to 100% (solid-to-water, w/v) harvested almost the same removal efficiency of about 49% after 6 weeks incubation. So, high pulp density was recommended for bioremediation of the polluted soil because the higher pulp density meant lower water consumption, smaller bioreactor volume and cheaper remediation cost.

3.6. TOC removal dynamics under the optimum remediation conditions

The dynamics analysis of TOC removal is important for regulation and control of the bioremediation process. For this purpose, under the optimum conditions (4% of acetone addition, 50% of pulp density, pH 9.0 and 30 °C) linear relationship between TOC at certain time and incubation time (T) was established as zero order reaction, the linear relationship between In (TOC) and incubation time (T) was set up as first order reaction, the linear relationship between TOC^{-1} and incubation time (T) was plotted as second order reaction [29]. It was found that the second order reaction fits well the experimental data owing to the greatest R^2, although the other models can also describe the dynamics.

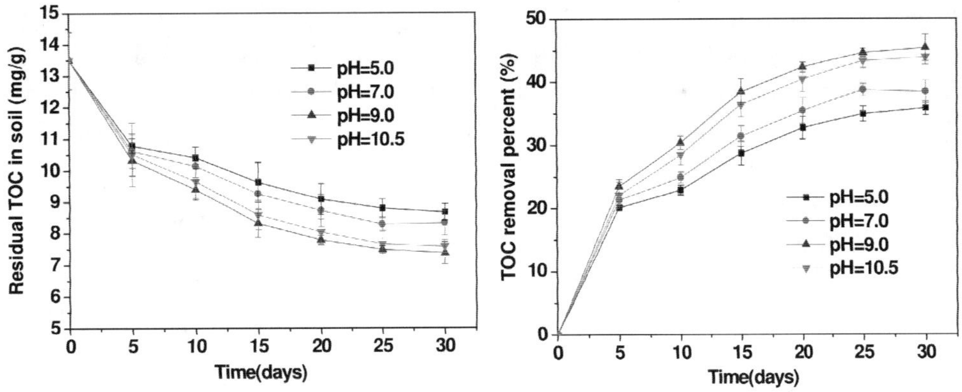

Figure 2. Time-course for TOC removal percent and residual TOC concentration in the soil of bioslurry under different initial pH.

3.7. Mechanisms exploration of remediation improvement under optimum conditions

For exploring the remediation mechanisms, the variation of both pH and ORP under optimum conditions was compared with that under average conditions. It was found that final pH value of both bioreactors approached the same due to the great buffer capacity of the soil (Figure 3). Different from pH, the ORP of both bioreactors kept fluctuation during the bioremediation process (Figure 3), indicating that the oxidation-reduction reaction occurred although the intermediates and details were unknown. However, greater variation range of the ORP with the optimum conditions suggested stronger oxidation-reduction reaction, and the lower ORP values might imply more reducing ambient originated from addition of acetone.

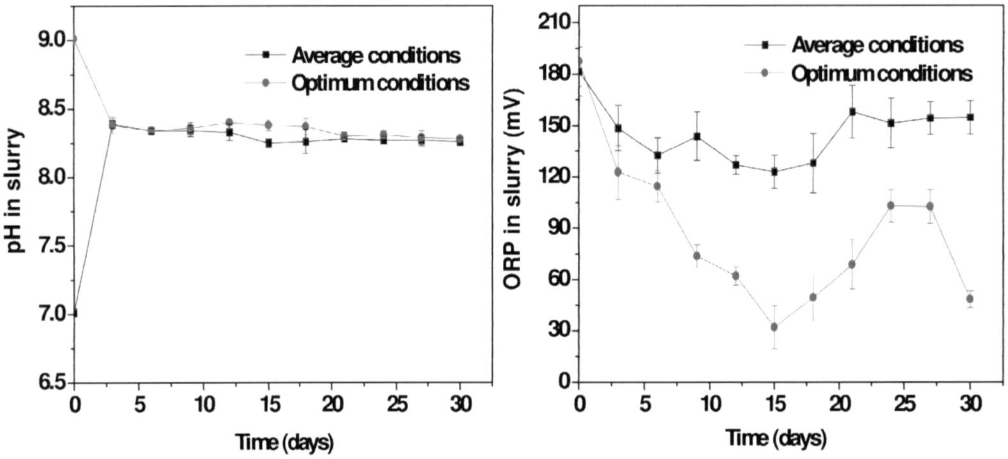

Figure 3. Comparison of both pH and ORP variation as functions of remediated time in slurry between the optimum and average conditions.

In aerobic bioremediation, the nitro group maybe release from the nitro-explosives in the form of NO_2^- which further transforms into NO_3^-. So, the variation of NO_2^- and NO_3^- of the slurry reactor under optimum conditions was compared with that under average conditions for investigating the bioremediation mechanisms (Figure 4). During bioremediation, removal of nitro group from the explosives caused increase of dose of NO_2^- in liquid phase; whereas oxidation of NO_2^- into NO_3^- under aerobic conditions resulted in decrease of concentration of NO_2^-, so there was a dynamic balance with NO_2^-. Like NO_2^-, a dynamic balance also occurred with NO_3^- through generation of NO_3^- from oxidation of NO_2^- and consumption of NO_3^- for growth of microorganisms. Under the average conditions, both NO_2^- and NO_3^- set at dynamic balance; therefore, the concentrations of both NO_2^- and NO_3^- kept almost unchanged over the period of bioremediation (Figure 4). However, the case was completely different with the optimum conditions. On one hand, the stronger growth of the intrinsic bacteria consumed much more NO_3^-, leading to continuous decline of dose of NO_3^- (Figure 4). On the other hand, the higher activity of the intrinsic bacteria achieved faster removal of nitro group from the explosives, resulting to accumulation of NO_2^- in liquid solution (Figure 4).

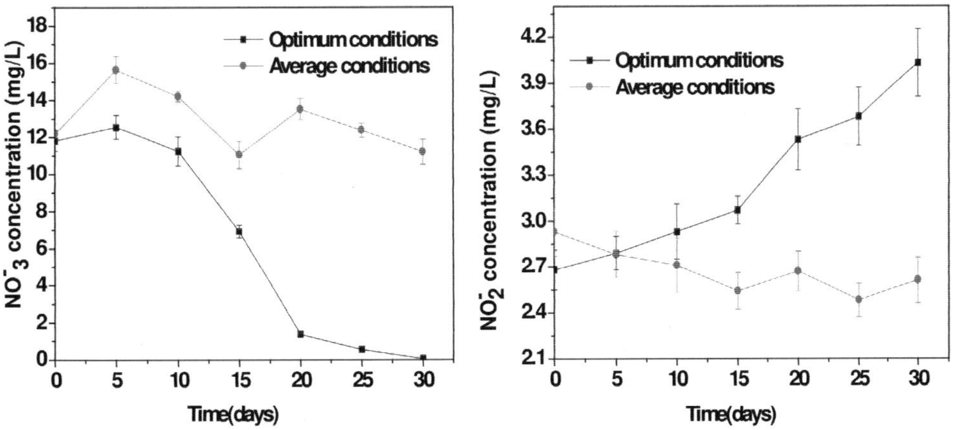

Figure 4. Comparison of both NO_2^- and NO_3^- concentration variation as functions of remediated time in slurry between the optimum and average conditions.

Under the average conditions, the liquid phase TOC decreased from 3404 to 3144 mg/L (i.e. 260 mg/L in removal amount) and the solid phase TOC dropped from 1022 to 104 mg/L (i.e. 918 mg/L in removal amount) after 30 days treatment, respectively, working together to achieve a 26.6% of total TOC removal of slurry from 4426 to 3248 mg/L (Figure 5). The results demonstrated that the solid bioremediation played a more important role than the liquid remediation under the average conditions, although the liquid phase covered 77% of the total TOC. In contrast with the average conditions, the optimum condition witnessed 48.9% of TOC removal from 4500 to 2300 mg/L in the liquid phase because of release of organic matters from the solid phase by acetone (Figure 5), leaving the solid phase free of bioremediation.

3.8. Toxicity testing of remediated slurry under optimum and average conditions

Both of the remediated slurries under the optimum conditions or average conditions witnessed a decrease of toxicity accompanied by bioremediation (Figure 6) indicating that the low-cost and simple aerobic biostimulation was substantially efficient for remediation of the explosive-polluted soil. With the average conditions, the dilution factor for 50% of luminescence inhibition decreased from 3.62 to 2.12, toxicity fell by 35% (Figure 6). In contrast, with the optimum conditions the dilution factor dropped from 3.43 to 1.32, toxicity decline by 62% (Figure 6). Although the optimum conditions achieved a lower toxicity than the average conditions due to the higher TOC removal, the further studies was needed for complete removal of toxicity of the polluted soil by supplement of organic carbon source for co-metabolism.

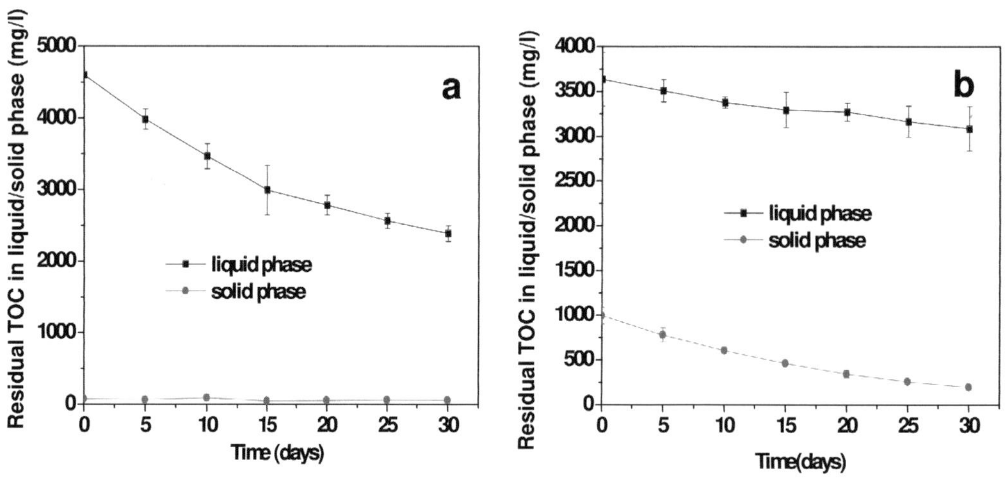

Figure 5. Variation of TOC in both liquid and solid phases as function of remediated time in optimum conditions (a) and average conditions (b)

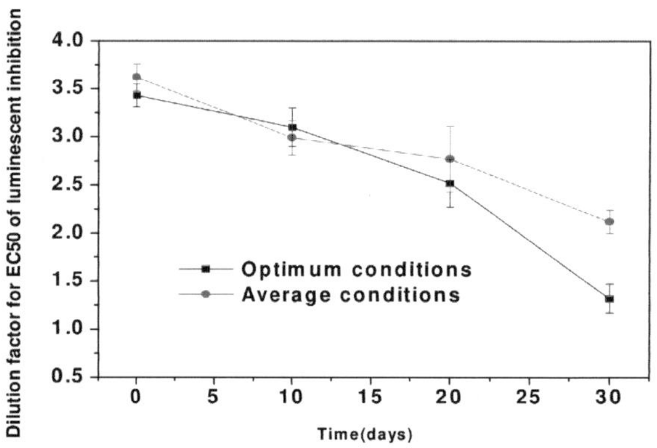

Figure 6. Comparison of toxicity change of the explosive-polluted soil with remediation progress under the optimum and average conditions

4. Conclusion

The maximum removal efficiencies of TOC increased from 25% to 38.4% when dose of acetone increased from 0% to 4%. The initial pH 9.0 achieved the maximum TOC removal of 45.4% from 38.4% at pH 7.0. The removal efficiency of TOC further rose from 45.4% to 49.1 when temperature went up from 25°C to 30 °C. The second order reaction fits well TOC removal dynamics under the optimum conditions. Solid bioremediation played a more important role than the liquid remediation under the average conditions; whereas the optimum condition witnessed complete TOC removal in the liquid phase, causing toxicity decline by 62%.

Acknowledgements

We really appreciate the Shandong Fund of Sciences and Technology for Environment Protection and Basic Research Fund of BIT for providing us the financial support. We also appreciate professor Ma

Mei (Research Center for Environmental Sciences and Ecology, CAS) for the kind help in toxicity test.

References

[1] Jerger D E, Woodhull P 2009 *Applications and costs for biological treatment of explosives-contaminated soils in the U.S. I*n: Spain J C, Hughes J B, Knackmuss H J (eds) Biodegradation of nitroaromatic compounds and explosives(Lewis, New York) pp 395–423

[2] Lewis T A, Newcombe D A, Crawford R L, 2004 Bioremediation of soils contaminated with explosives *J. Environ. Manage.* **70** 291-307

[3] Ayoub K, van Hullebusch E D, Cassir M, Bermond A 2010. Application of advanced oxidation processes for TNT removal: A review. *J. Hazard. Mater.* **178** 10–28.

[4] Esteve Núñez A, Caballero A, Ramos J L 2001 Biological degradation of 2,4,6-trinitrotoluene. *Microbiol. Mol. Biol.* Rev. **65** 335-352

[5] N Samotokin B, Marmiroli M (eds) Advanced science and technology for biological trinitrotoluene. Microbiol *Mol. Biol. Rev* 65 335–352

[6] Gerth A, Hebner A 2007 *Risk assessment and remediation of military and ammunition sites. In: Marmiroli decontamination of sites affected by chemical and radiological nuclear agents NATO science series* vol 75 Springer, Heidelberg pp 45–57

[7] Stenuit B A, Agathos S N, 2010 Microbial 2,4,6-trinitrotoluene degradation: could we learn from (bio)chemistry for bioremediation and vice versa? *Appl Microbiol Biotechnol* **88** 1043–1064

[8] Dillewijn P V, Caballero A, Paz J A, Gonzaalez M M., Oliva J M, Ramos J M, 2007 Bioremediation of 2,4,6-trinitrotoluene under field conditions *Environ. Sci. Technol.* **41** 1378-1383

[8] In B H, Park J S, Namkoong W, Hwang E Y, Kim J D 2008 Effect of co-substrate on anaerobic slurry phase bioremediation of TNT-contaminated soil Korean J. Chem. Eng. **25** 102-107.

[9] Dalgren K E, Waara S, Düker A, von Kronhelm T, van Hees P A W 2009 Anaerobic bioremediation of a soil with mixed contaminants: explosives degradation and influence on heavy metal distribution, monitored as changes in concentration and toxicity. *Water Air Soil Pollut.* **202** 301–313

[10] Newcombe D A, Crawford R L 2007 Transformation and fate of 2,4,6-trinitrotoluene (TNT) in anaerobic bioslurry reactors under various aeration schemes: implications for the decontamination of soils. *Biodegradation* **18** 741–754

[11] Sheibani G, Naeimpoor F, Hejazi P 2011 Statistical factor-screening and optimization in slurry phase bioremediation of 2,4,6-trinitrotoluene contaminated soil *J. Hazard. Mater.* **188** 1–9

[12] Muter O, Potapova K, Limane B, Sproge K., Jakobsone I, Cepurnieks G, Bartkevics V 2012 The role of nutrients in the biodegradation of 2,4,6-trinitrotoluene in liquid and soil. *J. Environ. Manage.* **98** 51 -55

[13] Barreto-Rodrigues M, Silva F T, Paiva T C B 2009 Combined zero-valent iron and fenton processes for the treatment of Brazilian TNT industry wastewater *J. Hazard. Mater.* **165** 1224–1228

[14] Maloneya S W, Adriana N R, Hickeyb R F, Heineb R L 2002 Anaerobic treatment of pink water in a fluidized bed reactor containing GAC *J. Hazard. Mater.* **92** 77–88

[15] Zhao Q L, Ye Z F, Zhang M H 2010 Treatment of 2,4,6-trinitrotoluene (TNT) red water by vacuum distillation *Chemosphere* **80** 947–950

[16] Ciecone V J, 1982 ADA115218

[17] Patterson J 1976 PB265042

[18] Spanggord R J 1978 sADA059434

[19] Park C, Kim T H, Kim S, Lee J, Kim S W 2003 Bioremediation of 2,4,6-trinitrotoluene contaminated soil in slurry and column reactor *J. Biosci. Bioeng.* **96** 429–433

[20] Popesku J T, Singh A, El-Alawi Y, Ward O P 2006 Trinitrotoluene removal in a soil slurry and soil box systems by an oil-degrading mixed bacterial culture. World *J. Microbiol. Biotechnol.* **22** 1075–1081

[21] Robles-González I, Fava F, Poggi-Varaldo H M 2008 A review on slurry bioreactors for bioremediation of soils and sediments *Microb. Cell Fact.* **7** 5–21

[22] Ma M, Tong Z, Wang Z, Zhu W 1999 Acute Toxicity Bioassay Using the Freshwater Luminescent Bacterium Vibrio-qinghaiensis sp Nov. Q67. Bull. *Environ. Contam. Toxicol,* **62** 247-253

[23] Karvelas M, Katsoyiannis A, Samara C 2003 Occurrence and fate of heavy metals in the wastewater treatment process *Chemosphere* **53** 1201–1210

[24] Bremner J M., Mulvaney C S 1982 Nitrogen—total. In A. L. Page, R. H. Miller R H, & D. R. Keeney (Eds.), *Methods of soil analysis.* Part 2 (2nd ed.). Agron Monogr Vol. 9 Madison (WI: ASA and SSSA) ppS 595–624

[25] Wang et al 2010 A rapid selection strategy for an anodophilic consortium for microbial fuel cells *Journal of Bioresource technology*

[26] Guo X J, Xin B P, Ma X T, Xia Y T, Chen S, Yang Y P 2009 Biodegradation of Trinitrotoluene by a Bacterial Consortium Containing Klebsiella sp. and Burkholderia sp.. Chinese Journal of Catalysis, 30, 1261-1268

[27] El Fantroussi S, Agathos S N 2005 Is bioaugmentation a feasible strategy for pollutant removal and site remediation? Curr. Opin. *Microbiol* **8** 268–275

[28] Van Dillewijn P, Wittich R M, Caballero A, Ramos J L 2008 Type II hydride transferases from different microorganisms yield nitrite and diarylamines from polynitroaromatic compounds. *Appl. Environ. Microbiol* **74** 6820–6823

[29] Alberty R A, Silbey R J 2001 Physical chemistry (third edition). John Wiley & Sons, Inc. pp 656–703

Journal of Physics: Conference Series **439** (2013) 012048 doi:10.1088/1742-6596/439/1/012048

Dynamic Corneal Surface Mapping with Electronic Speckle Pattern Interferometry

S Iqbal*[1] and M M S Gualini[2]

[1] Institute of Industrial Control Systems, Rawalpindi, Pakistan
[2] Polytechnico di Torino, Turin, Italy

E-mail: drsalmaniqbal@yahoo.com

Abstract. In view of the fast advancement in ophthalmic technology and corneal surgery, there is a strong need for the comprehensive mapping and characterization techniques for corneal surface. Optical methods with precision non-contact approaches have been found to be very useful for such bio measurements. Along with the normal mapping approaches, elasticity of corneal surface has an important role in its characterization and needs to be appropriately measured or estimated for broader diagnostics and better prospective surgical results, as it has important role in the post-op corneal surface reconstruction process. Use of normal corneal topographic devices is insufficient for any intricate analysis since these devices operate at relatively moderate resolution. In the given experiment, Pulsed Electronic Speckle Pattern Interferometry has been utilized along with an excitation mechanism to measure the dynamic response of the sample cornea. A Pulsed ESPI device has been chosen for the study because of its micron-level resolution and other advantages in real-time deformation analysis. A bovine cornea has been used as a sample in the subject experiment. The dynamic response has been taken on a chart recorder and it is observed that it does show a marked deformation at a specific excitation frequency, which may be taken as a characteristic elasticity parameter for the surface of that corneal sample. It was seen that outside resonance conditions the bovine cornea was not that much deformed. Through this study, the resonance frequency and the corresponding corneal deformations are mapped and plotted in real time. In these experiments, data was acquired and processed by FRAMES plus computer analysis system. With some analysis of the results, this technique can help us to refine a more detailed corneal surface mathematical model and some preliminary work was done on this. Such modelling enhancements may be useful for finer ablative surgery planning. After further experimentation, this technique can possibly be developed for in-vivo experiments on animals and humans and then may prospectively be matured for future clinical usage.

1. Introduction

Corneal surgery has seen tremendous progress in the recent years, with millions of patients benefiting from it worldwide. On its forefront are the procedures involving laser ablation for corneal correction including PRK and LASIK. Although the results statistics from these procedures are impressive, yet in many of the cases the ultimate vision correction achieved misses the targeted correction level. In this regards, it is obvious that the planned laser application parameters inevitably depend on the parameters measured and the measured values obtained. Looking at the eye measurement technology in terms of corneal mapping and topography, most early generation machines are excessively rigid, not allowing adaptation to the intricacies of the corneal changes [1]. Use of topographic devices like videokeratometers has also proven insufficient since these devices operate at resolutions in tens of microns, while ablation depth may be less than few microns as total and for single-shot removal may even go to sub-micron range. White light interferometers have better resolution of around 10 micron but it may still not be sufficient.

The corneal surface reconstruction process after ablation well depends on multiple parameters including the corneal tissue elasticity, as confirmed by the use of techniques to modify the intra-ocular pressure or IOP during the post-op period. For the target of obtaining optimal results from laser ophthalmic surgery, it has been observed that more attention is required on the adjustment of the necessary ablative laser power according to eye parameters. In view of having a broader estimation of corneal characteristics and looking at the earlier development of topographers and videokeratormaters, it is found that the corneal elasticity is generally a neglected parameter in these studies.

1.1. Pulsed ESPI
Thus an improvement in PRK procedure may be brought by improving the eye metrology involved and combining wavefront analysis with an effective method for tissue elasticity measurement [1-4]. The use of a Twymann-Green Interferometer (TGI) for mapping of corneal surfaces under normal conditions is proposed, which may also be extended to its use on elastically-strained cornea under static stress conditions. Such system can go to better resolution and should prove to be an attractive alternative [1, 5]. For mapping of the behavior of cornea under dynamic stresses, an innovative laser device or Pulsed ESPI camera may be used. Pulsed ESPI is very good candidate for this purpose as its resolution goes to sub-micron ranges. Through the study given here, the measurement of corneal resonance frequency and the corresponding deformations are determined in real time, probably for the first time. In order to further establish our results, cornea-like structure was simulated for resonance conditions and comparable results were obtained with its oscillation analysis.

1.2. Some Advantages
This Pulsed ESPI approach has been used in this study because of the many advantages offered by this technique in real-time deformation analysis. Pulsed ESPI has seen good popularity as a reliable and accurate investigation method compared to conventional Holography and other schemes [3, 6]. Here, instead of using photographic films, Pulsed ESPI can directly record images in the PC through a CCD camera. Pulsed ESPI images can be obtained with very short light pulses at high repetition rates, which make the method insensitive to mechanical vibrations. Pulse widths and repetition rates can be adjusted conveniently to cover quite a wide range of applications. Due to the short recording time (few nanoseconds) Pulsed ESPI is insensitive to disruptive factors like low frequency vibrations or shocks transmitted to the apparatus.

2. Materials and Methods
A Pulsed ESPI camera was used in this work for the measurement of corneal deformation under dynamic stresses in real time. A normal bovine cornea was used for the study, which is probably not the best choice to emulate a human cornea, since other smaller animals may have eyes which are more comparable. But along with the reasons of availability, it does offer quite a wide surface for the analysis, so that the fringes can be conveniently displayed and the tip of a shaker could easily be applied without significantly disturbing the field of view. This research should pave way to prospective in-vivo experiments on human cornea after further experimentation. The experiments were conducted at Steinbichler Optotechnik GmbH in Neubeuren, Germany.

2.1. The Equipment
In these experiments, the equipment used had a high resolution ranging between $\lambda/30$ and $\lambda/10$ at 20 pixels fringe width. The camera resolution was 1280x1024 pixels. The ruby laser with a wavelength of 632.8 nm and pulse separation ranging from 2 to 800 μ sec was used. Polytec laser vibrometer was used to determine the resonance frequency. Data was acquired and processed by the powerful software FRAMES plus release 5.0, developed by Steinbichler Optotechnik GmBH.

For alignment, a He-Ne laser was applied to the optical system in eye-safe conditions for the operator, before the actual use of the high power ruby laser. Part of the beam was injected into a single-mode, polarization-maintaining optical fiber, which carried the reference beam signal directly onto the CCD camera surface.

For inducing strain to the bovine cornea, a mechanical vibrator in the shape of a rod connected with a motor was used. When the motor was electrically driven, the tip of rod would poke the sample with its tip moving back and forth using a displacement of a fraction of millimeter. The oscillating frequency of the tip was variable over wide range and Polytec laser vibrometer was used to measure its frequency. The bovine cornea was supported by aluminum film and polystyrene.

In our experiment, the ruby laser light was widely diffused to cover significant part of cornea. At first, the back-scattered signal was found to be very poor due to the high transparency of the bovine cornea. This problem was overcome by spraying the corneal surface with fine white powder film in order to ensure high reflectivity and uniformity of the back-scattered signal. This approach may appear a limitation for an in-vivo application on human cornea, but in that case a comparatively better reflectivity to the ruby laser is expected, as already reported by other authors [3, 6].

In the beginning of experiment, some static measurements were performed using a TGI setup in order to map the static condition of bovine cornea. First the cornea was mapped without deformation and then static and dynamic stresses were induced to the bovine corneal surface with mechanical vibrator. Figure 1 & 2 show the setup for Pulsed ESPI measurements showing sample, camera and vibration mechanism.

2.2. *Results Verification*
In order to establish our results further, simulation of cornea-like structure was also carried out and the results were compared with the experimental results. ANSys mechanical analysis software (version 13) was used for this purpose. Physical corneal structure was simulated in the shape of a hollow semi-sphere with diameter of 31mm and thickness of 1mm. A more detailed model could be built but this was sufficient for our target. The material used was soft rubber-like material with normal polymer properties and Young's Modulus was made selectable in few MPa range.

Simulation was done at ambient environmental conditions. First its modes were found out and then full harmonic analysis with frequency ranging up to 500 Hz was carried out at 100 different steps. A small sinusoidal vertical force was used as the input for the simulation. For the analysis at varying material conditions, Young's Modulus in its material properties was given different values and corresponding change in its resonance conditions was observed. Figure 3 shows the simulated semi-sphere.

Figure 1. Sample and camera

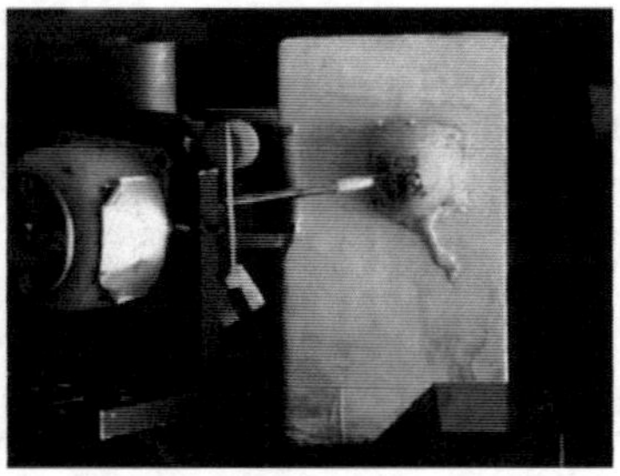

Figure 2. Vibration setup

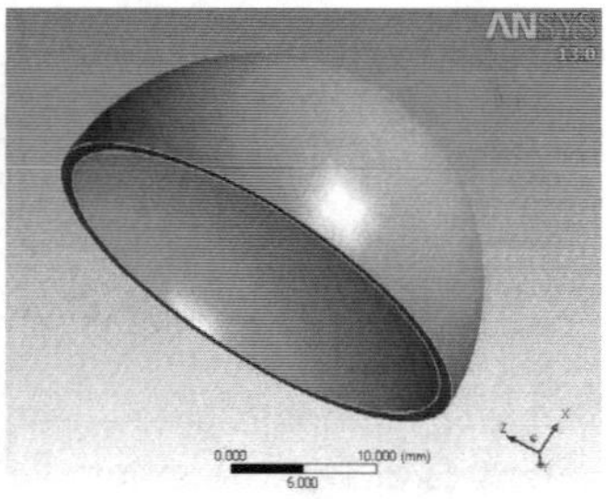

Figure 3. Simulation model

3. Results And Discussion
Firstly, the direct but static mapping of cornea with the help of modified Twymann-Green Interferometer was obtained. Though it gave impressive output, the target here was to obtain the mapping of the cornea under dynamic stress applications which is necessary to obtain the elastic behavior of the sample. Pulsed ESPI was able to provide this important information to us.

Static measurements of the sample cornea were taken using a TGI setup in order to map the surface in normal conditions. After mapping of cornea without deformation, a specific static stress was induced to the bovine corneal surface. The resulting corneal deformation mapped by the TGI device and obtained after the fringe analysis from the software is shown in figure 4.

3.1. *Experimental Results*
Here, figure 5 shows the amplitude response versus the frequency spectrum of the bovine cornea surface shacked with the tip synchronized to the laser vibrometer. The graph of the amplitude versus frequency spectrum was directly obtained using the vibrometer. As it is visible on the spectrum,

resonance frequency is clearly visible at 175 Hz. Thus, the tip vibration frequency was fixed at resonance conditions and the surface deformation at this resonance frequency was detected using Pulsed ESPI setup. As can be seen in figure 6, the cornea looks evidently deformed. Contrast to this, there is no apparent measurable deformation at frequencies offset from the resonance, as visible in figure 7. The bumpy surface seen in the maps is may be due to the unfiltered surface calculations as generated by the software, which are shown in the original form. It partly may also be coming from the powder coating given to cornea for better reflection.

It was found out that the cornea surface as used in this experiment exhibits a resonance frequency which can very well give information about its elastic parameters. This information of having a specific resonance frequency can be characteristic to a specific cornea or the eye. Also that Pulsed ESPI may be a promising method for the conduction of such elasticity-related investigations on ophthalmic or biological sample [7, 8].

3.2. Simulation Results

The results of simulating this experiment with a cornea-like structure in mechanical simulation software were also found to be supportive to the experiment. A hollow shell with different elasticity parameters of Young's Modulus was simulated for detailed vibration analysis built-in to ANSys software. It was found that the sample does exhibit a response with specific vibration nodes. Notably, a marked peak was found in the amplitude-frequency graph.

It was shown that as the Young's Modulus of the simulated shell was increased, the resonance condition given by the peak shifted towards higher frequency, but the general shape of the amplitude-frequency graph was not largely changed. Thus, it was found that the resonance frequency does shift with the changes given to elasticity of the cornea-like shell, which is the characteristic measurement and result of our study too. The results are shown in figures 8, 9 and 10. The real-time deformation of cornea as estimated under the resonance conditions is shown in figure 11. Deformation shape may well look different from the experimental one as the stressing here was from the top of the structure while it was from one side in the experimental setup.

3.3. Future Possibilities

Looking at the earlier researches for the development of topographers and videokeratormaters, it is found that the corneal elasticity is generally neglected parameter in these studies. In fact, wavefront analysis in the normal devices does not give any information about the critical parameter of corneal elasticity. Using Pulsed ESPI to extend the studies in this direction opens up another avenue which will pave way for other stress-related studies on biological samples with this technique.

Preliminary studies on animals can be useful to determine all the process parameters and functions needed for elasticity determination. As an implementation, treating patients with Excimer laser and then monitoring the corneal evolution during the post-op time can be compared with the calculated figures of forecasted cornea surface. This procedure will enable us to refine a mathematical model so that it can be used in vivo conditions for humans [9].

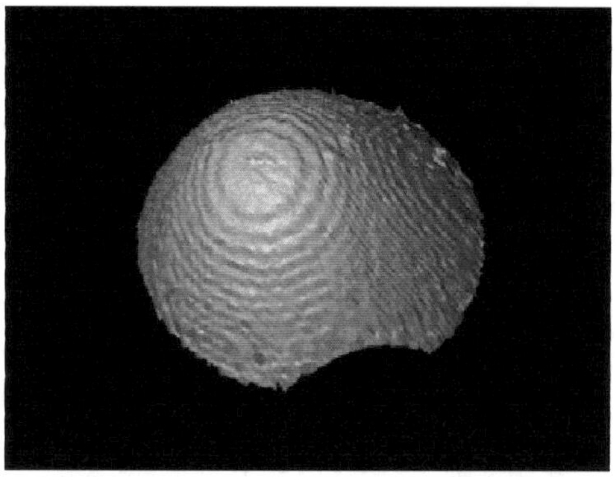

Figure 4. Static deformation from TGI

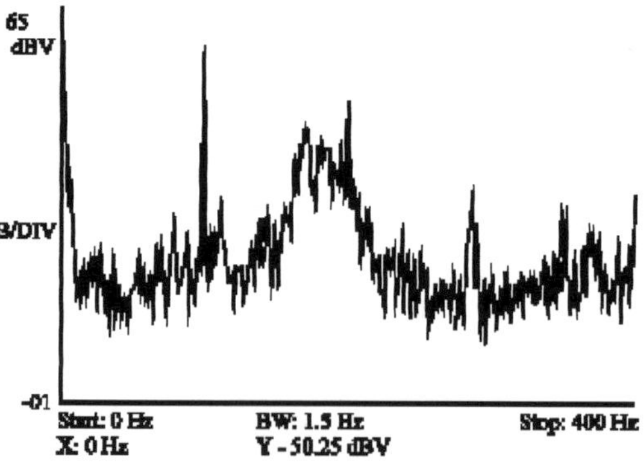

Figure 5. Vibration vs. frequency spectrum

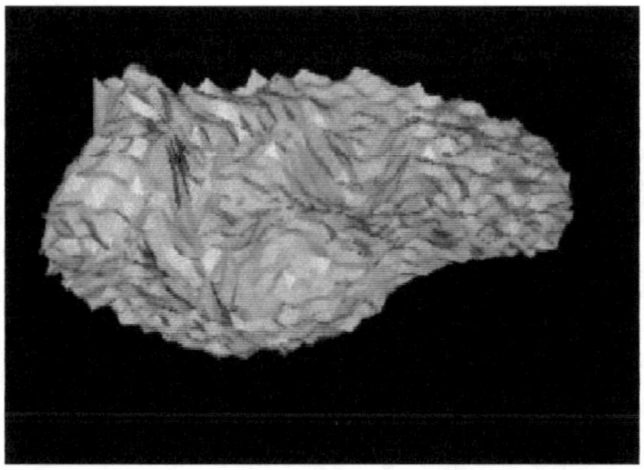

Figure 6. Deformation of cornea at resonance

Figure 7. No deformation outside resonance

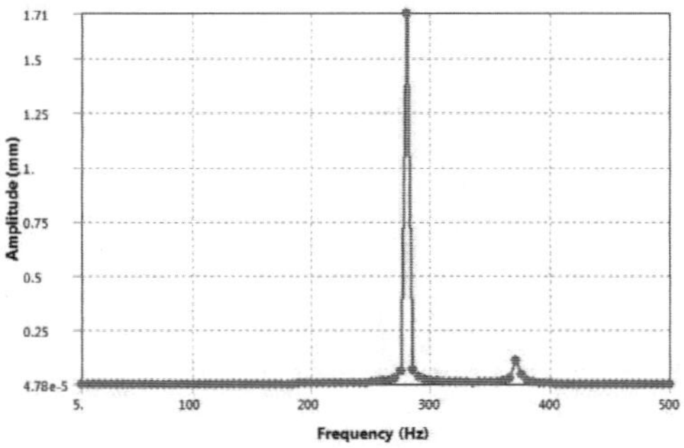

Figure 8. Resonance at Modulus of 1 MPa

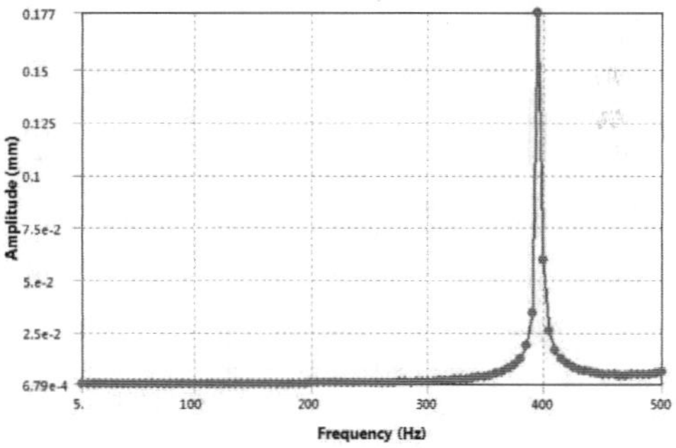

Figure 9. Resonance at Modulus of 2 MPa

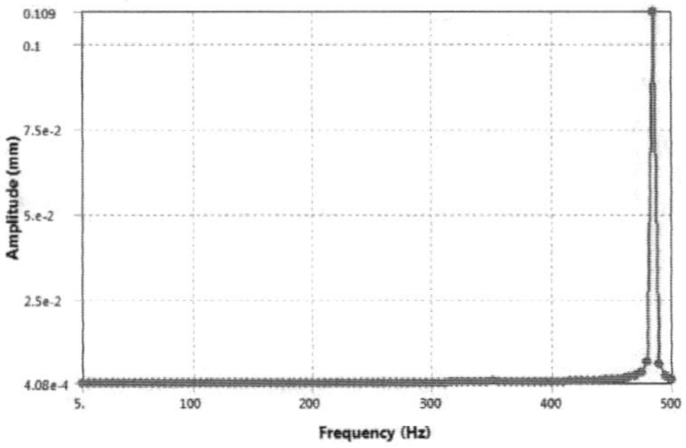

Figure 10. Resonance at Modulus of 3 MPa

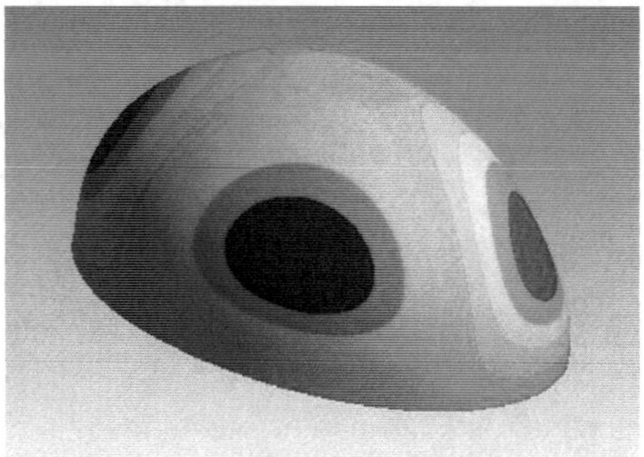

Figure 11. Deformation contour map

4. Conclusions

A useful Pulsed ESPI approach has been used in this study for dynamic corneal surface analysis because of its advantages in real-time deformation analysis including good resolution and insensitivity to mechanical vibrations. Here, a bovine cornea sample was mapped first at rest and then under static and dynamic stress conditions. To start with, mapping of cornea at rest and under static stress conditions was done by modified Twymann-Green interferometer. Then Pulsed ESPI was utilized to measure and map the deformations of the bovine cornea under a dynamic stress and to find the intrinsic resonance frequency of the bovine cornea where it does show marked deformation. It was then noticed that outside resonance condition the bovine cornea was not much deformed. The results of the experimental work were also checked by simulating the oscillatory behaviour of a cornea-like structure in mechanical simulation software; and a shift in resonance conditions with changes in elasticity was also confirmed there. Such study is apparently reported for the first time.

This technique can be used for the systematic studies and investigations on the elastic behavior of cornea. Information of the cornea elasticity can be very well utilized for fine-tuning the final ablative procedures in ophthalmic surgery. As already seen, normal optical methods may not be sufficient for establishing the elastic parameters and thus may not give optimal Targeted Correction. Thus stress analysis should help in these procedures and also in predicting the post-op time evolution of the corneal tissue. It is proposed that Pulsed ESPI can be used in combination with normal wave front analysis devices to achieve better results compared to their isolated use.

Using Pulsed ESPI to extend the studies in this direction opens up another avenue which will pave way for other stress-related studies on biological samples with this technique. The technique needs to be worked upon and to be refined with further experiments and then to be developed for in-vivo experiments on animals and humans.

Acknowledgments
The authors are particularly grateful to all the staff of Steinbichler Optotechnik GmbH for their support in handling the corneal sample and establishing the experimental setup. Help from staff of IICS, Rawalpindi, Pakistan is also acknowledged for the sample simulation work. Lastly, authors are grateful for the continued support from their families.

References
[1] Gualini M, Iqbal S, Khan W and Sixt W 2008 ESPI and PulsESPI Applied to Ophthalmology Using Modified Twymann-Green Interferometer *J. Appl. Sci.* **8** 677
[2] Drescher J et al 1999 Noncontact measurement of intraocular pressure using a modified Michelson interferometer *Proc. Conf. on Ophthalmic Technol. IX* (San Jose / SPIE) p 104
[3] Kasprzak H, Forster W and Von Bally G 1994 Holographic measurement of changes of the central corneal curvature due to intraocular pressure differences *Optical Eng.* **33** 198.
[4] Von Bally G 1982 *Optics in Biomedical Sciences* (New York: Springer-Verlag)
[5] Licznerski T, Kasprzak H Kowalik W 1999 Application of twyman-green interferometer for evaluation of in vivo breakup characteristic of the human tear film *J. Biomed. Optics* **4** 176
[6] Friedlander M et al 1991 Holographic interferometry of the corneal surface *Proc. SPIE Interferometry and Optical Pattern Recog. in Biomed.* (Los Angeles) p 229
[7] Reiss S 2003 Ophthalmology techniques get a boost *Biophotonics Int.* **10** 52
[8] Tyrer J 2001 Optical Interferometry analyzes the body *Laser Focus World* **37** 95
[9] Garcia D, Van de Pol C, Klein S and Barsky B 1999 Wavefront coherence area for predicting visual acuity of Post-PRK and Post-PARK refractive surgery patients *Proc. SPIE Conf. on Ophthalmic Technol. IX* (California) p 303

Design, modeling and simulation of MEMS-based silicon Microneedles

F Amin[1] and S Ahmed[1,2]
[1]Advanced Electronics Laboratory, Faculty of Engineering & Technology, International Islamic University, Islamabad - 44000, Pakistan.
[2] Centre for Emerging Sciences, Engineering & Technology, Islamabad-44000, Pakistan.

E-mail: faisal.msee226@iiu.edu.pk

Abstract: The advancement in semiconductor process engineering and nano-scale fabrication technology has made it convenient to transport specific biological fluid into or out of human skin with minimum discomfort. Fluid transdermal delivery systems such as Microneedle arrays are one such emerging and exciting Micro-Electro Mechanical System (MEMS) application which could lead to a total painless fluid delivery into skin with controllability and desirable yield. In this study, we aimed to revisit the problem with modeling, design and simulations carried out for MEMS based silicon hollow out of plane microneedle arrays for biomedical applications particularly for transdermal drug delivery. An approximate 200 μm length of microneedle with 40 μm diameter of lumen has been successfully shown formed by isotropic and anisotropic etching techniques using MEMS Pro design tool. These microneedles are arranged in size of 2 x 4 matrix array with center to center spacing of 750 μm. Furthermore, comparisons for fluid flow characteristics through these microneedle channels have been modeled with and without the contribution of the gravitational forces using mathematical models derived from Bernoulli Equation. Physical Process simulations have also been performed on TCAD SILVACO to optimize the design of these microneedles aligned with the standard Si-Fabrication lines.

1. Introduction

The area of miniaturization and micro fabrication in the past few years changes tremendously in numerous engineering and scientific fields due to extensive preponderance progress in micro electromechanical systems (MEMS) and Microsystems. The BioMEMS based devices increase its efficiency and robustness with the advancement of technology in both micromachining tools and design structure [1]. BioMEMS as a whole at micro scale integrates many biomedical systems and biological sciences disciplines and applications. Sampling and extraction of biomolecules and fluids, drug delivery, therapeutics and diagnostics etc are the major areas of biomedical industry where MEMS based microstructures have been successfully implemented and more research is still in progress to connect miniaturized products with the biological environment [2]. Biocompatibility and broad protocols of experiment procedures are primarily required for the manufacture of biomedical microstructures [3].

Patients have been delivering drugs through hypodermic needles for about 160 years [4]. For the transdermal drug delivery and transdermal blood extraction hollow needles have played a very vital role since they were manufactured in 18th century and later shortly first injection was then administered [4].In 1991, the foremost out of-plane microneedle array, consisting of 100 microneedles with a length of 1.5 mm was reported [5]. Microneedle arrays offer great application and advantage over conventional needles.Microneedles have been developed for different applications with different variety of materials, fabrication techniques, design, dimensions and tips shape etc.

For the very purpose of the extraction of biological fluid and drug delivery for skin permeability, the BioMEMS based microneedle array technology provides many advantages and applications over conventional needles, including (i) The length of microneedle can be controlled for the precise parenteral injection of therapeutic agents, by considering skin anatomy; (ii) It is minimally invasive, due to smaller diameters at the tip; (iii) Microneedles are arranged in arrays in order to transport large volume of fluid and penetrate invivo the skin over large distributed area rather than using single needle [6]; (iv) Provide painless drug delivery due to the fact that microneedles do not reach the nerves deep below the skin;(v) Single wafer is being used for the fabrication of arrays of microneedles due to its small size which tends to be a good design metrics including reliability, biocompatibility, fabrication cost, accuracy and etc [7];(iv) BioMEMS based microneedles array devices also find its application tremendously in biomedical sciences such as: Treatment of cancer;DNA transportation; Introducing biomolecules in living cell; and Diabetes.

In this work, an approach has been made to design and simulate fabrication steps for microneedle arrays for biomedical applications that would be hollow and out of plane in structure.

2. Types, Shapes and Materials of Microneedles

Based on the fabrication process techniques, microneedles are generally classified into two types that are: Out of plane microneedles and In-plane microneedles. In out of plane microneedles, the structure of the flow channel and shaft is normal and protrudes out of the surface of the wafer substrate [8]. In in-plane microneedles, the structure of the flow channel and shaft is parallel and in plane to surface of the wafer substrate [8].

With reference to the physical structure microneedles are also comprises of two types that are [9]: Hollow microneedles and Solid microneedles.Hollow microneedles are the microneedles that direct the flow of biological fluid and drugs through its internal lumen. Solid microneedles are the microneedles in which either the lumen surface is coated with particles of drugs and applied to the skin or since the base of solid microneedles is non dissolving, it is sequentially removed after the dissolution of microneedles [9].

The factors such as skin permeability and penetration along with the strong biocompatibility require an appropriate shape and selection of materials of the microneedles which is very critical during the process of fabrication and design. Any specific applications particularly in biomedical sciences, the selection of both shapes and materials is very important simultaneously. The shape of microneedles can be categorized into two types with respect to the Overall shape and Tip shape. Table 1 summarizes various designs of microneedle shapes and materials [10].

Table 1. Lists of Shapes and Materials of Microneedles [10].

Overall shape	Tip shape	Materials
Pyramidal	Tapered	SU-8
Spear	Beveled	Silicon
Candle	Snake fang	Metals
Cylindrical	Volcano	Polymers
Spike	Canonical	Silicon dioxide

3. Skin Anatomy and Microneedle Design Considerations

The human skin is the most multifunctional and largest organ of the body that interacts with the surrounding world [11]. The superficial epithelium layer epidermis is approximately 50-150 μm thick which largely consists of cells called keratinocytes [12] which are formed in the basal layer of epidermis [11]. The major barrier for drug delivery is the stratum corneum, about 10 to 20 μm thick, [13] is the outermost layer of the epidermis adherently consisting of stacked cornified cells of hexagonally flattened shape [11, 12, 13]. The deepest layers of the stratum corneum contain lamellar granules that act as major part of the permeability layer [11, 13]. The dermis is the second thick layer distributed with connective tissues [11]. The dermis mainly consists of elastin, network of blood vessels and capillaries and collagen matrix [11, 13]. The hypodermis also called subcutaneous fatis a fatty layer which is attached loosely to the superficial dermis [11, 13]. So, the structure of skin offers the great opening for the transdermal drug delivery as well as drug absorption [14]. The microneedle should therefore not reach the dermis but it may tend to penetrate the stratum corneum into epidermis of the skin. This suggests that for transdermal drug delivery, the length of microneedle fabricated must be in range of 150 – 250 μm size, as devised in other such experiments [14].

4. Modeling Experiments – Hollow Microneedle fluidic analysis

It is very significant to develop appropriate numerical model for microneedles with the characteristics of fluid flow and to account for "lumen dimensions" as the liquid drugs invivo into the skin through these arrays of microneedle. Therefore; mathematical models are developed and reported in literature [6] to evaluate and comprehend the working performance of microneedle arrays theoretically.

4.1 Flow Characteristics through Microneedle Arrays

Bernoulli equation (BE) [15], which is one of the most useful equation for fluid mechanics has been utilized in the form of Modified Bernoulli equation (equation 1) as reported in some studies [6], with the gravitational forces are ignored in calculations and analysis. We also considered the contributions from gravitational force and re-named the modified equation as Extended modified Bernoulli equation (equation 2) for ready reference, as shown below:

$$P_1 - P_2 + \rho g(Z_1 - Z_2) + \rho\left(\frac{V_1^2 - V_2^2}{2}\right) = \rho\frac{V^2}{2}\frac{fL}{D} + \Sigma K\frac{\rho V^2}{2} \tag{1}$$

$$P_1 - P_2 + \rho g(Z_1 - Z_2) + \rho\left(\frac{V_1^2 - V_2^2}{2}\right) = \rho\frac{V^2}{2\,g}\frac{fL}{D} + \Sigma K\frac{\rho V^2}{2\,g} \tag{2}$$

In equations (1) and (2), P_1 is inlet pressure, P_2 is outlet pressure, $V_1$is inlet velocity, V_2 is outlet velocity, Z_1 is inlet height, Z_2 is outlet height and ρ is fluid density [15]. The first term on the right side of the above equations describes the flow in the microchannel and second term on right side tells about the minor losses encounter during travelling via fluid particle [15].g = 9.81 m/s^2 is the gravitational force.

4.1.1 Analysis for pressure drop

Properties such as microneedle geometry, viscosity of fluid, microneedle density and flow rate are mainly acquired by the pressure drop in order to allow the flow of fluid through microneedles, as previously reporte [16]. The above mention equations (1) and (2) are exploited for analysis of pressure drop dynamics for an individual needle.

4.1.1.1 Pressure Drop Analysis from Modified Bernoulli Equation

An individual microneedle has been modeled using water as a "model liquid" as shown in figure 1[6] with Modified Bernoulli Equation, which is used to depict fluid flow through microneedle. In case of considered needle, the channel with a "constant" cross section area, equation (1) can be further reduced to:

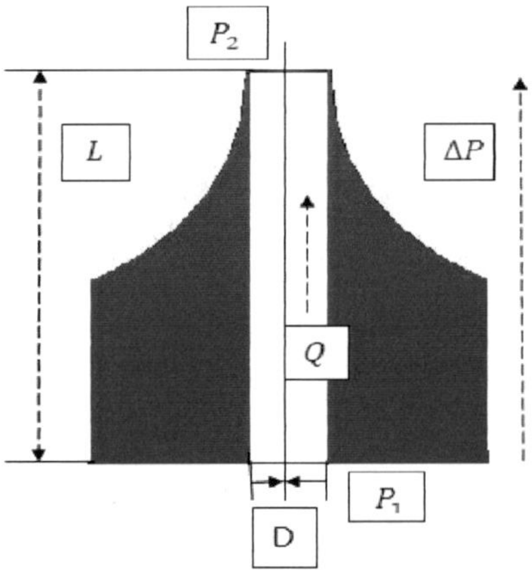

Figure 1. Characteristics of Liquid Flow Through Microneedle[6]

$$\Delta P = \mu \frac{128}{\pi}\frac{QL}{D^4} + \rho \frac{8\,(K_1 + K_2)}{\pi^2}\frac{Q^2}{D^4} \tag{3}$$

Equation (3) tells that inside a circular tube the viscous shear force of Poiseuille flow induced the pressure drop [17] and secondly corresponds to inertia effects along with the minor losses K_1& K_2 at the entrance and exit of the needle respectively [18]. Furthermore, the equation (3) illustrates that the contribution from the gravitational force 'g' are neglected. In the above calculations Gravitational forces were not considered because in previous studies it was assumed that at the time of flow, they might be negligible being relatively small [6]. This, however, is revisited in our studies conducted during the design and modeling of such microneedle devices, governed by equation 4 below:

$$\Delta P = \mu \frac{128}{\pi\, g}\frac{QL}{D^4} + \rho \frac{8\,(K_1 + K_2)}{\pi^2\, g}\frac{Q^2}{D^4} \tag{4}$$

This equation tells about the influence of gravitational forces 'g' along with the pressure drop inside a circular duct of Poiseuille flow due to the viscous shearforce which in turn also corresponds to inertia effects along with the minor losses K_1&K_2under the influence of gravitational forces 'g', which may be of greater importance when using the devices on altitudes and/or deviations from a standard "g" value systems.

4.2 Analysis for Microchannel Diameter
Diameter is one the very important parameter for the design of microneedles because it describes the "channel inner" from where the fluid flows and injected into the skin. This parameter is dealt with full appreciation in our design. Again, two approaches for microchannel diameter design have been employed by using the equations (3) and (4) which in turn obtained the following two equations that are,

$$D = \left(\mu \frac{128}{\pi \Delta P} QL + \rho \frac{8\,(K_1 + K_2)}{\pi^2 \Delta P} Q^2 \right)^{0.25} \quad (5)$$

$$D = \left(\mu \frac{128}{\pi\, g\, \Delta P} QL + \rho \frac{8\,(K_1 + K_2)}{\pi^2\, g\, \Delta P} Q^2 \right)^{0.25} \quad (6)$$

Equation (5) illustrates microchannel diameter under the pressure drop dynamics where the gravitational forces "g" are ignored, whereas equation (6) illustrates micochannel diameter under pressure drop dynamics with the gravitational force contribution considered.

The summary of the underlying assumption for our design are as follows:

For equations (3), (4), (5) and (6), $f = \frac{64}{Re}$ is the friction factor which has been used for laminar flow. Standard macroscopic values for an edged inlet (K_1= 0.5) and for an exit (K_2= 1.0) were chosen in line with similar studies conducted [6] to model the flow characteristics of the L = 550 μm long microneedle lumen [6]. μ = 1.006e^{-3} Pa s and ρ = 1000 kg/m^3 were assumed as viscosity and density respectively as standard values for the model [6, 15] and g = 9.81 m/s^2 is the gravitational force which is subject to change on standard calculations for altitudes.

5. Design Experiments I – Design and Fabrication processes of Microneedles

5.1 Experimental tool: MEMS PRO and Process Steps (Design Strategy)
The simulations for design and fabrication processes of microneedles were carried out on MEMS PRO design tool which is a computer aided design tool provided by the SOFTMEMS Company that is reliably utilized for the scalable design, modeling and analysis of Micro Electro Mechanical Systems (MEMS) [19].

The properties of the needles are generally reflected by the structure of microneedles [20]. The primary considerations for our proposed design of microneedles are:

(a) That during the insertion into skin they must not break or bend. This demands a needle with sharp or pointed tip along with a wide base must be taken into account;
(b) That during the injection period, there are some lateral forces that tends to act on the microneedles, which in turn produce bending moments on the needle body. This demands that the microneedles must have such good design that they withstand with all certain forces;
(c) That one needs to organize the microneedles in the form of arrays because by designing them in arrays the fluid obtained would be in large volume and also be able to have a penetration onto the skin over a large distributed area [21]. For successful resulted microneedle, it must have pointed tips, hollow and out of plane shape.

In order to sketch out the 200 μm long hollow out of plane symmetric microneedle array for biomedical applications, simulations for fabrication process on process simulator MEMS Pro were executed and illustrated in the following steps:

STEP 1: The design process started with initial substrate of silicon wafer with 550 μm thickness shown in figure 2(a), having <100> orientation and polished on top and bottom side.

STEP 2: The second step is to deposit mask material on the top (surface side) and bottom (bulk side) of the silicon wafer shown in figure 2(b) and for this purpose 6 μm thickness of silicon dioxide (SiO_2) is deposit by applying a process of Low Pressure Chemical Vapor Deposition (LPCVD).

STEP 3: Photoresist is then deployed on the bottom side of the wafer and by applying photolithography, process pattern was transferred on the oxide layer (SiO_2) for the hole of 40 μm in diameter shown in figure 2(c).

STEP 4: The anisotropic etching which accomplish by Deep Reactive Ion etching (DRIE) is applied to form the needle channels deep down the silicon substrate to a depth of 556 μm from bottom side stopping close to the top oxide layer as shown in figure 2(d).

STEP 5: 4μm thickness of silicon nitride by applying a process of LPCVD is deposited on the top and bottom along with the channels shown in figure 2(e) which will act as a protector during the next steps of etching including isotropic etching of silicon.

STEP 6: On the front side of the wafer photoresist was deployed and circular pattern was transferred by applying photolithography process on the silicon dioxide and silicon nitride masks shown in figure 2(f) which could be utilize in the next step to create the needle shaft.

STEP 7: Now to craft the outer shape of the needle, isotropic underetched is performed asshown in figure 2(g). The isotropic undercut parameter is primarily utilized to optimize the etching of silicon beneath the circular masks horizontally. Here the silicon nitride plays its critical role and protects the side walls of channels. The obtained outer shape of the needle has a 200μm long shaft with a wide base.

STEP 8: In the last step the remaining silicon dioxide and silicon nitride are stripped resulting in a final structure of microneedle as shown in figure 2(h).

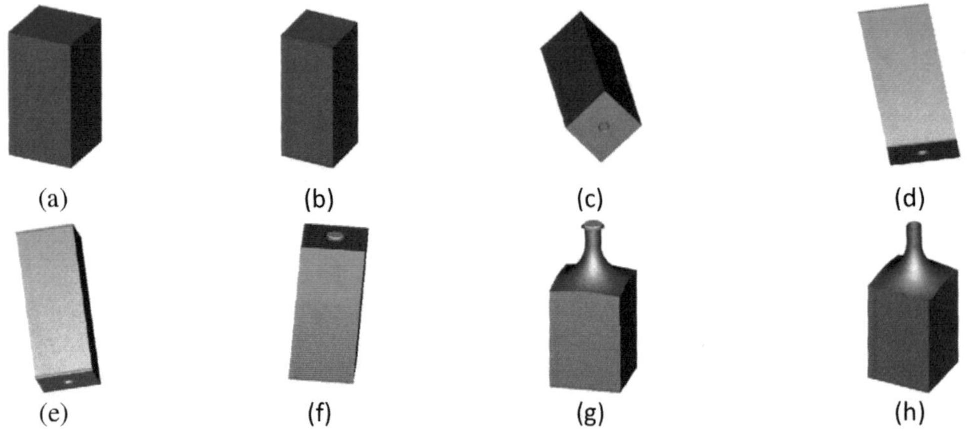

Figure 2. Fabrication Processes Steps for Hollow Out of Plane Microneedles (a) Silicon Wafer (b) SiO_2 Deposition (c) Backside Etching of SiO_2 (d) DRIE for Needle Channel (e) Depostion of Si_3N_4 (f) Pattern transfer for Front Side (g) Isotropic Etching for Needle Shaft (h) Removal of SiO_2 and Si_3N_4.

5.2 Microneedle Design Dimensions

The modeled symmetric and pointed tip hollow out of plane microneedles are shown in figure 3(a) and figure 3(b) respectively,consisting of following main design specifications.(i) Silicon wafer thickness = 550 µm, (ii) Needle length = 200 µm, (iii) Lumen diameter = 40 µm, (iv) Center to center spacing between needles = 750 µm, (v) Array size = 2 x 4.

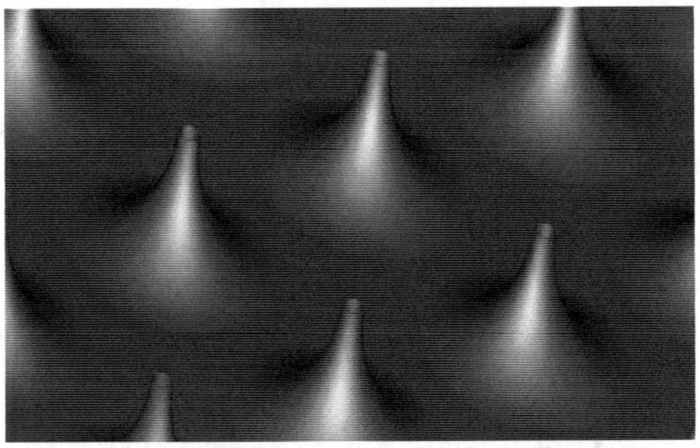

Figure 3(a). Symmetric microneedle arrays

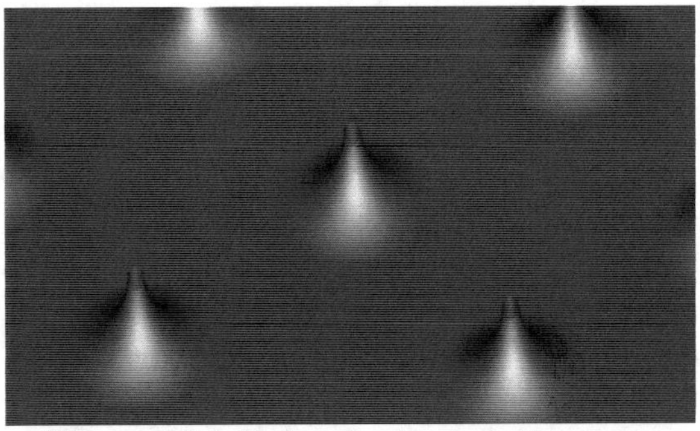

Figure 3(b). Pointed microneedle arrays

6. Design Experiments II - Process selection optimization for the fabrication of microneedles

6.1 Experimental Tool: TCAD SILVACO

Process simulator ATHENA given by SILVACO TCAD has been used for the fabrication of microneedles to gain optimized process selection parameters. This tool offers process simulation and modeling of semiconductor devices with a variety of modeling options [22].

6.2 Broader Process Parameters Steps (SILVACO) Specific to Si-CMOS fabrication (Fab) Line

Since the SILVACO is primarily based on Si-CMOS processes, therefore the processes fabrication steps for microneedles simulated is slightly different from the processes fabrication steps desiged in MEMS PRO design tool. This in turn facilitates the device design engineer to make the fabrication and batch production of MEMS based microneedles easy, cost effective, reliable and flexible on standard Si-CMOS fab lines, without altering the experimental set up a great deal. The processes used in SILVACO are invoked by simple physical process parameters rather than using machine parameters. The process steps for the optimized fabrication of microneedles are provided in the following steps. The selection of the optimized process parameters and final design shapes (refer to figures 3(a) to 3(h)) are based on numerous routines run on the simulator and in close proximity to the design maintained in MEMS PRO, detailed in section 5.

STEP 1: The first step is the selection of appropriate substrate or material and preferred material is silicon as shown in figure 4(a).

STEP 2: Background doping is then carried out in the silicon wafer with p-type boron impurity to modulate its electrical properties and also to provide a region for preferential etch.

STEP 3: Now in order to craft the outer shape of the microneedle at a height of 200 µm shown in figure 4(b) geometrical etching of silicon is performed by invoking arbitrary points.

STEP 4: In this step silicon dioxide is thermally grown by dry oxidation in order to protect the side walls and top of the needle shown in figure 4(c).

STEP 5: Following the thermal oxidation it is then annealed to strengthen the sidewalls by activating the silicon atoms and also to create a sharp tip of the microneedle during subsequent oxidizing process as shown in figure 4(d).

STEP 6: Now the grown silicon dioxide is selectively geometric etched away as shown in figure 4(e). The remaining portion is covering the needle top.

STEP 7: Silicon nitride (Si_3N_4) is then conformally deposited shown in figure 4(f) mainly to cover the entire channel of the needle and top of the needle tip.

STEP 8: Nitride is now annealed as can be seen in figure 4(g) to harden the channel of the microneedle.

STEP 9: Now the silicon nitride is selectively geometric etched away as shown in figure 4(h) remaining on top of silicon dioxide.

STEP 10: As a last process, minor steps were performed including mirroring the structure and stripping of remaining silicon dioxide (SiO_2) and silicon nitride (Si_3N_4), as can be seen in figure 4(i). In the end microneedle with sharp tip shape is formed as shown in contours of the figure 4(j).

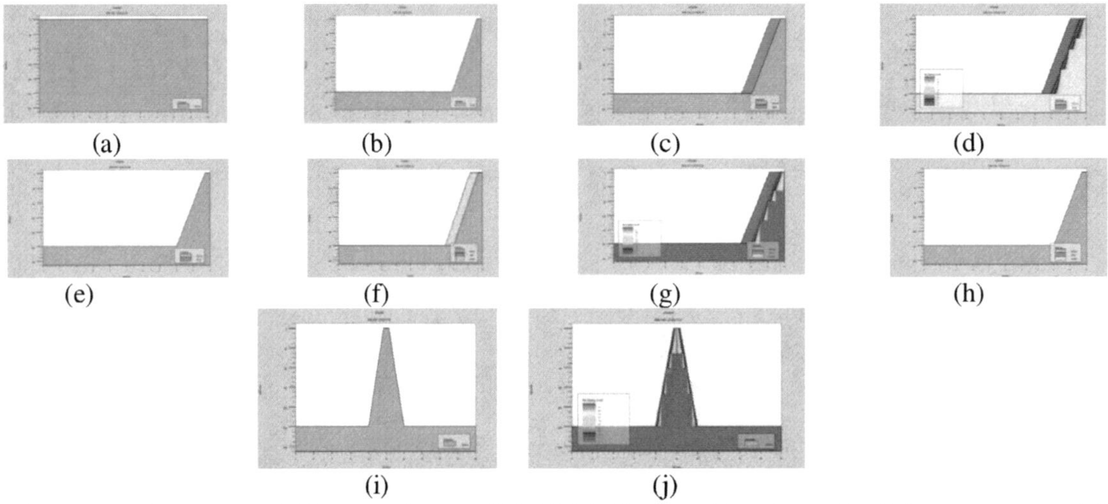

Figure 4. Process Simulation of Microneedles on SILVACO Process Simulator, (a) Silicon substrate (b) Geometrical etching for microneedle outer shape (c) Deposition of SiO_2 (d) Annealing of SiO_2 (e) Geometrical etching of SiO_2 (f) Deposition of Si_3N_4 (g) Annealing of Si_3N_4 (h) Geometric etching of Si_3N_4 (i) Final shape of microneedle (j) Contour of final shape microneedle with sharp tip.

7. Discussion on salient features of results

The design and simulation results are detailed at length in sections 6 and 7. Based on the design assessments made and modeling of the underlying mechanics deliberated in section 4, we provide an insight to these results in this section and discuss them in line with the objectives set forth for this study:

The microstructure needles were designed to inject or extract out the drug or fluid from skin via stratum corneum. Mathematical models were re-visited in form of Modified Bernoulli Equations (1) and Extended Modified Bernoulli Equation (2) which are further reduced to equation (3) and equation (4) respectively. These revised models simulated for fluid flow characteristics through needles. The physical relationships between several parameters such as flow rate, pressure drop and needle diameters were also developed. Modified Bernoulli Equation was re-visited in such areas where gravitational forces are either considered as negligible or ignored, whereas in Extended Modified Bernoulli Equation gravitational forces are taken into consideration, for applications of devised microneedles on ambients where the contribution of gravitation forces are appreciable.

7.1 The contribution of gravitational forces

The gravitational forces were mainly not considered during previous studies of microneedles [18] and treated as negligible during the numerical and experimental analysis of microneedles. In our study, the gravitational forces (g) were taken into account during the numerical analysis in order to see its effects on design and flow characteristics of microneedles. During the comparison of two models, it was found that gravitational force has appreciable impact on geometric aspects of the design and subsequently on its flow characteristics. Gravity is known to have effects and impact on human body as well as on other biological entities, therefore the gravitational forces may become a highly considerable parameter during the design of biomedical systems when used in special environments such as altitudes. This has been a missing link so far, largely in the literature and very little is known on this account. One very recent study published to address this issue partially [23] claimed that during the design of microneedle arrays for biomedical systems, gravitational forces may also be taken into consideration for the transport of drug formulation [23].

7.1.1 Relationship between Pressure Drop and Flow Rate

The pressure drop as a function of flow rate, diameter as a function of flow rate and diameter as function of pressure drop with microneedle geometry has been simulated in our study for an individual microneedle using water as a model liquid with the help of equations (3) and (4). Standard macroscopic values for an edged inlet (K_1= 0.5) and for an exit (K_2= 1.0) were chosen to model the flow characteristics. Other parameters are: L = 550 μm long microneedle lumen. $\mu = 1.006e^{-3}$ Pa s and $\rho = 1000$ kg/m^3 were assumed as viscosity and density respectively and g = 9.81 m/s^2 is the standard gravitational value.

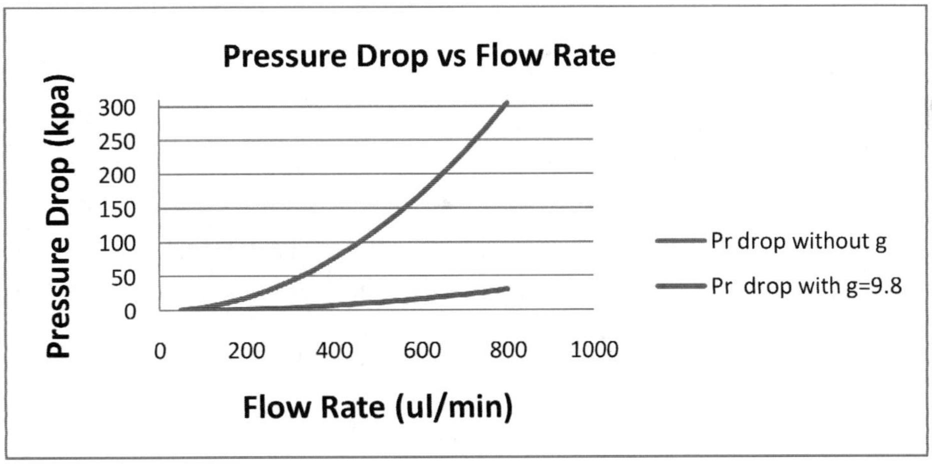

Figure 5. Variation between Pressure Drop and Flow Rate

Figure 5 illustrates the relationship between pressure drop and flow rate simulated at a fixed 40 μm channel diameter of the designed microneedles. From this result, one can notice that the diameter of channel is determined by desired flow rate and saturated pressure drop. The curves show that as the flow rate increases, the pressure drop also increases initially for both conditions of "g". The pressure drop however decreases dramatically in case of gravitational force accounted for as compared to the pressure drop without g contribution. It is due to the physical implication of gravity on the viscous and dense medium and directly impacts the behaviour of pressure dynamics due to a certain maintained flow rate within the device.

7.1.2 Relationship between microneedle diameter and pressure drop

Figure 6 discusses the relationship between the variations in diameter and pressure drop simulated with constant flow rate at Q = 400 µl/min and 800 µl/min. As the pressure drop increases, the diameter gets decreased in both the conditions i-e with and without taking the contribution of 'g' into account. The curves of diameter vs pressure drop are saturated after the drop of pressure reaches a certain value. The diameter of lumen increased with the increased flow rate at certain values of pressure drop which is true for both the conditions. In this graph the curves are becoming flat with the increase of the pressure drop. The region of flattened lines illustrates that the diameter is not very sensitive to the variation of the high pressure drop. Hence a smaller diameter is required for the higher pressure drop at the desired flow rate in the microneedles.

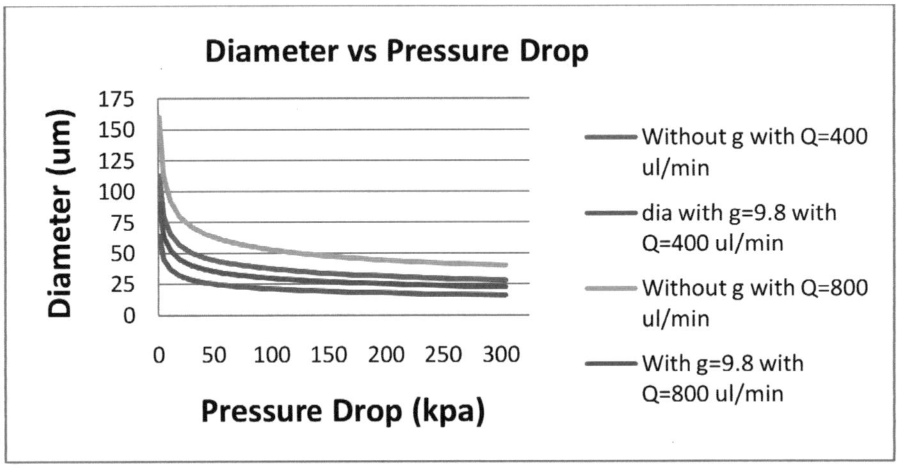

Figure 6. Variation between Diameter and Pressure Drop

It can be noticed that by comparing this graph for both conditions of 'g', the diameter gets much reduced in case where value of "g" is accounted for as compared to the diameter modeled without "g". This also suggests that at a certain value of pressure drop, the increase in the flow rate makes the diameter of lumen increased.

We gathered from these experiments that the designed diameter of the microneedles can be determined at the saturated region of diameter vs pressure drop curves. The slope of the graph presents the variation of the diameter with pressure drop. After the pressure drop reaches a certain value, the variation of diameter with pressure drop is negligible, which is true for both the conditions of gravitational forces.

Further, it can be gathered from the results presented in Figure 6 that the velocity of liquid flow also goes to a decrease with the increase in the diameter, which results in lower pressure drop across a microneedle. Since the water has low viscosity so the flow rate will increase readily. This is true for both conditions of "g". However, the values become smaller for all above parameters when they are compared with the condition where experiments are modeled without taking the contribution of "g" into account.

7.1.3 Relationship between diameter of the microneedle and flow rate

Figure 7 illustrates the relationship between the variation in the diameter and flow rate simulated with two separate constant pressure drop at 75.9 kpa and 304 kpa. The diameter gets increased with decrease of pressure drop at certain desired flow rate for both conditions of "g". This means that the pressure that is needed to flow the fluid rises dramatically when channel diameter decreases. In figure 7, it can further be seen that a very small decrease in the diameter reduces the flow drastically. This is due to the fact that cross section area (which is directly proportional to the flow rate) gets impacted for both conditions of "g". By comparing the value of diameter with and without the consideration of value of "g", it is found that by including the value of "g", the diameter becomes much smaller at higher pressure drop at a desired flow rate, which suggests that there is higher velocity experienced at that particular point.

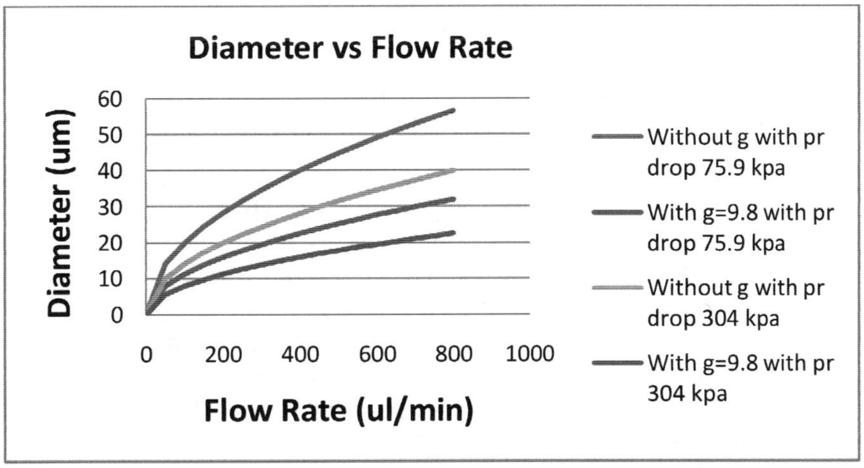

Figure 7. Relationship between Diameter and Flow Rate

7.2 The Geometry of microneedle & Physical Processes

The pressure drop is mainly dependent on the geometry of microneedle in combination with fluid viscosity and density for a smooth fluid transportation through microneedle channel. Therefore appropriate selection of needle geometry leads to an effective biomedical microsystem device. The Deep Reactive Ion Etching (DRIE) is suggested to be performed to create the microneedle lumen. The outside shape of microneedle is suggested to be created by isotropic undercut of silicon dioxide and silicon nitride masks.

The geometry of the design is adjusted in such away to provide a solution using which a process engineer may fabricate MEMS-based microneedles with the following characteristics:

(a) A miniaturized cross section (controlled by diameter modeling)
(b) A substantial capillary flow rate (controlled by the modeling of pressure drop, contribution of gravitational force and optimized diameter)
(c) Full strength not to break during the skin penetration (controlled by physical process optimization)

Physical process based simulations carried out as shown in Figure 4 are aimed to optimize the structure and design of the microneedles. The purpose of the process simulation of microneedles is to predict and manipulate the design structure knowledge that has been obtained after applying specified process steps which was not possible to obtain from the MEMS PRO design tool (refer to figure 3 and its constituents, for example). Since the ATHENA simulator is primarily based on the CMOS fabrication process, the dimensions and design used for microneedles in these simulations are complex, relevant to the fab-less solutions and cost effective in nature. The selection of parameters (refer to figure 4 and its constituents and steps elaborated in section 6) is purely iterative in behaviour and provides a lot of insight while optimizing the design of MEMS-based microneedles by the optimization of both the modeling and physical process parameters. Our study is somewhat different as the utility of both the powerful design and modeling tools for the design and consequent fabrication of MEMS-based microneedles is focused on (a) the cost effective Si-CMOS fab specific solutions, and (b) the physical process parameters rather than machine parameters in TCAD SILVACO environment, which makes it viable for design optimization studies and applications rather than an 'embedded-only' solution in a mechanical application.

8. Conclusion

This work was focused to the design and fabrication process of "hollow out of plane" silicon based microneedle arrays for biomedical applications in MEMS environment. The purpose of such microneedles was to offer a design solution which would in turn control physical parameters for a successful painless epidermal drug delivery and cause no trauma at the site. The design of microneedles presented in this work is carried out on the MEMS PRO design in conjunction with TCAD SILVACO process simulation tool. The MEMS PRO design provided the basic geometrical definition of the device. With the help of MEMS PRO, it was found that Bernoulli Equation (and its revisited versions) had been a good approximation for liquid flow through microneedle lumens. TCAD SILVACO was utilized to improve the specific design in line with standard Si-CMOS fab lines.A trade-off process optimization was thus achieved between fab-less and in-fab solutions for MEMS-based microneedles for effective applications in biomedical industry. The design in this work has been specific to transdermal drug delivery so the dimension of microneedle shaft was set to be 200 um long which might prove to be consequently sufficient to easily penetrate the skin barrier stratum corneum and reach the epidermis. An appropriate selection of microneedle diameter was needed that could deal with the design limitations. With large value of diameters, the microneedle channel gotwidened making the base of microneedles wider. But for pointed microneedles (our design) with small diameter, sharp tips were created intentionally. Therefore diameter ranging between 35 μm to 45 μm was found suitable for the microchannel created with both types of microneedles.

Acknowledgements

The authors would like to acknowledge Islamic Development Bank, Jeddah for awarding the grant to S. Ahmed for the establishment of Advanced Electronics Laboratory in International Islamic University, where Electronic Device Design Suite is accessed to perform the design, modeling and simulation experiments. The partnership of Lawrence Berkeley National Laboratory, Berkeley, USA is also appreciated for re-run and test checks of some parts of this work.

References

[1] Bhandari R, Negi S and Solzbacher F 2008 *Electronic Components and Technology Conf.* Florida pp 1306 – 09
[2] Bashir R 2004 *Advanced Drug Delivery Reviews* **56** 1565 – 86
[3] Sharaf R, Aggarwal P, Kaler K V I S and Badawy W 2003 *Proc. of the Inl. Conf. on MEMS, NANO and Smart Systems* Banff Alta pp 32 – 35
[4] Gerstel M.S and Place V.A 1976 US 3 964 482 patent
[5] Campbell P K, Jones K E, Huber R J, K. Horch W and Normann R A 1991 *IEEE Transactions on Biomedical Engineering.* **38** pp 758-68
[6] Stoeber B and Liepmann D 2005 *IEEE. J. Microelectromechanical. Systems* **14** 472-79
[7] Yadav J D, Vaidya K A, Kulkarni P R and Raut R A 2011 *Int. J. Pharma and Bio Sciences.* **2** 684-08
[8] Bodhale D W, Nisar A and Afzulpurkar N 2010 *Proc. of the 3rd. Int. Conf. on the Development of Biomedical Engineering*. Vietnam pp 84 – 88
[9] Ashraf M W, Tayyaba S, Afzulpurkar N, Nisar A, Bohez E L J, and Tuantranont A 2010 *6th annual Conf. on Automation Science and Engineering* Toronto pp 258 – 62
[10] Ashraf M W, Tayyaba S, Afzulpurkar N, Tuantranont A, Nisar A and Bohez E L J 2010 *Int. Conf. on Electrical Engineering/Electronics Computer Telecommunications and Information Technology* Chiang Mai pp 950 - 53

[11] Hendriks F M 2001*Mechanical behaviour of human skin in vivo* - a literature review Nat.Lab. Unclassified Report 820 *Philips Research Laboratories*
[12] Roxhed N, Gasser T C, Griss P, Holzapfel G A, Stemme G 2007 *J. of Microelectromechanical Systems* **16** 1429 – 40
[13] Ebling F J G, Eady R A J, and Leigh I M. 1992 *Textbook of Dermatology Anatomy and organization of human skin.* ed Champion et al (New York: Blackwell Scientific)
[14] Ma B, Gan Z and Liu S 2005 6th Int. Conf. on Electronic Packaging Technology Shenzhen pp 1 - 5
[15] Janna W S 1993 *Design of Fluid Thermal Systems* (Boston: PWS Kent)
[16] Ashraf M W, Tayyaba S , Nisar A , Afzulpurkar N, Bodhale D W, Lomas T, Poyai A and Tuantranont A 2010 *J. Cardiovascular Engineering* **10** 91-108
[17] Batchelor G.K 1967 *An Introduction to fluid dynamics* (Cambridge: Cambridge University Press)
[18] Stoeber B and Liepmann D 2000 1st Annual Int. EMBS Special Topic Conf. on Microtechnologies in Medicine & Biology France pp 224 – 28
[19] Softmems 2009 *MEMS Pro v7.0 data sheet* (Santa Clara CA: Softmems)
[20] Zhang P, Dalton C and Jullien G A 2009 *J. Microsystem Technologies* **15** 1073 – 82
[21] Zhang P and Jullien G A 2003 *Nanotech technical Proc. of the Nanotechnology Conf. and Trade Show 1* pp 510 – 13
[22] Silvaco International 2008 *ATHENA user's manual release notes* (Santa Clara CA: Silvaco)
[23] VadimV Y 2010 US 7,658,728 B2 Patent

Journal of Physics: Conference Series **439** (2013) 012050 doi:10.1088/1742-6596/439/1/012050

Laser induced breakdown spectroscopy (LIBS) as a rapid tool for material analysis

T Hussain[1] and M A Gondal [2]

[1]National Institute of Vacuum Science and Technology (NINVAST), P.O. Box 3125, Islamabad, Pakistan
[2]Laser Research Laboratory, Physics Department, King Fahd University of Petroleum & Minerals (KFUPM), Box 5047, Dhahran 31261, Saudi Arabia

Corresponding e-mail: thminhas7@yahoo.com

Abstract. Laser induced breakdown spectroscopy (LIBS) is a novel technique for elemental analysis based on laser-generated plasma. In this technique, laser pulses are applied for ablation of the sample, resulting in the vaporization and ionization of sample in hot plasma which is finally analyzed by the spectrometer. The elements are identified by their unique spectral signatures. LIBS system was developed for elemental analysis of solid and liquid samples. The developed system was applied for qualitative as well as quantitative measurement of elemental concentration present in iron slag and open pit ore samples. The plasma was generated by focusing a pulsed Nd:YAG laser at 1064 nm on test samples to study the capabilities of LIBS as a rapid tool for material analysis. The concentrations of various elements of environmental significance such as cadmium, calcium, magnesium, chromium, manganese, titanium, barium, phosphorus, copper, iron, zinc etc., in these samples were determined. Optimal experimental conditions were evaluated for improving the sensitivity of developed LIBS system through parametric dependence study. The laser-induced breakdown spectroscopy (LIBS) results were compared with the results obtained using standard analytical technique such as inductively couple plasma emission spectroscopy (ICP). Limit of detection (LOD) of our LIBS system were also estimated for the above mentioned elements. This study demonstrates that LIBS could be highly appropriate for rapid online analysis of iron slag and open pit waste.

1. Introduction

Laser Induced Breakdown Spectroscopy (LIBS) is a useful technique that requires optical access to the sample surface and can be used as a fast method for the detection and quantification of the contaminants in various categories of samples. It is an emerging technique for elemental determination in a wide range of environmental samples, metallurgical, metallic and non-metallic, sewage sludge, liquid, aerosol, gases and biological samples etc [1-7]. Several elements can be measured with this technique simultaneously even at parts per million concentration level. In this technique, laser pulses are applied for ablation of the samples, resulting in the vaporization and ionization of sample in hot plasma which is finally analyzed by the spectrometer. The elements are identified by their unique spectral signatures [8-10].

Recent developments towards improving analytical capabilities of this technique have led to more applications of LIBS. However, still more efforts are required to improve the precision and the sensitivity of this technique. Some improvement in methodology is needed for its wide applications in terms of its efficiency and limit of detection [11-13].

The quantification and characterization of large variety of samples of various natures require special technique like LIBS. Some capabilities of LIBS as a rapid tool for analysis of sample are pollution monitoring in air, soil, sewage sludge, ground and wastewater, determination of ore composition during mining, chemical analysis of explosive materials, forensics and biochemical applications, detection of radioactive contamination and hazardous materials. This technique can also be used for rapid identification of metals and alloys during recycling of scrap materials, on-line compositional analysis of the molten metals, on-line compositional analysis of liquid glass for process control (iron, lead etc), depth-profiling and compositional analysis of surface coatings (galvanized steel, plastic coating and heavy-metals in paint), on-line monitoring of particulates in air (stack emission monitoring etc.) [14].

Some of the potential benefits of this technique include, that sample analysis cost by LIBS is low as compared to other conventional analytical techniques and can analyze large number of samples in short span of time. It is capable of doing multi elemental analysis simultaneously. All state of matter (solid, liquid and gases) can be analyzed. LIBS can be used for analysis of extremely hard materials that are difficult to digest or dissolve, e.g. ceramics and superconductors. In LIBS minimal sample preparation is required which reduces worker exposure to hazardous radioactive or toxic substances. LIBS requires small amount (μgm) of sample, and element detection limit can be achieved in the range of ppm/ppb. In addition to its potential benefits there are some inherent disadvantages of this technique: It is difficult to obtain suitable standards for LIBS analysis because of matrix effect. For quantitative analysis, the sample matrix must closely match the matrix of the standards. The detection limits for solids are generally not as good as established by others techniques [15-16].

The detection and quantification of heavy metals within liquid samples are pertinent to industrial processing, environmental monitoring and waste treatment. In the last decade laser-induced breakdown spectroscopy has become a prime area of scientific and technological interest for researchers in the various fields of science. To-date, laser-induced breakdown spectroscopy (LIBS) has been tested with limited success for liquid samples. Dissolved gases, particulate material and bubbles produced by prior laser pulses can lead to misfocusing of the laser beam and can also serve as breakdown sites prior to the laser beam focus and additionally, the high local density within the liquid leads to rapid quenching which prohibits temporally selective detection, high collision rates which broaden spectral transitions and confines the plasma emission spatially rendering spatially selective detection. Due to these inherent problems, in LIBS analysis of liquids has not been widely successful [17].

In this study the LIBS was applied for characterization of iron & steel plant slag and open pit ore samples. By considering various needs for optimum LIBS measurement a sample chamber and sample rotating system was developed and fabricated. In this system the solid sample is placed in the aluminum cup and sample is rotated by a step motor. The important parameters such as laser energy, distance between fibre optics and target material, sample rotation speed and delay time were optimized. The optimized parameters were tested for the preparation of the calibration curves for Mg, Pb, Cu, Ca, Cr, Cd, S, Mn, P, K and other element of interest. Limits of detection (LOD) of our system were also estimated for these elements. The application of LIBS for the analysis of iron and steel plant slag waste residue is very innovative and environmentally significant. Similarly the use of LIBS for the analysis of waste generated from open pit mine are of diverse in nature. Wastes produced by open pit mine are in large quantity and most of their constituents are hazardous. Similarly iron slag is produced annually as a by-product of iron in the steel production industry. Steel plants normally use electric arc furnaces to produce steel from steel scrap. During melting of scarp steel, a slag forms on the top surface of the molten metal and is removed. Steel plant slag is a hard, dense material. Slag output in iron and steel production depend on the feed material chemistry and the type of furnace. Typically for an ore feed of 60% to 66% iron blast furnace, slag production ranges from about 220 to 370 kilograms per metric ton of pig iron produced. Lower grade ores produce much higher slag fraction. Steel slag output is typically 20% by mass of the steel output as per U.S. geological survey report. Indeed, the steel industry slag has been classified as hazardous waste and has been of great concern to the U.S Environmental Protection Agency [18].

Iron scraps material collected from various locations, contain variety of contaminants. The contaminants present in scrap material can be harmful to the environment and therefore proper analysis of such slag is important. The primary hazardous components of iron slag are zinc, lead, cadmium, nickel and chromium [18]. Classification and identification of slag has been the subject of most government environmental agencies in the world and different research laboratories are using different techniques for elemental identification of slag waste. Due to huge demands of iron and steel to fulfil the needs of the construction industry, automobile industry, and various other industrial sectors, environmental issues related to iron slag are now becoming important and gaining high priority. There are numerous health risk involved to the steel factory workers at workplace and surrounding environment due to solid waste generated from the iron and steel mill industry.

The second sample under investigation was open pit ore sample. The reason to select this kind of sample is that, from an environmental prospect, the acidity and metal concentrations originating from the open pits are extremely toxic. Changes in surface water and groundwater due to waste-dumps at open pits and mining sites could cause significant environmental effect. For example, under certain conditions, surplus water discharged from mine dump into the nearby streams disrupts riparian ecosystems, can effect local and regional groundwater qualities. Due to these environmental concerns, the characterization and analysis of ore samples from open pit mining are very important and needs rapid detection. We tested our LIBS spectrometer for the analysis of these ore samples.

2. Materials and Methods

Two different types of samples iron slag and open pit ore sample were analyzed using LIBS technique to study its capabilities as a rapid tool for material analysis. The details about each sample collection and its preparation are as fallow: - Iron slag samples were collected from a major steel manufacturing plant situated in Jubail, Saudi Arabia. These samples were grinded in spectro mill ball pestle impact grinder (Model 1100-11, Chemplex Industries, Inc. 10707, USA) to suitable mesh size for preparation of pellets. The slag samples under investigation were pressed to pellets shape by applying a load of 8.6× 107 Nm-2 for half an hour by placing the homogenous mixture in a die. These pellets have diameter of 20 mm and thickness of 10 mm. While for testing on ICP the samples were digested with aqua regia, ashed, dissolved in nitric acid and analyzed using inductively coupled plasma emissions spectroscopy. The second sample under investigation was the ore sample. The ore samples tested in this study were collected from an open pit of the Al-Khnaiguiyah Zn-Cu mine, located at GPS: N, 24° 16' 10.2"; E, 45° 05' 24.5". For calibration purposes, the elements used were Pb, Cu, Cr, Ca, S, Mg, Zn, Ti, Sr, Ni, Si, Fe, Al, Ba, Na and Zr. All these elements were obtained in powder form with high purity 99.99%. For the construction of calibration curves, stoichiometrically different samples comprising these metals were prepared. Metals in powder form were mixed with the matrix material KBr in a ball milling apparatus in order to ensure good mixing and homogeneity.

For the analysis of ore samples using the ICP technique, the ore samples were ground to powder form, digested by concentrated HNO3 (99.99%, Fisher Scientific, USA) and were left over night. The resulting residue was then dried at 500ºC. The dried samples were further treated with HNO_3 and the resulting solution was analyzed for poisonous metals using the inductively coupled plasma spectrometer. For LIBS, ore samples under investigation were prepared in the dimensions 4 cm x 6 cm and a thickness of 2 cm. These samples were filed, polished and were placed in to LIBS ablation chamber for analysis.

The experimental setup applied for the analysis of slag and ore sample is presented in Figure 2. Nd: YAG Laser (Spectra Physics, Model GCR100) was employed for production of plasma spark at the sample surface. This laser can deliver maximum pulse energy of one joule with a pulse width of 8 ns and operate at a 10 Hz pulse repetition rate in Q-switched mode. A pulse from this Nd-YAG laser at fundamental wavelength 1064 nm was focused by a convex lens having focal length 30 mm onto the surface of samples to create a plasma spark in the sample. The pulse energy utilized in these experiments was in the range of 20–120 mJ. The laser energy was measured with a calibrated energy meter (Ophir Model 300) for the study of dependence of LIBS signal on incident laser energy. The spectrometer used was Ocean Optics LIBS 2000+ system and a locally-designed sample chamber as shown in Figure 1, along with the OOLIBS software.

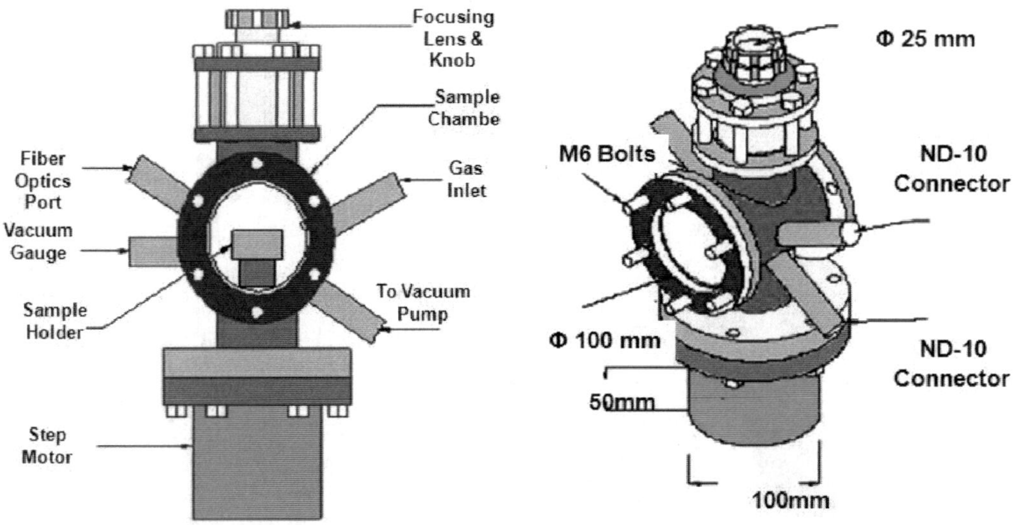

Figure 1. Locally designed sample chamber

The light from the plasma spark is collected by a collimating lens using UV graded fused silica 1 meter, multimode sampling fiber with SMA connector and is transferred to LIB2000 + Spectrometer (Ocean Optics). Our LIBS 2000+ has four spectrometers modules to provide high resolution (FWHM 0.1 nm) in the 200 nm to 620 nm wavelength region. The detector has a gated CCD camera having 14,336 pixels. This makes it possible to measure a LIBS spectrum over broad spectral range (200–620 nm) simultaneously with high spectral resolution (0.1 nm). The emission is observed at a 45° angle to the laser pulse. Prior to the application of our LIBS technique on samples under investigation the spectrometer was calibrated for each trace element present in the sample. The detection limit for various elements under investigation was also measured. Detection limit here means the lowest concentration that can be detected with LIBS. The limit of detection (LOD) was estimated using the equation [19].

$$\text{Limit of detection (LOD)} = \frac{2\sigma b}{S}$$

Where σb is the standard deviation of the background (noise level) and S is calibration sensitivity (S = slope of the calibration curve) which is the ratio of the line intensity to the concentration. The standard deviation of the mean value was calculated from these measurements. All kinds of electronic noise (e.g., CCD dark current) were measured separately and were subtracted from the measured spectra prior to the analysis of samples.

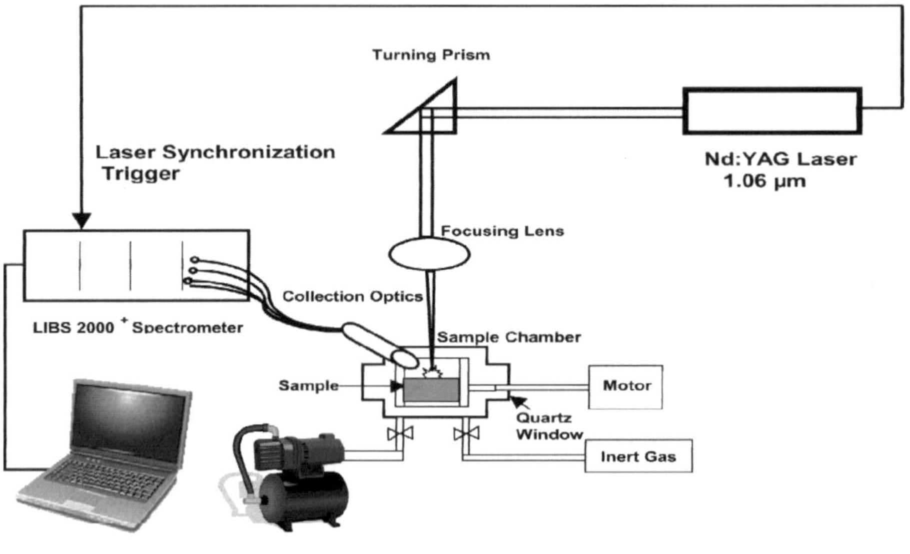

Figure 2. Schematic diagram of the LIBS experimental setup applied for the analysis of samples

2.1. Optimization of various parameters of LIBS
Before going to perform experiments on actual samples the LIBS system developed at our laboratory was optimized for various parameters involved. The optimization of signal intensity Vs laser pulse energy for Mg 518.36 nm line, signal Intensity Vs time delay for Mg 518.36 nm line, variation of the intensity of the Mg 518.36 nm emission line with the change in the position of collecting lens from the sample, effect of lens distance on focusing signal intensity and sample rotation speed are shown in the following diagrams.

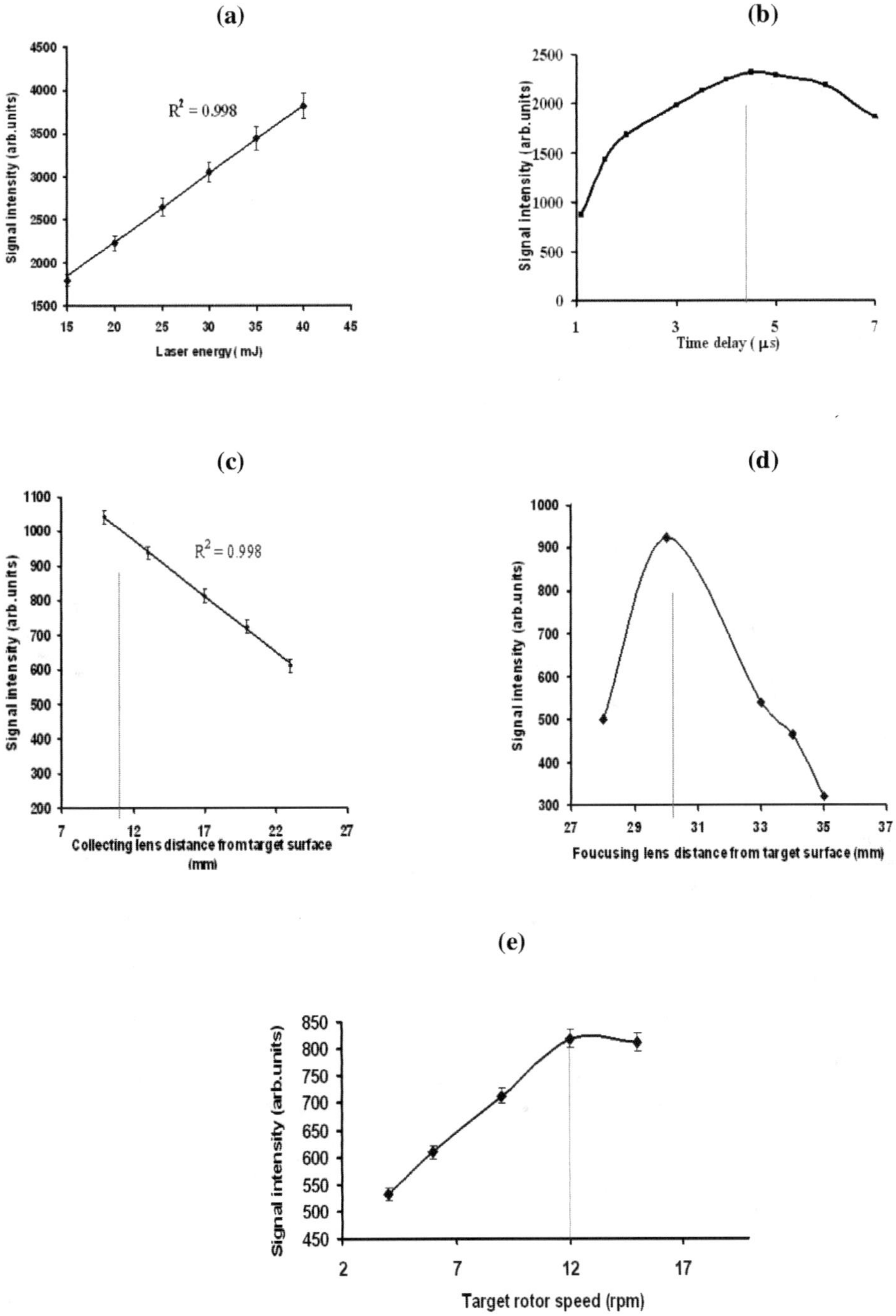

Figure 3.
(a) Signal Intensity Vs Laser pulse energy for Mg 518.36 nm line.
(b) Signal Intensity Vs Time delay for Mg 518.36 nm line.
(c) Variation of the intensity of the Mg 518.36 nm emission line with the change in the position of collecting lens from the sample.
(d) Effect of lens distance on focusing signal intensity.
(e) Sample Rotation Speed.

3. Results and Discussion

3.1. Analysis of Iron and Steel plant slag

Figure- 4 depicts the emission spectra of iron slag samples recorded at 4.5 μs delay time in 200-620 nm region. Here the laser pulse energy was 20 mJ. The selected lines of iron were used to study the effect of different experimental parameters and to compare the signal intensities with the calibration curves of the elements under investigation. Elements detected in slag samples and comparison of concentration detected with LIBS and ICP is presented in Table 1. From Table 1, it is clear that Ba, Cr, and S are present in very high concentrations. For example in sample under investigation, the concentration of Ba = 564 ppm, Cr = 1866 ppm and S = 1397 ppm while the maximum permissible limits set by EPA and other regulatory authorities for these elements are: Ba = 2, Cr = 1 and S = 800 ppm respectively. The health risk associated with barium to workers in iron and steel mill are breathing dust, eating soil or drinking water that is polluted with barium and harm to skin due to contact with soil or scrap. Similarly we detected chromium concentrations also higher than the maximum permissible limit although a limited amount of chromium is an essential nutrient that helps the body to use sugar, protein and fat. However, breathing a high level of chromium can cause irritation to the nose. During several industrial processes (such as iron slag formation) sulfur compounds are added to the environment that is hazardous to animals, as well as humans. These sulphur compounds are unwanted because of their unpleasant smells and are highly toxic, and they can have different effects on human health [20- 23]. For slag samples the precision of LIBS measurements depend on the complexity of the sample, homogeneity, and the reproducibility of the laser shots.

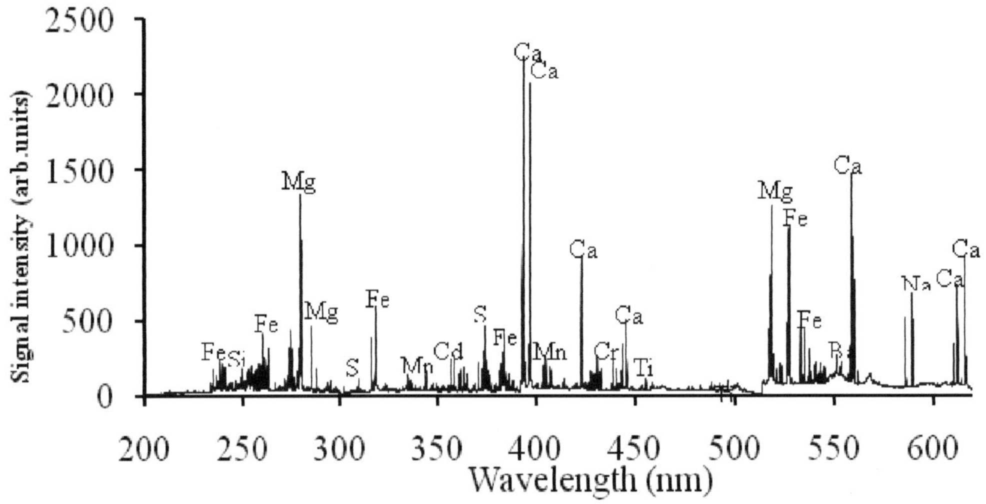

Figure 4. Typical spectrum of slag sample in 200 to 620 nm region recorded at delay time of 4.5 μs and laser pulse energy of 20 mJ

Typical RSD values for LIBS are in the range 1–10% and for slag samples, the precision was 2%. The relative accuracy (RA) is calculated using equation 1

$$RA = \frac{|d| + SD \times t_0.975n^{-1/2}}{M} \quad (1)$$

In above equation d is the difference between the LIBS measurement and the ICP (standard method). SD is the standard deviation of LIBS measurement, M is the measurement from standard method, n is the number of measurements and t 0 .975 is the t value at 2.5% error confidence. As it can be seen from Table 1 the relative accuracy is in the range of 0.001 – 0.049, which is quite acceptable for any good instrument.

Table 1. Comparison of LIBS with ICP, relative accuracy, limit of detection and delay times for different elements detected in slag sample

Element	Wavelength (nm) (Striganov and Sventitski, 1968)	Slag Sample		Relative accuracy (RA)	Standard deviation (S.D) LIBS	LOD LIBS (ppm)	Delay time (μs)
		LIBS (ppm)	ICP (ppm)				
Ba	553.5	564.06	553.0	0.020	0.098	14.0	5.5
Ca	393.3	188700	188500	0.001	0.122	11.0	5.0
Cd	361.2	54.4	52.0	0.049	0.129	10.0	4.0
Cr	427.3	1866.6	1830	0.020	0.091	7.0	5.5
Fe	526.9	281520	276000	0.028	0.132	6.0	4.5
Mg	518.2	80194	79400	0.010	0.115	9.0	5.0
Mn	403.4	16564	16400	0.012	0.095	6.0	4.5
Na	588.9	1131.2	1120	0.015	0.119	5.0	4.0
P	438.5	2747.2	2720	0.018	0.125	13.0	4.5
S	373.8	1397.4	1370	0.020	0.128	16.0	4.5
Si	288.1	79285	78500	0.017	0.130	10.0	4.0
Ti	498.17	5635.8	5580	0.013	0.135	9.0	5.0

3.1.1. Calibration Curves for Elements under Investigation

In order to test the homogeneity of our samples, several LIBS measurements were performed at different locations at the surface of the pellets. Four samples of known concentration such as 100,000, 10,000, 1000 and 100 ppm of elements under investigation were prepared in the matrix of KBr and LIBS spectra were recorded for these four concentrations of each element. All these spectra were recorded with an average of 20 laser shots, at three different locations on the sample surface. The calibration curves for Cd and Cr are depicted in Figures 5 and 6.

Thus in this study, slag samples collected from the local iron and steel plant were characterized qualitatively as well as quantitatively for elemental detection. Ba, Ca, Cd, Cr, Fe, Mg, Mn, Na, P, S, and Si were detected in reasonable abundance in these samples. The detected concentration of some of the elements such as Ba, Cr and S was higher than the permissible safe limits.

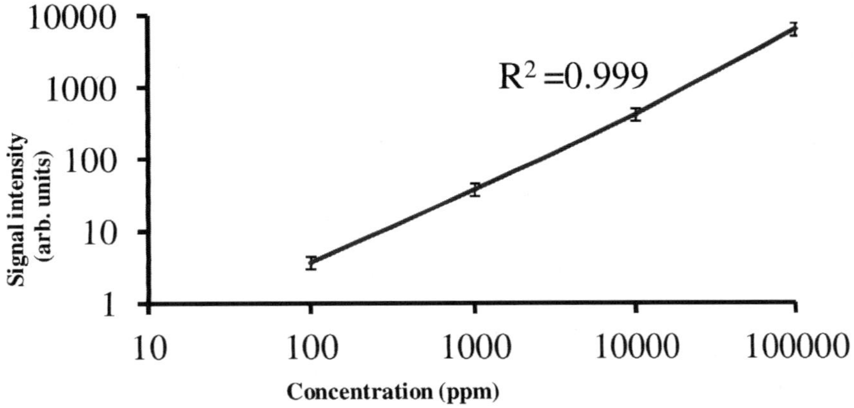

Figure 5. Signal intensity versus Cd concentration at 4.5 μs time delay and laser pulse energy 20 mJ. Error bars denote 95% confidence limits and are based on triplicate measurements.

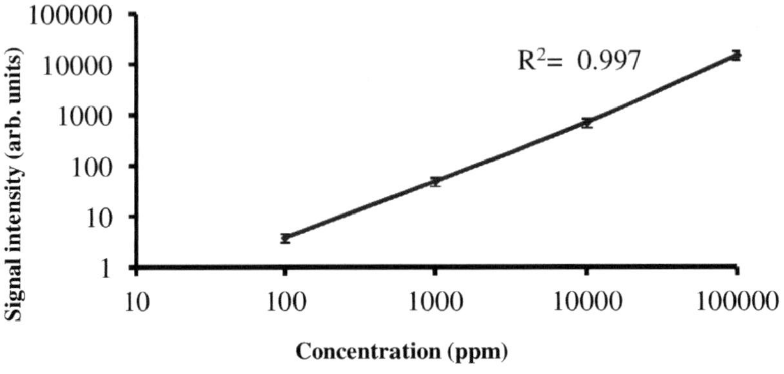

Figure 6. Signal intensity versus Cr concentration at 4.5 μs time delay and laser pulse energy 20 mJ. Error bars denote 95% confidence limits and are based on triplicate measurements.

Calibration curves for elements under investigation were also estimated by the total intensity of a spectral line 'I' from an excited atom or ion in homogenous and optically thin plasma for a transition from state j→i. It is given by [24].

$$I = h\nu_{ji} A_{ji} N_j \quad (2)$$

Where ν_{ji} is the frequency of the transition from state j→i, A_{ji} the Einstein coefficient for spontaneous emission, N_j the population of the upper level j and h is the Planks constant.
If the energy levels concerned are in local thermodynamic equilibrium (LTE), then population density of upper level N_j is related to ground level number density (N) by Boltzmann's equation [29].

$$N_j = Ng_j Q^{-1} \exp\left(-\frac{E_j}{KT}\right) \quad (3)$$

In above equation g_j and E_j are the statistical weight and energy of upper level j, respectively and Q is the partition function of relevant species, K is the Boltzmann's constant and T is the electron temperature of the plasma. So from above two equations, we get:

$$I = h\nu_{ji} A_{ji} Ng_j Q^{-1} \exp\left(-\frac{E_j}{KT}\right) \quad (4)$$

According to equation 4, one can determine the population density of the relevant species (atoms or ions) for an element in plasma from a measurement of the absolute intensity of corresponding transition and knowledge of the excitation temperature and atomic constants. However it is not possible to derive a theoretical expression relating N and concentration of that element in solid sample under investigation. Hence, in practice, an empirical relationship is sought between the observed line intensity (I_j) and corresponding concentration (C), i.e.:

$$I_j = F(C) \quad (5)$$

Here F(C) is called analyte calibration function and the graph between the LIBS intensity of a line of specific element and the concentration is called the calibration curve. Such curve can be drawn by measurements of reference samples of known concentrations termed as standard samples and can be used to determine the elemental concentration in unknown sample like our solid samples. A plot of intensity I_{ji} as a function of elemental concentration n_j yields the calibration curve. If the dependence of LIBS signal intensity on the elemental concentration is linear then we can determine the unknown concentration of particular element at any LIBS signal intensity I_{ji} provided that the same experimental parameters (time delay, incident laser energy, collecting fiber optics and incident focusing optics distance, target rotation speed, atmospheric pressure) are selected.

3.2. Analysis of ore samples

Figure 7 shows the emission spectrum of the ore sample at time delay of 4 μs recorded in 200 nm to 620 nm spectral region. In this case the laser pulse energy was 25 mJ. The LIBS spectra of Fe, Cr, Mn, Si, and Al were recorded in 200 to 620 nm spectral region to detect the lines of each element. The Cu lines were detected at wavelength of 515.3 and 521.8 nm. The Mg line was detected at 516.7 nm while other lines detected for Ca, Si and Fe are listed in Table 2.

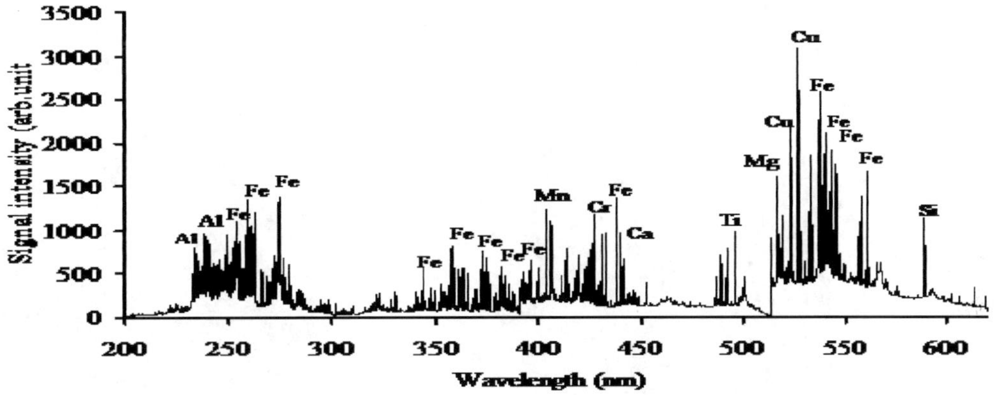

Figure 7. Emission spectrum of ore sample in 200-620 nm spectral region

Table 2. Concentration of the elements detected in ore sample

Element	Wavelength (nm)	LOD (ppm)	Concentration LIBS (ppm)	Concentration ICP (ppm)	Delay time (μs)
Cu	521.82	14.0	805	716	5.0
Zn	330.6	12.0	145	125	4.5
Ca	442.5	6.0	1215	1110	5.0
Ti	498.17	3.5	73	69.6	4.5
Fe	537.14	5.0	597,000	591,000	5.0
Si	594.85	10.0	81,150	80,600	4.0
Mg	516.73	11.0	3478	3380	4.5
Mn	404.13	2.0	1150	1060	5.5
Cr	427.48	4.0	72.1	60	5.5
Al	237.31	10.0	10,205	10,100	4.5

The major elements detected in the sample are copper, chromium, calcium, magnesium, zinc, titanium, silicone, iron, aluminum and manganese are shown in Figure 7. All the spectral lines for the above-mentioned elements recorded with our LIBS setup were identified using the NIST atomic spectral data base and also using the reference [24]. It is clear from the Table 2 that concentration of iron, aluminum and calcium is relatively high which is due to natural geological composition of area. Reasonable amounts of Cu, Zn and Mg are also present in the sample. The concentration of aluminum and chromium in the ore sample are 10,205 ppm and 72.1 ppm which are quite high as compared to permissible safe limits of 30 ppm and 1 ppm set by Environmental Regulation Standards by Royal Commission Jubail, Saudi Arabia. These high concentrations can effect the nearby surface water resources and soil through run off during rain.

Journal of Physics: Conference Series **439** (2013) 012050 doi:10.1088/1742-6596/439/1/012050

Plasma plume images of laser induced plasma recorded at ore sample at laser pulse energies of 25, 30, 35 and 40 mJ are shown in Figure 8.

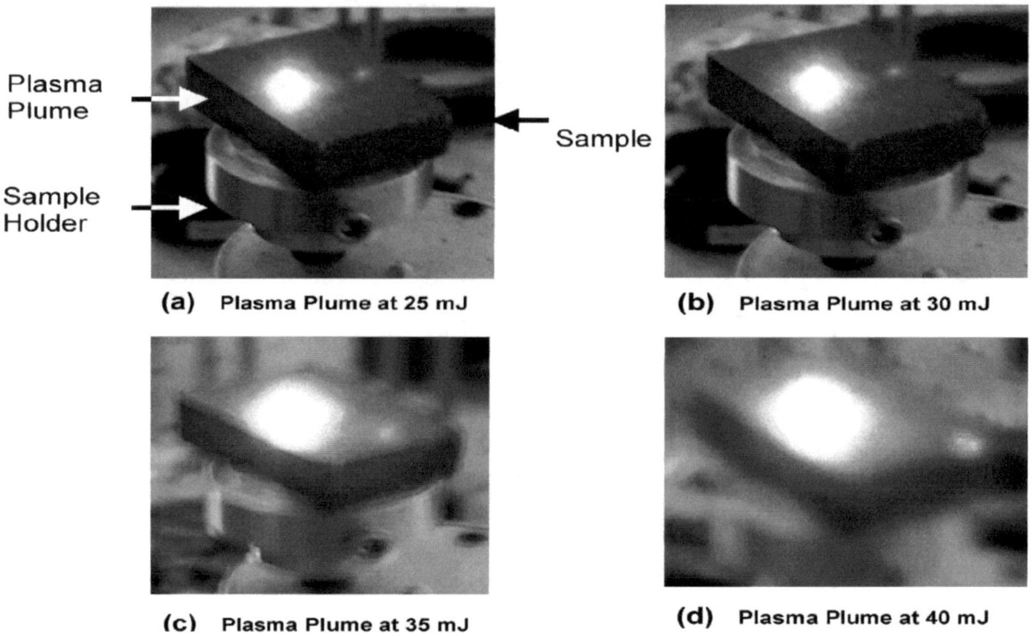

Figure 8. Images of laser-induced plasma plume produced on the ore sample at different laser energies.

The concentrations of trace metals estimated with our LIBS setup are in close agreement with the results achieved with the inductively coupled plasma spectrometry (ICP). The experience gained through this work can be useful for the development of a portable system for rapid online analysis of solid waste from open pits, mineral processing units and geological rock analysis.

4. Conclusions

In this study, the ability of LIBS as a rapid tool for material analysis was discussed briefly. Two samples slag from iron and steel plant and open pit ore sample was tested. Slag samples collected from the local iron and steel plant were characterized qualitatively as well as quantitatively for elemental detection. Ba, Ca, Cd, Cr, Fe, Mg, Mn, Na, P, S, and Si were detected in reasonable abundance in these samples. The detected concentration of some of the elements such as Ba, Cr and S was higher than the permissible safe limits. Similarly the second sample under investigation was the open pit ore sample. The concentration of various trace elements (Cu, Cr, Ca,Mg,Zn, Ti, Si,Fe and Al) present in the ore sample collected from the Al-Khnaiguiyah open pit mine, west of Riyadh in Saudi Arabia, was measured to study the capabilities of LIBS. The sensitive lines for above mentioned elements in both sample iron slag and ore were identified using standard data published by NIST for the elemental analysis and the calibration curves were effective in quantifying the trace metal concentration in samples. The concentrations of trace metals estimated with our LIBS setup agree with another well-established standard method such as the inductively coupled plasma spectrometry (ICP). The experience gained through this work can be useful for the development of a portable system for rapid online analysis of solid waste from open pits, mineral processing units, geological rock and iron slag analysis.

Acknowledgements
The support by the Physics Department of King Fahd University of Petroleum and Minerals is gratefully acknowledged. Authors offer their sincere thanks to all who co-operated during the experiments.

References

[1] Ying M, Xia Y, Sun Y, Lu Q, Zhao M and Liu X 2003 Study of the plasma produced from laser ablation of a KTP crystal *Appl. Surf. Science* **207** 227-235

[2] Russell S H, Frank C D, Andrzej W M, Kevin L M, Roy A W and Patrick D F 2005 Laser-induced breakdown spectroscopy (LIBS) – an emerging field-portable sensor technology for real-time, in-situ geochemical and environmental analysis *Environment Analysis* **5** 21-28

[3] Lazic V,Colao F, Fantoni R, and Spizzicchino V 2005 Laser-induced breakdown spectroscopy in water: Improvement of the detection threshold by signal processing *Spectrochimica Acta Part B* **60** 1002-1013

[4] Gondal M A, Hussain T, Yamani Z H and. Bakry A H 2007 Study of hazardous metals in iron slag waste using laser-induced breakdown spectroscopy *J. Environment Science and Health Part A* **42 (6)** 767-775

[5] Gondal M A, Hussain T, Yamani Z H and Baig M A 2007 The role of various binding materials for trace elemental analysis of powder samples using laser-induced breakdown spectroscopy *Talanta* **72(2)** 642-649

[6] Gondal M A, Hussain T, Ahmad Z and Bakry A 2007 Detection of contaminants in ore samples using laser-induced break down spectroscopy *Journal of Environmental Sciences and Health Part A* **42** 879-887

[7] Hussain T and Gondal M A 2008 Detection of toxic metals in waste water from dairy products plant using laser-induced breakdown spectroscopy *Bulletin of Environmental Contamination & Toxicology* **80(6)** 561-565

[8] Gondal M A, Hussain T, Yamani Z H and Ahmad Z 2007 Determination of toxic metals in petroleum, cultivated land and ore samples using laser-induced breakdown spectroscopy *Bull. Environ. Contam. Toxicol* **78** 270-274

[9] Hussain T, Gondal M A, Yamani Z H and Baig M A 2007 Measurement of nutrients in green house soil with laser-induced breakdown spectroscopy *Environmental Monitoring and Assessment* **124** 131-139

[10] Gondal M A, Hussain T, Yamani Z H and Baig M A 2006 Detection of heavy metals in Arabian crude oil residue using laser-induced breakdown spectroscopy *Talanta* **69** 1072 -1078

[11] Gondal M A, Hussain T and Yamani Z H 2008 Optimization of the LIBS parameters for detection of trace metals in petroleum products *Energy Sources Part A-Recovery, Utilization and Environmental Effects* **30** 441-451

[12] Radziemski L J and Cremers D A 1989 *Laser-induced plasma and applications* (New York: Marcel Dekker) pp 295-325

[13] Natrellla M G 1963 *Experimental Statistics NBS Handbook* vol 91 (National Institute of Standards and Technology, Gaithessburg, MD)

[14] Cho H H, Kim Y J, Soo JoY, Arai N and Lee Y 2001 Application of laser-induced breakdown spectrometry for direct determination of trace elements in starch-based flours *J. Anal. At. Spectrum* **16** 622-627

[15] Radziemski L J 2002 From LASER to LIBS, the path of technology development *Spectrochimica.Acta part B Atom Spectrosc* **57** 1109-1113

[16] Rusak D A, Castle B C, Smith B W and Winefordner J D 1998 Recent trends and the future of laser-induced breakdown spectroscopy. *Trends in Anal. Chem.* **17** 453-461

[17] Cremers D A, Radziemski L J and. Loree T R 1984 Spectrochemical analysis of liquids using the laser spark *Appl. Spectrosc.* **38 (5)** 721-729

[18] 1998 *World Bank Handbook Pollution Prevention and Abatement, Iron and Steel manufacturing; Draft technical background document* (Washington, DC: environment

department world bank) pp 327-331

[19] Telle H H, Beddows D C S, Morris G W and Samek O 2001 Sensitive and selective spectrochemical analysis of metallic samples: the combination of laser-induced breakdown spectroscopy and laser-induced fluorescence spectroscopy *Spectrochim Acta Part B* **56** 947–960

[20] Lee P N and.Corning D M 1993 *Experimental* toxicology: The basic issues (Cambridge: The Royal Society of Chemistry) pp 405-440

[21] 1999 (Royal Commission: Jubail Industrial City-Saudi-Arabian Environmental Regulation Standards) and Greenberg A E 1992 *Standard Methods for the examination of water and wastewater 18th ed* , ed Clescerl L and Eatom A D,(American Public Health Association: Washington- D.C)

[22] Abdellatif G and Imam H 2002 A study of laser plasma parameters at different laser wavelength *Spectrochimica Acta Part B* **57** 1155-1165

[23] Song K, Lee Y I and Sneddon J 1997 Application of laser–induced breakdown spectroscopy *Applied Spectroscopy Reviews* **32** 183-235

[24] Striganov A and Sventitski N 1968 *Handbook Table of Spectral Lines of Neutral and Ionized Atoms* (New York: Plenum)

Laser etching rates of sapphire glass using metal targets

A H Hamdani[1*], A Nasir[2], S Sarwar[2], A Ansar[1], R Akhter[1], M Aslam[1]

[1] Pakistan Institute of Lasers and Optics, P. O. Box 505, Rawalpindi, Pakistan
[2] Department of Electrical Engineering, International Islamic University, Islamabad, Pakistan

[*] *Coressponding author: pilo786@yahoo.com*

Abstract. Q-switched Nd:YAG laser with an output power of 40 watts has been used to study LIPAA etch rates of Sapphire glass in air with different target materials including stainless steel, brass and mild steel. Experiments showed that an ablation depth of 50μm has been achieved in sapphire at a zero target-substrate gap at ambient pressure. The etch depth versus laser scan speed and laser power have also been studied. An etched depth of 200μm is achieved at a scan speed of 20mm/sec with stainless steel as target.

Keywords: Pulsed Nd:YAG Laser, LIPAA process, Laser Ablation, Sapphire glass

1. Introduction

Micro-fabrication on transparent materials such as quartz and sapphire glasses has been growing since last two decades due to various technological developments in optoelectronic industry. Specially, development of high brightness blue LED's has further accelerated the development of micro machining on sapphire substrates such as scribing, deep etching, dicing etc. Laser Induced Plasma Assisted Ablation (LIPAA) has emerged as an effective technique resulting in high etch rates (~ few hundreds of nanometers). This process has been studied by various scientific research groups. Y.Handa et al used single pulse second harmonic of Nd:YAG laser and femto second laser for dual pulse irradiation for LIPAA studies on soda glass[1-2]. The reported etch rate is around 10nm/pulse. Zhang. J et al used silver target for ablation of quartz and pyrex glasses using first, second and third harmonics of 1064nm laser to find etch rates of these glasses. The achieved ablation rate was 50nm/pulse at a laser fluence of 7J/cm^2[3-4]. Zhang. J et al also reported ablation of quartz with stainless steel as target with fourth harmonic of Nd:YAG laser at 266nm[5]. The Laser Induced Back side Wet Etching (LIBWE) of sapphire glass has been reported by K.Zimmer et al [6]. Y.Handa has also reported LIPAA etching of sapphire using second harmonic of Nd:YAG laser[7]. However high etch rate LIPAA parameters for sapphire glass are still lacking.

We presented parametric study of LIPAA process for quartz in air and under vacuum in our prior works [8-9] with etch rates as high as 700nm/pulse. Now we present results of ablation depths in sapphire glass obtained during LIPAA with high fluence fundamental harmonic of Nd:YAG at 1064nm using different target materials. The high etch rates reported in this paper can be useful for fast dicing, engraving and scribing of sapphire substrates for various applications including thin film solar cell, blue LED manufacturing etc.

2. Experimental

The experimental setup for this study is shown in Figure 1 below. The laser source is an actively Q-switched 40W Nd:YAG laser (1.064μm) operating at 200Hz-50kHz with a galvano type laser scanning head.

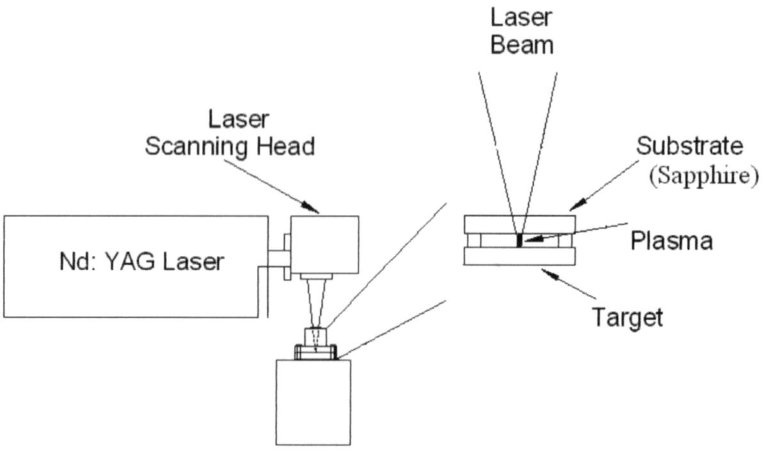

Figure 1. Experimental setup for LIPAA process

The laser is focused on target surface with a spot size of 225µm in a scan area of $3x4mm^2$ with varying scan speeds. The gap between target and substrate was varied from 0 to 540 µm with an increment of 60µm. The ablation depth was measured by a precision mechanical dial and averaged over 20 number of scans. The laser fluence was $65J/cm^2$ at focus spot for 40W output power.

3. Results and Discussion

Figure 2 shows the results of ablation depth for Sapphire glass using stainless steel as target material. It can be seen that a maximum ablation depth of 50µm is attained at a zero target-substrate gap at a scan speed of 100mm/sec. The ablation depth drops to 5µm at maximum gap size of 540µm. The maximum and minimum ablation rates are found to be 746 nm/pulse at zero gap and 74nm/pulse at maximum gap.

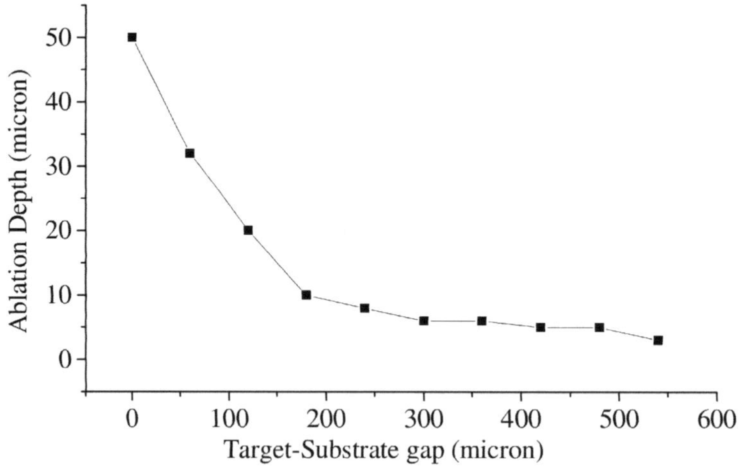

Figure 2. Ablated depth of Sapphire for stainless steel target

Figure 3 shows the results of ablation depth for Sapphire using mild steel target. The maximum ablation depth of 40µm is produced at zero gap which gives a maximum ablation rate of 595 nm/pulse and a minimum of 45nm/pulse at a gap size of 0 and 540 µm, respectively.

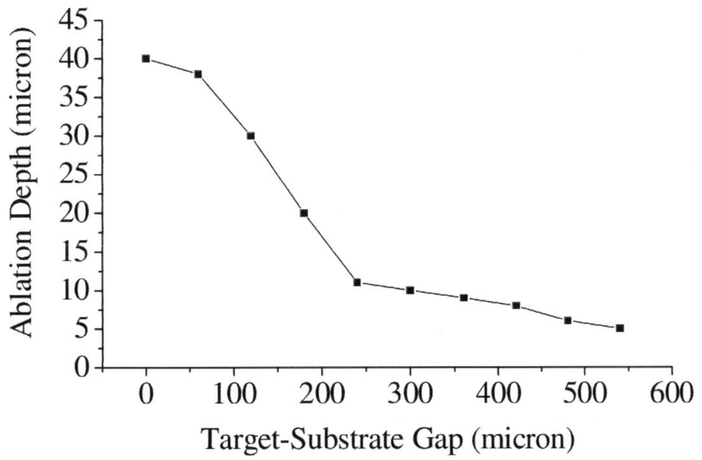

Figure 3. Ablated depth of Sapphire with mild steel target at varying gaps

The results for ablation depth in Sapphire glass using non ferrous material such as brass are shown in Figure 4. The maximum ablation depth of 30μm is attained with brass target at zero gap which corresponds to a maximum ablation rate of 446nm/pulse and a minimum of 74nm/pulse at a gap size of 0 and 540 μm respectively.

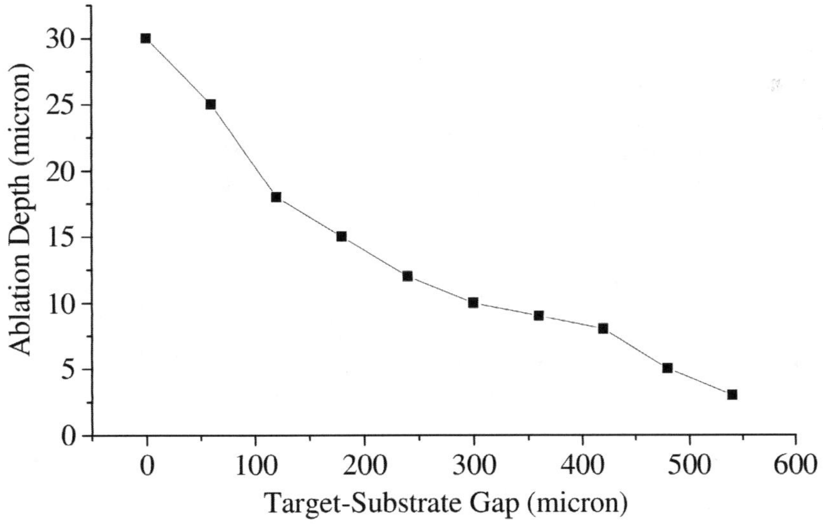

Figure 4. Ablated depth of Sapphire with brass target at varying gaps

The comparison of Figure 2, Figure 3 and Figure 4 shows that overall ablation depths are higher in case of ferrous metals. This can be attributed to the fact that laser absorption of ferrous material at 1.064μm is higher as compared to non-ferrous materials. The amount and energy of the plasma generated at the surface of ferrous materials is significantly higher than that formed on non-ferrous metals. The healthy plasma imparts more kinetic energy on the glass surface resulting in deeper etching.

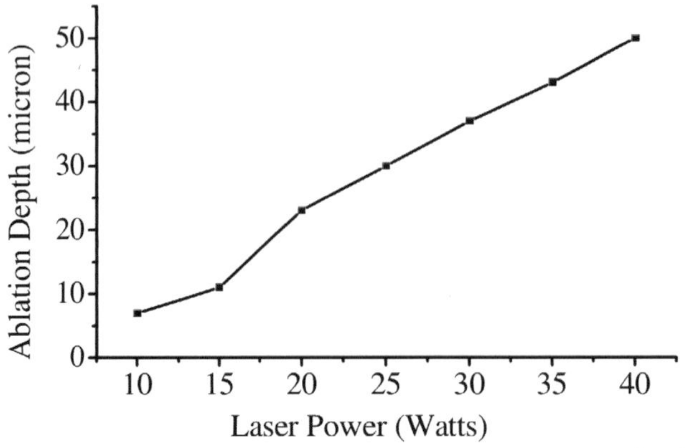

Figure 5. Ablation depth versus laser power at zero target-substrate gap for Sapphire-stainless steel

Since the maximum ablation depth and etch rates were obtained for Sapphire with stainless steel target, hence variation of ablation depth with laser power was studied for this case.
Figure 5 shows that ablated depth increases with laser power which can be attributed to increased plasma formation at target surface and hence more ablation at substrate surface. A maximum ablation depth of 50μm is obtained at 40Watts of average power at 0μm gap size.

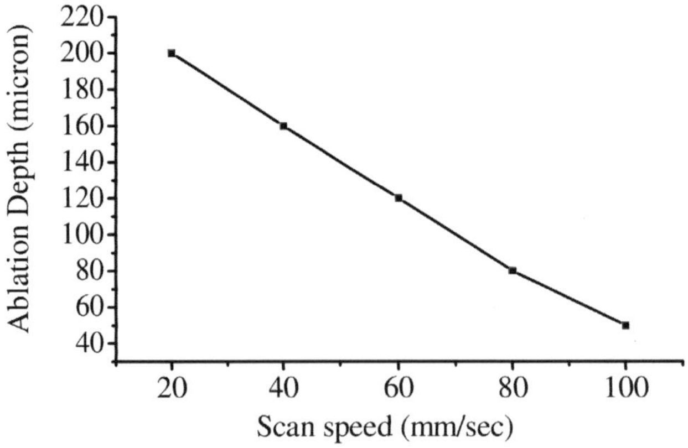

Figure 6. Ablation depth in Sapphire with varying scan speed

The etched depth in Sapphire glass was also studied as a function of scan speed at zero gap and stainless steel target. The results are shown in Figure 6 where lower the scan speed results in higher etch depths. The 0.2 mm thick sapphire plate was cut at a scan speed of 20m/sec in 20 scans.

4. Conclusion

The present studies on ablation rates of sapphire glass using different target materials have shown that ferrous metals such as stainless steel and mild steel produce maximum etch depths at zero target-substrate gap. The maximum ablation depth of 50μm at an etch rate of 744nm/pulse has been measured in stainless steel. The ablation depth at different scan speeds has also been measured for practical substrate cutting applications. Sapphire has shown a near linear relationship between laser power and ablated depth in available laser power regime (0-40Watts) which shows that further increase in cutting rates are expected at higher fluence.

Acknowledgement
Authors are grateful to management and staff of Pakistan Institute of Laser and Optics for their continued support during completion of this work.

References

1. Lee J M et al, 2000 *Appl. Phys.* **A 70** 561
2. Sugioka K and Midorikawa K, 2001 *RIKEN Review* **32** 36
3. Jie Z et al. 1998 *Applied Phys.* A. **67** 499
4. Hanada Y et al. 2004 *Applied Phys.* **A. 79** 1001
5. Z. Jie et al. 1999 *Applied Phys.* **A. 69** S879
6. Zimmer K et al. 2008 *Laser Chemistry*, 2008 (article ID170632)
7. Hanada et al 2005 *Journal of High Temperature Society*, **30(2)** 105
8. Hamdani A H et al. 2010 *Key Engineering Materials*, **442** 172
9. Hamdani A H et al. 2012 *Key Engineering Materials*, **261** 510-511

Computation and modeling of Ionospheric Plasma Pressure for F_2 layer at Pakistan air space

[1]M Ayub Khan Yousuf Zai and [2]Khusro Mian

[1]Solar-Terrestral and Atmospheric research center, department of Applied Physics University Karachi, Pakistan.
[2]Institute of Space and Planetary Astrophysics (ISPA) University of Karachi, Pakistan

Corresponding Email: ayubzai@yahoo.com

Abstract: This paper deals with the computation of Ionospheric plasma pressure for F_2 layer at Pakistan air space. The physical processes of plasmas have been studied extensively by now both in laboratory as well as in space environment as in the present study. This paper deals with the computation of Ionospheric plasma pressure for F2 layer at Pakistan air space. Modeling plasma pressure has been a part in terms of instabilities in the ionosphere. We use stochastic approach to model and analyze the stability properties of plasma. These models are used to sustain our investigations of instabilities due to plasma pressure at Pakistan air space.

Key words: Plasma pressure, Plasma instabilities, Stochastic process Stability properties.

1. Introduction

Gauss 1839 speculated on the existence of electrical phenomena in the upper atmosphere, and Balfour Stewart in 1883 postulated a conducting layer as the site of the currents that cause the small daily variations of the earth's magnetic field. This conducting layer now called ionosphere to have been discovered by Marconi when he sent radio signals across the Atlantic in 1901.This ionosphere forms space plasma that could be defined as the quasi neutral gas of charged particles that can exhibit collective behavior and also follow the three criteria:

(a) $\lambda_D << L$
(b) $N_D >>> 1$
(c) $\omega_t > 1$

It has been known that magnetohydrodynamics (MHD) explains the process that deals with the motion of a highly conducting fluid like in the present case ionospheric plasma due to presence of magnetic field. In formulating this description of plasma comprising the momentum equation, Debye shielding, mean free path or thermal gyro-radius etc. Our assumption that the time scales of interest are long compared to the microscopic particle motion and that of spatial scales with respect to Debye length and mean free path [1-5]

For a fluid like ionospheric plasma the particles are tied with the field lines. In the magnetic intensity B the magnetic stresses are equivalent to a tension $B^2/ 4\pi$ and hydrostatic pressure $B^2/ 8\pi$. The magnetic pressure is balanced by the fluid pressure (plasma pressure) so that the lines of force behave as taut string under tension. The magnetic force in the form of

$\frac{1}{c} J \times B$ on the right hand side of the momentum equation and the momentum equation to

Journal of Physics: Conference Series **439** (2013) 012052 doi:10.1088/1742-6596/439/1/012052

$$\rho\,(\frac{d}{d_t} + u.\nabla\,) = -\ \nabla(p + \frac{B^2}{8\pi}) + \frac{1}{4\pi}(B.\nabla)B$$

It is evident from the above expression that $\frac{B^2}{8\pi}$ indicates magnetic pressure and has dimension of kinetic pressure. The ratio of plasma pressure to the magnetic field pressure is represented by the symbol β where β = p / B2/ 8 π. If β << 1 then the plasma is cold and if β >>1 then the plasma is considered as warm and the plasma currents may become important.

The plasma gas may be created, either by the natural ionization of the upper atmosphere at Pakistan air space. The Ionospheric plasma is coupled formation by the earth magnetic field. It is possible to obtain a state of equilibrium for plasma in presence of field, that interact with plasma, we have already shown that there can exist a simple equilibrium configuration in which plasma pressure is balanced by magnetic pressure. Now given perturbation to this system, it is important to know whether this equilibrium is stable or unstable. Numerical models are used to maintain our examine instabilities due to plasma pressure. The particles have maxwellian velocity distributions and the states are not perfectly in thermodynamic equilibrium. Albeit they are in the equilibrium in the sense that all the forces are in balance and a time-independent solution is possible. The free energy that is present can cause waves to be self-excited equilibrium then an unstable one. Thus an instability is always a motion that decreases the free energy and brings the plasma closer to true thermodynamic equilibrium. Instabilities are classified according to the kind of free energy available to derive them.

2. Plasma Pressure

We have known the plasma instability in which the effect of plasma pressure has been considered. We considered warm plasma the elastic waves exist, where a plasma gradient can build up where plasma pressure is considerable.

The plasma pressure is product of some constant, plasma concentration and temperature

$$p\ = \gamma\ nKT$$

Where $\gamma = 5/3$ for long wave, the distance traversed by an electron during one wave period much smaller than the wavelength and thus the effect of temperature perturbations will be negligible. Plasma pressure can not be ignored. The plasma gradient term in the momentum equation has to be included. Even where there are no obvious drive force such an electric or a gravitational field, plasma is not in perfect thermodynamics equilibrium as long as it is confined. The plasma pressure tends to make the plasma expand and the expansion energy can drive instability. This type of free energy is always present in any finite plasma, and the resulting waves are called universal instability.[6-7]

3. Methodology

Our purpose of this study to construct a mathematical model and predict plasma pressure at upper atmosphere. The technique that has become the standard for estimating AR models, was proposed by making successive approximations through various stages, Identification, estimation, and diagnostic checking. It can be used for forecasting for the period the process for developing this model begins by examining whether the time series under study is stationary non-stationary. [8-9]

The data collected from SUPARCO for virtual height in km and critical frequency in MHz. Figure 1 shows time plot of the plasma pressure for Ionospheric layers for Pakistan air space. The residual analysis has been done and plot is shown in figure 2.

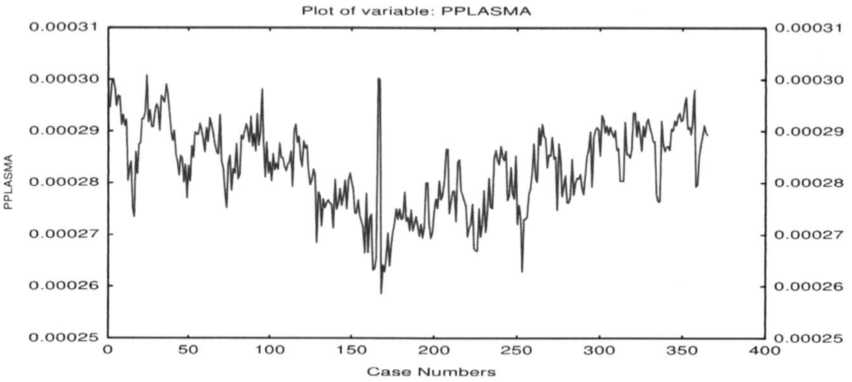

Figure 1. The temporal plot for plasma pressure of Ionospheric F_2 layer

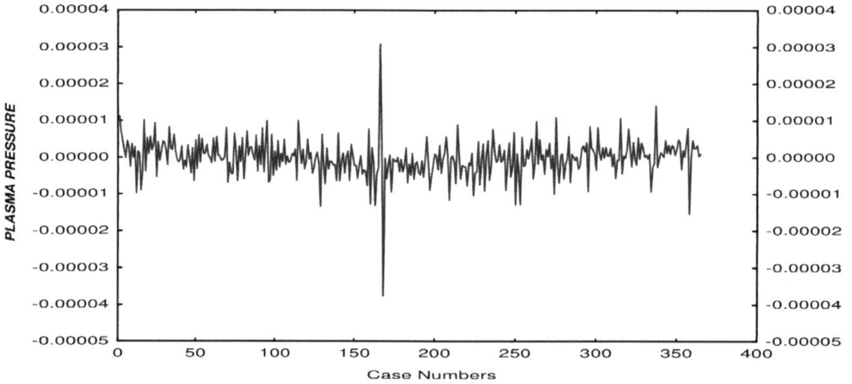

Figure 2. Residual analysis for plasma pressure

4. Test of Hypothesis

Mean of plasma pressure of Ionospheric layer is 0.000283. the standard deviation is known to be 0.000008 a random sample of month of Jan 30 days plasma pressure points are taken and the mean of plasma pressure is computed as 0.000291 and level of significance value α = 0.05 then test of hypothesis is accept H_0 or reject H_1.

H_0: $\mu \leq 0.000283$ (μ =0.000283)

H_1: μ >0.000283

$\alpha = 0.05$

Test of statistic since standard deviation of population σ =0.000008 known, therefore Z-statistics is suitable to use.

$$Z = \frac{\overline{X} - \mu}{\frac{\sigma}{\sqrt{n}}}$$, n=30, $\overline{X}$ =0.000291, σ =0.000008 and μ =0.000283

$$Z = \frac{0.000291 - 0.000283}{\frac{0.000008}{\sqrt{30}}}$$

$Z_{\text{Calculated Value}} = 5.477$, critical region at $\alpha = 0.05$ for one tailed test is Z> 1.65

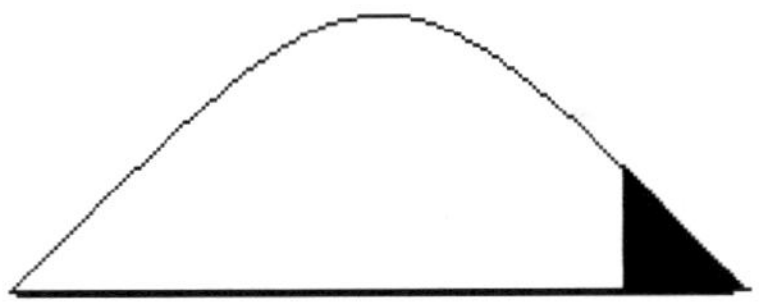

Figure 1: Right tailed test $H_1 : \mu > \mu_0$

Conclusion since the calculated value of Z falls in critical rejoin, therefore we reject null hypothesis H_0

$$|Z_{\text{Calculated Value}}| = 5.477 \text{ and } |Z_{\text{Tabulated Value}}| = 1.65$$

$$|Z_{\text{Calculated Value}}| > |Z_{\text{Tabulated Value}}|, \text{ then we reject } H_0$$

5. Model Fitting and diagnostics

We have used the computed plasma pressure for ionosphereic F_2 layer data that has been explained in last portion and fitted model AR(1) is suitable model whose parametric values s shown in table 1 and figure.2 that manifests the residual plot for AR(1) model as given below:

$$Y_t = 0.77\, Y_{t-1} + 6.4\text{x}10^{-5} + e_t \qquad (1)$$

Table: 1 Model parametric values for AR (1) model of plasma pressure Ionospheric F_2 layer

Variable	Constant	P(1)
Coefficient	0.000283	0.773277
Standard Error	0.000001	0.033709
t-statistics	239.7984	22.9398
Probability	0.00	0.00
Lower 95 percent	0.000281	0.706988
Upper 95 percent	0.000286	0.839566

Journal of Physics: Conference Series **439** (2013) 012052 doi:10.1088/1742-6596/439/1/012052

6. Forecasting

Equation (1) shows prediction equation that is used to forecast series opf plasma pressure values as shown in table 2 and the figure.3 manifests the forecasts values.

Table 2: Forecast values of plasma pressure for ionosphereic F_2 layer

No of forecasting	Forecast
366	0.000288
367	0.000287
368	0.000286
369	0.000286
370	0.000285
371	0.000285
372	0.000284
373	0.000284
374	0.000284
375	0.000284

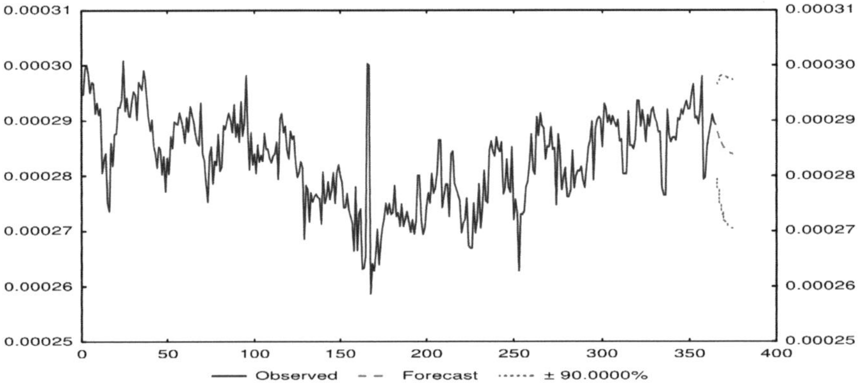

Figure 3. Forecasts of Ionospheric data showing random process plasma pressure ionospheric F_2 layer a first order AR

7. Validity of the constructed model

This model can be validated using the figures 4 and 5 that depict the autocorrelation function of the random process for the plasma pressure at Ionospheric F_2 layer and the partial autocorrelation function of the data of the same random process in the plasma pressure for ionospheric F_2 layer

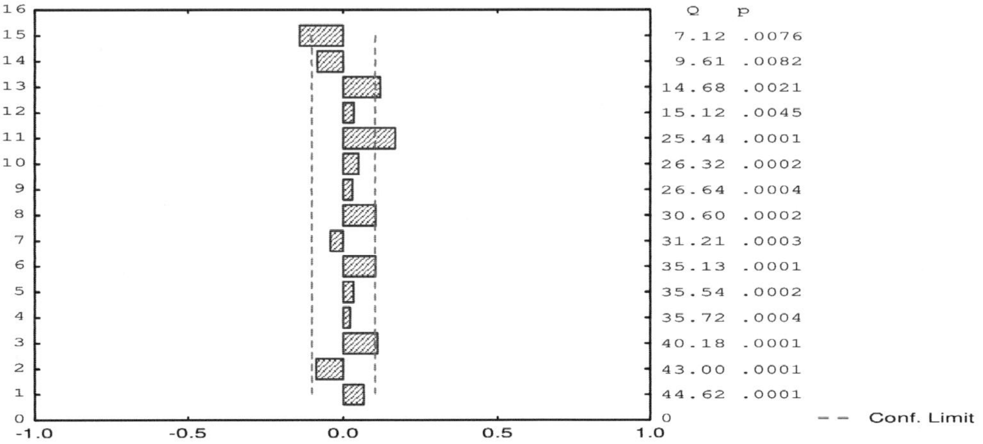

Figure 4. autocorrelation function showing a random process the plasma pressure for Ionospheric F_2 layer a first order AR

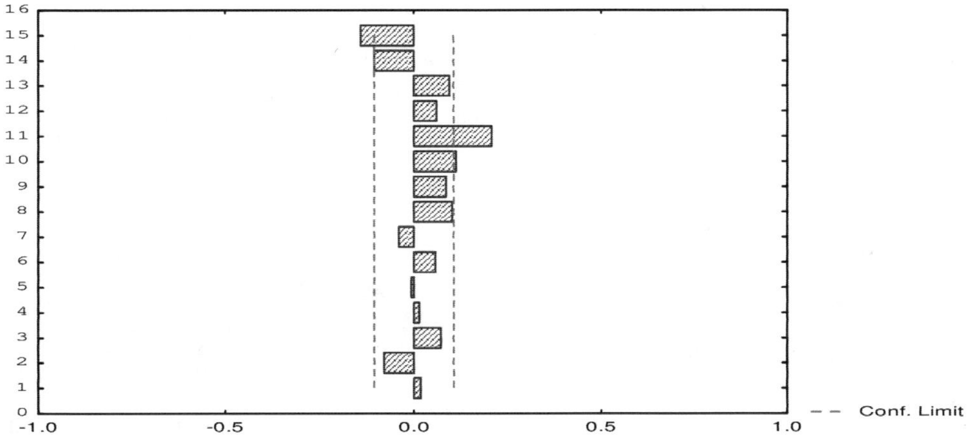

Figure 5. The plot showing the partial autocorrelation coefficient of the data that exhibits a random process in the plasma pressure for ionospheric F_2 layer a first order AR

8. Conclusion

In this communication we have computed plasma pressure and explored an idea to quantify the plasma pressure and its instability in the ionospheric layers at this region. Forecast values have been obtained and more useful results were reported from the Analysis of autoregressive (AR) model of first order and the higher orders could also be verified based on the availability of data sets. The variation of the predicted values of plasma pressure depend on the ionospheric conditions

Acknowledgement
The authors are grateful to the officials of Space and Upper Atmosphere Research Commission SUPARCO for providing us the Ionospheric data.

References

[1] ARMEL PICQUENARD, "Radio Wave Propagation", the Macmillan Press Ltd, 1974 U K.

[2] Chen M C, "Introduction to Plasma Physics", Plenum Publishing Corporation, 1929 New York

[3] Das A C, "Space Plasma Physics "An Introduction, Narosa Publishing House, 2004.

[4] John H. Seinfeld, Spyros N. Pandis "Atmospheric Chemistry and Physics" (From Air Pollution to Climate Change) 2nd Ed., Published by John Wiley & Sons, Inc., Hoboken, 2006 New Jersey.

[5] H Rishbeth, "Introduction to Ionospheric Physics ", Academic Press, 1969, New York and London.

[6] R W Schunk, "Ionospheres Physics, Plasma Physics, and Chemistry", Published by the Press Syndicate of the University of Cambride, 2000 United Kingdom.

[7] Clare Smith, Environmental Physics, 2001 by Routledge, USA and Canada

[8] Chatfield C., The analysis of time series: an introduction, 4/e, Chapman & Hall, 1989 London

[9] S Makridakis, Forcasting Methods and application 2nd Ed, 1983, John Wiley and sons, Inc. Canada

[10] Allan H. Murphy and Richard W. Katz, Probability, Statistics, and decision making in the Atmospheric Sciences, 1985 by West view Press, Ine Boulder and London